CANADIAN EDITION

Earth

An Introduction to Physical Geology

Edward J. Tarbuck
Illinois Central College, Emeritus

Frederick K. Lutgens
Illinois Central College, Emeritus

Cameron J. Tsujita
University of Western Ontario

PEARSON

Prentice
Hall

Toronto

National Library of Canada Cataloguing in Publication

Tarbuck, Edward J.
Earth : an introduction to physical geology / Edward J. Tarbuck,
Frederick K. Lutgens, Cameron J. Tsujita. — Canadian ed.

ISBN 0-13-121724-0

1. Physical geology—Textbooks. I. Lutgens, Frederick K
II. Tsujita, Cameron J. (Cameron James), 1966- III. Title.

QE28.2.T37 2005   550   C2003-906480-8

ISBN 0-13-121724-0

Vice President, Editorial Director: Michael J. Young
Senior Acquisitions Editor: Kelly Torrance
Marketing Manager: Marlene Olsavsky
Senior Developmental Editor: Paul Donnelly
Production Editor: Judith Scott
Copy Editor: Jennifer Therriault
Production Coordinator: Janette Lush
Photo and Permissions Research: David Dillon and Susan Wallace-Cox
Page Layout: Alicia Countryman
Art Director: Mary Opper
Interior and Cover Design: Anthony Leung
Cover Image: Gettyimage

1 2 3 4 5   09 08 07 06 05
Printed and bound in the USA.

Brief Contents

GEODe III CD-ROM

A copy of GEODe III CD-ROM is packaged with each copy of EARTH: AN INTRODUCTION TO PHYSICAL GEOLOGY, Seventh Edition. This dynamic learning aid reinforces key geologic concepts by using tutorials, animations and interactive exercises.

Unit Introduction

A. A View of Earth
B. Earth's Layered Structure
C. Features of the Continents
D. Floor of the Ocean

Unit Earth Materials

A. Minerals
   1. Introduction
   2. Major Mineral Groups
   3. Properties Used to Identify Minerals
   4. Mineral Identification
B. Mineral Review
C. Rock Cycle
D. Igneous Rocks
   1. Introduction
   2. Igneous Textures
   3. Mineral Composition of Igneous Rocks
   4. Naming Igneous Rocks
E. Sedimentary Rocks
   1. Introduction
   2. Types of Sedimentary Rocks
   3. Interpreting Environments
F. Metamorphic Rocks
   1. Introduction
   2. Agents of Metamorphism
   3. Textural and Mineralogical Changes
   4. Common Metamorphic Rocks
G. Rock Review

Unit External Processes

A. External vs. Internal Processes
B. Hydrologic Cycle
C. Running Water
   1. Stream Characteristics
   2. Review—Valleys and Stream-Related Features
D. Groundwater
   1. Groundwater and its Importance
   2. Springs and Wells
E. Glaciers
   1. Introduction
   2. Budget of a Glacier
   3. Reviewing Glacial Features
F. Deserts
   1. Distribution and Causes of Dry Lands
   2. Common Misconceptions About Deserts
   3. Review of Landforms and Landscapes
G. Coastal Processes
   1. Waves and Beaches
   2. Wave Erosion

Unit Internal Processes

A. Plate Tectonics
   1. Introduction
   2. Plate Boundaries
B. Crustal Deformation
   1. Introduction
   2. Mapping Geologic Structures
   3. Folds
   4. Faults and Fractures (Joints)
   5. Review
C. Earthquakes
   1. What is an Earthquake?
   2. Seismology
   3. Locating an Earthquake
D. Igneous Activity
   1. The Nature of Volcanic Eruptions
   2. Materials Extruded During an Eruption
   3. Volcanoes
   4. Intrusive Igneous Activity

Unit Geologic Time

A. Geologic Time Scale
B. Relative Dating
C. Radiometric Dating

This GEODe III icon appears throughout the book wherever a text discussion has a corresponding activity on the CD-ROM.

Contents

 This icon for the GEODe III CD-ROM appears whenever a text discussion has a corresponding GEODe III activity.

xiv   Contents

Preface

Earth is a very small part of a vast universe, but it is our home. Like Goldilocks, on Earth we enjoy just the right combination of conditions and ingredients necessary to maintain life and support our activities. The science of geology is particularly important in our quest for understanding how planet Earth works.

Media reports keep us aware of the geological forces at work on our planet. News stories graphically portray the violent force of a volcanic eruption, the devastation generated by a strong earthquake, and the large numbers left homeless by landslides and flooding. Such events, and many others as well, are destructive to life and property, and we must always be prepared to deal with them. To comprehend and prepare for such events requires an awareness of how science is done and the scientific principles that influence our planet, its rocks, mountains, atmosphere, and oceans.

Geology can also yield clues to help us understand, prevent, and repair environmental problems. Just as an auto mechanic must have a good working knowledge of the interactions among the parts in a car in order to fix it, the knowledge of our planet is critical to our well being and, is indeed, vital to our survival.

The Canadian edition of *Earth: An Introduction to Physical Geology* is a university-level text that is intended to be a meaningful, non-technical primer for students taking their first course in geology. In addition to being informative and up-to-date, a major goal of *Earth* is to meet the need of students for a readable and user-friendly text, a book that is an effective resource for learning the basic principles and concepts of geology.

Distinguishing Features

Readability

The language of the book is straightforward and written to be understood by the layperson. Clear, readable discussions with a minimum of technical language are the rule. The frequent headings and subheadings help students follow discussions and identify the important ideas presented in each chapter. Large portions of the text were substantially rewritten in an effort to make the material less repetitive, more understandable, and more relevant to a Canadian audience.

Illustrations and Photographs

Geology is a highly visual science. Therefore, photographs and artwork are a very important part of an introductory book. *Earth* contains hundreds of photographs, of which over 100 are new to the Canadian edition. These have been carefully selected to aid understanding, add realism, and contribute a Canadian perspective to geological concepts.

The illustrations in *Earth* are renowned for their quality. The original art program was carried out by Dennis Tasa, a gifted artist and respected geological illustrator. The Canadian edition also features many pieces of new or substantially redesigned line art. The new art illustrates ideas and concepts more clearly and realistically than ever before. Most of the new illustrations produced for the Canadian edition were custom-designed by Cam Tsujita, with the help of the outstanding graphics team at Pearson Education Canada.

Focus on Learning

When a chapter has been completed, three useful devices help students review. First, the Chapter Summary recaps all the major points. Next is a checklist of Key Terms with page references. Learning the language of geology helps students learn the material. This is followed by Review Questions that help students examine their knowledge of significant facts and ideas. Each chapter closes with a reminder to visit the Companion Website for *Earth* (**http://www.pearson.ca/tarbuck**). It contains many excellent opportunities for review and exploration.

 ## Earth as a System

An important occurrence in modern science, particularly in geology, has been the realization that Earth is a giant multidimensional system. Our planet consists of many distinct but interacting parts. The Earth system responds to changes in any of its constituent parts, often in ways that are neither obvious nor immediately apparent. Although it is not possible to study the entire system at once, it is possible to develop an awareness and appreciation for the concept and for many of the system's important interrelationships. Therefore, the theme of "Earth as a System," recurs at appropriate places throughout the book. It is a thread that weaves through the chapters and helps tie them together.

New and revised special-interest boxes relate to "Earth as a System." To remind the reader of this important theme, the small icon you see at the beginning of this section is used to mark these boxes.

 ## People and the Environment

Because knowledge about our planet and how it works is necessary to our survival and well-being, the treatment of environmental issues plays an important part of *Earth*. Such discussions serve to illustrate the relevance and application of geological knowledge. This theme is given considerable prominence in the Canadian edition. The text integrates a great deal of information about the relationship between people and the natural environment and explores applications of geology to understanding and solving problems that arise from these interactions. In addition to many basic text discussions, several of the text's special-interest boxes involve the "People and the Environment" theme and are quickly recognized by the distinctive icon you see at the beginning of this section.

 ### Canadian Profile

Many of the people, places, issues, and ideas that play pivotal roles in understanding physical geology also happen to originate in Canada. Of course, this theme pervades the Canadian edition, but we have also highlighted certain topics for discussion in special-interest boxes in almost every chapter. You can identify these in the text by the small icon you see at the beginning of this paragraph.

Maintaining a Focus on Basic Principles and Instructor Flexibility

The main focus of the Canadian edition is to foster a basic understanding of physical geology. As much as possible, we have attempted to provide the reader with a sense of the observational techniques and reasoning processes that constitute the discipline of geology.

The organization of the text remains intentionally traditional. Following the overview of geology in the introductory chapter, we turn to a discussion of Earth materials and the related processes of volcanism and weathering. Next a discussion of a most basic topic, geologic time, is followed by an examination of the geological work of gravity, water, wind and ice in modifying and sculpting landscapes. After this look at external processes, we examine Earth's internal structure and the processes that deform rocks and give rise to mountains. Finally, the text concludes with an important chapter on resources and a step beyond the physical confines of our planet to view Earth in larger-scale context of the solar system. This particular organization was selected largely to accommodate the study of minerals and rocks in the laboratory, which usually comes early in an introductory-level geology course.

Realizing that some instructors prefer to structure their courses differently, each chapter is self-contained, so that chapters may be taught in a different sequence. Thus, the instructor who wishes to discuss earthquakes, plate tectonics and mountain building prior to dealing with erosional processes may do so without difficulty. We also chose to introduce the principles of plate tectonics in the first chapter so that this basic but important theory could subsequently be elaborated in appropriate places throughout the text.

More About the Canadian Edition

The Canadian edition of *Earth* represents a robust revision of its U.S. predecessor. Every part of the book was examined carefully with the goals of keeping topics current, addressing Canadian perspectives and issues, and improving the clarity of text discussions.

Here are some examples of what is new in the Canadian edition of *Earth:*

- Over 200 new photos, maps, and illustrations.

- Canadian examples used judiciously throughout, while retaining the best of the U.S. and international examples.

- New or substantially revised "People and the Environment" boxes, including features on: "Asbestos: What Are the Risks?" (Ch. 2) and "Return to the Dust Bowl?" (Ch. 5).

- New or substantially revised "Understanding Earth" boxes, including features on: "A Closer Look at Facies" (Ch. 6) and "Index Fossils and Ecology of Organisms" (Ch. 8).

- New to the Canadian edition, the "Canadian Profile" box focuses on notable geologists, geological features, and geological events that are unique to Canada. Topics include: "Sir William Logan: Canada's Premier Geologist" (Ch. 1); "The Burgess Shale" (Ch. 6); "The Walkerton Tragedy: Geology Forms the Link" (Ch. 11); "Snowball Earth: Canadian Cryospheric Controversy" (Ch. 12); "The Okanagan Valley: A Canadian Desert" (Ch. 13); "Earthquakes in Canada" (Ch. 16); "LITHOPROBE: Probing the Depths of Canada" (Ch. 17); "The Grand Banks Earthquake and Turbidity Current" (Ch. 18); "John Tuzo Wilson: Canada's Sponsor of Plate Tectonics" (Ch. 19); "Some Important Events in the History of Canada's Petroleum Industry" (Ch. 21); and "Fall and Recovery of the Tagish Lake Meteorite: A Messenger from the Early Solar System" (Ch. 22)

- Chapter 5, Weathering and Soil, includes coverage of the Canadian Soils Classification System, as well as a section on paleosols.

- Chapter 7, Metamorphism and Metamorphic Rocks, has been effectively rewritten and condensed to communicate more clearly the processes and products of metamorphism. Concepts are progressively built on one another, culminating in the concept of metamorphic facies and the significance of this concept with respect to plate tectonic processes.

- Chapter 10, Running Water, has been reformatted for a more logical flow of concepts, and now includes a Canadian-based account of the Red River Flood of 1997 as well as some interesting facts about the history of the Niagara River. In addition, examples of ancient river deposits are provided to give the reader a sense of how processes observed in modern rivers are used to interpret features of ancient river systems.

- Chapter 13, Deserts and Wind, now also includes information on polar deserts.

- Chapter 14, Shorelines, has been substantially revised, with greater emphasis on depositional features of coastlines and the sedimentary characteristics of ancient shoreline deposits.

- Chapter 20, Mountain Building and the Evolution of Continents, has been extensively rewritten to allow the step-by-step explanation of how, where, and why mountain building occurs, with the concept of isostasy as a common thread. This approach allows the reader to then understand mountain building episodes in the context of the Wilson Cycle, a concept given less prominence in U.S. editions of this text.

- Chapter 21, Mineral and Energy Resources, bears little resemblance to its U.S. counterpart. Substantial changes were made to this chapter to provide better explanations of how the many types of mineral and fossil fuel deposits came to be, to highlight environmental issues surrounding the extraction and use of natural resources, and more importantly, to make the chapter more relevant to Canadian issues. We view this chapter as perhaps the most important of all the chapters in this text, and much effort went into its revision. All the maps and most of the figures in this chapter are new and are focused on Canada.

The Teaching and Learning Package

We have prepared an excellent supplements package to accompany the text. This package includes the traditional supplements that students and professors have come to expect from authors and publishers, as well as some new kinds of supplements that involve electronic media.

For the Student

Geode III CD-ROM.   Each copy of *Earth*, Canadian Edition, comes with GEODe III, by Ed Tarbuck, Fred Lutgens, and Dennis Tasa of Tasa Graphic Arts, Inc. GEODe III is a dynamic program that reinforces key concepts by using animations, tutorials, and interactive exercises. A special GEODe III icon appears throughout the book wherever a text discussion has a corresponding GEODe III activity. This special offering gives students two valuable products (GEODe III and the textbook) for the price of one.

Companion Web Site.   This site, created specifically for the text, contains numerous review exercises (from which students get immediate feedback), exercises to expand one's understanding of geology, and resources for further exploration. It provides an excellent platform from which to start using the Internet for the study of geology. Please visit the site at **http://www.pearsoned.ca/tarbuck**.

For the Professor

Instructor's Resource CD-ROM.   This valuable aid provides quick and easy access to a wealth of valuable teaching tools, including the following:

- Instructor's Manual

- Test Generator. Contains more than 1600 questions in multiple-choice, true/false, and short-answer format, a selection of which you can link to actual figures and photos from the text.

- Customizable PowerPoint® lecture presentations with Digital Image Gallery

- The Prentice Hall Geoscience Animations. A library of more than 30 Flash animations for physical geography, meteorology, physical geology, and Earth sciences. Created through a uniq[ue] collaboration among four of Prentice Hall's le[ad]ing geoscience authors, these animations [make] visual and accessible some key physical pro[cesses]

Acknowledgments

Writing or revising a textbook requires [...] and cooperation of many individuals [...] approached by Pearson Education Ca[nada ...] prospect of adapting *Earth* for Canadi[an ...] at the chance, partly because I was [...] quality of previous editions. E[d ...] Frederick K. Lutgens, and their U[...] have indeed set a high standar[d ...] geology textbooks and I am h[...] involved in the production [...]

The excellent graphics and user-friendly explanations of geological processes provided an excellent framework on which to build the current text.

I am indebted to many friends and colleagues who enthusiastically contributed material to the Canadian edition of *Earth*. I am also deeply thankful to the many professors who influenced my academic philosophy during my academic career and enhanced my interests beyond my own area of expertise in paleontology. Dr. Steve Hicock, a glacial geologist at the University of Western Ontario, is an individual who I particularly respect as a teacher. For this reason, I asked Steve for suggestions on how to adapt Chapter 12 (Glaciers and Glaciation) to better reflect the diversity of glacial features in Canada. The result is a product of Steve's enthusiasm for the project, and nearly all the improvements made to the chapter I owe to him.

Many other colleagues generously contributed material for the new "Canadian Profile" boxes found throughout the textbook. Many thanks are extended to these individuals:

W. Glen E. Caldwell
Rick Cheel
Claudia Cochrane
Matt Devereux
David Eaton
Alessandro Forte
Brian Hart
Steve Hicock
Phil McCausland

Richard Léveillé
Mike Powell
Rob Schincariol
Gordon Southam
Keith Tinkler
Vic Tyrer
Gordon Winder
Grant Young

Several colleagues and other individuals were also very generous in providing new photographs for this edition of *Earth* and in many cases allowed me to use multiple images in their personal collections. I would particularly like to thank Christopher Collom, Catherine Hickson, Steve Hicock, Bob Hodder, Alfred Lenz, Mike Schering, Kristin Singh, Guy Plint, Denis Tetreault, and Grant Young for their cooperation in this regard. Although too numerous to mention here, several other people contributed individual images that enhance the tangibility of the concepts covered in the textbook, and I am grateful for their permission to showcase their photographic work.

Any project of this scale requires the work of many individuals to ensure that all of the logistical tasks are accomplished. Matt Devereux, Bob Hodder, Erica Tsujita, Anna Sergueeva, and Marie Schell were particularly helpful in ensuring that the quality of this textbook reached a consistently high standard through their assistance in initial editorial work. I specially thank David Dillon for the tremendous amount of effort he invested and in the acquisition of photo permissions.

Special thanks go to those colleagues who prepared in-depth reviews for the Canadian edition. Their critical comments and thoughtful input helped guide our work and clearly strengthened the text. I wish to thank:

Kevin Ansdell, University of Saskatchewan
Dileep Athaide, Capilano College
S.A. Balzer, University of Alberta
Elliott Burden, Memorial University
Mary Lou Bevier, University of British Columbia
Ward Chesworth, University of Guelph
Rick Cheel, Brock University
Brewster Conant, Wilfrid Laurier University
Christopher Collom, Mount Royal College
Jaroslav Dostal, Saint Mary's University
Wayne Haglund, Mount Royal College
Charles Henderson, University of Calgary
S.R. Hicock, University of Western Ontario
Ian Hutcheon, University of Calgary
Joyce Lundberg, Carleton University
Anne Marie Ryan, Dalhousie University
Robert J. Ryan, Dalhousie University
Stuart Sutherland, University of British Columbia

I also want to acknowledge the team of professionals at Pearson Education Canada. Thanks are due to my Acquisitions Editors: Leslie Carson, who brought me on board, and Kelly Torrance, for her help in bringing this textbook to fruition. An enormous thanks is awarded to Paul Donnelly, Senior Development Editor, who did a remarkable job of coordinating this very ambitious endeavour and kept the project on an even keel at all times, even when it felt like the boat was tipping. On a personal level, I also thank Paul for his patience with my inexperience as a textbook author, for absorbing my frequent bouts of frustration that accompanied the writing of the text, and for tolerating my bad jokes. I'd also like to express my gratitude to the Production team, led by Judith Scott and Janette Lush, for their outstanding work.

Lastly, heartfelt thanks are extended to my family. My father (James Tsujita), mother (Lilly Tsujita), and sister (Kimberly Tsujita) have provided unwavering support throughout my academic journey. Many thanks are extended to my wife, Erica Tsujita, and my two sons Jeremy and Casey, who persevered through my long work hours and frequent absences from home during the preparation of this textbook. It is their love and support that keeps me going during these crazy periods of my life. Hang in there—I'll be home soon!

Cam Tsujita

A Great Way to Learn and Instruct Online

The Pearson Education Canada Companion Website is easy to navigate and is organized to correspond to the chapters in this textbook. Whether you are a student in the classroom or a distance learner you will discover helpful resources for in-depth study and research that empower you in your quest for greater knowledge and maximize your potential for success in the course.

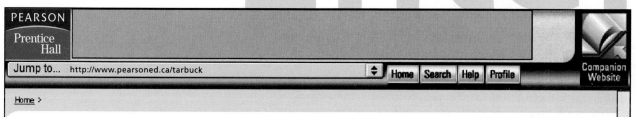

[www.pearsoned.ca/tarbuck] Enter

PEARSON
Prentice
Hall

Companion
Website

Jump to...   http://www.pearsoned.ca/tarbuck   ⬍ | Home | Search | Help | Profile |

Home >

PH Companion Website

Earth: An Introduction to Physical Geology, Canadian Edition,
by Tarbuck, Lutgens, and Tsujita

Student Resources

The modules in this section provide students with tools for learning course material.
These modules include:
- Concept Review (Multiple Choice Questions)
- Visualizing Geology (Labelling Questions)
- Thinking Geologically (Essay Questions)
- Quick Quiz (True/False Questions)
- Field Trips (Destinations)

In the quiz modules students can send answers to the grader and receive instant feedback on their progress through the Results Reporter. Coaching comments and references to the textbook may be available to ensure that students take advantage of all available resources to enhance their learning experience.

Instructor Resources

Links to our Instructor Central Site provide password-protected access to the Instructor's Manual, PowerPoint Presentations, and more.

Chapter 1

An Introduction to Geology

The glacier-capped peaks of the Canadian
Cordillera attest to past and present geological
processes that have shaped the Canadian
landscape. (Photo by Mike Schering)

The spectacular eruption of a volcano, the terror brought by an earthquake, the magnificent scenery of a mountain valley, and the destruction created by a landslide are all subjects for the geologist (Figure 1.1). The study of geology deals with many fascinating and practical questions about our physical environment. What forces produce mountains? Will there soon be another great earthquake in California? What was the Ice Age like? Will there be another? How are ore deposits formed? Should we look for water here? Is strip mining practical in this area? Will oil be found if a well is drilled at that location?

The Science of Geology

The subject of this text is **geology**, from the Greek *geo*, "Earth" and *logos*, "discourse." It is the science that pursues an understanding of planet Earth. Geology is traditionally divided into two broad areas—physical and historical. **Physical geology**, which is the primary focus of this book, examines the materials composing Earth and seeks to understand the many processes that operate beneath and upon its surface. The aim of **historical geology**, on the other hand, is to understand the origin of Earth and its development through time. Thus, it strives to establish an orderly chronological arrangement of the multitude of physical and biological changes that have occurred in the geologic past. The study of physical geology logically precedes the study of Earth history because we must first understand how Earth works before we attempt to unravel its past.

To understand Earth is challenging because our planet is a dynamic body with many interacting parts and a complex history. Throughout its long existence, Earth has been changing. In fact, it is changing as you read this page and will continue to do so in the future. Sometimes the changes are rapid and violent, as when landslides or volcanic eruptions occur. Conversely, change can take place so slowly that it goes unnoticed during a lifetime. Scales of size and space also vary greatly among the phenomena that geologists study. Sometimes they must focus on phenomena that are submicroscopic, and at other times they must deal with features that are continental or global in scale.

Geology is perceived as a science that is done outdoors, and rightly so. A great deal of geology is based on observations and experiments conducted in the field. But geology is also done in the laboratory where, for example, the study of various Earth materials provides insights into many basic processes. Frequently geology requires an understanding and application of knowledge and principles from physics, chemistry, and biology. Geology is a science that seeks to expand our knowledge of the natural world and our place in it. This basic philosophy has endured since the time of the pioneers of geology, including the knighted Canadian geologist, Sir William Logan (Box 1.1).

Geology, People, and the Environment

The primary focus of this book is to develop an understanding of basic geological principles, but along the way, we will explore numerous important relationships

◆ **Figure 1.1** California's eastern Sierra Nevada. Geologists study the process that created these majestic peaks. (Photo by Galen Rowell/Mountain Light Photography)

Canadian Profile
Sir William Logan: Canada's Premier Geologist

BOX 1.1

C. Gordon Winder*

In a 1998 list of the one hundred most important Canadians in history, a **geologist** was cited as the number one scientist. William Edmond Logan (1798-1875) was born in Montréal, the third child, and second son, of Scottish immigrants. His elementary education started in Montréal and finished in Edinburgh. In 1816, he registered for medicine at the University of Edinburgh, and completed the first year. He had the highest mark in his geometry class of 200 students and was awarded a brass octant, engraved with his name.

In 1817, Logan started working in London for his uncle, a commodity broker. He travelled to Northern Ireland, Paris, Spain (to view copper deposits), and Rome, where his diary records types of building stone. He was reading the available geology books and continued with lessons in geometry, languages, painting, and flute.

In 1830, Logan's uncle purchased a copper smelter at Swansea, Wales, and Logan was appointed business manager. Coal was mined around Swansea, but the supply was sporadic. Logan started mapping the area's geology. He used precise surveying instruments and his map proved to be so superior that the British Geological Survey adopted it as its official map. On the basis of upright petrified tree trunks and stumps in the strata, he proposed a theory for the origin of coal, a theory that is still valid today.

In the early 1840s, the government of Canada decided to establish a geological survey. Logan applied for the directorship and was appointed on April 20, 1842 and headquartered in Montreal. In 1843, he measured the rocks with coal at Joggins, Nova Scotia. In 1844, he examined the Gaspé peninsula, looking for coal. His collections of rock chips and fossils were wrapped in paper; he counted paces and used a Rochon micrometer to measure distance, a compass for direction, and mercury barometers for elevation; and he recorded his observations in leatherbound notebooks during the evenings beside a campfire.

His accommodation was a blanket thrown over saplings and he was assisted by aboriginal people who moved the camp and constructed birch bark canoes. He ate salt pork, sea biscuits, berries, fish, birds, and wild game, and drank tea with sugar. On one occasion, local people, puzzled by this aimlessly wandering mumbler, attempted to deliver him to an insane asylum.

In subsequent years, Logan traversed the Ottawa river to Lake Temiskaming; surveyed the areas around Lakes Superior and Huron for copper and iron deposits; and studied the complex geology of the Eastern Townships, where copper deposits were being explored. No coal deposit was ever discovered.

Logan trained his assistants himself. Alexander Murray was a retired naval officer who mapped southern Ontario (including the Petrolia oil fields) and later the area north of Lake Huron. Sterry Hunt was a university-trained chemist, who analyzed rocks, minerals, and water. Elkanah Billings, a lawyer and amateur paleontologist, identified fossils. There were two mapmakers and a handyman, who was once mistaken as the Survey Director for his elegant dress. The Director, commonly dressed in field clothes, looked like a handyman.

Logan drew attention to the potential mineral wealth of Canada at international exhibitions in London (1851) and Paris (1855). The Canadian rock and mineral specimens were the largest and most spectacular at the exhibitions. In January 1856, he was knighted by Queen Victoria and was the first individual born in Canada to be so honoured. He received numerous other awards in his lifetime.

As the Geological Survey director, Logan did administrative work, including writing field reports, cost accounting, answering letters (including some from swindlers) and sending requests to the government for funds. Occasionally, the government reduced the funding and Logan, convinced the Geological Survey was essential, paid the additional costs with his own money.

◆ **Figure 1.A**   Sir William Logan. (Reproduced with the permission of the Minister of Public Works and Government Services Canada, 2004 and courtesy of Natural Resources Canada, Geological Survey of Canada.)

Logan wrote the four copies of all reports, scientific and administrative, by hand. He worked day and night in the Survey office and had a folding chair that, with blankets, served as a bed—some people wondered if he ever slept. His worn out field boots lined the walls.

In 1863, a 983-page text, *Geology of Canada* was published and included a small atlas with a coloured geologic map of Canada, including parts of Newfoundland, the Maritime Provinces, the adjacent areas of the United States, and areas west of Lake Superior. In 1869, a large scale map was published; all copies were hand-coloured, probably by Logan, which contributed to eyestrain.

On November 30, 1869, Logan retired as the Geological Survey director and went to Llechryd, Wales to live with his sister Elizabeth. Each summer he returned to the Eastern Townships to examine the complex geology. He died in Wales on June 22, 1875, and is buried in the Anglican churchyard at Cilgerran, Wales. His name is on Canada's highest mountain, Mount Logan, 5959 metres, located in the southwest corner of the Yukon and featured on the front cover of this textbook.

*C. Gordon Winder is a Professor Emeritus in the Department of Earth Sciences, The University of Western Ontario

between people and the natural environment. Many of the problems and issues addressed by geology are of practical value to people.

Natural hazards are a part of living on Earth. Every day they adversely affect millions of people worldwide and are responsible for staggering damages (Figure 1.2). Among the hazardous Earth processes studied by geologists are volcanoes, floods, earthquakes, and landslides. Of course, geologic hazards are simply *natural* processes. They become hazards only when people try to live where these processes occur (Figure 1.3).

Resources represent another important focus of geology that is of great practical value to people. They include water and soil, a great variety of metallic and nonmetallic minerals, and energy. Together they form the very foundation of modern civilization. Geology deals not only with the formation and occurrence of these vital resources but also with maintaining supplies and with the environmental impact of their extraction and use.

Complicating all environmental issues is rapid world population growth and everyone's aspiration to a better standard of living. Earth is now gaining about 100 million people each year. This means a skyrocketing demand for resources and a growing pressure for people to dwell in environments having significant geologic hazards.

Not only do geologic processes have an impact on people but we humans can dramatically influence geologic processes as well. For example, river flooding is natural, but the magnitude and frequency of flooding can be changed significantly by human activities such as clearing forests, building cities, and constructing dams. Unfortunately, natural systems do not always adjust to artificial changes in ways that we can anticipate. Thus, an alteration to the environment that was intended to benefit society often has the opposite effect.

At appropriate places throughout this book, you will have the opportunity to examine different aspects of our relationship with the physical environment. It will be rare to find a chapter that does not address some aspect of natural hazards, environmental issues, or resources. Significant parts of some chapters provide the basic geologic knowledge and principles needed to understand environmental problems. Moreover, a number of the book's special-interest boxes focus on geology, people, and the environment by providing case studies or highlighting topical issues, particularly those of Canadian interest.

◆ **Figure 1.2** Two geologic hazards are represented in this image. On January 13, 2001, a magnitude 7.6 earthquake caused considerable damage in El Salvador. The damage pictured here was caused by a landslide triggered by the earthquake. As many as 1000 people were buried under 8 metres of landslide debris. (Photo by Reuters/STR/Archive Photos)

◆ **Figure 1.3** This is an image of Italy's Mt. Vesuvius in September 2000. This major volcano is surrounded by the city of Naples and the Bay of Naples. In AD 79, Vesuvius explosively erupted, burying the towns of Pompeii and Herculanaeum in volcanic ash. Will it happen again? Geologic hazards are *natural* processes. They only become hazards when people try to live where these processes occur. (Image courtesy of NASA)

Some Historical Notes About Geology

The nature of our Earth—its materials and processes—has been a focus of study for centuries. Writings about such topics as fossils, gems, earthquakes, and volcanoes date back to the early Greeks, more than 2300 years ago.

Certainly the most influential Greek philosopher was Aristotle. Unfortunately, Aristotle's explanations about the natural world were not based on keen observations and experiments. Instead, they were arbitrary pronouncements. He believed that rocks were created under the "influence" of the stars and that earthquakes occurred when air crowded into the ground, was heated by central fires, and escaped explosively. When confronted with a fossil fish, he explained "a great many fishes live in the earth motionless and are found when excavations are made."

Although Aristotle's explanations may have been adequate for his day, they unfortunately continued to be expounded for many centuries, thus thwarting the acceptance of more up-to-date accounts. Frank D. Adams states in *The Birth and Development of the Geological Sciences* (New York: Dover, 1938) that "throughout the Middle Ages Aristotle was regarded as the head and chief of all philosophers; one whose opinion on any subject was authoritative and final."

Catastrophism   In the mid-1600s, James Ussher, Anglican Archbishop of Armagh, Primate of all Ireland, published a major work that had immediate and profound influence. A respected scholar of the Bible, Ussher constructed a chronology of human and Earth history in which he determined that Earth was only a few thousand years old, having been created in 4004 B.C. Ussher's treatise earned widespread acceptance among Europe's scientific and religious leaders, and his chronology was soon printed in the margins of the Bible itself.

During the seventeenth and eighteenth centuries the doctrine of **catastrophism** strongly influenced people's thinking about Earth. Briefly stated, catastrophists believed that Earth's landscapes had been shaped primarily by great catastrophes. Features such as mountains and canyons, which today we know take great periods of time to form, were explained as having been produced by sudden and often worldwide disasters produced by unknown causes that no longer operate. This philosophy was an attempt to force-fit the rates of Earth processes to the then-current ideas on the age of Earth.

The relationship between catastrophism and the age of Earth has been summarized as follows:

That the earth had been through tremendous adventures and had seen mighty changes during its obscure past was plainly evident to every inquiring eye; but to concentrate these changes into a few brief millenniums required a tailor-made philosophy, a philosophy whose basis was sudden and violent change.*

The Birth of Modern Geology   Modern geology began in the late 1700s when James Hutton, a Scottish physician and gentleman farmer, published his *Theory of the Earth*. In this work, Hutton put forth a fundamental principle that is a pillar of geology today: **uniformitarianism**. It simply states that the *physical, chemical, and biological laws that operate today have also operated in the geologic past*. This means that the forces and processes that we observe presently shaping our planet have been at work for a very long time. Thus, to interpret ancient rocks, we must first understand present-day processes and their results. This idea is summarized in the statement "the present is the key to the past."

Prior to Hutton's *Theory of the Earth*, no one had effectively demonstrated that geological processes occur over extremely long periods of time. However, Hutton persuasively argued that forces that appear small could, over long spans of time, produce effects that were just as great as those resulting from sudden catastrophic events. Unlike his predecessors, Hutton carefully cited verifiable observations to support his ideas.

For example, when he argued that mountains are sculpted and ultimately destroyed by weathering and the work of running water, and that their wastes are carried to the oceans by processes that can be observed, Hutton said, "We have a chain of facts which clearly demonstrates... that the materials of the wasted mountains have traveled through the rivers"; and further, "There is not one step in all this progress... that is not to be actually perceived." He then went on to summarize this thought by asking a question and immediately providing the answer: "What more can we require? Nothing but time."

Today the basic tenets of uniformitarianism are just as viable as in Hutton's day. Indeed, we realize more strongly than ever that the present gives us insight into the past and that the physical, chemical, and biological laws that govern geological processes remain unchanging through time. However, we also understand that the doctrine should not be taken too literally. To say that geological processes in the past were the same as those occurring today is not to suggest that they always had the same relative importance or that they operated at precisely the same rate. Moreover, some important geologic processes are not currently observable, but evidence that they occur is well established. For example, we know that Earth has experienced impacts from large meteorites even though we have no human witnesses. Such events altered Earth's crust, modified its climate, and strongly influenced life on the planet.

The acceptance of uniformitarianism meant the understanding of a very old age for Earth. Although Earth processes vary in intensity, they still take a long time to create or destroy major landscape features (Figure 1.4).

For example, geologists have established that mountains once existed in portions of the present-day Maritimes. Today the region consists of low hills and plains. Erosion (processes that wear land away) gradually destroyed these peaks. Estimates indicate that the North American continent is being lowered at a rate of about 3 centimetres per 1000 years. At this rate it would take 100 million years for water, wind, and ice to lower mountains that were 3000 metres high.

But even this time span is relatively short on the time scale of Earth history, for the rock record contains evidence that shows Earth has experienced many cycles of mountain building and erosion. Concerning the ever-changing nature of Earth through great expanses of geologic time, Hutton made a statement that was to become his most famous. In concluding his classic 1788 paper published in the *Transactions of the Royal Society of Edinburgh*, he stated, "The results, therefore, of our present enquiry is, that we find no vestige of a beginning—no prospect of an end." A quote from William L. Stokes sums up the significance of Hutton's basic concept:

In the sense that uniformitarianism implies the operation of timeless, changeless laws or principles, we can say that nothing in our incomplete but extensive knowledge disagrees with it.*

In the chapters that follow, we will be examining the materials that compose our planet and the processes that modify it. It is important to remember

*H. E. Brown, V. E. Monnett, and J. W. Stovall, *Introduction to Geology* (New York: Blaisdell, 1958).

Essentials of Earth History (Englewood Cliffs, New Jersey: Prentice Hall, 1966), p. 34.

◆ **Figure 1.4** Geologist Dr. Gerd Westermann contemplates how weathering and erosion have sculpted the spectacular hoodoos of the Drumheller area in southern Alberta. Some geologic processes act so slowly that changes may not be visible during an entire human lifetime.
(Photo by C. Tsujita)

that, although many features of our physical landscape may seem to be unchanging over the decades we observe them, they are nevertheless changing, but on time scales of hundreds, thousands, or even many millions of years.

Geologic Time

 Geologic Time
↳ Relative Dating

Although Hutton and others recognized that geologic time is exceedingly long, they had no methods to accurately determine the age of Earth. However, in 1896 radioactivity was discovered. Using radioactivity for dating was first attempted in 1905 and has been refined ever since. Geologists are now able to assign fairly accurate dates to events in Earth history.*

---

*Chapter 8 is devoted to a much more complete discussion of geologic time.

For example, we know the dinosaurs became extinct about 65 million years ago. Today the age of Earth is put at about 4.5 billion years.

Relative Dating and the Geologic Time Scale

During the nineteenth century, long before the advent of radiometric dating, a geologic time scale was developed using principles of relative dating. **Relative dating** means that events are placed in their proper sequence or order without knowing their age in years. This is done by applying principles such as the **law of superposition** (*super* = over, *positum* = to place), which states that in layers of sedimentary rocks or lava flows, the youngest layer is on top and the oldest is on the bottom (assuming that nothing has turned the layers upside down, which sometimes happens). Arizona's Grand Canyon provides a fine example in which the oldest rocks are located in the inner gorge and the youngest rocks are found on the rim (Figure 1.5). The law of superposition establishes the *sequence*

◆ **Figure 1.5**   These rock layers were exposed by the downcutting of the Colorado River as it created the Grand Canyon. Relative ages of these layers can be determined by applying the law of superposition. The youngest rocks are on top, and the oldest are at the bottom.
(Photo by Marc Muench/David Muench Photography, Inc.)

of rock layers—not their numerical ages. Today such a proposal appears to be elementary, but 300 years ago it amounted to a major breakthrough in scientific reasoning by establishing a rational basis for relative time measurements.

Fossils, the remains or traces of prehistoric life, were also essential to the development of the geologic time scale (Figure 1.6). Fossils are the basis for the **principle of fossil succession**, which states that *fossil organisms succeed one another in a definite and determinable order, and therefore any time period can be recognized by its fossil content.* This principle was laboriously worked out over decades by collecting fossils from countless rock layers around the world. Once established, it allowed geologists to identify rocks of the same age in widely separated places and to build the geologic time scale shown in Figure 1.7.

Notice that units having the same designations do not necessarily extend for the same number of years. For example, the Cambrian period lasted about 50 million years, whereas the Silurian period spanned

only about 26 million years. As we will emphasize again in Chapter 8, this situation exists because the basis for establishing the time scale was not the regular rhythm of a clock, but the changing character of life forms through time. Specific dates were added long after the time scale was established. A glance at Figure 1.7 also reveals that the Phanerozoic eon is divided into many more units than earlier eons even though it encompasses only about 12 percent of Earth history. The meagre fossil record for these earlier eons is the primary reason for the lack of detail on this portion of the time scale. Without abundant fossils, geologists lose their primary tool for subdividing geologic time.

The Magnitude of Geologic Time

The concept of geologic time is new to many non-geologists. People are accustomed to dealing with increments of time that are measured in hours, days, weeks, and years. Our history books often examine events over spans of centuries, but even a century is difficult

Chapter 13:

Deserts and Winds   365

Chapter 14:

Shorelines   385

A   B

◆ **Figure 1.6** **A.** Armoured head region of a primitive Devonian fossil fish from Quebec. **B.** An Ordovician trilobite from the Georgian Bay area of Ontario. (Photo A courtesy of Terry Ciotka, Canada Fossils Inc. Photo B by Denis Tetreault)

to appreciate fully. For most of us, someone or something that is 90 years old is *very old*, and a 1000-year-old artifact is *ancient*.

By contrast, those who study geology must routinely deal with vast time periods—millions or billions (thousands of millions) of years. When viewed in the context of Earth's 4.5-billion-year history, a geologic event that occurred 100 million years ago may be characterized as "recent" by a geologist, and a rock sample that has been dated at 10 million years may be called "young."

An appreciation for the magnitude of geologic time is important in the study of geology because many processes are so gradual that vast spans of time are needed before significant changes occur.

How long is 4.5 billion years? If you were to begin counting at the rate of one number per second and continued 24 hours a day, 7 days a week and never stopped, it would take about two lifetimes (150 years) to reach 4.5 billion! Another interesting basis for comparison is as follows:

Compress, for example, the entire 4.5 billion years of geologic time into a single year. On that scale, the oldest rocks we know date from about mid-March. Living things first appeared in the sea in May. Land plants and animals emerged in late November and the widespread swamps that formed the Pennsylvanian coal deposits flourished for about four days in early December. Dinosaurs became dominant in mid-December, but disappeared on the 26th, at about the time the Rocky Mountains were first uplifted. Human-like creatures appeared sometime during the evening of December 31st, and the most recent continental ice sheets began to recede from the Great Lakes area and from northern Europe about 1 minute and 15 seconds before midnight on the 31st. Rome ruled the Western world for 5 seconds

from 11:59:45 to 11:59:50. Columbus discovered America 3 seconds before midnight, and the science of geology was born with the writings of James Hutton just slightly more than one second before the end of our eventful year of years.*

The foregoing is just one of many analogies that have been conceived in an attempt to convey the magnitude of geologic time. Although helpful, all of them, no matter how clever, only begin to help us comprehend the vast expanse of Earth history.

The Nature of Scientific Inquiry

All science is based on the assumption that the natural world behaves in a consistent and predictable manner that is comprehensible through careful, systematic study. The overall goal of science is to discover the underlying patterns in nature and then to use this knowledge to make predictions about what should or should not be expected, given certain facts or circumstances. For example, by knowing how oil reservoirs form, geologists are able to predict the most favourable sites for exploration and how to avoid regions having little or no potential.

The development of new scientific knowledge involves some basic logical processes that are universally accepted. To determine what is occurring in the natural world, scientists collect scientific "facts" through observation and measurement (Figure 1.8). Because some error is inevitable, the accuracy of a particular measurement or observation is always open to question. Nevertheless, these data are essential to science and serve as the springboard for the development of scientific theories.

*Don L. Eicher, *Geologic Time*, 2nd ed. (Englewood Cliffs, New Jersey: Prentice Hall, 1978), pp. 18–19. Reprinted by permission.

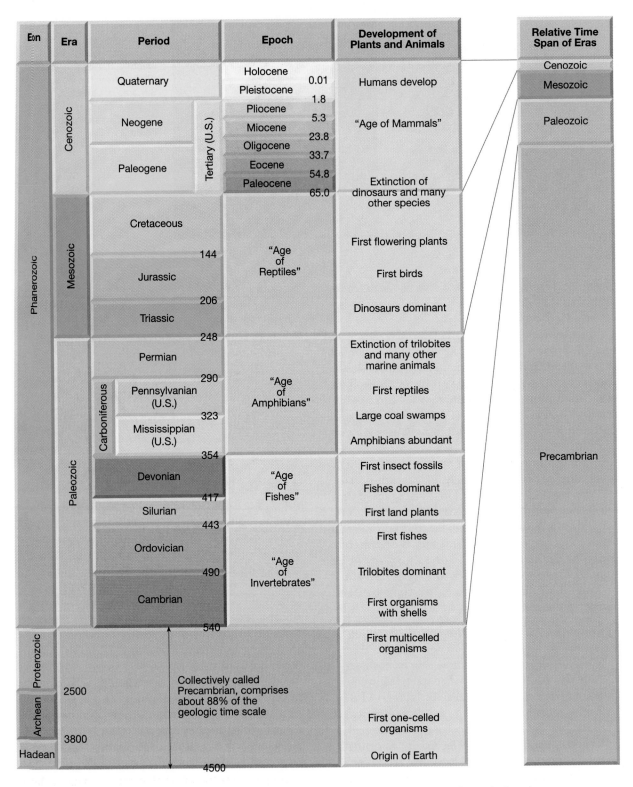

◆ **Figure 1.7**   The geologic time scale. Numbers on the time scale represent time in millions of years before the present. These dates were added long after the time scale had been established using relative dating techniques. The Precambrian accounts for more than 88 percent of geologic time. The Mississippian and Pennsylvanian are internationally recognized as sub-periods except in the USA where the names originated.
(Modified after the Geological Society of America)

Hypothesis

Once facts have been gathered and principles have been formulated to describe a natural phenomenon, investigators try to explain how or why things happen in the manner observed. They often do this by constructing a tentative (or untested) explanation, which is called a scientific *hypothesis* or *model*. (The term *model*, although often used synonymously with hypothesis, is a less precise term because it is sometimes used to describe a scientific theory as well.) It is best if an investigator can formulate more than one hypothesis to explain a given set of observations. If an individual scientist is unable to devise multiple models, others in the scientific community will almost always develop alternative explanations. A spirited debate frequently ensues. As a result, extensive research is conducted by proponents of opposing models, and the results are made available to the wider scientific community in scientific journals.

Before a hypothesis can become an accepted part of scientific knowledge, it must pass objective testing and analysis. (If a hypothesis cannot be tested, it is not scientifically useful, no matter how interesting it might seem.) The verification process requires that *predictions* be made based on the model being considered and that the predictions be tested by comparing them against objective observations of nature. Put another way, hypotheses must fit observations other than those used to formulate them in the first place. Those hypotheses that fail rigorous testing are ultimately discarded. The history of science is littered with discarded hypotheses. One of the best known is the Earth-centred model of the universe—a proposal that was supported by the apparent daily motion of the Sun, Moon, and stars around Earth. As the mathematician Jacob Bronowski so ably stated, "Science is a great many things, but in the end they all return to this: Science is the acceptance of what works and the rejection of what does not."

Theory

When a hypothesis has survived extensive scrutiny and when competing models have been eliminated, a hypothesis may be elevated to the status of a scientific **theory**. In everyday language we may say, "that's only a theory." But a scientific theory is a well-tested and widely accepted view that the scientific community agrees best explains certain observable facts.

Theories that are extensively documented are held with a very high degree of confidence. Theories of this stature that are comprehensive in scope have a special status. They are called **paradigms** because they explain a large number of interrelated aspects of the natural world. For example, the theory of plate

◆ **Figure 1.8**   This scientist is gathering ice samples in Antarctica.
(Joyce Photographics/Photo Researchers, Inc.)

tectonics is a paradigm of the geological sciences that provides the framework for understanding the origin of mountains, earthquakes, and volcanic activity. In addition, plate tectonics explains the evolution of the continents and the ocean basins through time—a topic we will consider later in this chapter.

Scientific Methods

The process just described, in which researchers gather facts through observations and formulate scientific hypotheses and theories, is called the *scientific method*. Contrary to popular belief, the scientific method is not a standard recipe that scientists apply in a routine manner to unravel the secrets of our natural world. Rather, it is an endeavour that involves creativity and insight. Rutherford and Ahlgren put it this way: "Inventing hypotheses or theories to imagine how the world works and then figuring out how they can be put to the test of reality is as creative as writing poetry, composing music, or designing skyscrapers."*

---

*F. James Rutherford and Andrew Ahlgren, *Science for All Americans* (New York: Oxford University Press, 1990), p. 7.

There is not a fixed path that scientists always follow that leads unerringly to scientific knowledge. Nevertheless, many scientific investigations involve the following steps: (1) the collection of scientific facts through observation and measurement; (2) the development of one or more working hypotheses or models to explain these facts; (3) development of observations and experiments to test the hypotheses; and (4) the acceptance, modification, or rejection of the model based on extensive testing.

Other scientific discoveries may result from purely theoretical ideas, which stand up to extensive examination. Some researchers use high-speed computers to simulate what is happening in the "real" world. These models are useful when dealing with natural processes that occur on very long time scales or take place in extreme or inaccessible locations. Still other scientific advancements are made when something totally unexpected happens during an experiment. These serendipitous discoveries are more than pure luck, for as Louis Pasteur said, "In the field of observation, chance favours only the prepared mind."

Scientific knowledge is acquired through several avenues, so it might be best to describe the nature of scientific inquiry as the methods of science rather than the scientific method. In addition, it should always be remembered that even the most compelling scientific theories are still simplified explanations of the natural world.

Students Sometimes Ask...

In class you compared a hypothesis to a theory. How is each one different from a scientific law?

A scientific *law* is a basic principle that describes a particular behaviour of nature that is generally narrow in scope and can be stated briefly—often as a simple mathematical equation. Because scientific laws have been shown time and time again to be consistent with observations and measurements, they are rarely discarded. Laws may, however, require modifications to fit new findings. For example, Newton's laws of motion are still useful for everyday applications (NASA uses them to calculate satellite trajectories), but they do not work at velocities approaching the speed of light. For these circumstances, they have been supplanted by Einstein's theory of relativity.

A View of Earth

Introduction
↳ A View of Earth

A view of Earth, such as the one in Figure 1.9A, provided the *Apollo 8* astronauts and the rest of humanity with a unique perspective of our home. Seen from space, Earth is breathtaking in its beauty and startling

A   B

◆ **Figure 1.9   A.** View that greeted the *Apollo 8* astronauts as their spacecraft emerged from behind the Moon. **B.** Africa and Arabia are prominent in this image of Earth taken from *Apollo 17*. The tan cloud-free zones over the land coincide with major desert regions. The band of clouds across central Africa is associated with a much wetter climate that in places sustains tropical rain forests. The dark blue of the oceans and the swirling cloud patterns remind us of the importance of the oceans and the atmosphere. Antarctica, a continent covered by glacial ice, is visible over the South Pole.
(Photo A. NASA Headquarters; Photo B. NASA/Science Source/Photo Researchers, Inc.)

in its solitude. Such an image reminds us that our home is, after all, a planet—small, self-contained, and fragile.

As we look more closely at our planet from space, it becomes apparent that Earth is much more than rock and soil (Figure 1.9B). In fact, the most conspicuous features are not the continents but swirling clouds suspended above the surface and the vast global ocean. These features emphasize the importance of air and water to our planet.

The closer view of Earth from space, shown in Figure 1.9B, helps us appreciate why the physical environment is traditionally divided into three major parts: the water portion of our planet, the hydrosphere; Earth's gaseous envelope, the atmosphere; and, of course, the solid Earth.

It needs to be emphasized that our environment is highly integrated and not dominated by rock, water, or air alone. Rather, it is characterized by continuous interactions as air comes in contact with rock, rock with water, and water with air. Moreover, the biosphere, which is the totality of all plant and animal life on our planet, interacts with each of the three physical realms and is an equally integral part of the planet. Thus, Earth can be thought of as consisting of four major spheres: the hydrosphere, atmosphere, solid Earth, and biosphere.

The interactions among Earth's four spheres are uncountable. Figure 1.10 provides us with one easy-to-visualize example. The shoreline is an obvious meeting place for rock, water, and air. Ocean waves created by the drag of air moving across open water break against the rocky shore. The force of the water can be powerful, and the erosional work that is accomplished can be great.

Hydrosphere

Earth is sometimes called the *blue* planet. Water more than anything else makes Earth unique. The **hydrosphere** is a dynamic mass of water that is continually on the move, evaporating from the oceans to the atmosphere, precipitating to the land, and running back to the ocean again. The global ocean is certainly the most prominent feature of the hydrosphere, blanketing nearly 71 percent of Earth's surface to an average depth of about 3800 metres. It accounts for about 97 percent of Earth's water. However, the hydrosphere also includes the fresh water found underground and in streams, lakes, and glaciers. Moreover, water is an important component of all living things.

Although these latter sources constitute just a tiny fraction of the total, they are much more important than their meagre percentage indicates. In addition to providing the fresh water that is so vital to life on

◆ **Figure 1.10** The shoreline is one obvious meeting place for rock, water, and air. The rocks that dominate this scene were smoothed by ocean waves that were created by the force of moving air. The calm conditions apparent in this photograph of Peggy's Cove, Nova Scotia, are deceptive, for the force of the water can be powerful during storms, and the erosional work that is accomplished can be great. (Photo by Mike Schering)

land, streams, glaciers, and groundwater are responsible for sculpting and creating many of our planet's varied landforms.

Atmosphere

Earth is surrounded by a life-giving gaseous envelope called the **atmosphere**. Compared with the solid Earth, the atmosphere is thin and tenuous. One half lies below an altitude of 5.6 kilometres, and 90 percent occurs within just 16 kilometres of Earth's surface. By comparison, the radius of the solid Earth (distance from the surface to the centre) is about 6400 kilometres! Despite its modest dimensions, this thin blanket of air is an integral part of the planet. It not only provides the air that we breathe but also protects us from the Sun's intense heat and dangerous ultraviolet radiation. The energy exchanges that continually occur between the atmosphere and Earth's surface and between the atmosphere and space produce the effects we call weather and climate.

If, like the Moon, Earth had no atmosphere, our planet would not only be lifeless but many of the processes and interactions that make the surface such a dynamic place could not operate. Without weathering and erosion, the face of our planet might more closely resemble the lunar surface, which has not changed appreciably in nearly 3 billion years.

Biosphere

The **biosphere** includes all life on Earth. It is concentrated near the surface in a zone that extends from the ocean floor upward for several kilometres into the atmosphere. Plants and animals depend on the physical environment for the basics of life. However, organisms do more than just respond to their physical environment. Through countless interactions life forms help maintain and alter their physical environment. Without life, the makeup and nature of the solid Earth, hydrosphere, and atmosphere would be very different.

Solid Earth

Lying beneath the atmosphere and the oceans is the solid Earth. Much of our study of the solid Earth focuses on the more accessible surface features. Fortunately, many of these features represent the outward expressions of the dynamic behaviour of Earth's interior. By examining the most prominent surface features and their global extent, we can obtain clues to the dynamic processes that have shaped our planet. A first look at the structure of Earth's interior and at the major surface features of the solid Earth will come later in the chapter.

Earth as a System

As we study Earth, it becomes clear that our planet can be viewed as a dynamic body with many separate but interacting parts or *spheres*. The hydrosphere, atmosphere, biosphere, and solid Earth and all of their components can be studied separately. However, the parts are not isolated. Each is related in some way to the others to produce a complex and continuously interacting whole that we call the *Earth system*.

A **system** is a group of interacting, or interdependent, parts that form a complex whole. Most of us hear and use the term frequently. We may service our car's cooling *system*, make use of the city's transportation *system*, and be a participant in the political *system*. A news report might inform us of an approaching weather

system. Further, we know that Earth is just a small part of a large system known as the *solar system*.

The parts of the Earth system are linked, so a change in one part can produce changes in any or all of the other parts. For example, when a volcano erupts, lava from Earth's interior may flow out at the surface and block a nearby valley. This new obstruction influences the region's drainage system by creating a lake or causing streams to change course. The large quantities of volcanic ash and gases that can be emitted during an eruption might be blown high into the atmosphere and influence the amount of solar energy that can reach Earth's surface. The result could be a drop in air temperatures over the entire hemisphere.

Where the surface is covered by lava flows or a thick layer of volcanic ash, existing soils are buried. This causes the soil-forming processes to begin anew to transform the new surface material into soil (Figure 1.11). The soil that eventually forms will reflect the interactions among many parts of the Earth system—the volcanic parent material, the type and rate of weathering, and the impact of biological activity. Of course, there would also be significant changes in the biosphere. Some organisms and their habitats would be eliminated by the lava and ash, whereas new settings for life, such as the lake, would be created. The potential climate change could also impact sensitive life forms.

The Earth system is characterized by processes that vary on spatial scales from fractions of millimetres to thousands of kilometres. Time scales for Earth's processes range from milliseconds to billions of years. As we learn about Earth, it becomes increasingly clear that despite significant separations in distance or time, many processes are connected and that a change in one component can influence the entire system.

The Earth system is powered by energy from two sources. The Sun drives external processes that occur in the atmosphere, hydrosphere, and at Earth's surface. Weather and climate, ocean circulation, and erosional processes are driven by energy from the Sun. Earth's interior is the second source of energy. Heat remaining from when our planet formed, and heat that is continuously generated by radioactive decay, powers the internal processes that produce volcanoes, earthquakes, and mountains.

Humans are *part of* the Earth system, a system in which the living and nonliving components are entwined and interconnected. Therefore, our actions produce changes in all of the other parts. When we burn gasoline and coal, build seawalls along the shoreline, dispose of our wastes, and clear the land, we cause other parts of the system to respond, often in unforeseen ways. Throughout this book you will learn about many of Earth's subsystems: the hydrologic system, the tectonic (mountain-building) system, and the rock

◆ **Figure 1.11**   When Mount St. Helens erupted in May 1980 the area shown here was buried by a volcanic mudflow. Now plants are re-established and new soil is forming. (Jack W. Dykinga Associates)

cycle, to name a few. Remember that these components *and we humans* are all part of the complex interacting whole we call the Earth system.

The Rock Cycle: Part of the Earth System

 Earth Materials
↳ The Rock Cycle

The Earth system has a nearly endless array of cycles in which matter is recycled over and over again. Viewed over long time spans, the rocks of the solid Earth are constantly forming, changing, and reforming. The cycle that involves the processes by which one type of rock changes to another is called the **rock cycle**. This cycle allows us to view many of the interrelationships among different parts of the Earth system. It helps us understand the origin of igneous, sedimentary, and metamorphic rocks and to see that each type is linked to the others by the processes that act upon and within the planet. Learn the rock cycle well; you will be examining its interrelationships in greater detail throughout this book.

The Basic Cycle

We will begin our look at the rock cycle at the bottom of Figure 1.12. *Magma* is molten material that forms in certain environments of Earth's interior where temperatures and pressures are such that rock melts. Once formed, magma migrates upward into Earth's outer layer or crust. Eventually magma cools and solidifies. This process, called *crystallization*, may occur either beneath the surface or, following a volcanic eruption, at the surface. In either situation the resulting rocks are called **igneous rocks** (*ignis* = fire, *ous* = full of).

If igneous rocks are exposed at the surface, they will undergo *weathering*, in which the day-in and day-out influences of the atmosphere slowly disintegrate and decompose rocks. The materials that result are often moved downslope by gravity before being picked up and transported by any of a number of erosional agents—running water, glaciers, wind, or waves. Eventually these particles and dissolved substances are deposited as **sediment**. Although most sediment ultimately comes to rest in the ocean, other sites of deposition include river floodplains, desert basins, swamps, and dunes.

Next the sediments undergo *lithification*, a term meaning "conversion into rock." Sediment is usually lithified into **sedimentary rock** when compacted by the weight of overlying layers or when cemented as percolating water fills the pores with mineral matter.

If the resulting sedimentary rock is buried deep within Earth and involved in the dynamics of mountain building, or intruded by a mass of magma, it will

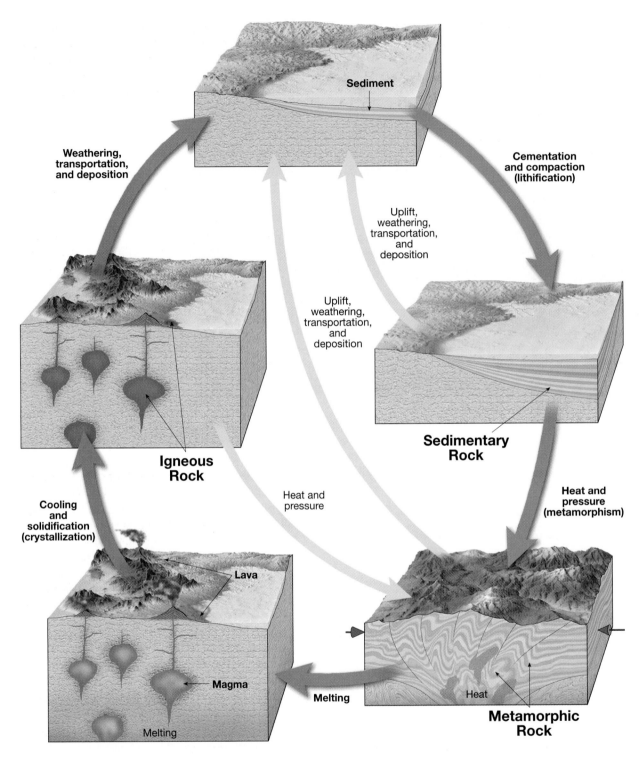

Weathering, transportation, and deposition

Sediment

Cementation and compaction (lithification)

Uplift, weathering, transportation, and deposition

Uplift, weathering, transportation, and deposition

Igneous Rock

Sedimentary Rock

Heat and pressure

Heat and pressure (metamorphism)

Cooling and solidification (crystallization)

Lava

Magma

Melting

Heat

Metamorphic Rock

Melting

◆ **Figure 1.12**   Viewed over long spans, rocks are constantly forming, changing, and reforming. The rock cycle helps us understand the origin of the three basic rock groups. Arrows represent processes that link each group to the others.

be subjected to great pressures and intense heat. The sedimentary rock will react to the changing environment and turn into the third rock type, **metamorphic rock** (*meta* = change, *morph* = form). When metamorphic rock is subjected to additional pressure changes or to still higher temperatures, it will melt, creating magma, which will eventually crystallize into igneous rock.

Processes driven by heat from Earth's interior are responsible for creating igneous and metamorphic rocks. Weathering and erosion—external processes powered by a combination of energy from the Sun and gravity—produce the sediment from which sedimentary rocks form.

Alternative Paths

The paths shown in the basic cycle are not the only ones that are possible. To the contrary, other paths are just as likely to be followed as those described in the preceding section. These alternatives are indicated by the blue arrows in Figure 1.12.

Igneous rocks, rather than being exposed to weathering and erosion at Earth's surface, may remain deeply buried. Eventually these masses may be subjected to the strong compressional forces and high temperatures associated with mountain building. When this occurs, they are transformed directly into metamorphic rocks.

Metamorphic and sedimentary rocks, as well as sediment, do not always remain buried. Rather, overlying layers may be stripped away, exposing the once-buried rock. When this happens, the material is attacked by weathering processes and turned into new raw materials for sedimentary rocks.

Although rocks might seem to be unchanging masses, the rock cycle clearly shows that they are not. The changes, however, take time—great amounts of time.

The Face of Earth

Introduction
 ↳ Features of the Continents
 ↳ Floor of the Ocean

The two principal divisions of Earth's surface are the continents and the ocean basins (Figure 1.13). A significant difference between these two areas is their relative levels. The average elevation of the continents above sea level is about 840 metres, whereas the average depth of the oceans is about 3800 metres. Thus, the continents stand, on average, 4640 metres above the level of the ocean floor.

Although land and sea meet at the shoreline, this is not the boundary between the continents and the ocean basins. Rather, along most coasts a gently sloping platform of continental material, called the **continental shelf**, extends seaward from the shore. A glance at Figure 1.13 shows that the width of the continental shelf is variable. For example, it is broad along the East and Arctic coasts of North America but relatively narrow along the Pacific margin of the continent.

The boundary between the continents and the deep-ocean basins lies along the **continental slope**, which is a steep dropoff that extends from the outer edge of the continental shelf to the floor of the deep ocean (Figure 1.13). Using this as the dividing line, we find that about 60 percent of Earth's surface is represented by ocean basins and the remaining 40 percent by continents.

Continents

The most prominent topographic features of the continents are linear mountain belts. Although the distribution of mountains may appear to be random, this is not the case. When the youngest mountains are considered, we find that they are located principally in two zones. The circum-Pacific belt (the region surrounding the Pacific Ocean) includes the mountains of the western Americas and continues into the western Pacific in the form of volcanic island arcs. Island arcs are active mountainous regions composed largely of deformed volcanic rocks and include the Aleutian Islands, Japan, the Philippines, and New Guinea.

The other major mountain belt extends eastward from the Alps through Iran and the Himalayas and then dips southward into Indonesia. Careful examination of mountainous terrains reveals that most are places where thick sequences of rocks have been squeezed and highly deformed, as if placed in a gigantic vise.

Older mountains are also found on the continents. Examples include the Appalachians in eastern North America and the Urals in Russia. Their once lofty peaks are now worn low, the result of millions of years of erosion. Still older are the stable continental interiors. Within these stable interiors are areas known as **shields**, extensive and relatively flat expanses composed largely of crystalline material. Radiometric ages of some samples approach 4 billion years. Even these oldest-known rocks exhibit evidence of enormous forces that have folded and deformed them.

Ocean Basins

A diversity of features occur on the ocean-basin floor, including linear chains of volcanoes, deep canyons, and large plateaus. The topography is nearly as varied as that on the continents.

The ocean basins contain one of the most prominent topographic features on Earth: the **oceanic ridge system**. In Figure 1.13 the Mid-Atlantic Ridge and the East Pacific Ridge are parts of this system. This broad elevated feature forms a continuous belt that winds for more than 70,000 kilometres around the globe in a manner similar to the seam of a baseball. Rather than consisting of highly deformed rock, such as most of the mountains found on the continents,

◆ **Figure 1.13**   Major physical features of the continents and ocean basins. The diversity of features on the ocean floor is as varied as on the continents.

the oceanic ridge system consists of layer upon layer of igneous rock that has been fractured and uplifted.

The ocean floor also contains extremely deep depressions that are occasionally more than 11,000 metres deep. Although these deep-ocean **trenches** are relatively narrow and represent only a small fraction of the ocean floor, they are very significant features. Some trenches are located adjacent to young mountains that flank the continents. For example, in Figure 1.13 the Peru–Chile trench off the west coast of South America parallels the Andes Mountains. Other trenches parallel linear island chains called volcanic island arcs.

What is the connection, if any, between the young, active mountain belts and the oceanic trenches? What is the significance of the enormous ridge system that extends through all the world's oceans? What forces crumple rocks to produce majestic mountain ranges? These questions must be answered if we are to understand the dynamic processes that shape our planet.

Early Evolution of Earth

Recent earthquakes caused by displacements of Earth's crust, and lavas erupted from active volcanoes, represent only the latest in a long line of events by which our planet has attained its present form and structure. The geologic processes operating in Earth's interior can be best understood when viewed in the context of much earlier events in Earth history.

Origin of Planet Earth

Earth is one of nine planets that, along with several dozen moons and numerous smaller bodies, revolve around the Sun. The orderly nature of our solar system leads most researchers to conclude that Earth and the other planets formed at essentially the same time and from the same primordial material as the Sun. The **nebular hypothesis** suggests that the bodies of our solar system evolved from an enormous rotating cloud called the *solar nebula*, composed mostly of hydrogen and helium, with a small percentage of the heavier elements (Figure 1.14).

About 5 billion years ago this huge cloud of gases and minute particles began to gravitationally contract. (What triggered the collapse is not known.) As this slowly spiralling nebula contracted, it rotated faster and faster for the same reason ice skaters do when they draw their arms toward their bodies. Eventually the inward pull of gravity came into balance with the outward force caused by the rotational motion of the nebula (Figure 1.14). By this time the once vast cloud had assumed a flat disk shape with a large concentration of material at its centre called the *protosun* (pre-Sun).

During the collapse, gravitational energy was converted to thermal energy (heat), causing the temperature of the inner portion of the nebula to dramatically rise. However, at distances beyond the orbit of Mars, the temperatures probably remained quite low. Here, at −200°C, the tiny particles of the nebula were likely covered with a thick layer of ices made of water, carbon dioxide, ammonia, and methane. The disk-shaped cloud also contained appreciable amounts of the lighter gases, hydrogen and helium.

The formation of the Sun marked the end of the period of contraction and thus the end of gravitational heating. Temperatures in the region where the inner planets now reside began to decline. The decrease in temperature caused those substances with high melting points to coalesce (join together). Materials such as iron and nickel and the elements of which the rock-forming minerals are composed—silicon, calcium, sodium, and so forth—formed metallic and rocky clumps that orbited the Sun. Repeated collisions caused these fragments to coalesce into larger asteroid-sized bodies, which in a few tens of millions of years accreted into the four inner planets we call Mercury, Venus, Earth, and Mars. Not all of these clumps of matter were incorporated into the *protoplanets* (pre-planets). Those that remained in orbit are called *meteorites* when they survive an impact with Earth.

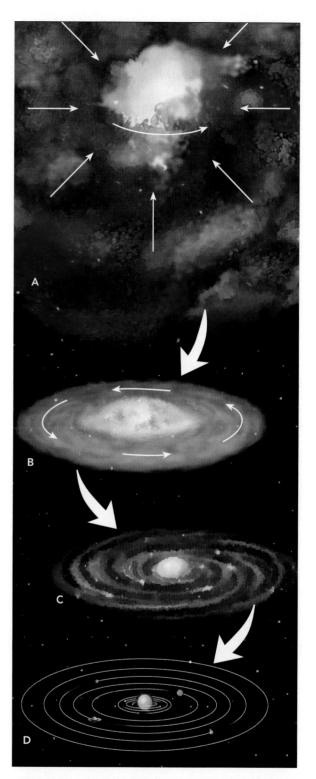

◆ **Figure 1.14** Nebular hypothesis. **A.** A huge rotating cloud of dust and gases (nebula) begins to contract. **B.** Most of the material is gravitationally swept toward the centre, producing the Sun. However, owing to rotational motion, some dust and gases remain orbiting the central body as a flattened disk. **C.** The planets begin to accrete from the material that is orbiting within the flattened disk. **D.** In time most of the remaining debris was either collected into the nine planets and their moons or swept out into space by the solar wind.

As more and more material was swept up by the protoplanets, the high-velocity impact of nebular debris caused the temperature of these bodies to rise. Because of their relatively high temperatures and weak gravitational fields, the inner planets were unable to accumulate much of the lighter components of the nebular cloud. The lightest of these, hydrogen and helium, were eventually whisked from the inner solar system by the solar winds. However, Earth, Mars, and Venus were able to retain various amounts of the heavier gases, including water vapour and carbon dioxide. Mercury, the smallest of the inner planets, and Earth's moon, were unable to retain even the heaviest of these gases.

At the same time that the inner planets were forming, the larger, outer planets (Jupiter, Saturn, Uranus, and Neptune), along with their extensive satellite systems, were also developing. Because of the frigid temperatures existing far from the Sun, the fragments from which these planets formed contained a high percentage of ices—water, carbon dioxide, ammonia, and methane. The two most massive planets, Jupiter and Saturn, had a surface gravity sufficient to attract and hold even the lightest elements—hydrogen and helium.

Formation of Earth's Layered Structure

As material accumulated to form Earth (and for a short period afterward), the high-velocity impact of nebular debris and the decay of radioactive elements caused the temperature of our planet to steadily increase. During this time of intense heating, Earth became hot enough that iron and nickel began to melt. Melting produced liquid blobs of heavy metal that sank toward the centre of the planet. This process occurred rapidly on the scale of geologic time and produced Earth's dense iron-rich core.

The early period of heating resulted in another process of chemical differentiation, whereby melting formed buoyant masses of molten rock that rose toward the surface where they solidified to produce a primitive crust. These rocky materials were enriched in oxygen and "oxygen-seeking" elements, particularly silicon and aluminum, along with lesser amounts of calcium, sodium, potassium, iron, and magnesium. In addition, some heavy metals such as gold, lead, and uranium, which have low melting points or were highly soluble in the ascending molten masses, were scavenged from Earth's interior and concentrated in the developing crust. This early period of chemical segregation established the three basic divisions of Earth's interior—the iron-rich *core*, the thin *primitive crust*, and Earth's largest layer, called the *mantle*, which is located between the core and crust.

An important consequence of this early period of chemical differentiation is that large volumes of gases were allowed to escape from Earth's interior, as happens today during volcanic eruptions. By this process a primitive atmosphere gradually evolved. It is on this planet, with this atmosphere, that life as we know it came into existence.

Following the events that established Earth's basic internal structure, the primitive crust was lost to erosion and other geologic processes, so we have no direct record of its makeup. When and exactly how the continental crust—and thus Earth's first landmasses—came into existence is still a matter of much debate. One position contends that since there are precious few rocks approaching 4 billion years of age, there could not have been continental crust before that time. These researchers propose that the continental crust formed gradually over a vast period of geologic time. However, other geologists contend that the available evidence indicates that nearly all of the continental crust formed in a relatively short span beginning as early as 4 billion years ago. (The oldest rocks yet discovered are isolated fragments found in the Northwest Territories that have radiometric dates of about 4 billion years.)

Whichever view is correct, at some point the continents began to take on a composition and structure similar to that of today. However, as you will see later in this chapter, Earth is an evolving planet whose continents and ocean basins have continually changed shape, and even location, over at least the last 3 billion years.

Earth's Internal Structure

Introduction
↳ Earth's Layered Structure

In the preceding section you learned that the segregation of material that began early in Earth's history resulted in the formation of three layers defined by their chemical composition—the crust, mantle, and core. In addition to these compositionally distinct layers, Earth can be divided into layers based on physical properties. The physical properties used to define such zones include whether the layer is solid or liquid and how weak or strong it is. Knowledge of both types of layered structures is essential to our understanding of basic geologic processes, such as volcanism, earthquakes, and mountain building (Figure 1.15).

Layers Defined by Composition

Crust   The **crust**, Earth's comparatively thin, rocky outer skin, is generally divided into oceanic and continental crust. The oceanic crust is roughly 7 kilometres thick and composed of dark igneous rocks called *basalt*. By contrast, the continental crust averages 35–40 kilometres thick but may exceed 70 kilometres in some mountainous regions. Unlike the oceanic crust, which has a relatively homogeneous chemical composition, the continental crust consists of many rock types. The upper crust has an average composition of a *felsic rock* called *granodiorite*, whereas the composition of the lowermost continental crust is *mafic* (see Chapter 3). Continental rocks have an average density of about 2.7 g/cm^3, and some have been discovered that are 4 billion years old. The rocks of the oceanic crust are younger (180 million years or less) and more dense (about 3.0 g/cm^3) than continental rocks.*

Mantle   Over 82 percent of Earth's volume is contained in the **mantle**, a solid, rocky shell that extends to a depth of 2900 kilometres. The boundary between the crust and mantle represents a marked change in chemical composition. The dominant rock type in the uppermost mantle is *peridotite*, which has a density of 3.3 g/cm^3. At greater depth peridotite changes by assuming a more compact crystalline structure and hence a greater density.

Core   The composition of the **core** is thought to be an iron-nickel alloy with minor amounts of oxygen, silicon, and sulphur—elements that readily form compounds with iron. At the extreme pressure found in the core, this iron-rich material has an average density of nearly 11 g/cm^3 and approaches 14 times the density of water at Earth's centre.

Layers Defined by Physical Properties

Earth's interior is characterized by a gradual increase in temperature, pressure, and density with depth. Estimates put the temperature at a depth of 100 kilometres at between 1200°C and 1400°C, whereas the temperature at Earth's centre may exceed 6700°C. Clearly, Earth's interior has retained much of the energy acquired during its formative years, despite the fact that heat is continuously flowing toward the surface, where it is lost to space. The increase in pressure with depth causes a corresponding increase in rock density.

The gradual increase in temperature and pressure with depth affects the physical properties and hence the mechanical behaviour of Earth materials. When a substance is heated, its chemical bonds weaken and its mechanical strength (resistance to deformation) is reduced. If the temperature exceeds the melting point

---

*Liquid water at sea level has a density of 1 g/cm^3; therefore, the density of basalt is three times that of water.

Students Sometimes Ask...

How do we know about the internal structure of Earth? You might suspect that the internal structure of Earth has been sampled directly. However, the deepest mine in the world (the Western Deep Levels mine in South Africa) is only about 4 kilometres deep, and the deepest drilled hole in the world (completed in the Kola Peninsula of Russia in 1992) goes down only about 12 kilometres. In essence, humans have never penetrated beneath the crust!

The internal structure of Earth is determined by using indirect observations. Every time there is an earthquake, waves of energy (called *seismic waves*) penetrate Earth's interior. Seismic waves change their speed and are bent and reflected as they move through zones having different properties. An extensive series of monitoring stations around the world detects and records this energy. The data are analyzed and used to work out the internal structure of Earth's interior. For more about this technique, see Chapter 17, "Earth's Interior."

---

of an Earth material, the material's chemical bonds break and melting ensues. If temperature were the only factor that determined whether a substance melted, our planet would be a molten ball covered with a thin, solid outer shell. However, pressure also increases with depth and tends to increase rock strength. Furthermore, because melting is accompanied by an increase in volume, it occurs at higher temperatures at depth because of greater confining pressure. Thus, depending on the physical environment (temperature and pressure), a particular Earth material may behave like a brittle solid, deform in a puttylike manner, or even melt and become liquid.

Earth can be divided into five main layers based on physical properties and hence mechanical strength—the *lithosphere, asthenosphere, mesosphere (lower mantle), outer core,* and *inner core.*

Lithosphere and Asthenosphere.   Based on physical properties, Earth's outermost layer consists of the crust and uppermost mantle and forms a relatively cool, rigid and brittle shell. Although this layer is composed of materials with markedly different chemical compositions, it tends to act as a unit that exhibits rigid behaviour—mainly because it is cool and thus strong. This layer, called the **lithosphere** (*sphere of rock*), averages about 100 kilometres in thickness but may be more than 250 kilometres thick below the older portions of

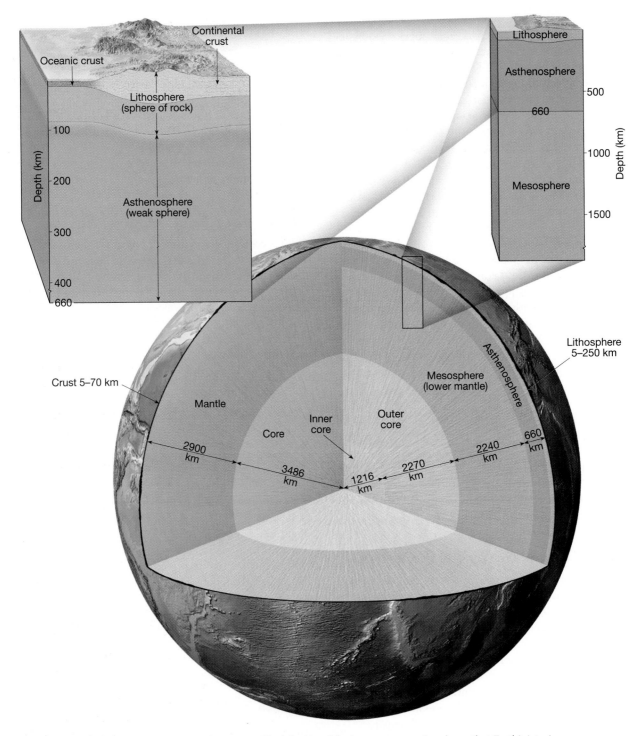

◆ **Figure 1.15** Views of Earth's layered structure. The left side of the large cross section shows that Earth's interior is divided into three different layers based on compositional differences—the crust, mantle, and core. The right side of the large cross section depicts the five main layers of Earth's interior based on physical properties and hence mechanical strength—the lithosphere, asthenosphere, mesosphere, outer core, and inner core. The block diagrams above the large cross section show an enlarged view of the upper portion of Earth's interior.

the continents (Figure 1.15). Within the ocean basins the lithosphere is only a few kilometres thick along the oceanic ridges but increases to perhaps 100 kilometres in regions of older and cooler oceanic crust.

Beneath the lithosphere, in the upper mantle (to a depth of about 660 kilometres) lies a soft, comparatively weak layer known as the **asthenosphere** ("weak sphere"). The top portion of the asthenosphere has a temperature/pressure regime that results in a small amount of melting. Within this very weak zone the lithosphere is mechanically detached from the layer below. The result is that the lithosphere is able to move independently of the asthenosphere, as considered in the next section.

It is important to emphasize that the strength of various Earth materials is a function of both their composition and of the temperature and pressure of their environment. You should not get the idea that the entire lithosphere behaves like a brittle solid similar to rocks found on the surface. Rather, the rocks of the lithosphere get progressively hotter and weaker (more easily deformed) with increasing depth. At the depth of the uppermost asthenosphere, the rocks are close enough to their melting temperatures (some melting may actually occur) that they are very easily deformed. Thus, the uppermost asthenosphere is weak because it is near its melting point, just as hot wax is weaker than cold wax.

Mesosphere or Lower Mantle. Below the zone of weakness in the uppermost asthenosphere, increased pressure counteracts the effects of higher temperature, and the rocks gradually strengthen with depth. Between the depths of 660 kilometres and 2900 kilometres a more rigid layer, called the **mesosphere** (*middle sphere*) or **lower mantle**, is found. Despite their strength, the rocks of the mesosphere are still very hot and capable of very gradual flow.

Inner and Outer Core. The core, which is composed mostly of an iron-nickel alloy, is divided into two regions that exhibit very different mechanical strengths. The **outer core** is a *liquid layer* 2270 kilometres thick. It is the convective flow of metallic iron within this zone that generates Earth's magnetic field. The **inner core** is a sphere having a radius of 1216 kilometres. Despite its higher temperature, the material in the inner core is stronger (because of immense pressure) than the outer core and behaves like a *solid*.

Dynamic Earth

Internal Processes
↳ Plate Tectonics

Earth is a dynamic planet! If we could go back in time a few hundred million years, we would find the face of our planet dramatically different from what we see today. There would be no Great Lakes, Rocky Mountains, or prairies. Moreover, we would find continents having different sizes and shapes and located in different positions than today's landmasses. In contrast, over the past few billion years the Moon's surface has remained essentially unchanged—only a few craters have been added.

The Theory of Plate Tectonics

Within the past several decades a great deal has been learned about the workings of our dynamic planet. This period has seen an unequalled revolution in our understanding of Earth. The revolution began in the early part of the twentieth century with the radical proposal of *continental drift*—the idea that the continents moved about the face of the planet. This proposal contradicted the established view that the continents and ocean basins are permanent and stationary features on the face of Earth. For that reason, the notion of drifting continents was received with great scepticism and even ridicule. More than 50 years passed before enough data were gathered to transform this controversial hypothesis into a sound theory that wove together the basic processes known to operate on Earth. The theory that finally emerged, called the **theory of plate tectonics** (*tekton* = to build), provided geologists with the first comprehensive model of Earth's internal workings.*

According to the plate tectonics model, Earth's rigid outer shell (*lithosphere*) is broken into numerous slabs called **plates**, which are in motion and are continually changing shape and size. As shown in Figure 1.16, seven major lithospheric plates are recognized. They are the North American, South American, Pacific, African, Eurasian, Australian, and Antarctic plates. Intermediate-size plates include the Caribbean, Nazca, Philippine, Arabian, Cocos, and Scotia plates. In addition, over a dozen smaller plates have been identified, but are not shown in Figure 1.16. Note that several large plates include an entire continent plus a large area of seafloor (for example, the South American plate). However, none of the plates are defined entirely by the margins of a single continent.

The lithospheric plates move relative to each other at a very slow but continuous rate that averages about 5 centimetres a year. This movement is ultimately driven by the unequal distribution of heat within Earth. Hot material found deep in the mantle moves slowly upward and serves as one part of our planet's internal convective system. Concurrently, cooler, denser slabs of lithosphere descend back into the mantle, setting Earth's rigid outer shell in motion. Ultimately, the titanic, grinding movements of Earth's lithospheric plates generate earthquakes, create volcanoes, and deform large masses of rock into mountains.

---

*Plate tectonics can be defined as the composite of various ideas explaining the observed motion of Earth's lithosphere through the mechanisms of subduction and seafloor spreading, which in turn generate Earth's major features, including continents and ocean basins.

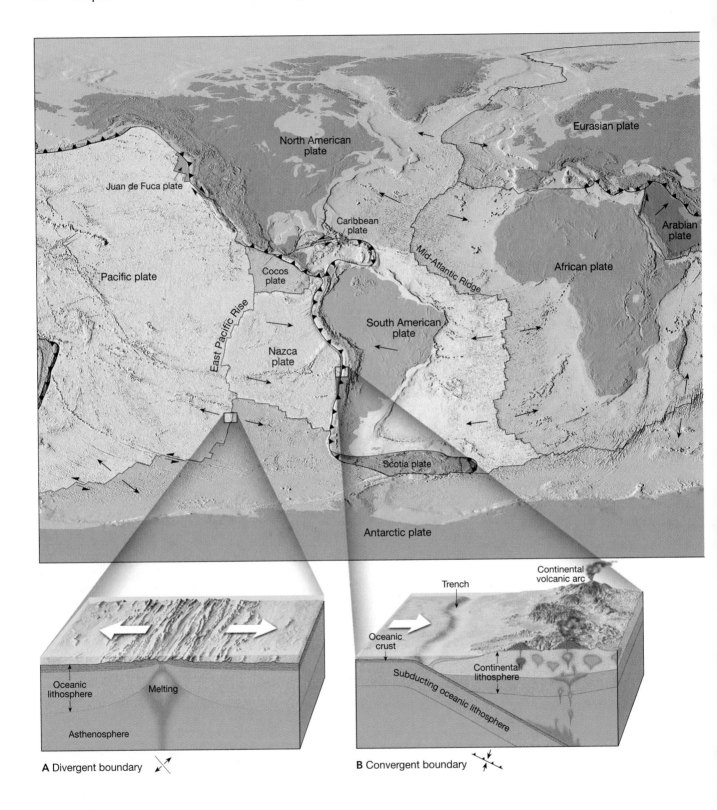

A Divergent boundary

B Convergent boundary

Plate Boundaries

Lithospheric plates move as coherent units relative to all other plates. Although the interiors of plates may experience some deformation, all major interactions among individual plates (and therefore most defor-mation) occurs along their *boundaries*. In fact, the first attempts to outline plate boundaries were made using locations of earthquakes. Later work showed that plates are bounded by three distinct types of boundaries, which are differentiated by the type of relative movement they exhibit. These boundaries are depicted at the bottom of Figure 1.16 and are briefly described here:

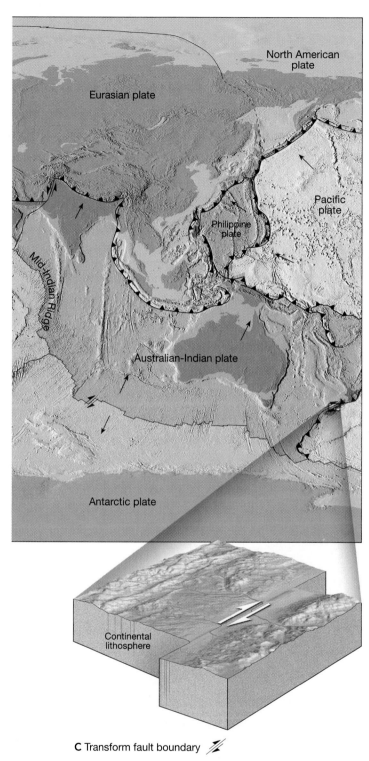

C Transform fault boundary

◆ **Figure 1.16** Mosaic of rigid plates that constitute Earth's outer shell.
(After W. B. Hamilton, U.S. Geological Survey)

1. **Divergent boundaries**—where plates move apart, resulting in upwelling of material from the mantle to create new seafloor (Figure 1.16A).

2. **Convergent boundaries**—where plates move together, resulting in the subduction (consumption) of oceanic lithosphere into the mantle (Figure 1.16B). Convergence can also result in the collision of two continental margins to create a major mountain system.

3. **Transform fault boundaries**—where plates grind past each other without the production or destruction of lithosphere (Figure 1.16C).

If you examine Figure 1.16, you can see that each large plate is bounded by a combination of these boundaries. Movement along one boundary requires that adjustments be made at the others.

Divergent Boundaries. Plate spreading (divergence) occurs mainly at the mid-ocean ridge. As plates pull apart, the fractures created are immediately filled with molten rock that wells up from the asthenosphere below (Figure 1.17). This hot material slowly cools to hard rock, producing new slivers of seafloor. This happens again and again over millions of years, adding thousands of square kilometres of new seafloor.

This mechanism has created the floor of the Atlantic Ocean during the past 160 million years and is appropriately called **seafloor spreading** (Figure 1.17). The rate of seafloor spreading varies considerably from one spreading centre to another. Spreading rates of only 2.5 centimetres per year are typical in the North Atlantic, whereas much faster rates (20 centimetres) have been measured along the East Pacific Rise. Even the most rapid rates of spreading are slow on the scale of human history. Nevertheless, the slowest rate of lithosphere production is rapid enough to have created all of Earth's ocean basins over the last 200 million years. In fact, none of the ocean floor that has been dated exceeds 180 million years in age.

Along divergent boundaries where molten rock emerges, the oceanic lithosphere is elevated, because it is hot and occupies more volume than do cooler rocks. Worldwide, this ridge extends for over 70,000 kilometres through all major ocean basins. As new lithosphere is formed along the oceanic ridge, it is slowly yet continually displaced away from the zone of upwelling along the ridge axis. Thus, it begins to cool and contract, thereby increasing in density. This thermal contraction accounts for the greater ocean depths that exist away from the ridge.

In addition, cooling causes the mantle rocks below the oceanic crust to strengthen, thereby adding to the plate thickness. Stated another way, the thickness of oceanic lithosphere is age-dependent. The older (cooler) it is, the greater its thickness.

Convergent Boundaries.   Although new lithosphere is constantly being added at the oceanic ridges, the planet is not growing in size—its total surface area remains constant. To accommodate the newly created lithosphere, older oceanic plates return to the mantle along *convergent boundaries*. As two plates slowly converge, the leading edge of one slab is bent downward, allowing it to slide beneath the other. The surface expression produced by the descending plate is an ocean *trench*, like the Peru–Chile trench illustrated in Figure 1.17.

Plate margins where oceanic crust is being consumed are called **subduction zones**. Here, as the subducted plate moves downward, it enters a high-temperature, high-pressure environment. Some subducted materials, as well as more voluminous amounts of the asthenosphere located above the subducting slab, melt and migrate upward into the overriding plate. Occasionally this molten rock may reach the surface, where it gives rise to explosive volcanic eruptions like Mount St. Helens in 1980 (see Figure 4.4). However, much of this molten rock never reaches the surface; rather, it solidifies at depth and acts to thicken the crust.

Whenever continental lithosphere moves toward an adjacent slab of oceanic lithosphere, the continental plate—being less dense—remains "floating," while the denser oceanic lithosphere sinks into the asthenosphere (Figure 1.17). The classic convergent boundary of this type occurs along the western margin of South America where the Nazca plate descends beneath the adjacent continental block. Here subduction along the Peru–Chile trench gave rise to the Andes Mountains, a linear chain of deformed rocks capped with numerous volcanoes—a number of which are still active.

The simplest type of convergence occurs where one oceanic plate is thrust beneath another. Here subduction results in the production of molten rock in a manner similar to that in the Andes, except volcanoes grow from the floor of the ocean rather than on a continent. If this activity is sustained, it will eventually build a chain of volcanic structures that emerge from the sea as a *volcanic island arc* (see Figure 4.34, upper left). Most volcanic island arcs are found in the Pacific Ocean, as exemplified by the Aleutian, Mariana, and Tonga islands.

As we saw earlier, when an oceanic plate is subducted beneath continental lithosphere, an Andean-type mountain range develops along the margin of the continent. However, if the subducting plate also contains continental lithosphere, continued subduction eventually brings the two continents together (Figure 1.18). Whereas oceanic lithosphere is relatively dense and sinks into the asthenosphere, continental lithosphere is buoyant, which prevents it from being subducted to any great depth. The result is a collision between the two continental blocks (Figure 1.18). Such a collision

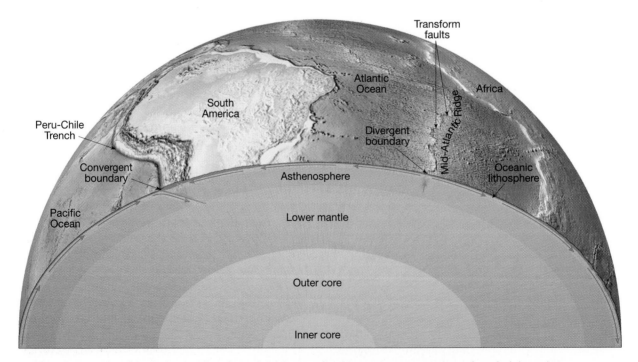

◆ **Figure 1.17**   View of Earth showing the relationship between divergent, convergent, and transform fault boundaries.

occurred when the subcontinent of India "rammed" into Asia and produced the Himalayas—the most spectacular mountain range on Earth. During this collision, the continental crust buckled, fractured, and was generally shortened and thickened. In addition to the Himalayas, several other major mountain systems, including the Alps, Appalachians, and Urals, formed during continental collisions along convergent plate boundaries.

Transform Fault Boundaries.
Transform fault boundaries are located where plates grind past each other without either generating new lithosphere or consuming old lithosphere. These faults form in the direction of plate movement and were first discovered in association with offsets in oceanic ridges (see Figure 1.17).

Although most transform faults are located along oceanic ridges, some slice through the continents. Two examples are the earthquake-prone San Andreas Fault of California and the Alpine Fault of New Zealand. Along the San Andreas Fault the Pacific plate is moving toward the northwest, relative to the adjacent North American plate (see Figure 1.16). The movement along this boundary does not go unnoticed. As these plates pass, strain builds in the rocks on opposite sides of the fault. Occasionally the rocks adjust, releasing energy in the form of a great earthquake of the type that devastated San Francisco in 1906.

Changing Boundaries.
Although the total surface area of Earth does not change, individual plates may diminish or grow in area depending on the distribution of convergent and divergent boundaries. For example, the Antarctic and African plates are almost entirely bounded by spreading centres and hence are growing larger. By contrast, the Pacific plate is being subducted along much of its perimeter and is therefore diminishing in area. At the current rate, the Pacific would close completely in 300 million years—but this is unlikely because changes in plate boundaries will probably occur before that time.

New plate boundaries are created in response to changes in the forces acting on the lithosphere. For example, a relatively new divergent boundary is located in Africa, in a region known as the East African Rift Valleys (see Figure 4.34, lower right). If spreading continues in this region, the African plate will split into two plates, separated by a new ocean basin. At other locations plates carrying continental crust are moving toward each other. Eventually these continents may collide and be sutured together. Thus, the boundary that once separated these plates disappears and two plates become one.

As long as temperatures within the interior of our planet remain significantly higher than those at the surface, material within Earth will continue to circulate. This internal flow, in turn, will keep the rigid outer shell of Earth in motion. Thus, while Earth's internal heat engine is operating, the positions and shapes of the continents and ocean basins will change and Earth will remain a dynamic planet.

In the remaining chapters we will examine in more detail the workings of our dynamic planet in light of the theory of plate tectonics.

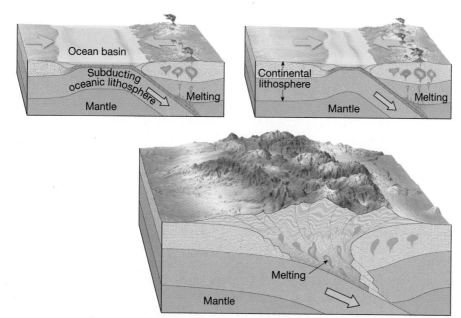

◆ **Figure 1.18** When two plates containing continental lithosphere collide, complex mountains are formed. The formation of the Himalayas represents a relatively recent example.

Chapter Summary

- *Geology* means "the study of Earth." The two broad areas of the science of geology are (1) *physical geology*, which examines the materials composing Earth and the processes that operate beneath and upon its surface; and (2) *historical geology*, which seeks to understand the origin of Earth and its development through time.

- The relationship between people and the natural environment is an important focus of geology. This includes natural hazards, resources, and human influences on geologic processes.

- During the seventeenth and eighteenth centuries, *catastrophism* influenced the formulation of explanations about Earth. Catastrophism states that Earth's landscapes have been developed primarily by great catastrophes. By contrast, *uniformitarianism*, one of the fundamental principles of modern geology advanced by *James Hutton* in the late 1700s, states that the physical, chemical, and biological laws that operate today have also operated in the geologic past. The idea is often summarized as "the present is the key to the past." Hutton argued that processes that appear to be slow-acting could, over long spans of time, produce effects that were just as great as those resulting from sudden catastrophic events.

- Using the principles of *relative dating*, the placing of events in their proper sequence or order without knowing their age in years, scientists developed a geologic time scale during the nineteenth century. Relative dates can be established by applying such principles as the *law of superposition* and the *principle of fossil succession*.

- All science is based on the assumption that the natural world behaves in a consistent and predictable manner. The process by which scientists gather facts and formulate scientific *hypotheses* and *theories* is called the *scientific method*. To determine what is occurring in the natural world, scientists often (1) collect facts, (2) develop a scientific hypothesis, (3) construct experiments to test the hypothesis, and (4) accept, modify, or reject the hypothesis on the basis of extensive testing. Other discoveries represent purely theoretical ideas that have stood up to extensive examination. Still other scientific advancements have been made when something totally unexpected happened during an experiment.

- Earth's physical environment is traditionally divided into three major parts: the solid Earth; the water portion of our planet, the *hydrosphere*; and Earth's gaseous envelope, the *atmosphere*. In addition, the *biosphere*, the totality of life on Earth, interacts with each of the three physical realms and is an equally integral part of Earth.

- Although each of Earth's four spheres can be studied separately, they are all related in a complex and continuously interacting whole that we call the *Earth system*. Changing one part of the Earth system can produce changes in any or all of the other parts.

- The two sources of energy that power the Earth system are (1) the Sun, which drives the external processes that occur in the atmosphere, hydrosphere, and at Earth's surface, and (2) heat from Earth's interior that powers the internal processes that produce volcanoes, earthquakes, and mountains.

- The *rock cycle* is one of the many cycles of the Earth system in which matter is recycled. The rock cycle is a means of viewing many of the interrelationships of geology. It illustrates the origin of the three basic rock types (sedimentary, igneous, and metamorphic) and the role of various geologic processes in transforming one rock type into another.

- Two principal divisions of Earth's surface are the continents and ocean basins. The *continental shelf* and *continental slope* mark the continent–ocean basin transition. Major continental features include *mountains* and *shields*. Important zones on the ocean floor are *trenches* and the extensive *oceanic ridge system*.

- The *nebular hypothesis* describes the formation of the solar system. The planets and Sun began forming about 5 billion years ago from a large cloud of dust and gases. As the cloud contracted, it began to rotate and assume a disk shape. Material that was gravitationally pulled toward the centre became the *protosun*. Within the rotating disk, small centres, called *protoplanets*, swept up more and more of the cloud's debris. Because of their high temperatures and weak gravitational fields, the inner planets were unable to accumulate and retain many of the lighter components. Because of the very cold temperatures existing far from the Sun, the large outer planets consist of huge amounts of lighter materials. These gaseous substances account for the comparatively large sizes and low densities of the outer planets.

- Earth's internal structure is divided into layers based on differences in chemical composition and on the basis of changes in physical properties. Compositionally, Earth is divided into a thin outer *crust*, a solid rocky *mantle*, and a dense *core*. Based on physical properties, the layers of Earth are (1) the *lithosphere*—the cool, rigid outermost layer that averages about 100 kilometres thick, (2) the *asthenosphere*, a relatively weak layer located in the mantle beneath the lithosphere, (3) the more rigid *mesosphere*, where rocks are very hot and capable of very gradual flow, (4) the liquid *outer core*, where Earth's magnetic field is generated, and (5) the solid *inner core*.

- The theory of *plate tectonics* provides a comprehensive model of Earth's internal workings. It holds that Earth's rigid outer lithosphere consists of several segments called *plates* that are slowly and continually in motion relative to each other. Most earthquakes, volcanic activity, and mountain building are associated with the movements of these plates.

- The three distinct types of plate boundaries are (1) *divergent boundaries*—where plates move apart; (2) *convergent boundaries*—where plates move together, causing one to go beneath the other, or where plates collide, which occurs when the leading edges are made of continental crust; and (3) *transform fault boundaries*—where plates slide past each other.

Review Questions

1. Geology is traditionally divided into two broad areas. Name and describe these two subdivisions.

2. Briefly describe Aristotle's influence on the science of geology.

3. How did the proponents of catastrophism perceive the age of Earth?

4. Describe the doctrine of uniformitarianism. How did the advocates of this idea view the age of Earth?

5. About how old is Earth?

6. The geologic time scale was established without the aid of radiometric dating. What principles were used to develop the time scale?

7. How is a scientific hypothesis different from a scientific theory?

8. List and briefly describe the four "spheres" that constitute our environment.

9. What are the two sources of energy for the Earth System?

10. Using the rock cycle, explain the statement "One rock is the raw material for another."

11. Briefly describe the events that led to the formation of the solar system.

12. List and briefly describe Earth's compositional divisions.

13. Contrast the asthenosphere and the lithosphere.

14. With which type of plate boundary is each of the following associated: subduction zone, San Andreas fault, seafloor spreading, and Mount St. Helens?

Key Terms

asthenosphere (p. 22)
atmosphere (p. 13)
biosphere (p. 14)
catastrophism (p. 5)
continental shelf (p. 17)
continental slope (p. 17)
convergent boundary
   (p. 25)
core (p. 22)
crust (p. 20)
divergent boundary (p. 25)

fossil succession,
   principle of (p. 8)
geologist (p. 3)
geology (p. 2)
historical geology (p. 2)
hydrosphere (p. 13)
igneous rock (p. 15)
inner core (p. 23)
lithosphere (p. 22)
lower mantle (p. 23)
mantle (p. 22)
mesosphere (p. 23)

metamorphic rock (p. 16)
nebular hypothesis (p. 19)
oceanic ridge system
   (p. 17)
outer core (p. 23)
paradigm (p. 11)
physical geology (p. 2)
plate (p. 23)
plate tectonics, theory of
   (p. 23)
relative dating (p. 7)
rock cycle (p. 15)

seafloor spreading (p. 25)
sediment (p. 15)
sedimentary rock (p. 15)
shield (p. 17)
subduction zone (p. 26)
superposition, law of (p. 7)
system (p. 14)
theory (p. 11)
transform fault boundary
   (p. 25)
trench (p. 18)
uniformitarianism (p. 6)

Web Resources

The *Earth* Web site uses the resources and flexibility of the Internet to aid in your study of the topics in this chapter. Written and developed by geology instructors, this site will help improve your understanding of geology. Visit **http://www.pearson.ca/tarbuck** and click on the cover of the text to find:

■ Online review quizzes.

■ Web-based critical thinking and writing exercises.

■ Links to chapter-specific Web resources.

■ Internet-wide key-term searches.

http://www.pearson.ca/tarbuck

Chapter 2

Matter and Minerals

Dolomite crystals, Nararre, Spain.
(Photo by E. R. Degginger/Photo
Researchers, Inc.)

arth's crust and oceans are the source of a wide variety of useful and essential minerals (Figure 2.1). In fact, practically every manufactured product contains materials obtained from minerals. Most people are familiar with the common uses of many basic metals, including aluminum in beverage cans, copper in electrical wiring, and gold and silver in jewellery. But some people are not aware that pencil "lead" contains the greasy-feeling mineral graphite and that baby powder comes from the mineral talc. Moreover, many do not know that drill bits impregnated with diamonds are employed by dentists to drill through tooth enamel, or that the common mineral quartz is the source of silicon for computer chips. As the mineral requirements of modern society grow, the need to locate additional supplies of useful minerals also grows, and becomes more challenging as well.

In addition to the economic uses of rocks and minerals, all of the processes studied by geologists are in some way dependent on the properties of these basic Earth materials. Events such as volcanic eruptions, mountain building, weathering and erosion, and even earthquakes involve rocks and minerals. Consequently, a basic knowledge of Earth materials is essential to the understanding of all geologic phenomena.

Minerals: The Building Blocks of Rocks

 Earth Materials
    ↳ Minerals

We begin our discussion of Earth materials with an overview of **mineralogy** because minerals are the building blocks of rocks. Geologists define **minerals** as any naturally occurring inorganic solids that possess an orderly internal structure and a definite chemical composition. Thus, for any Earth material to be considered a mineral, it must exhibit the following characteristics:

1. It must occur naturally.
2. It must be inorganic.
3. It must be a solid.
4. It must possess an orderly internal structure: that is, its atoms must be arranged in a definite pattern.
5. It must have a definite chemical composition that may vary within specified limits.

When geologists use the term *mineral*, only substances that meet these criteria are considered minerals. Consequently, synthetic diamonds and other useful materials produced by chemists are not

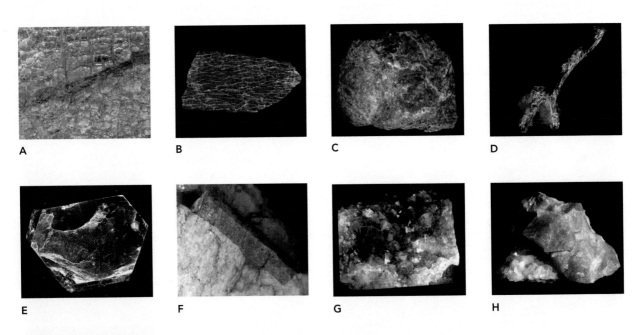

A   B   C   D

E   F   G   H

◆ **Figure 2.1**   Canadian Mineral Samples. **A.** Aragonite (as fossil mother of pearl) from Lethbridge, Alberta **B.** Microcline (variety Amazonite) from Lac Sairs, Québec **C.** Sodalite from Bancroft, Ontario **D.** Gold (on quartz) from Louvricourt Mines, Québec **E.** Phlogopite from Baffin Island, Nunavut; **F.** Beryl (variety: Emerald) from Quadeville, Ontario **G.** Quartz (variety: Amethyst) from Loon, Ontario **H.** Chabazite from Wasson's Bluff, Nova Scotia **I.** Gypsum (variety: Selenite) from Swift Current Creek, Saskatchewan
(Photos A-G by C. Tsujita; H and I by Arnum Walter; specimens B-I courtesy of the Department of Earth Sciences, the University of Western Ontario)

I

considered minerals. Further, the gemstone *opal* is classified as a *mineraloid*, rather than a mineral, because it lacks an orderly internal structure.

Rocks, on the other hand, are more loosely defined. Simply, a **rock** is any solid mass of mineral or mineral-like matter that occurs naturally as part of our planet. A few rocks are composed almost entirely of one mineral. A common example is the sedimentary rock *limestone*, which is composed of impure masses of the mineral calcite. However, most rocks, like the common rock granite (shown in Figure 2.2), occur as aggregates of several kinds of minerals. Here, the term *aggregate* implies that the minerals are joined in such a way that the properties of each mineral are retained. Note that you can easily distinguish the mineral constituents of the sample of granite shown in Figure 2.2.

A few rocks are composed of nonmineral matter. These include the volcanic rocks *obsidian* and *pumice*, which are noncrystalline glassy substances, and *coal*, which consists of solid organic debris.

Although this chapter deals primarily with the nature of minerals, keep in mind that most rocks are simply aggregates of minerals. Because the properties of rocks are determined largely by the chemical compo-

sition and internal structure of those minerals contained within them, we will first consider these Earth materials. Subsequent chapters describe the major rock types.

Students Sometimes Ask...

Are the minerals you talked about in class the same as those found in dietary supplements?

Not ordinarily. From a geologic perspective, a mineral must be a *naturally occurring* crystalline solid. Minerals found in dietary supplements are manufactured inorganic compounds that contain *elements* needed to sustain life. These dietary minerals typically contain elements that are metals—calcium, potassium, phosphorus, magnesium, and iron—as well as trace amounts of a dozen other elements, such as copper, nickel, and vanadium. Although these two types of "minerals" are different, they are related. The sources of the elements used to make dietary supplements are in fact the naturally occurring minerals of Earth's crust. It should also be noted that vitamins are *organic compounds* produced by living organisms, not *inorganic compounds*, like minerals.

Granite
(Rock)

Quartz
(Mineral)

Hornblende
(Mineral)

Feldspar
(Mineral)

◆ **Figure 2.2** Most rocks are aggregates of several kinds of minerals.

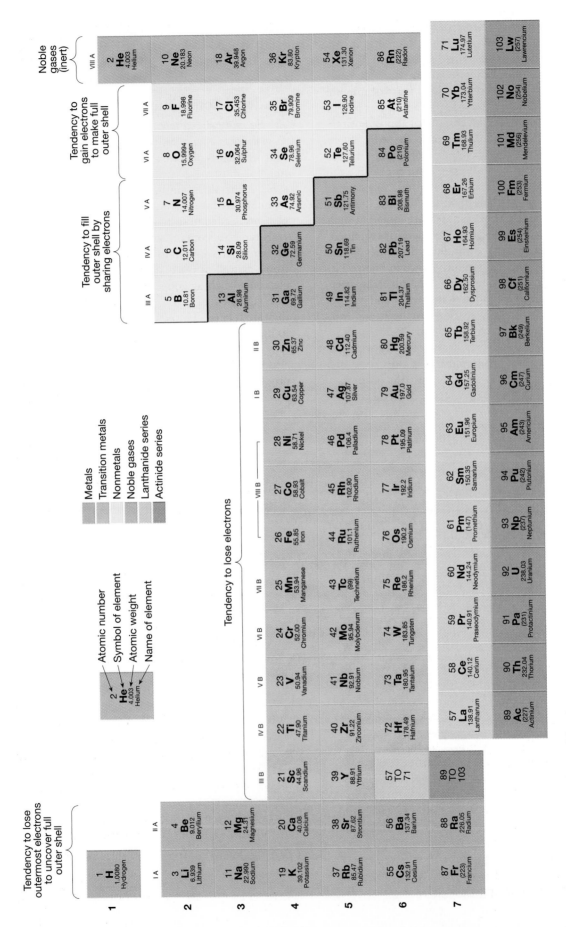

◆ **Figure 2.3**   Periodic Table of the Elements.

The Composition of Minerals

Each of Earth's nearly 4000 minerals is uniquely defined by its chemical composition and internal structure. In other words, every sample of the same mineral contains the same elements joined together in a consistent, repeating pattern. We will first review the basic building blocks of minerals, the *elements*, and then examine how elements bond together to form mineral structures.

At present, 112 elements are known. Of these, only 92 are naturally occurring (Figure 2.3). Some minerals, such as gold and sulphur, consist entirely of one element. But most minerals are a combination of two or more elements joined to form a chemically stable compound. To understand better how elements combine to form molecules and compounds, we must first consider the **atom** (*a* = not, *tomos* = cut), the smallest part of matter that still retains the characteristics of an element. It is this extremely small particle that does the combining.

Atomic Structure

Two simplified models illustrating basic atomic structure are shown in Figure 2.4. Note that atoms have a central region, called the **nucleus** (*nucleos* = a little nut), which contains very dense **protons** (particles with positive electrical charges) and equally dense **neutrons** (particles with neutral electrical charges). Surrounding the nucleus are very light particles called **electrons**, which travel at high speeds and are negatively charged. For simplicity, we often diagram atoms showing the electrons in orbits around the nucleus, like the orbits of the planets around the Sun. However, electrons *do not* travel in the same plane like planets. Further, because of their rapid motion, electrons create spherically shaped negatively charged zones around the nucleus called **energy levels**, or **shells**. Hence, a more realistic picture of the atom can be obtained by envisioning cloudlike shells of fast-moving electrons surrounding a central nucleus (Figure 2.4B). As we shall see, an important fact about these shells is that each can accommodate a specific number of electrons.

The number of protons found in an atom's nucleus determines the **atomic number** and name of the element. For example, all atoms with six protons are carbon atoms, those with eight protons are oxygen atoms, and so forth. Because atoms have the same number of electrons as protons, the atomic number also equals the number of electrons surrounding the nucleus (Table 2.1). Moreover, because neutrons have no charge, the positive charge of the protons is exactly balanced by the negative charge of the electrons.

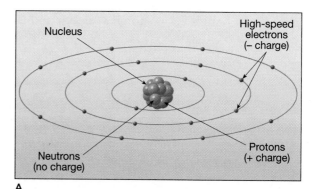

A

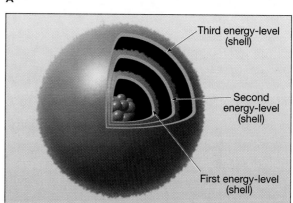

B

◆ **Figure 2.4** Two models of the atom. **A.** A very simplified view of the atom, which consists of a central nucleus, consisting of protons and neutrons, encircled by high-speed electrons. **B.** Another model of the atoms showing spherically shaped electron clouds (energy level shells). Note that these models are not drawn to scale. Electrons are minuscule in size compared to protons and neutrons, and the relative space between the nucleus and electron shells is much greater than illustrated.

Consequently, atoms are electrically neutral. Thus, an **element** is a large collection of electrically neutral atoms, all having the same atomic number.

The simplest element, hydrogen, is composed of atoms that have only one proton in the nucleus and one electron surrounding the nucleus. Each successively heavier atom has one more proton and one more electron, in addition to a certain number of neutrons (Table 2.1). Studies of electron configurations have shown that each electron is added in a systematic fashion to a particular energy level or shell. In general, electrons enter higher energy levels after lower energy levels have been filled to capacity.*

The first principal shell holds a maximum of two electrons, while each of the higher shells holds eight or more electrons. As we shall see, it is generally the outermost electrons, also referred to as **valence electrons**, that are involved in chemical bonding.

---

*This principle holds for the first 18 elements.

TABLE 2.1   **Atomic Number and Distribution of Electrons**

Element	Symbol	Atomic Number	Number of Electrons in Each Shell			
			1st	2nd	3rd	4th
Helium	He	2	2			
Lithium	Li	3	2	1		
Beryllium	Be	4	2	2		
Boron	B	5	2	3		
Carbon	C	6	2	4		
Nitrogen	N	7	2	5		
Oxygen	O	8	2	6		
Fluorine	F	9	2	7		
Neon	Ne	10	2	8		
Sodium	Na	11	2	8	1	
Magnesium	Mg	12	2	8	2	
Aluminum	Al	13	2	8	3	
Silicon	Si	14	2	8	4	
Phosphorus	P	15	2	8	5	
Sulphur	S	16	2	8	6	
Chlorine	Cl	17	2	8	7	
Argon	Ar	18	2	8	8	
Potassium	K	19	2	8	8	1
Calcium	Ca	20	2	8	8	2

Bonding

Elements combine with each other to form a wide variety of more complex substances. The strong attractive force linking atoms together is called a *chemical bond*. When chemical bonding joins two or more elements together in definite proportions, the substance is called a **compound**. Most minerals are chemical compounds.

Why do elements join together to form compounds? From experimentation it has been learned that the forces holding the atoms together are electrical. Further, it is known that chemical bonding results in a change in the electron configuration of the bonded atoms. As we noted earlier, it is the valence electrons (outer-shell electrons) that are generally involved in chemical bonding. Other than the first shell, which contains two electrons, *a stable configuration occurs when the valence shell contains eight electrons.* Only the so-called noble gases, such as neon and argon, have a complete outermost electron shell. Hence, the noble gases are the least chemically reactive, and thus their designation as "inert." However, all other atoms seek a valence shell containing eight electrons like the noble gases.

The octet rule, literally "a set of eight," refers to the concept of a complete outermost energy level. Simply, the **octet rule** states that atoms combine to form compounds and molecules in order to obtain the stable electron configuration of the noble gases. To satisfy the octet rule, an atom can gain, lose, or share electrons with one or more atoms (see Figure 2.5). The result of this process is the formation of an electrical "glue" that bonds the atoms. In summary, *most atoms are chemically reactive and bond together in order to achieve the stable noble-gas configuration while retaining overall electrical neutrality.*

Ionic Bonds. Perhaps the easiest type of bond to visualize is an **ionic bond**. In ionic bonding, one or more valence electrons are transferred from one atom to another. Simply, one atom gives up its valence electrons, and the other uses them to complete its outer shell. A common example of ionic bonding is sodium (Na) and chlorine (Cl) joining to produce sodium chloride (common table salt). This is shown in Figure 2.5. Notice that sodium gives up its single outer electron to chlorine. As a result, sodium achieves a stable configuration having eight electrons in its outermost shell. By acquiring the electron that sodium loses, chlorine—which has seven valence electrons—completes its outermost shell. Thus, through the transfer of a single electron, both the sodium and chlorine atoms have acquired the stable noble-gas configuration.

Once electron transfer takes place, atoms are no longer electrically neutral. By giving up one electron, a neutral sodium atom (11 protons/11 electrons) becomes *positively charged* (11 protons/10 electrons). Similarly, by acquiring one electron, the neutral chlorine

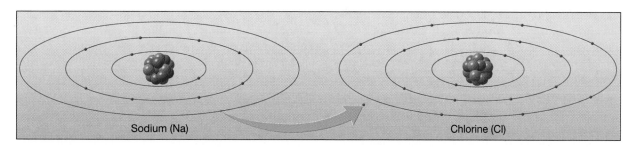

◆ **Figure 2.5** Chemical bonding of sodium and chlorine through the transfer of the lone outer electron from a sodium atom to a chlorine atom. The result is a positive sodium ion (Na$^+$) and a negative chloride ion (Cl$^-$). Bonding to produce sodium chloride (NaCl) is due to electrostatic attraction between the positive and negative ions. In this process note that both the sodium and chlorine atoms have achieved the stable noble-gas configuration (eight electrons in their outer shell).

atom (17 protons/17 electrons) becomes *negatively charged* (17 protons/18 electrons). Atoms such as these, which have an electrical charge because of the unequal numbers of electrons and protons, are called **ions**. (An atom that picks up an extra electron and becomes negatively charged is called an *anion*. An atom that loses an electron and becomes positively charged is called a *cation*.)

We know that particles (ions) with like charges repel, and those with unlike charges attract. Thus, an *ionic bond* is the attraction of oppositely charged ions to one another, producing an electrically neutral compound. Figure 2.6 illustrates the arrangement of sodium and chloride ions in ordinary table salt. Notice that salt consists of alternating sodium and chloride ions, positioned in such a manner that each positive ion is attracted to and surrounded on all sides by negative ions, and vice versa. This arrangement maximizes the attraction between ions with unlike charges while min-

imizing the repulsion between ions with like charges. Thus, *ionic compounds consist of an orderly arrangement of oppositely charged ions assembled in a definite ratio that provides overall electrical neutrality.*

The properties of a chemical compound are *dramatically different* from the properties of the elements comprising it. For example, chlorine is a green, poisonous gas that is so toxic it was used as a chemical weapon during World War I. Sodium is a soft, silvery metal that reacts vigorously with water and, if held in your hand, could burn it severely. Together, however, these atoms produce the compound sodium chloride (table salt), a clear crystalline solid that is essential for human life. This example also illustrates an important difference between a rock and a mineral. Most *minerals* are *chemical compounds* with unique properties that are very different from the elements that comprise them. A *rock*, on the other hand, is a *mixture* of minerals, with each mineral retaining its own identity.

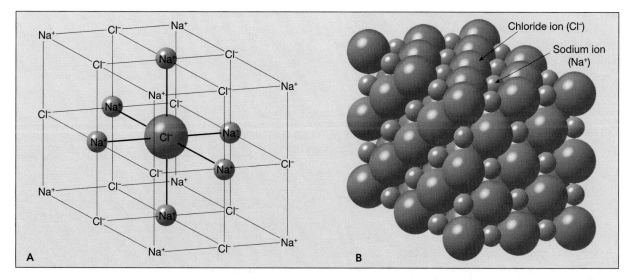

◆ **Figure 2.6** Schematic diagrams illustrating the arrangement of sodium and chloride ions in table salt.
A. Structure has been opened up to show arrangement of ions. **B.** Actual ions are closely packed.

Covalent Bonds. Not all atoms combine by transferring electrons to form ions. Other atoms *share* electrons. For example, the gaseous elements oxygen (O_2), hydrogen (H_2), and chlorine (Cl_2) exist as stable molecules consisting of two atoms bonded together, without a complete transfer of electrons.

Figure 2.7 illustrates the sharing of a pair of electrons between two chlorine atoms to form a molecule of chlorine gas (Cl_2). By overlapping their outer shells, these chlorine atoms share a pair of electrons. Thus, each chlorine atom has acquired, through cooperative action, the needed eight electrons to complete its outer shell. The bond produced by the sharing of electrons is called a **covalent bond**.

A common analogy may help you visualize a covalent bond. Imagine two people at opposite ends of a dimly lit room, each reading under a separate lamp. By moving the lamps to the centre of the room, they are able to combine their light sources so each can see better. Just as the overlapping light beams meld, the shared electrons that provide the "electrical glue" in covalent bonds are indistinguishable from each other. The most common mineral group, the silicates, contains the element silicon, which readily forms covalent bonds with oxygen.

Other Bonds. As you might suspect, many chemical bonds are actually hybrids. They consist to some degree of electron sharing, as in covalent bonding, and to some degree of electron transfer, as in ionic bonding. Furthermore, both ionic and covalent bonds may occur within the same compound. This occurs in many silicate minerals, where silicon and oxygen atoms are covalently bonded to form the basic building block common to all silicates. These structures in turn are ionically bonded to metallic ions, producing various electrically neutral chemical compounds.

Another chemical bond exists in which valence electrons are free to migrate from one ion to another. The mobile valence electrons serve as the "electrical glue." This type of electron sharing is found in metals such as copper, gold, aluminum, and silver and is called **metallic bonding**. Metallic bonding accounts for the high electrical conductivity of metals, the ease with which metals are shaped, and numerous other special properties of metals.

Isotopes and Radioactive Decay

Subatomic particles are so incredibly small that a special unit was devised to express their mass. A proton or a neutron has a mass just slightly more than one *atomic mass unit*, whereas an electron is only about one two-thousandth of an atomic mass unit. Thus, although electrons play an active role in chemical reactions, they do not contribute significantly to the mass of an atom.

The **mass number** of an atom is simply the total of its neutrons and protons in the nucleus. Atoms of the same element always have the same number of protons but commonly have varying numbers of neutrons. This means that an element can have more than one mass number. These variants of the same element are called **isotopes** of that element.

For example, carbon has three well-known isotopes. One has a mass number of 12 (carbon-12), another has a mass number of 13 (carbon-13), and the third, carbon-14, has a mass number of 14. All atoms of the same element must have the same number of protons (atomic number), and carbon always has six. Hence, carbon-12 must have six protons plus *six* neutrons to give it a mass number of 12, whereas carbon-14 must have six protons plus *eight* neutrons to give it a mass number of 14. The *average* atomic mass of any random sample of carbon is much closer to 12 than 13 or 14 (12.011, as in Figure 2.3), because carbon-12 is the more abundant isotope. This average is called **atomic weight**.*

---

*The term *weight* as used here is a misnomer that has been sanctioned by long use. The correct term is atomic *mass*.

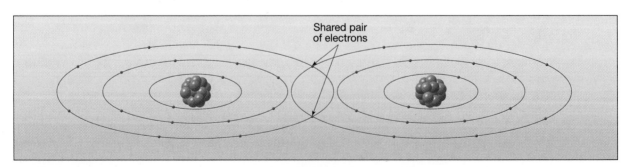

◆ **Figure 2.7**   Illustration of the sharing of a pair of electrons between two chlorine atoms to form a chlorine molecule. Notice that by sharing a pair of electrons, both chlorine atoms have eight electrons in their valence shell.

Note that in a chemical sense, all isotopes of the same element are nearly identical. To distinguish among them would be like trying to differentiate individual members from a group of similar objects, all having the same shape, size, and colour, with some being only slightly heavier. Further, different isotopes of an element are generally found together in the same mineral.

Although the nuclei of most atoms are stable, some elements do have isotopes in which the nuclei are unstable. Unstable isotopes, such as carbon-14, disintegrate through a process called **radioactive decay**. During radioactive decay unstable nuclei spontaneously break apart, giving off subatomic particles and/or electromagnetic energy similar to X-rays. The rate at which the unstable nuclei decay is steady and measurable, thus making such isotopes useful "clocks" for dating the events of Earth history. A discussion of radioactive decay and its applications in dating past geological events can be found in Chapter 8.

The Structure of Minerals

Earth Materials
↳ Minerals

A mineral is composed of an ordered array of atoms chemically bonded together to form a particular crystalline structure. This orderly packing of atoms is reflected in the regularly shaped objects we call crystals (see chapter-opening photo).

What determines the particular crystalline structure of a mineral? For those compounds formed by ions, the internal atomic arrangement is determined partly by the charges on the ions, but even more so by the size of the ions involved. To form stable ionic compounds, each positively charged ion is surrounded by the largest number of negative ions that will fit, while maintaining overall electrical neutrality, and vice versa. Figure 2.8 shows some ideal arrangements for various-sized ions.

Let's examine the geometric arrangement of sodium and chloride ions in the mineral halite (Figure 2.9). We see in Figure 2.9A that the sodium and chloride ions pack together to form a cubic-shaped internal structure. Also note that the orderly arrangement of ions found at the atomic level is reflected on a much larger scale in the cubic-shaped halite crystals shown in Figure 2.9B. Like halite, all samples of a particular mineral contain the same elements, joined together in the same orderly arrangement.

Although it is true that every specimen of the same mineral has the same internal structure, some *elements* are able to join together in more than one way.

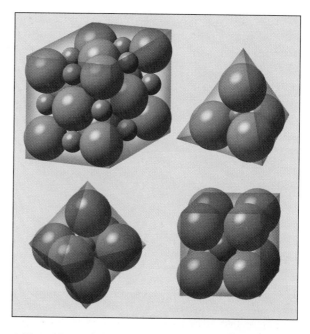

◆ **Figure 2.8** Ideal geometrical packing for various-sized positive and negative ions.

Thus, two minerals with totally different properties may have exactly the same chemical composition. Minerals of this type are said to be **polymorphs** (*poly* = many, *morph* = form). Graphite and diamond are particularly good examples of polymorphism because they both consist exclusively of carbon yet have drastically different properties. Graphite is the soft grey material of which pencil lead is made, whereas diamond is the hardest known mineral. The differences between these minerals can be attributed to the conditions under which they were formed. Diamonds form at depths approaching 200 kilometres, where extreme pressures produce the compact structure shown in Figure 2.10A. Graphite, on the other hand, forms under low pressure and temperature and consists of sheets of carbon atoms that are widely spaced and weakly held together (Figure 2.10B). Because these carbon sheets will easily slide past one another, graphite makes an excellent lubricant.

Scientists have learned that by heating graphite under high pressure, they can produce diamonds. Although synthetic diamonds are generally not gem quality, they have many industrial uses. The transformation of one polymorph to another is called a *phase change*. In nature, certain minerals go through phase changes as they move from one environment to another. For example, when rocks are carried to greater depths by a subducting plate, the mineral *olivine* changes to a more compact form called *spinel*.

Two other minerals with identical chemical compositions ($CaCO_3$) but different crystal forms

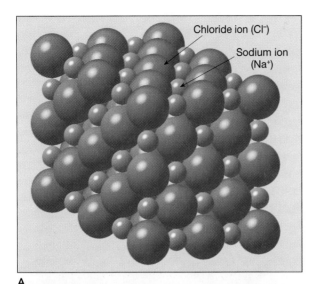

B

◆ **Figure 2.9**   The structure of sodium chloride. **A**. The orderly arrangement of sodium and chloride ions in the mineral halite. **B**. The orderly arrangement at the atomic level produces regularly shaped crystals.
(Photo by M. Claye/Jacana Scientific Control/Photo Researchers, Inc.)

Students Sometimes Ask...

Are there any artificial materials harder than diamonds?
Yes, but you won't be seeing them anytime soon. A hard form of carbon nitride (C_3N_4), described in 1989 and synthesized in a laboratory shortly thereafter may be harder than diamond but hasn't been produced in large enough amounts for a proper test. In 1999, researchers discovered that a form of carbon made from fused spheres of 20 and 28 carbon atoms—relatives of the famous "buckyballs"—also could be as hard as a diamond. These materials are expensive to produce, so diamonds continue to be used as abrasives and in certain kinds of cutting tools. Synthetic diamonds, produced since 1955, are now widely used in these industrial applications.

are calcite and aragonite. Calcite forms mainly through biochemical processes and is the major constituent of the sedimentary rock limestone. Aragonite is commonly deposited by hot springs and is also an important constituent of pearls and the shells of some marine organisms. Because aragonite changes to the more stable crystalline structure of calcite, it is rare in rocks older than 50 million years. Diamond is also somewhat unstable at Earth's surface, but (fortunately for jewellers) its rate of change to graphite is extremely slow.

Physical Properties of Minerals

Minerals are solids formed by inorganic processes. Each mineral has an orderly arrangement of atoms (crystalline structure) and a definite chemical composition, which give it a unique set of physical properties. Because the internal structure and chemical composition of a mineral are difficult to determine without the aid of sophisticated tests and apparatuses, the more easily recognized physical properties are frequently used in identification. A discussion of some diagnostic physical properties follows.

Crystal Form

Most people think of a crystal as a rare commodity, when in fact most inorganic solid objects are composed of crystals. The reason for this misconception is that most crystals do not exhibit good crystal form. The **crystal form** is the external expression of a mineral that reflects the orderly internal arrangement of atoms. Figure 2.11A illustrates the characteristic form of the iron-bearing mineral pyrite.

Generally, whenever a mineral is permitted to form without space restrictions, it will develop individual crystals with well-formed crystal faces. Some crystals such as those of the mineral quartz have a very distinctive crystal form that can be helpful in identification (Figure 2.11B). However, most of the time, crystal growth is interrupted because of competition for space, resulting in an intergrown mass of crystals, none of which exhibits its crystal form.

Lustre

Lustre is the appearance or quality of light reflected from the surface of a mineral. Minerals that have the appearance of metals, regardless of colour, are said to have a *metallic lustre*. Minerals with a *nonmetallic lustre* are described by various adjectives, including vitreous (glassy), pearly, silky, resinous, and earthy (dull). Some minerals appear partially metallic in lustre and are said to be submetallic.

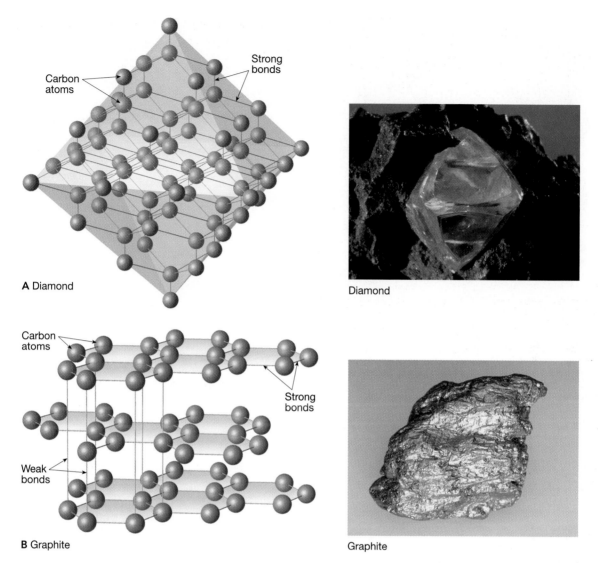

◆ **Figure 2.10** Comparing the structures of diamond and graphite. Both are natural substances with the same chemical composition—carbon atoms. Nevertheless, their internal structure and physical properties reflect the fact that each formed in a very different environment. **A.** All carbon atoms in diamond are covalently bonded into a compact, three-dimensional framework, which accounts for the extreme hardness of the mineral. (Photo courtesy of Smithsonian Institution) **B.** In graphite the carbon atoms are bonded into sheets that are joined in a layered fashion by very weak electrical forces. These weak bonds allow the sheets of carbon to readily slide past each other, making graphite soft and slippery, and thus useful as a dry lubricant.
(A., photographer Dane Pendland, courtesy of Smithsonian Institution; B., E. J. Tarbuck)

Colour

Although **colour** is an obvious feature of a mineral, it is often unreliable as a diagnostic property. Slight impurities in the common mineral quartz, for example, give it a variety of colours, including pink, purple (amethyst), white, and even black (see Figure 2.25, p. 53).

Streak

Streak is the colour of a mineral in its powdered form and is obtained by rubbing the mineral across a piece of unglazed porcelain known as a *streak plate*.

Although the colour of a mineral may vary from sample to sample, the streak usually does not and is therefore the more reliable property.

Hardness

One of the most useful diagnostic properties is **hardness**, a measure of the resistance of a mineral to abrasion or scratching. This property is determined by rubbing a mineral of unknown hardness against one of known hardness, or vice versa. A numerical value can be obtained by using the **Mohs scale** of hardness, which consists of 10 minerals arranged in order from 1 (softest) to 10 (hardest), as shown in Table 2.2.

A

B

◆ **Figure 2.11**   Crystal form is the external expression of a mineral's orderly internal structure.
A. Pyrite, commonly known as "fool's gold," often forms cubic crystals. They may exhibit parallel lines
(striations) on the faces. (Photo courtesy of E. J. Tarbuck)
B. Quartz sample that exhibits well-developed hexagonal crystals with pyramidal-shaped ends.
(Photo by Breck P. Kent)

TABLE 2.2	**Mohs Scale of Hardness**	
Relative Scale	**Mineral**	**Hardness of Some Common Objects**
Hardest 10	Diamond	
9	Corundum	
8	Topaz	
7	Quartz	
6	Potassium Feldspar	
5	Apatite	5.5 Glass, Pocketknife
4	Fluorite	
3	Calcite	3 Copper Penny
2	Gypsum	2.5 Fingernail
Softest 1	Talc	

Any mineral of unknown hardness can be compared to these or to other objects of known hardness. For example, a fingernail has a hardness of 2.5, a copper penny 3, and a piece of glass 5.5. The mineral gypsum, which has a hardness of 2, can be easily scratched with your fingernail. On the other hand, the mineral calcite, which has a hardness of 3, will scratch your fingernail but will not scratch glass. Quartz, the hardest of the common minerals, will scratch glass.

Cleavage

Within the crystal structure of a mineral, some bonds are weaker than others. These bonds are where a mineral will break when it is stressed. **Cleavage** (*kleiben* = carve) is the tendency of a mineral to break along planes of weak bonding. Not all minerals have definite planes of weak bonding, but those that possess cleavage can be identified by the distinctive smooth surfaces that are produced when the mineral is broken.

The simplest type of cleavage is exhibited by the micas (Figure 2.12). Because the micas have weak bonds in one direction, they cleave to form thin, flat sheets. Some minerals have several cleavage planes, which produce smooth surfaces when broken, while others exhibit poor cleavage and still others, such as quartz, have no cleavage at all. When minerals break evenly in more than one direction, cleavage is described by the *number of planes* exhibited and the *angles at which they meet* (Figure 2.13).

Do not confuse cleavage with crystal form. When a mineral exhibits cleavage, it will break into pieces *that each have the same geometry*. By contrast, the quartz crystals shown in Figure 2.11B do not have cleavage. If broken, they fracture into shapes that do not resemble each other or the original crystals.

◆ **Figure 2.13** Smooth surfaces produced when a mineral with cleavage is broken. The sample on the left (fluorite) exhibits four planes of cleavage (eight sides), whereas the other two samples exhibit three planes of cleavage (six sides). Also notice that the mineral in the centre (halite) has cleavage planes that meet at 90° angles, whereas the mineral on the right (calcite) has cleavage planes that meet at 75° angles. (Photo courtesy of E. J. Tarbuck)

◆ **Figure 2.12** The thin sheets shown here were produced by splitting a mica (muscovite) crystal parallel to its perfect cleavage. (Photo by Breck P. Kent)

Fracture

Minerals that do not exhibit cleavage when broken, such as quartz, are said to **fracture**. Those that break into smooth curved surfaces resembling broken glass have a *conchoidal fracture* (Figure 2.14). Others break into splinters or fibres, but most minerals fracture irregularly.

Specific Gravity

Specific gravity is a number representing the ratio of the mass of a mineral to the weight of an equal volume of water. For example, if a mineral weighs three times as much as an equal volume of water, its specific gravity is 3. With a little practice, you can estimate the specific gravity of minerals by hefting them in your hand. For example, if a mineral feels as heavy as the common rocks you have handled, its specific gravity will probably be somewhere between 2.5 and 3. Some metallic minerals have a specific gravity two or three times that of common rock-forming minerals. Galena, which is an ore of lead (Figure 2.15), has a specific gravity of roughly 7.5, whereas the specific gravity of 24-karat gold is approximately 20.

Other Properties of Minerals

In addition to the properties already discussed, some minerals can be recognized by other distinctive properties. For example, halite is ordinary salt, so it is quickly identified with your tongue. Thin sheets of mica will bend and elastically snap back. Gold is malleable and

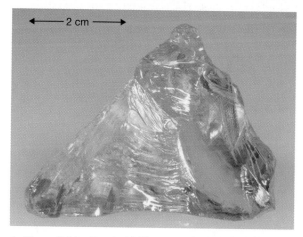

◆ **Figure 2.14** Conchoidal fracture. The smooth curved surfaces result when minerals break in a glasslike manner. (Photo by E. J. Tarbuck)

◆ **Figure 2.15** Galena is lead sulphide and, like other metallic ores, has a relatively high specific gravity. (Photo courtesy of E. J. Tarbuck)

can be easily shaped. Talc and graphite both have distinctive feels; talc feels soapy and graphite feels greasy. A few minerals, such as magnetite, have a high iron content and can be picked up with a magnet, while some varieties (lodestone) are natural magnets and will pick up small iron-based objects such as pins and paper clips.

Moreover, some minerals exhibit special optical properties. For example, when a transparent piece of calcite is placed over printed material, the letters appear twice. This optical property is known as *double refraction*. The streak of many sulphur-bearing minerals smells like rotten eggs.

One very simple chemical test involves placing a drop of dilute hydrochloric acid from a dropper bottle onto a freshly broken mineral surface. Carbonate minerals will effervesce (fizz) with hydrochloric acid. This test is useful in identifying the mineral calcite, which is a common carbonate mineral.

In summary, a number of special physical and chemical properties are useful in identifying certain minerals. These include taste, smell, elasticity, malleability, feel, magnetism, double refraction, and chemical reaction to hydrochloric acid. Remember that every one of these properties depends on the composition (elements) of a mineral and its structure (how the atoms are arranged).

Mineral Groups

Earth Materials
  ↳ Minerals

Nearly 4000 minerals have been named, and several new ones are identified each year. Fortunately, for students who are beginning to study minerals, no more than a few dozen are abundant! Collectively, these few make up most of the rocks of Earth's crust and, as such, are classified as the *rock-forming minerals*. It is also interesting to note that *only eight elements* make up the bulk of these minerals and represent over 98 percent (by weight) of the continental crust (Table 2.3). These elements in order of abundance are oxygen (O), silicon (Si), aluminum (Al), iron (Fe), calcium (Ca), sodium (Na), potassium (K), and magnesium (Mg).

Silicon and oxygen combine to form the framework of the most common mineral group, the **silicates**. Perhaps the next most common mineral group is the carbonates, of which calcite is the most prominent member. Other common rock-forming minerals include gypsum and halite.

We will first discuss the most common mineral group, the silicates, and then consider some of the other prominent mineral groups.

TABLE 2.3   Relative Abundance of the Most Common Elements in the Continental Crust

Element	Approximate Percentage by Weight
Oxygen (O)	46.6
Silicon (Si)	27.7
Aluminum (Al)	8.1
Iron (Fe)	5.0
Calcium (Ca)	3.6
Sodium (Na)	2.8
Potassium (K)	2.6
Magnesium (Mg)	2.1
All others	1.7
Total	100

The Silicates

Every silicate mineral contains the elements oxygen and silicon. Moreover, except for a few minerals such as quartz, a silicate mineral includes one or more additional elements that are needed to produce electrical neutrality. These additional elements give rise to the great variety of silicate minerals and their varied properties (Box 2.1).

◆ **Figure 2.16** Two representations of the silicon-oxygen tetrahedron. **A.** The four large spheres represent oxygen ions, and the blue sphere represents a silicon ion. The spheres are drawn in proportion to the radii of the ions. **B.** An expanded view of the tetrahedron using rods to depict the bonds that connect the ions.

The Silicon–Oxygen Tetrahedron.   All silicates have the same fundamental building block, the **silicon–oxygen tetrahedron** (*tetra* = four, *hedra* = a base). This structure consists of four oxygen ions surrounding a much smaller silicon ion (Figure 2.16). The silicon–oxygen tetrahedron is a complex ion (SiO_4^{4-}) with a charge of −4.

In nature one of the simplest ways in which these tetrahedra join together to become neutral compounds is through the addition of positively charged ions (Figure 2.17). In this way, a chemically stable structure is produced, consisting of individual tetrahedra linked together by cations.

More Complex Silicate Structures.   In addition to cations providing the opposite electrical charge needed to bind the tetrahedra, the tetrahedra may link with other tetrahedra in a variety of configurations. For example, the tetrahedra may be joined to form *single chains*, *double chains*, or *sheet structures*, as shown in Figure 2.18. The joining of tetrahedra in each of these configurations results from the sharing of a different number of oxygen atoms between silicon atoms in adjacent tetrahedra.

To understand better how this sharing takes place, select one of the silicon ions (small blue spheres) near the middle of the single-chain structure shown in Figure 2.18A. Notice that this silicon ion is completely surrounded by four larger oxygen ions (you are looking *through* one of the four to see the blue silicon ion). Also notice that of the four oxygen ions, two are joined to other silicon ions, whereas the other two are not shared in this manner. *It is the linkage across the shared oxygen ions that joins the tetrahedra into a chain structure.* Now examine a silicon ion near the middle of the sheet structure and count the number of shared and unshared oxygen ions surrounding it (Figure 2.18C). The increase

in the degree of sharing accounts for the sheet structure. Other silicate structures exist, and the most common has all of the oxygen ions shared to produce a complex three-dimensional framework.

By now you can see that the ratio of oxygen ions to silicon ions differs in each of the silicate structures. In the isolated tetrahedron, there are four oxygen ions for every silicon ion. In the single chain the oxygen-to-silicon ratio is 3:1, and in the three-dimensional framework this ratio is 2:1. Consequently, as more of the oxygen ions are shared, the percentage of silicon in the structure increases. Silicate minerals are therefore described as having a "high" or "low" silicon content based on their ratio of oxygen to silicon. This difference in silicon content is important, as we shall see in a later chapter when we consider the formation of igneous rocks.

Most silicate structures, including single chains, double chains, or sheets, are not neutral chemical compounds. Thus, like the individual tetrahedra, they are all neutralized by the inclusion of metallic cations that bond them together into a variety of complex crystalline configurations. The cations that most often link silicate structures are those of the elements iron (Fe), magnesium (Mg), potassium (K), sodium (Na), aluminum (Al), and calcium (Ca).

Notice in Figure 2.17 that each of these cations has a particular atomic size and a particular charge. Generally, ions of approximately the same size are able to substitute freely for one another. For instance, ions of iron (Fe^{2+}) and magnesium (Mg^{2+}) are nearly the same size and substitute for each other without altering the mineral structure. This also holds true for calcium and sodium ions, which can occupy the same site in a crystalline structure. In addition, aluminum (Al^{3+}) often substitutes for silicon in the silicon–oxygen tetrahedron.

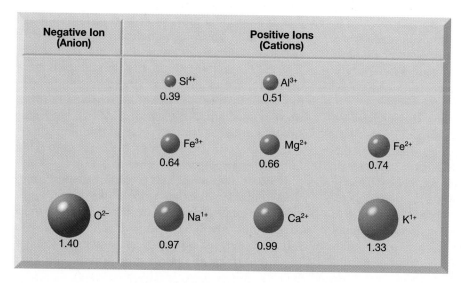

◆ **Figure 2.17** Relative sizes and electrical charges of ions of the eight most common elements in Earth's crust. These are the most common ions in rock-forming minerals. Ionic radii are expressed in angstroms (1 angstrom equals 10^{-8} cm).

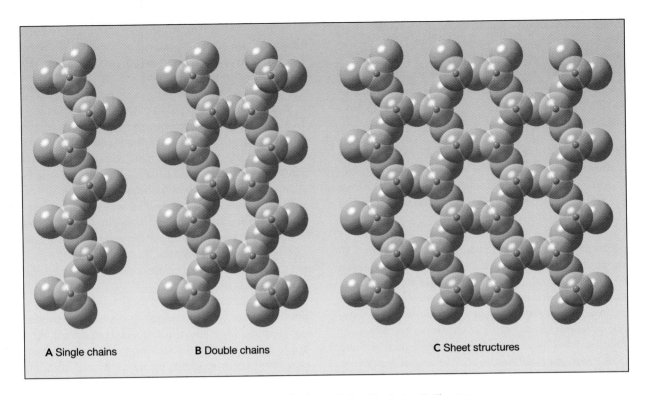

A Single chains   **B** Double chains   **C** Sheet structures

◆ **Figure 2.18**   Three types of silicate structures. **A.** Single chains. **B.** Double chains. **C.** Sheet structures.

Because of the ability of silicate structures to readily accommodate different cations at a given bonding site, individual specimens of a particular mineral may contain varying amounts of certain elements. A mineral of this type is often expressed by a chemical formula that uses parentheses to show the variable component. A good example is the mineral olivine, $(Mg, Fe)_2SiO_4$, which is magnesium/iron silicate. As you can see from the formula, it is the iron (Fe^{2+}) and magnesium (Mg^{2+}) cations in olivine that freely substitute for each other. At one extreme, olivine may contain iron without any magnesium (Fe_2SiO_4, or iron silicate), and at the other, iron is totally lacking (Mg_2SiO_4, or magnesium silicate). Between these end members, any ratio of iron to magnesium is possible. Thus, olivine, as well as many other silicate minerals, is actually a *family* of minerals with a range of composition between the two end members.

In certain substitutions, the ions that interchange do not have the same electrical charge. For instance, when calcium (Ca^{2+}) substitutes for sodium (Na^+), the structure gains a positive charge. In nature, one way in which this substitution is accomplished, while still maintaining overall electrical neutrality, is the simultaneous substitution of aluminum (Al^{3+}) for silicon (Si^{4+}). This particular double substitution occurs in a mineral called plagioclase feldspar. It is a member of the most abundant family of minerals found in Earth's

crust. The end members of this particular feldspar series are a calcium–aluminum silicate (anorthite, $CaAl_2Si_2O_8$), and a sodium-aluminum silicate (albite, $NaAlSi_3O_8$).

We are now prepared to review silicate structures in light of what we know about chemical bonding. An examination of Figure 2.17 shows that among the major constituents of the silicate minerals, only oxygen is an anion (negatively charged). Because oppositely charged ions attract (and similarly charged ions repel), the chemical bonds that hold silicate structures together form between oxygen and oppositely charged cations. Thus, cations arrange themselves so that they can be as close as possible to oxygen while remaining as far apart from each other as possible. Because of its small size and high charge (14), the silicon (Si) cation forms the strongest bonds with oxygen. Aluminum (Al) is more strongly bonded to oxygen than are calcium (Ca), magnesium (Mg), iron (Fe), sodium (Na), or potassium (K), but not as strongly as silicon. In many ways, aluminum plays a role similar to that of silicon by being the central ion in the basic tetrahedral structure.

Most silicate minerals consist of a basic framework composed of either a single silicon or a single aluminum cation surrounded by four negatively charged oxygen ions. These tetrahedra often link together to form a variety of other silicate structures (chains, sheets, etc.) through shared oxygen atoms. The other

People and the Environment
Asbestos: What Are the Risks?

Asbestos is a commercial term applied to a variety of silicate minerals that readily separate into thin, strong fibres that are highly flexible, heat resistant, and relatively inert. These properties make asbestos a desirable material for the manufacture of a wide variety of products including insulation, fireproof fabrics, cement, floor tiles, and car brake linings.

The mineral chrysotile, marketed as "white asbestos," belongs to the serpentine mineral group and accounts for the vast majority of asbestos sold commercially. Canada's only currently mined source of chrysotile occurs in eastern Quebec in the vicinity of Thetford Mines. Still, Canada is the world's top exporter and second largest producer of chrysotile. All other forms of asbestos are amphiboles and constitute only a minor proportion of asbestos used commercially. The two most common amphibole asbestos minerals are amosite and crocidolite, informally called "brown" and "blue" asbestos respectively.

Health concerns about asbestos have prevailed since the 1980s and stem largely from claims of high death rates attributed to asbestosis (lung scarring from asbestos fibre inhalation), mesothelioma (cancer of chest and abdominal cavity), and lung cancer among asbestos mine workers. The degree of concern generated by these claims has been well demonstrated by

◆ **Figure 2.A**   Chrysotile asbestos. This sample is a fibrous form of the mineral serpentine. Inset: Thetford Mines, Quebec, the source of all Canadian chrysotile asbestos. (Photo by E. J. Tarbuck)

the growth of an entire industry built around asbestos removal from buildings.

The stiff, straight fibres of brown and blue (amphibole) asbestos are known to readily pierce, and remain lodged in, the linings of human lungs. The fibres are physically and chemically stable and are not broken down in the human body. These forms of asbestos are therefore a genuine cause for concern. White asbestos, however, being a different mineral, has different properties. The curly fibres of white asbestos are readily expelled from the lungs and, if they are not expelled, can dissolve

within a year. Studies conducted on people living in the Thetford Mines area exposed to high levels of white asbestos dust suggest little, if any, difference in mortality rates from mesothelioma and lung cancer relative to the general public. Despite the fact that over 90 percent of all asbestos used commercially is white asbestos, a number of countries have moved to ban the use of asbestos in many applications, as the different mineral forms of asbestos are not distinguished. Still, the health hazards of chrysotile at low doses remain controversial in the eyes of health officials.

cations bond with the oxygen atoms of these silicate structures to create the more complex crystalline structures that characterize the silicate minerals.

Common Silicate Minerals

To reiterate, the silicates are the most abundant mineral group and have the silicate ion (SiO_4^{4-}) as their basic building block. The major silicate groups and common examples are given in Figure 2.19. The feldspars (*feld* = field, *spar* = mineral) are by far the most abundant silicate mineral, comprising over 50 percent of Earth's crust. Quartz, the second most abundant mineral in

the continental crust, is the only common mineral made completely of silicon and oxygen.

Notice in Figure 2.19 that each mineral *group* has a particular silicate *structure* and that a relationship exists between the internal structure of a mineral and the *cleavage* it exhibits. Because the silicon–oxygen bonds are strong, silicate minerals tend to cleave between the silicon–oxygen structures rather than across them. For example, the micas have a sheet structure and thus tend to cleave into flat plates (see Figure 2.12, p. 43). Quartz, which has equally strong silicon–oxygen bonds in all directions, has no cleavage, but fractures instead.

Most silicate minerals form (crystallize) as molten rock is cooling. This cooling can occur at or near Earth's surface (low temperature and pressure) or at great depths (high temperature and pressure). The environment during crystallization and the chemical composition of the molten rock determine to a large degree which minerals are produced. For example, the silicate mineral olivine crystallizes at high temperatures, whereas quartz crystallizes at much lower temperatures.

In addition, some silicate minerals form at Earth's surface from the weathered products of pre-existing silicate minerals. Still other silicate minerals are formed under the extreme pressures associated with mountain building. Each silicate mineral, therefore, has a structure and a chemical composition that *indicate the conditions under which it formed*. Thus, by carefully examining the mineral constituents of rocks, geologists can often determine the circumstances under which the rocks formed.

We will now examine some of the most common silicate minerals, which we divide into two major groups on the basis of their chemical makeup.

Mineral		Idealized Formula	Cleavage	Silicate Structure
Olivine		$(Mg, Fe)_2SiO_4$	None	Single tetrahedron 
Pyroxene group (Augite)		$(Mg,Fe)SiO_3$	Two planes at right angles	Single chains 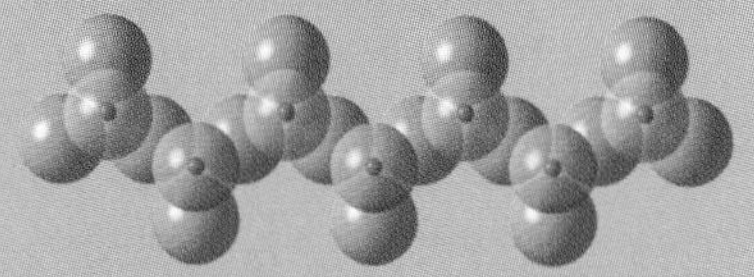
Amphibole group (Hornblende)		$Ca_2(Fe,Mg)_5Si_8O_{22}(OH)_2$	Two planes at 60° and 120°	Double chains 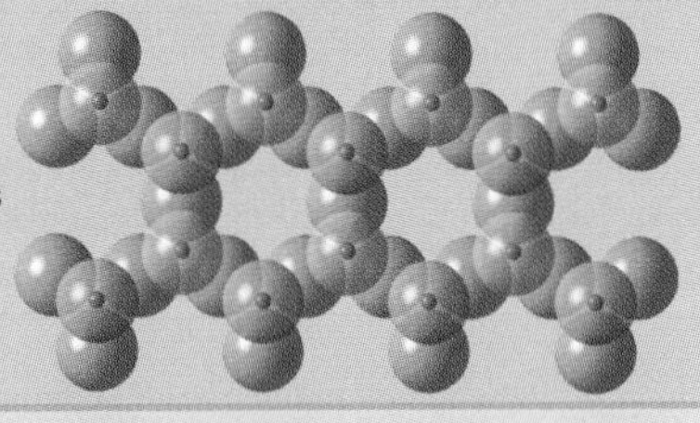
Micas	Biotite	$K(Mg,Fe)_3AlSi_3O_{10}(OH)_2$	One plane	Sheets 
	Muscovite	$KAl_2(AlSi_3O_{10})(OH)_2$		
Feld-spars	Orthoclase	$KAlSi_3O_8$	Two planes at 90°	Three-dimensional networks 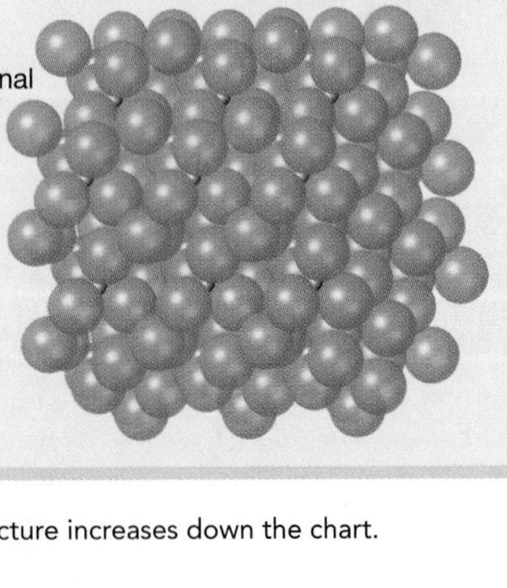
	Plagioclase	$(Ca,Na)AlSi_3O_8$		
Quartz		SiO_2	None	

◆ **Figure 2.19** Common silicate minerals. Note that the complexity of the silicate structure increases down the chart.

Ferromagnesian (Dark) Silicates. The **dark (or ferromagnesian) silicates** are those minerals containing ions of iron (iron = *ferro*) and/or magnesium in their structure. Because of their iron content, ferromagnesian silicates are dark in colour and have a greater specific gravity, between 3.2 and 3.6, than nonferromagnesian silicates. The most common dark silicate minerals are olivine, the pyroxenes, the amphiboles, the dark mica (biotite), and garnet.

The *olivine* family of high-temperature silicate minerals that are black to olive green in colour and have a glassy lustre and a conchoidal fracture. Rather than developing large crystals, olivine commonly forms small, rounded crystals that give rocks consisting largely of olivine a granular appearance. Olivine is composed of individual tetrahedra, which are bonded together by a combination of iron and magnesium ions positioned so as to link the oxygen atoms with the magnesium and iron atoms. Because the three-dimensional network generated in this fashion does not have its weak bonds aligned, olivine does not possess cleavage.

The *pyroxenes* are a group of complex minerals thought to be important components of Earth's mantle. The most common member, *augite*, is a black, opaque mineral with two directions of cleavage that meet at nearly a 90-degree angle. Its crystalline structure consists of single chains of tetrahedra bonded together by ions of iron and magnesium. Because the silicon–oxygen bonds are stronger than the bonds joining the adjacent silicate chain, augite cleaves parallel to the silicate chains. Augite is one of the dominant minerals in basalt, a common igneous rock of the oceanic crust and volcanic areas on the continents.

Hornblende is the most common member of a chemically complex group of minerals called *amphiboles* (Figure 2.20). Hornblende is usually dark green to black in colour, and except for its cleavage angles, which are about 60 degrees and 120 degrees, it is very similar in appearance to augite (Figure 2.21). The double chains of tetrahedra in the hornblende structure account for its particular cleavage. In a rock, hornblende often forms elongated crystals. This helps

◆ **Figure 2.20** Hornblende amphibole. Hornblende is a common dark silicate mineral having two cleavage directions that intersect at roughly 60° and 120°. (Photo courtesy of E. J. Tarbuck.)

← 5 cm →

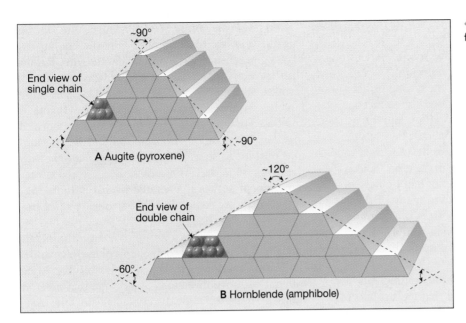

distinguish it from pyroxene, which forms rather blocky crystals. Hornblende is found predominantly in continental rocks, where it often makes up the dark portion of an otherwise light-coloured rock.

Biotite is the dark iron-rich member of the mica family. Like other micas, biotite possesses a sheet structure that gives it excellent cleavage in one direction. Biotite also has a shiny black appearance that helps distinguish it from the other dark ferromagnesian minerals. Like hornblende, biotite is a common constituent of continental rocks, including the igneous rock granite.

Garnet is similar to olivine in that its structure is composed of individual tetrahedra linked by metallic ions. Also like olivine, garnet has a glassy lustre, lacks cleavage, and possesses conchoidal fracture. Although the colours of garnet are varied, this mineral is most often brown to deep red. Garnet readily forms equidimensional crystals that are most commonly found in metamorphic rocks (Figure 2.22). When garnets are transparent, they may be used as gemstones.

Nonferromagnesian (Light) Silicates.   As the name implies, the **light (or nonferromagnesian) silicates** are generally light in colour and have a specific gravity of about 2.7, which is considerably less than the ferromagnesian silicates. As indicated earlier, these differences are mainly attributable to the presence or absence of iron and magnesium. The light silicates contain varying amounts of aluminum, potassium, calcium, and sodium rather than iron and magnesium.

Muscovite is a common member of the mica family. It is light in colour and has a pearly lustre. Like other micas, muscovite has excellent cleavage in one direction (see Figure 2.12, p. 43). In thin sheets, muscovite is clear, a property that accounts for its use as window "glass" during the Middle Ages. Because muscovite is very shiny, it can often be identified by the sparkle it gives a rock. If you have ever looked closely at beach sand, you may have seen the glimmering brilliance of the mica flakes scattered among the other sand grains.

◆ **Figure 2.22**   A deep-red garnet crystal embedded in a light-coloured, mica-rich metamorphic rock. (Photo by E. J. Tarbuck)

BOX 2.2

Understanding Earth
Gemstones

Precious stones have been prized since antiquity. But misinformation abounds regarding gems and their mineral makeup. This stems partly from the ancient practice of grouping precious stones by colour rather than mineral makeup. For example, *rubies* and red *spinels* are very similar in colour, but they are completely different minerals. Classifying by colour led to the more common spinels being passed off to royalty as rubies. Even today, with modern identification techniques, common *yellow quartz* is sometimes sold as the more valuable gemstone *topaz*.

Naming Gemstones

Most precious stones have common names that are different from their parent mineral. For example, *sapphire* is one of two names given to the mineral *corundum*. Minute amounts of foreign elements can produce vivid sapphires of nearly every colour (Figure 2.B). Traces of titanium and iron in corundum produce the most prized blue sapphires. When the mineral corundum contains a sufficient quantity of chromium, it exhibits a brilliant red colour, and the gem is called *ruby*. Further, if a specimen is not suitable as a gem, it simply goes by the mineral name *corundum*. Because of its hardness, corundum that is not of gem quality is often crushed and sold as an abrasive.

To summarize, when corundum exhibits a red hue, it is called *ruby*, but if it exhibits any other colour, the gem is called *sapphire*. Whereas corundum is the base mineral for two gems, quartz is the parent of more than a dozen gems. Table 2.A lists some well-known gemstones and their parent minerals.

What Constitutes a Gemstone?

When found in their natural state, most gemstones are dull and would be passed over by most people as "just another rock." Gems must be cut and polished by experienced professionals before their true beauty is displayed (Figure 2.B). Only those mineral specimens that

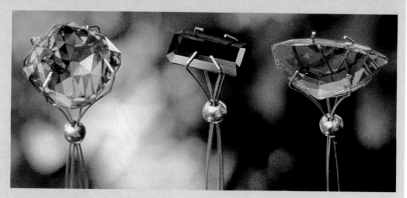

◆ **Figure 2.B**   Australian sapphires showing variation in cuts and colours. (Photo by Fred Ward, Black Star)

TABLE 2.A   Important Gemstones

Gem	Mineral Name	Prized Hues
Precious		
Diamond	Diamond	Colourless, yellows
Emerald	Beryl	Greens
Opal	Opal	Brilliant hues
Ruby	Corundum	Reds
Sapphire	Corundum	Blues
Semiprecious		
Alexandrite	Chrysoberyl	Variable
Amethyst	Quartz	Purples
Cat's-eye	Chrysoberyl	Yellows
Chalcedony	Quartz (agate)	Banded
Citrine	Quartz	Yellows
Garnet	Garnet	Reds, greens
Jade	Jadeite or nephrite	Greens
Moonstone	Feldspar	Transparent blues
Peridot	Olivine	Olive greens
Smoky quartz	Quartz	Browns
Spinel	Spinel	Reds
Topaz	Topaz	Purples, reds
Tourmaline	Tourmaline	Reds, blue-greens
Turquoise	Turquoise	Blues
Zircon	Zircon	Reds

are of such quality that they can command a price in excess of the cost of processing are considered gemstones.

Gemstones can be divided into two categories: precious and semiprecious. A *precious* gem has beauty, durability, size, and rarity, whereas a *semiprecious* gem generally has only one or two of these qualities. The gems traditionally held in highest esteem are diamonds, rubies, sapphires, emeralds, and some varieties of opal (Table 2.A). All other gemstones are classified as semiprecious. However, large high-quality specimens of semiprecious stones often command a very high price.

Today translucent stones with evenly tinted colours are preferred. The most favoured hues are red, blue, green, purple, rose, and yellow. The most prized stones are pigeon-blood rubies, blue sapphires, grass-green emeralds, and canary-yellow diamonds. Colourless gems are generally less than desirable except for diamonds that display "flashes of colour" known as *brilliance*.

The durability of a gem depends on its hardness; that is, its resistance to abrasion by objects normally encountered in everyday living. For good durability, gems should be as hard or harder than quartz as defined by the Mohs scale of hardness. One notable exception is opal, which is comparatively soft (hardness 5 to 6.5) and brittle. Opal's esteem comes from its "fire," which is a display of a variety of brilliant colours, including greens, blues, and reds.

It seems to be human nature to treasure that which is rare. In the case of gemstones, large, high-quality specimens are much rarer than smaller stones. Thus, large rubies, diamonds, and emeralds, which are rare in addition to being beautiful and durable, command the very highest prices.

Students Sometimes Ask...

I've seen garnet sandpaper at the hardware store. Is it really made of garnets?
Yes, and it's one of many things at the hardware store that's made of minerals! Hard minerals such as garnet (Mohs hardness = 6.5 to 7.5) and corundum (hardness = 9) make good abrasives. The abundance and hardness of garnets make them suitable for producing abrasive wheels, polishing materials, non-skid surfaces, and in sandblasting applications. Alternatively, those minerals that have low numbers on the Mohs hardness scale are commonly used as lubricants. For example, another mineral found in hardware stores is graphite (hardness = 1), which is used as an industrial lubricant (see Figure 2.10B).

◆ **Figure 2.23** Sample of the mineral orthoclase feldspar. (Photo courtesy of E. J. Tarbuck)

Feldspar, the most common mineral group, can form under a very wide range of temperatures and pressures, a fact that partially accounts for its abundance. All of the feldspars have similar physical properties. They have two planes of cleavage meeting at or near 90-degree angles, are relatively hard (6 on the Mohs scale), and have a lustre that ranges from glassy to pearly. As one component in a rock, feldspar crystals can be identified by their rectangular shape and rather smooth shiny faces (Figure 2.23).

The structure of feldspar minerals is a three-dimensional framework formed when oxygen atoms are shared by adjacent silicon atoms. In addition, one quarter to one half of the silicon atoms in the feldspar structure are replaced by aluminum atoms. The difference in charge between aluminum (+3) and silicon (+4) is made up by the inclusion of one or more of the following ions into the crystal lattice: potassium (+1), sodium (+1), and calcium (+2). Because of the large size of the potassium ion as compared to the size of the sodium and calcium ions, two different feldspar structures exist. *Orthoclase feldspar* is a common member of a group of feldspar minerals that contains potassium ions in its structure and is therefore often referred to as *potassium feldspar*. The other group, called *plagioclase feldspar*, contains both sodium and calcium ions that freely substitute for one another depending on the environment during crystallization.

Potassium feldspar is usually light cream to salmon pink in colour. The plagioclase feldspars, on the other hand, range in colour from white to medium grey. However, colour should not be used to distinguish these groups. The only sure way to distinguish the feldspars physically is to look for a multitude of fine parallel lines, called *striations* (*striat* = a streak). Striations are found on some cleavage planes of plagioclase feldspar but are not present on orthoclase feldspar (Figure 2.24).

Quartz is the only common silicate mineral consisting entirely of silicon and oxygen. As such, the term *silica* is applied to quartz, which has the chemical formula SiO_2. As the structure of quartz contains a ratio of two oxygen ions (O^{2-}) for every silicon ion (Si^{4+}), no other positive ions are needed to attain neutrality.

In quartz, a three-dimensional framework is developed through the complete sharing of oxygen by adjacent silicon atoms. Thus, all of the bonds in quartz are of the strong silicon–oxygen type. Consequently, quartz is hard, resistant to weathering, and does not have cleavage. When broken, quartz generally exhibits conchoidal fracture. In a pure form, quartz is clear and if allowed to solidify without interference, will form hexagonal crystals that develop pyramid-shaped ends (see Figure 2.11B, p. 42). However, like most other clear minerals, quartz is often coloured by the inclusion of various ions (impurities) and forms without developing good crystal faces. The most common varieties of quartz are milky (white), smoky (grey), rose (pink), amethyst (purple), and rock crystal (clear) (Figure 2.25).

Clay is a term used to describe a variety of complex minerals that, like the micas, have a sheet structure. The clay minerals are generally very fine grained and can only be studied microscopically. Most clay minerals originate as products of the chemical weathering of

← 2 cm →

◆ **Figure 2.24** These parallel lines, called striations, are a distinguishing characteristic of the plagioclase feldspars. (Photo by E. J. Tarbuck)

other silicate minerals. Thus, clay minerals make up a large percentage of the surface material we call soil. Because of the importance of soil in agriculture, and because of its role as a supporting material for buildings, clay minerals are extremely important to humans.

Students Sometimes Ask...

What's the difference between a carrot, a karat, and a carat?

Like many words in English, *carrot*, *karat*, and *carat* have the same sound but have different meanings. Such words are called *homonyms* (*homo* = same, *nym* = name). A *carrot* is the familiar orange crunchy vegetable, and *karat* and *carat* have to do with gold and gems, respectively. (Both *karat* and *carat* derive from the Greek word *keration* = carob bean, which early Greeks used as a weight standard.)

Karat is a term used to indicate the purity of gold, where 24 karats represent pure gold. Gold less than 24 karats is an alloy (mixture) of gold and another metal, usually copper or silver. For example, 14-karat gold contains 14 parts of gold (by weight) mixed with 10 parts of other metals.

Carat is a unit of weight used for precious gems such as diamonds, emeralds, and rubies. The size of a carat has varied throughout history, but early in the twentieth century it was standardized at 200 milligrams (0.2 gram). For example, a typical diamond on an engagement ring might range from one half to one carat, and the famous Hope Diamond at the Smithsonian Institution is 45.52 carats.

◆ **Figure 2.25** Quartz. Some minerals, such as quartz, occur in a variety of colours. These samples include crystal quartz (colourless), amethyst (purple quartz), citrine (yellow quartz), and smoky quartz (grey to black). (Photo courtesy of E. J. Tarbuck)

One of the most common clay minerals is *kaolinite*, which is used in the manufacture of fine chinaware and in the production of high-gloss paper such as that used in this textbook. Further, some clay minerals absorb large amounts of water, which allows them to swell to several times their normal size. These swelling clays have been used commercially in a variety of ingenious ways, including as an additive to thicken milkshakes in fast-food restaurants.

Important Nonsilicate Minerals

Other mineral groups can be considered scarce when compared to the silicates, although many are extremely important geologically and economically. Table 2.4 lists examples of several nonsilicate mineral groups of economic value. A discussion of a few of the more common nonsilicate, rock-forming minerals follows.

The carbonate minerals are much simpler structurally than are the silicates. This mineral group is composed of the carbonate ion (CO_2^{2-}), and one or more kinds of positive ions. The two most common carbonate minerals are *calcite*, $CaCO_3$ (calcium carbonate) and *dolomite*, $CaMg(CO_3)_2$ (calcium/magnesium carbonate). Because these minerals are similar both physically and chemically, they are difficult to distinguish from one another. Both have a vitreous lustre, a hardness between 3 and 4, and nearly perfect rhombic cleavage (see Figure 2.13, right side, p. 43). They can, however, be distinguished by using dilute hydrochloric acid. Calcite reacts vigorously with this acid, whereas dolomite reacts much more slowly. Calcite and dolomite are usually found together as the primary constituents in the sedimentary rocks limestone and dolostone. When calcite is the dominant mineral, the rock is called *limestone*, whereas *dolostone*

TABLE 2.4   Common Nonsilicate Mineral Groups

Group	Member	Formula	Economic Use
Oxides	Hematite	Fe_2O_3	Ore of iron, pigment
	Magnetite	Fe_3O_4	Ore of iron
	Corundum	Al_2O_3	Gemstone, abrasive
	Ice	H_2O	Solid form of water
	Chromite	$FeCr_2O_4$	Ore of chromium
	Ilmenite	$FeTiO_3$	Ore of titanium
Sulphides	Galena	PbS	Ore of lead
	Sphalerite	ZnS	Ore of zinc
	Pyrite	FeS_2	Sulphuric acid production
	Chalcopyrite	$CuFeS_2$	Ore of copper
	Bornite	Cu_5FeS_2	Ore of copper
	Cinnabar	HgS	Ore of mercury
Sulfates	Gypsum	$CaSO_4 \cdot 2H_2O$	Plaster
	Anhydrite	$CaSo_4$	Plaster
	Barite	$BaSO_4$	Drilling mud
Native elements	Gold	Au	Trade, jewelry
	Copper	Cu	Electrical conductor
	Diamond	C	Gemstone, abrasive
	Sulphur	S	Sulpha drugs, chemicals
	Graphite	C	Pencil lead, dry lubricant
	Silver	Ag	Jewellery, photography
	Platinum	Pt	Catalyst
Halides	Halite	$NaCl$	Common salt
	Fluorite	CaF_2	Used in steelmaking
	Sylvite	KCl	Fertilizer
Carbonates	Calcite	$CaCO_3$	Portland cement, lime
	Dolomite	$CaMg(CO_3)_2$	Portland cement, lime
	Malachite	$Cu_2(OH)_2CO_3$	Gemstone
	Azurite	$Cu_3(OH)_2(CO_3)_2$	Gemstone
Hydroxides	Limonite	$FeO(OH) \cdot nH_2O$	Ore of iron, pigments
	Bauxite	$Al(OH)_3 \cdot nH_2O$	Ore of aluminum
Phosphates	Apatite	$Ca_5(F,Cl,OH)(PO_4)_3$	Fertilizer
	Turquoise	$CuAl_6(PO_4)_4(OH)_8$	Gemstone

results from a predominance of dolomite. Limestone has numerous economic uses, including as road aggregate, as building stone, and as the main ingredient in portland cement.

Three other nonsilicate minerals found in sedimentary rocks are *halite sylvite*, and *gypsum*. These minerals are commonly found in thick layers, which are the last vestiges of ancient seas that have long since evaporated (Figure 2.26). Like limestone, both are important nonmetallic resources. Halite is the mineral name for common table salt (NaCl), whereas sylvite (KCl) is the principal mineral of potash. Gypsum ($CaSO_4 \cdot 2H_2O$), which is calcium sulphate with water bound into the structure, is the mineral of which plaster and other similar building materials are composed.

In addition, a number of other minerals are prized for their economic value (see Table 2.4 and Box 2.2). Included in this group are the ores of metals, such as hematite (iron), sphalerite (zinc), and galena (lead); the native (free-occurring, not in compounds) elements, including gold, silver, carbon (diamonds); and a host of others, such as fluorite, corundum, and uraninite (a uranium source).*

_____

*For more on the economic significance of these and other minerals, see Chapter 21.

◆ **Figure 2.26**   Thick beds of potash (potassium-rich salt deposit) in an underground mine in Saskatchewan. (Photo by Jisuo Jin)

Chapter Summary

- A *mineral* is a naturally occurring inorganic solid that possesses a definite chemical structure that gives it a unique set of physical properties. Most *rocks* are aggregates composed of two or more minerals.
- The building blocks of minerals are *elements*. An *atom* is the smallest particle of matter that still retains the characteristics of an element. Each atom has a *nucleus*, which contains *protons* (particles with positive electrical charges) and *neutrons* (particles with neutral electrical charges). Orbiting the nucleus of an atom in regions called *energy levels*, or *shells*, are *electrons*, which have negative electrical charges. The number of protons in an atom's nucleus determines its *atomic number* and the name of the element. An element is a large collection of electrically neutral atoms, all having the same atomic number.
- Atoms combine with each other to form more complex substances called *compounds*. Atoms bond together by gaining, losing, or sharing electrons with other atoms. In *ionic bonding*, one or more electrons are transferred from one atom to another, giving the atoms a net positive or negative charge. The resulting electrically charged atoms are called *ions*. Ionic compounds consist of oppositely charged ions assembled in a regular, crystalline structure that allows for the maximum attraction of ions, given their sizes. Another type of bond, the *covalent bond*, is produced when atoms share electrons.
- *Isotopes* are variants of the same element but with a different *mass number* (the total number of neutrons plus protons found in an atom's nucleus). Some isotopes are unstable and disintegrate naturally through a process called *radioactivity*.
- The properties of minerals include *crystal form*, *lustre*, *colour*, *streak*, *hardness*, *cleavage*, *fracture*, and *specific gravity*. In addition, a number of special physical and chemical properties (*taste*, *smell*, *elasticity*, *malleability*, *feel*, *magnetism*, *double refraction*, and *chemical reaction to hydrochloric acid*) are useful in identifying certain minerals. Each mineral has a unique set of properties that can be used for identification.
- Of the nearly 4000 minerals, no more than a few dozen make up most of the rocks of Earth's crust and, as such, are classified as rock-forming minerals. Eight elements (oxygen, silicon, aluminum, iron, calcium, sodium, potassium, and magnesium) make up the bulk of these minerals and represent over 98 percent (by weight) of Earth's continental crust.
- The most common mineral group is the *silicates*. All silicate minerals have the negatively charged *silicon–oxygen tetrahedron* as their fundamental building block. In some silicate minerals the tetrahedra are joined in chains (the pyroxene and amphibole groups); in others, the tetrahedra are arranged into sheets (the micas, biotite, and muscovite), or three-dimensional networks (the feldspars and quartz). The tetrahedra and various silicate structures are often bonded together by the positive ions of iron, magnesium, potassium, sodium, aluminum, and calcium. Each silicate mineral has a structure and a chemical composition that indicates the conditions under which it formed.
- The *nonsilicate* mineral groups, which contain several economically important minerals, include the *oxides* (e.g., the mineral hematite, mined for iron), *sulphides* (e.g., the mineral sphalerite, mined for zinc; and the mineral galena, mined for lead), *sulphates*, *halides*, and *native elements* (e.g., gold and silver). The more common nonsilicate rock-forming minerals include the *carbonate minerals*, calcite and dolomite. Two other nonsilicate minerals frequently found in sedimentary rocks are halite and gypsum.

Review Questions

1. Define the term *rock*.
2. List the three main particles of an atom and explain how they differ from one another.
3. If the number of electrons in a neutral atom is 35 and its mass number is 80, calculate the following:
   (a) the number of protons
   (b) the atomic number
   (c) the number of neutrons
4. What is the significance of valence electrons?
5. Briefly distinguish between ionic and covalent bonding.
6. What occurs in an atom to produce an ion?
7. What is an isotope?
8. Although all minerals have an orderly internal arrangement of atoms (crystalline structure), most mineral samples do not exhibit their crystal form. Why?
9. Why might it be difficult to identify a mineral by its colour?

10. If you found a glassy-appearing mineral while rock hunting and had hopes that it was a diamond, what simple test might help you make a determination?

11. Explain the use of corundum as given in Table 2.4 (p. 54) in terms of the Mohs hardness scale.

12. Gold has a specific gravity of almost 20. If a 25-litre pail of water weighs 25 kilograms, how much would a 25-litre pail of gold weigh?

13. Explain the difference between the terms *silicon* and *silicate*.

14. What do ferromagnesian minerals have in common? List examples of ferromagnesian minerals.

15. What do muscovite and biotite have in common? How do they differ?

16. Should colour be used to distinguish between orthoclase and plagioclase feldspar? What is the best means of distinguishing between these two types of feldspar?

17. Each of the following statements describes a silicate mineral or mineral group. In each case, provide the appropriate name:
    (a) the most common member of the amphibole group
    (b) the most common nonferromagnesian member of the mica family
    (c) the only silicate mineral made entirely of silicon and oxygen
    (d) a high-temperature silicate with a name that is based on its colour
    (e) one that is characterized by striations
    (f) one that originates as a product of chemical weathering

18. What simple test can be used to distinguish calcite from dolomite?

Key Terms

atom (p. 35)	energy-levels, or shells (p. 35)	mass number (p. 38)	proton (p. 35)
atomic number (p. 35)	ferromagnesian silicates (p. 49)	metallic bond (p. 38)	radioactive decay (p. 39)
atomic weight (p. 38)		mineral (p. 32)	rock (p. 33)
cleavage (p. 42)	fracture (p. 43)	mineralogy (p. 32)	shell (p. 35)
colour (p. 41)	hardness (p. 41)	Mohs scale (p. 41)	silicate mineral (p. 44)
compound (p. 36)	ion (p. 37)	neutron (p. 35)	silicon-oxygen tetrahedron (p. 45)
covalent bond (p. 38)	ionic bond (p. 36)	nonferromagnesian silicates (p. 50)	specific gravity (p. 43)
crystal form (p. 40)	isotope (p. 38)	nucleus (p. 35)	streak (p. 41)
dark silicates (p. 49)	light silicates (p. 50)	octet rule (p. 36)	valence electron (p. 35)
electron (p. 35)	lustre (p. 40)	polymorph (p. 39)	
element (p. 35)			

Web Resources

 The *Earth* Web site uses the resources and flexibility of the Internet to aid in your study of the topics in this chapter. Written and developed by geology instructors, this site will help improve your understanding of geology. Visit **http://www.pearson.ca/ tarbuck** and click on the cover of the text to find:

■ Online review quizzes.

■ Web-based critical thinking and writing exercises.

■ Links to chapter-specific Web resources.

■ Internet-wide key-term searches.

http://www.pearson.ca/tarbuck

"Smokey the Bear"

ha

Chapter 3
Igneous Rocks

snout

mouth

Geologists gather above "Smokey the Bear," part of a large, granitic igneous intrusion in Cathedral Provincial Park in south-central British Columbia.
(Photo by Peter Rotheisler)

Igneous rocks make up the bulk of Earth's crust. In fact, with the exception of Earth's core, the remaining solid portion of our planet is basically a huge mass of igneous rock that is partially covered by a thin veneer of sedimentary and metamorphic rocks. Consequently, a basic knowledge of igneous rocks is essential to our understanding of the structure, composition, and internal workings of our planet.

Magma: The Parent Material of Igneous Rock

In our discussion of the rock cycle, it was pointed out that **igneous rocks** (*ignis* = fire) form as molten rock cools and solidifies. Abundant evidence indicates that the parent material for igneous rocks, called *magma*, is formed by a process called *partial melting*. Partial melting occurs at various levels within Earth's crust and upper mantle to depths of perhaps 250 kilometres. We will explore the origin of magma later in this chapter.

Once formed, a magma body buoyantly rises toward the surface because it is less dense than the surrounding rocks. Occasionally molten rock breaks through, producing a volcanic eruption. Magma that reaches Earth's surface is called **lava**. The lava fountain shown in Figure 3.1 was produced when escaping gases propelled molten rock from a magma chamber. Sometimes magma is explosively ejected from a vent producing a catastrophic eruption. However, not all eruptions are violent; many volcanoes emit quiet outpourings of very fluid lava.

Igneous rocks that form when magma solidifies *at the surface* are classified as **extrusive** (*ex* = out, *trudere* = thrust), or **volcanic** (after the fire god Vulcan). Extrusive igneous rocks are abundant in western portions of the Americas, including the volcanic cones of the Cascade Range and the extensive lava flows of the Columbia Plateau. In addition, many oceanic islands, such as the Hawaiian chain, are composed almost entirely of volcanic igneous rocks.

Magma that loses its mobility before reaching the surface eventually crystallizes at depth. Igneous rocks that *form at depth* are termed **intrusive** (*in* = into, *trudere* = thrust), or **plutonic** (after Pluto, the god of the lower world in classical mythology). Intrusive igneous rocks would never be exposed at the surface if portions of the crust were not uplifted and the overlying rocks stripped away by erosion. (When a mass of crustal rock is exposed—not covered

◆ **Figure 3.1**   Recent eruption of Hawaii's Kilauea Volcano.
(Photo by Soames Summerhays/Science Source/Photo Researchers, Inc.)

Students Sometimes Ask...

Are lava and magma the same thing?

No, but their *composition* might be similar. Both are terms that describe molten or liquid rock: Magma exists beneath Earth's surface, and lava is molten rock that has reached the surface. That's the reason why they can be similar in composition: Lava is produced from magma, but it generally has lost materials that escape as a gas, such as water vapour.

with soil—it is called an *outcrop*. An outcrop of plutonic rock is called a *pluton*.) Exposures of intrusive igneous rocks occur in many places, including much of western and central British Columbia, the St. Lawrence region of Quebec (Mount Royal in Montréal), the highlands of the maritime provinces, and in the Sweetgrass Hills area of Montana just south of Alberta. Very ancient igneous intrusions are also exposed throughout the Canadian Shield, which includes much of Nunavut, the Northwest Territories, much of Manitoba, Ontario, and Quebec, and parts of Alberta and Saskatchewan.

The Nature of Magma

Magma is completely or partly molten material which on cooling solidifies to form an igneous rock. Most magmas consist of three distinct parts—a liquid component, a solid component, and a gaseous phase.

The liquid portion, called **melt**, is composed of mobile ions of those elements commonly found in Earth's crust. Melt is made up mostly of ions of silicon and oxygen, as well as lesser amounts of aluminum, potassium, calcium, sodium, iron, and magnesium.

The solid components (if any) in magma are silicate minerals that have already crystallized from the melt. As a magma body cools, the size and number of crystals increases. During the last stage of cooling, a magma body is mostly a crystalline solid with only minor amounts of melt.

Water vapour (H_2O), carbon dioxide (CO_2), and sulphur dioxide (SO_2) are the most common gases found in magma and are confined by the immense pressure exerted by the overlying rocks. These gaseous components, called **volatiles**, are dissolved within the melt. Volatiles are those materials that will readily vaporize (form a gas) at surface pressures. Volatiles remain part of the magma until it either moves near the surface (low-pressure environment), or until the magma body is essentially crystallized, at which time any remaining volatiles freely migrate away. These hot fluids play an important role in metamorphism and will be considered further in Chapter 7.

From Magma to Crystalline Rock

As magma cools, the ions in the melt begin to lose their mobility and arrange themselves into orderly crystalline structures. This process, called **crystallization**, generates various silicate minerals that reside within the remaining melt.

Before we examine how magma crystallizes, let us first examine how a simple crystalline solid melts. In any crystalline solid, the ions are arranged in a closely packed regular pattern. However, they are not without some motion. They exhibit a sort of restricted vibration about fixed points. As the temperature rises, the ions vibrate more rapidly and consequently collide with ever increasing vigour with their neighbours. Thus, heating causes the ions to occupy more space, which in turn causes the solid to expand. When the ions are vibrating rapidly enough to overcome the force of the chemical bonds, the solid begins to melt. At this stage the ions are able to slide past one another, and their orderly crystalline structure disintegrates. Thus, melting converts a solid consisting of tight, uniformly packed ions into a liquid composed of unordered ions moving randomly about.

In the process of crystallization, cooling reverses the events of melting. As the temperature of the liquid drops, the ions pack closer and closer together as they slow their rate of movement. When cooled sufficiently, the forces of the chemical bonds will again confine the ions to an orderly crystalline arrangement.

When magma cools, it is generally the silicon and oxygen atoms that link together first to form silicon-oxygen tetrahedra, the basic building blocks of the silicate minerals. As magma continues to lose heat to its surroundings, the tetrahedra join with each other and with other ions to form embryonic crystal nuclei. Slowly each nucleus grows as ions lose their mobility and join the crystalline network.

The earliest formed minerals have space to grow and tend to have better developed crystal faces than do the later ones that fill the remaining space. Eventually all of the melt is transformed into a solid mass of interlocking silicate minerals that we call an *igneous rock* (Figure 3.2).

As you will see later, the crystallization of magma is much more complex than just described. Whereas a single compound, such as water, crystallizes at a specific temperature, solidification of magma with its diverse chemistry spans a temperature range of 200°C. During crystallization, the composition of the melt continually changes as ions are selectively removed and incorporated into the earliest formed minerals. If the melt should separate from the earliest formed minerals, its composition will be different from that of the original magma. Thus, a single magma may generate rocks with widely differing compositions. As

A

B

◆ **Figure 3.2   A.** Close-up of interlocking crystals in a coarse-grained igneous rock. The largest crystals are about 1 centimetre in length. **B.** Photomicrograph of interlocking crystals in a coarse-grained igneous rock. (Photos by E. J. Tarbuck)

a consequence, a great variety of igneous rocks exist. We will return to this important idea later in the chapter.

Although the crystallization of magma is complex, it is nevertheless possible to classify igneous rocks based on their mineral composition and the conditions under which they formed. Their environment during crystallization can be roughly inferred from the size and arrangement of the mineral grains, a property called *texture*. Consequently, *igneous rocks are most often classified by their texture and mineral composition*. We will consider these two rock characteristics in the following sections.

Igneous Textures

  Earth Materials
    ↳ Igneous Rocks

The term **texture**, when applied to an igneous rock, is used to describe the overall appearance of the rock based on the size, shape, and arrangement of its interlocking crystals (Figure 3.3). Texture is an important characteristic because it reveals a great deal about the environment in which the rock formed. This permits geologists to make inferences about a rock's origin while working in the field where sophisticated equipment is not available.

Factors Affecting Crystal Size

Three factors contribute to the textures of igneous rocks: (1) *the rate at which magma cools;* (2) *the amount of silica present;* and (3) *the amount of dissolved gases in the magma*. Of these, the rate of cooling is the dominant factor, but like all generalizations, this one has numerous exceptions.

As a magma body loses heat to its surroundings, the mobility of its ions decreases. When cooling occurs rapidly—for example, in a thin lava flow—the ions quickly lose their mobility and readily combine to form crystals. This results in the development of numerous embryonic nuclei, all of which compete for the available ions. The result is a solid mass of many small, intergrown crystals. On the other hand, a very large magma body located at great depth will cool over a period of perhaps tens or hundreds of thousands of years. Initially, relatively few crystal nuclei form. Slow cooling permits ions to migrate freely until they eventually join one of the existing crystalline structures. Consequently, slow cooling promotes the growth of fewer but larger crystals.

At the opposite extreme, when molten material is quenched quickly, there may not be sufficient time for the ions to arrange into a crystalline network. Rocks that consist of unordered ions are referred to as **glass**.

Types of Igneous Textures

As you saw, the effect of cooling on rock textures is fairly straightforward. Rapid cooling tends to generate small crystals, whereas slow cooling promotes the growth of large crystals. We will consider the other two factors affecting crystal growth as we examine the major textural types.

Aphanitic (fine-grained) Texture. Igneous rocks that form at the surface or as small masses within the upper crust where cooling is relatively rapid possess a very fine-grained texture termed **aphanitic** (*a* = not, *phaner* = visible). By definition, the crystals that make up aphanitic rocks are so small that individual minerals

◆ **Figure 3.17** Outcrop of welded tuff interbedded with obsidian (black) near Shoshone, California. Tuff is composed mainly of ash-sized particles and may contain larger fragments of pumice or other volcanic rocks. (Photo by Breck P. Kent)

Origin of Magma

The origin of magma has been a controversial topic in geology almost from the very beginning of the science. How can magma form in Earth's mantle, which is composed essentially of solid rock? How do magmas of different compositions form? Why do volcanoes in the deep-ocean basins primarily extrude mafic lava, whereas those on the continental margins adjacent to oceanic trenches extrude mainly andesitic lava? These are some of the questions we will address in the following section.

Generating Magma from Solid Rock

Based on available scientific evidence, *Earth's crust and mantle are composed primarily of solid, not molten, rock*. Although the outer core is a fluid, its iron-rich material is very dense and remains deep within Earth. So, what is the source of magma that produces igneous activity?

Geologists conclude that magma originates when essentially solid rock, located in the crust and upper mantle, melts. The most obvious way to generate magma from solid rock is to raise the temperature above the rock's melting point.

Role of Heat.   What source of heat is sufficient to melt rock? Workers in underground mines know that temperatures get higher as they go deeper. Although the rate of temperature change varies from place to place, it *averages* between 20°C and 30°C per kilometre in the *upper* crust. This change in temperature with depth is known as the **geothermal gradient** (Figure 3.18). Estimates indicate that the temperature at a depth of 100 kilometres ranges between 1200°C and 1400°C.* At these high temperatures, rocks in the lower crust and upper mantle are near their melting points; they are very hot but still essentially solid.

There are several ways that enough additional heat can be generated within the crust or upper mantle to produce some magma. First, at subduction zones, friction generates heat as huge slabs of crust slide past one another. Second, crustal rocks are heated as they descend into the mantle during subduction. Third, hot mantle rocks can rise and intrude crustal rocks. Although all of these processes generate some magma, the quantities produced are relatively small and the distribution is very limited.

As you will see, the vast bulk of magma forms without the aid of an additional heat source. Rock that is near its melting point may begin to melt if the

---

*We will consider the heat sources for the geothermal gradient in Chapter 17.

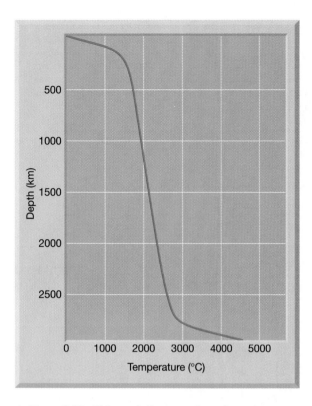

◆ **Figure 3.18**   This graph illustrates the estimated temperature distribution for the crust and mantle. Notice that temperature increases significantly from the surface to the base of the lithosphere (100 kilometres depth) and that the temperature gradient (rate of change) is much less in the mantle. Because the temperature difference between the top and bottom of the mantle is relatively small, geologists conclude that slow convective flow (hot material rising, and cool material sinking) must occur in the mantle.

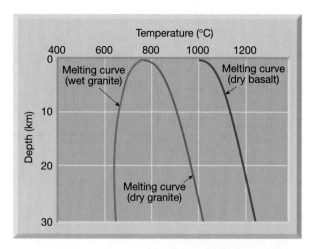

◆ **Figure 3.19**   Idealized melting temperature curves. These curves portray the minimum temperatures required to melt rock within Earth's crust. Notice that dry granite and dry basalt melt at higher temperatures with increasing depth. By contrast, the melting temperature of wet granite actually decreases as the confining pressure increases.

confining pressure drops or if fluids (volatiles) are introduced. We will now consider the roles that pressure and volatiles play in generating magma.

Role of Pressure.   If temperature were the only factor that determined whether or not rock melts, our planet would be a molten ball covered with a thin, solid outer shell. This, of course, is not the case. The reason is that pressure also increases with depth.

Melting, which is accompanied by an increase in volume, *occurs at higher temperatures at depth* because of greater confining pressure (Figure 3.19). Consequently, an increase in confining pressure causes a rise in the rock's melting temperature. Conversely, reducing confining pressure lowers a rock's melting temperature. When confining pressure drops enough, **decompression melting** is triggered. This may occur when rock *ascends* as a result of convective upwelling, thereby moving into zones of lower pressure. (Recall that even though the mantle is a *solid*, it does *flow* at very slow rates over time scales of millions of years.)

This process is responsible for generating magma along ocean ridges where plates are rifting apart (Figure 3.20).

Role of Volatiles.   Another important factor affecting the melting temperature of rock is its water content. Water and other volatiles cause rock to melt at lower temperatures. Further, the effect of volatiles is magnified by increased pressure. Consequently, "wet" rock buried at depth has a much lower melting temperature than does "dry" rock of the same composition and under the same confining pressure (see Figure 3.19). Therefore, in addition to a rock's composition, its temperature, depth (confining pressure), and water content determine whether it exists as a solid or liquid.

Volatiles play an important role in generating magma in regions where cool slabs of oceanic lithosphere descend into the mantle (Figure 3.21). As an oceanic plate sinks, both heat and pressure drive water from the subducting crustal rocks. These volatiles, which are very mobile, migrate into the wedge of hot mantle that lies above. This process is believed to lower the melting temperature of mantle rock sufficiently to generate some melt. Laboratory studies have shown that the melting point of basalt can be lowered by as much as 100°C by the addition of only 0.1 percent water.

When enough mantle-derived mafic magma forms, it will buoyantly rise toward the surface. In a continental setting, mafic magma may "pond" beneath crustal rocks, which have a lower density and are already near their melting temperature. This may result in some melting of the crust and the formation of a secondary, silica-rich magma.

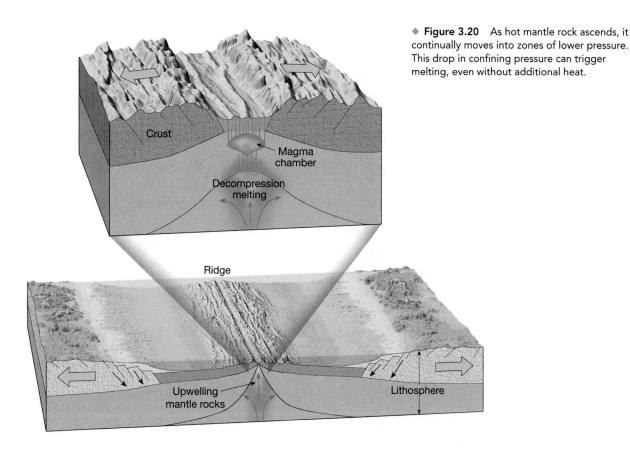

◆ **Figure 3.20** As hot mantle rock ascends, it continually moves into zones of lower pressure. This drop in confining pressure can trigger melting, even without additional heat.

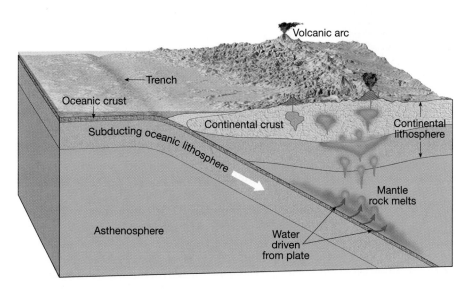

◆ **Figure 3.21** As an oceanic plate descends into the mantle, water and other volatiles are driven from the subducting crustal rocks. These volatiles lower the melting temperature of mantle rock sufficiently to generate melt.

In summary, magma can be generated under three sets of conditions: (1) *heat* may be added; for example, a magma body from a deeper source intrudes and melts crustal rock; (2) *a decrease in pressure* (without the addition of heat) can result in *decompression melting*; and (3) the *introduction of volatiles* (principally water) can lower the melting temperature of mantle rock sufficiently to generate magma.

How Magmas Evolve

Because a large variety of igneous rocks exists, it is logical to assume that an equally large variety of magmas must also exist. However, geologists have observed that a single volcano may extrude a number of lavas exhibiting quite different compositions (Figure 3.22). Data of this type led geologists to

◆ **Figure 3.22** Ash and pumice ejected during a large eruption of Mt. Mazama (Crater Lake). Notice the gradation from light-coloured, silica-rich ash near the base to dark-coloured rocks at the top. It is likely that prior to this eruption the magma began to segregate as the less-dense, silica-rich magma migrated toward the top of the magma chamber. The zonation seen in the rocks resulted because a sustained eruption tapped deeper and deeper levels of the magma chamber. Thus, this rock sequence is an inverted representation of the compositional zonation in the magma body; that is, the magma from the top of the chamber erupted first and is found at the base of these ash deposits and vice versa. (Photo by E. J. Tarbuck)

examine the possibility that magma might change (evolve) and thus become the parent to a variety of igneous rocks. To explore this idea, a pioneering investigation into the crystallization of magma was carried out by Canadian scientist N. L. Bowen in the first quarter of the twentieth century.

Bowen's Reaction Series and the Composition of Igneous Rocks

Recall that ice freezes at a single temperature whereas magma crystallizes through at least 200°C of cooling. In a laboratory setting Bowen and his coworkers demonstrated that as a mafic magma cools, minerals tend to crystallize in a systematic fashion based on their melting points. As shown in Figure 3.23, the first mineral to crystallize from a mafic magma is the ferromagnesian mineral olivine. Further cooling generates calcium-rich plagioclase feldspar as well as pyroxene, and so forth down the diagram.

During the crystallization process, the composition of the liquid portion of the magma continually changes. For example, at the stage when about a third of the magma has solidified, the melt will be nearly depleted of iron, magnesium, and calcium because these elements are constituents of the earliest-formed minerals. The removal of these elements from the melt will cause it to become enriched in sodium and potassium. Further, because the original mafic magma contained about 50 percent silica (SiO_2), the crystallization of the earliest-formed mineral, olivine, which is only about 40 percent silica, leaves the remaining melt richer in SiO_2. Thus, the silica component of the melt becomes enriched as the magma evolves.

Bowen also demonstrated that if the solid components of a magma remain in contact with the remaining melt, they would chemically react and evolve into the next mineral in the sequence shown in Figure 3.23. For this reason, this arrangement of minerals became known as **Bowen's reaction series** (see Box 3.3). As you will see, in some natural settings the earliest-formed minerals can be separated from the melt, thus halting any further chemical reaction.

The diagram of Bowen's reaction series in Figure 3.23 depicts the sequence that minerals crystallize from a magma of average composition under laboratory conditions. Evidence that this highly idealized crystallization model approximates what can happen in nature comes from the analysis of igneous rocks. In particular, we find that minerals that form in the same general temperature regime on Bowen's reaction series are found together in the same igneous rocks. For example, notice in Figure 3.23 that the minerals quartz, potassium feldspar, and muscovite, which are located in the same region of Bowen's diagram, are typically found together as major constituents of the plutonic igneous rock *granite*.

Magmatic Differentiation. Bowen demonstrated that minerals crystallize from magma in a systematic fashion. But how do Bowen's findings account for the great diversity of igneous rocks? It has been shown that, at one or more stages during crystallization, a

BOX 3.3

Understanding Earth
A Closer Look at Bowen's Reaction Series

Although highly idealized, Bowen's reaction series provides a visual representation of the order in which minerals crystallize from a magma of average composition (see Figure 3.23). This model assumes that the magma cools slowly at depth in an otherwise unchanging environment. Notice that Bowen's reaction series is divided into two branches—a discontinuous series and a continuous series.

Discontinuous Reaction Series. The upper left branch of Bowen's reaction series indicates that as a magma cools, olivine is the first mineral to crystallize. Once formed, olivine chemically reacts with the remaining melt to form pyroxene (see Figure 3.23). In this reaction, olivine, which is composed of individual silicon-oxygen tetrahedra, incorporates more silica into its structure, thereby linking its tetrahedra into single-chain structures of pyroxene. (Note: Pyroxene has a lower crystallization temperature than olivine and is more stable at lower temperatures.) As the magma body cools further, pyroxene crystals react with the melt to generate the double-chain structure of amphibole. This reaction continues until the last mineral in this series, biotite mica, crystallizes. In nature, these reactions do not usually run to completion, so that various amounts of each of the minerals in the series may exist at any given time, and some minerals, such as biotite, may never form.

This branch of Bowen's reaction series is called a *discontinuous reaction series* because at each step a different silicate structure emerges. Olivine, the first mineral in the sequence, is composed of isolated tetrahedra, whereas pyrox-

ene is composed of single chains, amphibole of double chains, and biotite of sheet structures.

Continuous Reaction Series. The right branch of the reaction series, called the *continuous reaction series*, illustrates that calcium-rich plagioclase feldspar crystals react with the sodium ions in the melt to become progressively more sodium-rich (see Figure 3.23). Here the sodium ions diffuse into the feldspar crystals and displace the calcium ions in the crystal lattice. Often, the rate of cooling occurs rapidly enough to prohibit a complete replacement of the calcium ions by sodium ions. In these instances, the feldspar crystals have calcium-rich interiors surrounded by zones that are progressively richer in sodium (Figure 3.C).

During the last stage of crystallization, after much of the magma has solidified, potassium feldspar forms. Finally, if the remaining melt has excess silica, the mineral quartz forms.

Testing Bowen's Reaction Series. During an eruption of Hawaii's Kilauea volcano in 1965, mafic lava poured into a pit crater, forming a lava lake that became a natural laboratory for testing Bowen's reaction series. When the surface of the lava lake cooled enough to form a crust, geologists drilled into the magma and periodically removed samples that were quenched to preserve the melt, and minerals that were growing within it. By sampling the lava at successive stages of cooling, a history of crystallization was recorded.

As Bowen's reaction series predicts, olivine crystallized early but later ceased to form, and was partly reabsorbed into the cooling melt. (In a larger magma

◆ **Figure 3.C** Photomicrograph of a zoned plagioclase feldspar crystal. After this crystal (composed of calcium-rich feldspar) solidified, further cooling resulted in sodium ions displacing calcium ions. Because replacement was not complete, this feldspar crystal has a calcium-rich interior surrounded by zones that are progressively richer in sodium. (Photo courtesy of E. J. Tarbuck)

body that cooled more slowly, we would expect most, if not all, of the olivine to react with the melt and change to pyroxene.) Most important, the melt changed composition throughout the course of crystallization. In contrast to the original mafic lava, which contained about 50 percent silica (SiO_2), the final melt contained more than 75 percent silica and had a composition similar to granite.

Although the lava in this setting cooled rapidly compared to rates experienced in deep magma chambers, it was slow enough to verify that minerals do crystallize in a systematic fashion that roughly parallels Bowen's reaction series. Further, had the melt been separated at any stage in the cooling process, it would have formed a rock with a composition much different from that of the original lava.

---

separation of the solid and liquid components of a magma can occur. One example is called **crystal settling**. This process occurs when the earlier-formed minerals are denser (heavier) than the liquid portion and sink toward the bottom of the magma chamber, as shown in Figure 3.24. When the remaining melt

solidifies—either in place or in another location if it migrates into fractures in the surrounding rocks—it will form a rock with a chemical composition much different from the parent magma (Figure 3.24). The formation of one or more secondary magmas from a single parent magma is called **magmatic differentiation**.

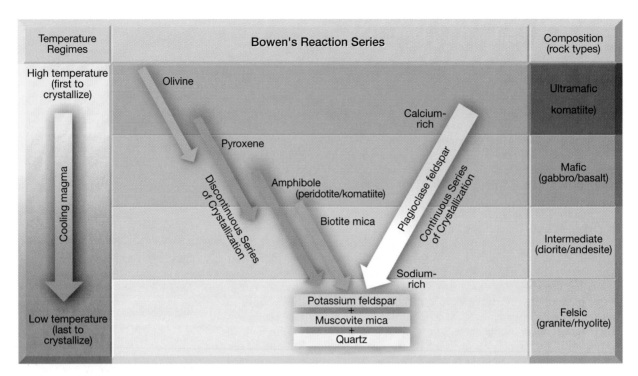

◆ **Figure 3.23** Bowen's reaction series shows the sequence in which minerals crystallize from a magma. Compare this figure to the mineral composition of the rock groups in Figure 3.8. Note that each rock group consists of minerals that crystallize in the same temperature range.

At any stage in the evolution of a magma, the solid and liquid components can separate into two chemically distinct units. Further, magmatic differentiation within the secondary melt can generate additional chemically distinct fractions. Consequently, magmatic differentiation and separation of the solid and liquid components at various stages of crystallization can produce several chemically diverse magmas and ultimately a variety of igneous rocks (see Figure 3.24).

Assimilation.   Bowen successfully demonstrated that through magmatic differentiation, a parent magma can generate several mineralogically different igneous rocks. However, more recent work indicates that magmatic differentiation cannot by itself account for the entire compositional spectrum of igneous rocks.

Once a magma body forms, its composition can change through the incorporation of foreign material. For example, as magma migrates upward, it may incorporate some of the surrounding host rock, a process called **assimilation** (Figure 3.25). This process may operate in a near-surface environment where rocks are brittle. As the magma pushes upward, stress causes numerous cracks in the overlying rock. The force of the injected magma is often sufficient to dislodge blocks of "foreign" rock and incorporate them into the magma body. In deeper environments,

the magma may be hot enough to simply melt and assimilate some of the surrounding host rock, which is near its melting temperature.

Magma Mixing.   Another means by which the composition of a magma body can be altered is called **magma mixing**. This process occurs whenever one magma body intrudes another (Figure 3.25). Once combined, convective flow may stir the two magmas and generate a fluid with an intermediate composition. Magma mixing may occur during the ascent of two chemically distinct magma bodies as the more buoyant mass overtakes the slower moving mass.

Partial Melting and Magma Formation

Recall that the crystallization of a magma occurs over a temperature range of at least 200°C. As you might expect, melting, the reverse process, spans a similar temperature range. As rock begins to melt, those minerals with the lowest melting temperatures are the first to melt. Should melting continue, minerals with higher melting points begin to melt and the composition of the magma steadily approaches the overall composition of the rock from which it was derived. Most often, however, melting is not complete. The incomplete melting of rocks is known as **partial melting**, a process that produces most, if not all, magma.

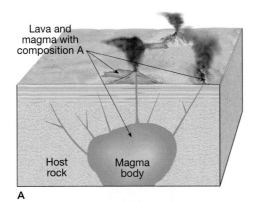

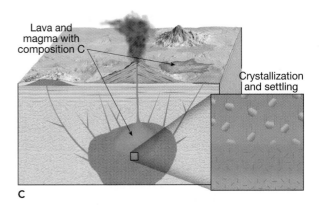

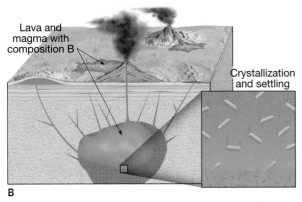

◆ **Figure 3.24**   Illustration of how a magma evolves as the earliest-formed minerals (those richer in iron, magnesium, and calcium) crystallize and settle to the bottom of the magma chamber, leaving the remaining melt richer in sodium, potassium, and silica (SiO_2). **A.** Emplacement of a magma body and associated igneous activity generates rocks having a composition similar to that of the initial magma. **B.** After a period of time, crystallization and settling changes the composition of the melt, while generating rocks having a composition quite different than the original magma. **C.** Further magmatic differentiation results in another more highly evolved melt with its associated rock types.

Notice in Figure 3.23 that rocks with a felsic composition are composed of minerals with the lowest melting (crystallization) temperatures—namely, quartz and potassium feldspar. Also note that as we move up Bowen's reaction series, the minerals have progressively higher melting temperatures and that olivine, which is found at the top, has the highest melting point. When a rock undergoes partial melting, it will form a melt that is enriched in ions from minerals with the lowest melting temperatures. The unmelted crystals are those of minerals with higher melting temperatures. Separation of these two fractions would yield a melt with a chemical composition that is richer in silica and nearer to the felsic end of the spectrum than the rock from which it was derived.

Formation of a Mafic Magma.   Most mafic magmas probably originate from partial melting of the ultra-mafic rock *peridotite*, the major constituent of the upper mantle. Mafic magmas that originate from direct melting of mantle rocks are called *primary* magmas because they have not yet evolved. Melting to produce these mantle-derived magmas may be triggered by a reduction in confining pressure (decompression melting). This can occur, for example, where mantle rock ascends as part of slow-moving convective flow at mid-ocean ridges (see Figure 3.20). Recall that mafic magmas may also be generated at subduction zones, where water driven from the descending slab

of oceanic crust promotes partial melting of mantle rocks (see Figure 3.21).

Because most mafic magma forms between about 50 and 250 kilometres below the surface, we might expect that this material would cool and crystallize at depth. However, as mafic magma migrates upward, the confining pressure steadily diminishes and reduces its melting temperature. As you will see in the next chapter, environments exist where mafic magmas ascend rapidly enough that the heat loss to the surrounding environment is offset by the drop in the melting temperature. Consequently, large outpourings of mafic lavas are common at Earth's surface. In some situations, however, mafic magmas that are comparatively dense will collect beneath crustal rocks and crystallize at depth.

Formation of Andesitic and Felsic Magmas.   If partial melting of mantle rocks generates mafic magmas, what is the source of the magma that generates intermediate and felsic rocks? Recall that intermediate and felsic magmas are not erupted from volcanoes in the deep-ocean basins; rather, they are found only within, or adjacent to, the continental margins (Figure 3.26). This is strong evidence that interactions between mantle-derived mafic magmas and more silica-rich components of Earth's crust generate these magmas. For example, as a mafic magma migrates upward, it may melt and assimilate some of the crustal rocks through

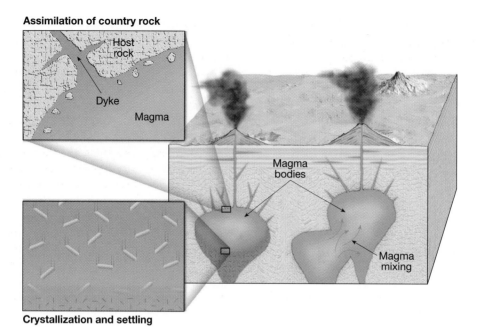

◆ **Figure 3.25**   This illustration shows three ways that the composition of a magma body may be altered: magma mixing; assimilation of host (country) rock; and crystallization and settling (magmatic differentiation).

◆ **Figure 3.26**   1996 eruption of Mount Ruapehu, Tongariro National Park, New Zealand. Volcanoes that border the Pacific Ocean are fed largely by magmas that have intermediate or felsic compositions. These silica-rich magmas often erupt explosively, generating large plumes of volcanic dust and ash. (Photo by Tui De Roy/Minden Pictures)

which it ascends. The result is the formation of a more silica-rich magma of intermediate or andesitic composition (between mafic and felsic).

Magma of intermediate composition may also evolve from a mafic magma by the process of magmatic differentiation. Recall from our discussion of Bowen's reaction series that as mafic magma solidifies, it is the silica-poor ferromagnesian minerals that crystallize first. If these iron-rich components are separated from the liquid by crystal settling, the remaining melt, now enriched in silica, will have a composition more akin to andesite. These evolved (changed) magmas are termed *secondary magmas*.

Felsic rocks are found in much too large a quantity to be generated solely from the magmatic differentiation of primary mafic magmas. Most likely they are the end product of the crystallization of an intermediate magma, or the product of partial melting of silica-rich continental rocks. The heat to melt crustal rocks often comes from hot mantle-derived mafic magmas that formed above a subducting plate within the crust.

Felsic melts are higher in silica and thus more viscous (thicker) than other magmas. Felsic magmas are also formed at lower temperatures than other magmas. Therefore, in contrast to mafic magmas that frequently produce vast outpourings of lava, felsic magmas usually lose their mobility before reaching the surface and tend to produce large plutonic structures. On those occasions when silica-rich magmas do reach the surface, explosive pyroclastic eruptions, such as those from Mount St. Helens, are the rule.

In summary, a large number of processes dominated by magmatic differentiation and partial melting, but also including magma mixing and contamination by crustal rocks, account for the great diversity of igneous rocks and the magmas from which they formed.

Chapter Summary

- *Igneous rocks* form when *magma cools* and solidifies. *Extrusive*, or *volcanic*, igneous rocks result when *lava* cools at the surface. Magma that solidifies at depth produces *intrusive*, or *plutonic*, igneous rocks.

- As magma cools, the ions that compose it arrange themselves into orderly patterns during a process called *crystallization*. Slow cooling results in the formation of rather large crystals. Conversely, when cooling occurs rapidly, the outcome is a solid mass consisting of tiny intergrown crystals. When molten material is quenched instantly, a mass of unordered atoms, referred to as *glass*, forms.

- Igneous rocks are most often classified by their *texture* and *mineral composition*.

- The texture of an igneous rock refers to the overall appearance of the rock based on the size and arrangement of its interlocking crystals. The most important factor affecting texture is the rate at which magma cools. Common igneous rock textures include *aphanitic*, with grains too small to be distinguished without the aid of a microscope; *phaneritic*, with intergrown crystals that are roughly equal in size and large enough to be identified with the unaided eye; *porphyritic*, which has large crystals (*phenocrysts*) interbedded in a matrix of smaller crystals (*groundmass*); and *glassy*.

- The mineral composition of an igneous rock is the consequence of the chemical makeup of the parent magma and the environment of crystallization. Igneous rocks are divided into broad compositional groups based on the percentage of dark and light silicate minerals they contain. *Felsic rocks* (e.g., granite and rhyolite) are composed mostly of the light-coloured silicate minerals potassium feldspar and quartz. Rocks of *intermediate* composition (e.g., andesite and diorite) contain plagioclase feldspar and amphibole. *Mafic rocks* (e.g., basalt and gabbro) contain abundant olivine, pyroxene, and calcium feldspar. They are high in iron, magnesium, and calcium, low in silicon, and are dark grey to black in colour.

- The mineral makeup of an igneous rock is ultimately determined by the chemical composition of the magma from which it crystallizes. N. L. Bowen discovered that as magma cools in the laboratory, those minerals with higher melting points crystallize before minerals with lower melting points. *Bowen's reaction series* illustrates the sequence of mineral formation within magma.

- During the crystallization of magma, if the earlier-formed minerals are denser than the liquid portion, they will settle to the bottom of the magma chamber during a process called *crystal settling*. Owing to the fact that crystal settling removes the earlier-formed minerals, the remaining melt will form a rock with a chemical composition much different from that of the parent magma. The process of developing more than one magma type from a common magma is called *magmatic differentiation*.

- Once a magma body forms, its composition can change through the incorporation of foreign material, a process termed *assimilation*, or by *magma mixing*.

- Magma originates from essentially solid rock of the crust and mantle. In addition to a rock's composition, its temperature, depth (confining pressure), and water content determine whether it exists as a solid or liquid. Thus, magma can be generated by *raising a rock's temperature*, as occurs when a hot mantle plume collects beneath crustal rocks. A *decrease in pressure* can cause *decompression melting*. Further, the *introduction of volatiles* (water) can lower a rock's melting point sufficiently to generate magma. Because melting is generally not complete, a process called *partial melting* produces a melt made of the lowest-melting-temperature minerals, which are higher in silica than the original rock. Thus, magmas generated by partial melting are nearer to the felsic end of the compositional spectrum than are the rocks from which they formed.

Review Questions

1. What is magma?

2. How does lava differ from magma?

3. How does the rate of cooling influence the crystallization process?

4. In addition to the rate of cooling, what two other factors influence the crystallization process?

5. The classification of igneous rocks is based largely on two criteria. Name these criteria.

6. The statements that follow relate to terms describing igneous rock textures. For each statement, identify the appropriate term.
   (a) Openings produced by escaping gases.
   (b) Obsidian exhibits this texture.
   (c) A matrix of fine crystals surrounding phenocrysts.
   (d) Crystals are too small to be seen without a microscope.
   (e) A texture characterized by two distinctly different crystal sizes.
   (f) Coarse-grained, with crystals of roughly equal size.
   (g) Exceptionally large crystals exceeding 1 centimetre in diameter.

7. Why are the crystals in pegmatites so large?

8. What does a porphyritic texture indicate about an igneous rock?

9. How are granite and rhyolite different? In what way are they similar?

10. Compare and contrast each of the following pairs of rocks:
    (a) granite and diorite
    (b) basalt and gabbro
    (c) andesite and rhyolite

11. How do tuff and volcanic breccia differ from other igneous rocks such as granite and basalt?

12. What is the geothermal gradient?

13. Describe the three conditions that are thought to cause rock to melt.

14. What is magmatic differentiation? How might this process lead to the formation of several different igneous rocks from a single magma?

15. Relate the classification of igneous rocks to Bowen's reaction series.

16. What is partial melting?

17. How does the composition of a melt produced by partial melting compare with the composition of the parent rock?

18. How are most mafic magmas generated?

19. Mafic magma forms at great depth. Why doesn't most of it crystallize as it rises through the relatively cool crust?

20. Why are rocks of intermediate (andesitic) and felsic (granitic) composition generally *not* found in the ocean basins?

Key Terms

andesitic (p. 66)
aphanitic texture (p. 62)
assimilation (p. 80)
basaltic (p. 66)
Bowen's reaction series
   (p. 78)
crystallization (p. 61)
crystal settling (p. 79)
decompression melting
   (p. 76)
extrusive (p. 60)

felsic (p. 66)
fragmental texture (p. 65)
geothermal gradient
   (p. 75)
glass (p. 62)
glassy texture (p. 64)
groundmass (p. 64)
granitic (p. 66)
igneous rocks (p. 60)
intermediate (p. 66)
intrusive (p. 60)

lava (p. 60)
mafic (p. 66)
magma (p. 61)
magma mixing (p. 80)
magmatic differentiation
   (p. 79)
melt (p. 61)
partial melting (p. 80)
pegmatite (p. 65)
pegmatitic texture (p. 65)
phaneritic texture (p. 64)

phenocryst (p. 64)
plutonic (p. 60)
porphyritic texture
   (p. 64)
porphyry (p. 64)
pyroclastic texture (p. 65)
texture (p. 62)
ultramafic (p. 67)
vesicular texture (p. 63)
volatiles (p. 61)
volcanic (p. 60)

Web Resources

 The *Earth* Web site uses the resources and flexibility of the Internet to aid in your study of the topics in this chapter. Written and developed by geology instructors, this site will help improve your understanding of geology. Visit **http://www.pearson.ca/tarbuck** and click on the cover of the text to find:

■ Online review quizzes.
■ Web-based critical thinking and writing exercises.
■ Links to chapter-specific Web resources.
■ Internet-wide key-term searches.

http://www.pearson.ca/tarbuck

Chapter 4

Volcanoes and Other Igneous Activity

Nature's fireworks. Twin fountains spew lava and volcanic bombs.
(Photo by George Shelley/The Stock Market)

◆ **Figure 4.1**   Before-and-after photographs show the transformation of Mount St. Helens caused by the May 18, 1980, eruption. The dark area in the "after" photo is debris-filled Spirit Lake, partially visible in the "before" photo. (Photos courtesy of U.S. Geological Survey)

On Sunday, May 18, 1980, the largest volcanic eruption to occur in North America in historic times transformed a picturesque volcano into a decapitated remnant (Figure 4.1). On this date in southwestern Washington State, Mount St. Helens erupted with tremendous force. The blast blew out the entire north flank of the volcano, leaving a gaping hole. In one brief moment, a prominent volcano whose summit had been more than 2900 metres above sea level was lowered by more than 400 metres.

The event devastated a wide swath of timber-rich land on the north side of the mountain. Trees within a 400-square-kilometre area lay intertwined and flattened, stripped of their branches and appearing from the air like toothpicks strewn about. The accompanying mudflows carried ash, trees, and water-saturated rock debris 29 kilometres down the Toutle River. The eruption claimed 59 lives, some dying from the intense heat and the suffocating cloud of ash and gases, others from being hurled by the blast, and still others from entrapment in the mudflows.

The eruption ejected nearly a cubic kilometre of ash and rock debris. Following the devastating explosion, Mount St. Helens continued to emit great quantities of hot gases and ash. The force of the blast was so strong that some ash was propelled more than 18,000 metres into the stratosphere. During the next few days, this very fine-grained material was carried around Earth by strong upper-air winds. Measurable deposits were reported well into southern Alberta, with crop damage into central Montana. Meanwhile, ash fallout in the immediate vicinity exceeded 2 metres in depth. The air over Yakima, Washington (130 kilometres to the east), was so filled with ash that residents experienced midnight-like darkness at noon.

Not all volcanic eruptions are as violent as the 1980 Mount St. Helens event. Some volcanoes, such as Hawaii's Kilauea volcano, generate relatively quiet outpourings of fluid lavas. These "gentle" eruptions are not without some fiery displays; occasionally fountains of incandescent lava spray hundreds of metres into the air (Figure 4.2). Such events, however, typically

Spirit Lake

pose minimal threat to human life and property, and the lava generally falls back into a lava pool.

Testimony to the quiet nature of Kilauea's eruptions is the fact that the Hawaiian Volcanoes Observatory has operated on its summit since 1912. Further, the longest and largest of Kilauea's eruptions began in 1983 and remains active, although it has received only modest media attention.

Why do volcanoes like Mount St. Helens (U.S.) and Mount Garibaldi (Canada, Figure 4.3) erupt explosively, whereas others like Kilauea are relatively quiet? Why do volcanoes occur in chains like the Aleutian Islands or the Cascade Range? Why do some volcanoes form on the ocean floor, while others occur on the continents? This chapter will deal with these and other questions as we explore the nature and movement of magma and lava.

The Nature of Volcanic Eruptions

 Internal Processes
↳ Igneous Activity

Volcanic activity is commonly perceived as a process that produces a picturesque, cone-shaped structure

◆ **Figure 4.2** Fluid mafic lava erupting from Kilauea Volcano, Hawaii.
(Photo by Douglas Peebles)

◆ **Figure 4.3** Mount Garibaldi, a stratovolcano located in southwestern British Columbia. Mount Garibaldi has been dormant since the last ice age and has suffered some erosion since that time.
(Photo by S. Irwin. Reproduced with the permission of the Minister of Public Works and Government Services Canada, 2004 and Courtesy of Natural Resources Canada, Geological Survey of Canada.)

that periodically erupts in a violent manner, like Mount St. Helens (Box 4.1). Although some eruptions are very explosive, many are not. What determines whether a volcano extrudes magma violently or "gently"? The primary factors include the magma's *composition*, its *temperature*, and the amount of *dissolved gases* it contains. To varying degrees, these factors affect the magma's mobility, or **viscosity** (*viscos* = sticky). The more viscous the material, the greater its resistance to flow. (For example, maple syrup is more viscous than water.) Magma associated with an explosive eruption may be five times more viscous than magma extruded in a quiescent manner.

Factors Affecting Viscosity

The effect of temperature on viscosity is easily seen. Just as heating maple syrup makes it more fluid (less viscous), the mobility of lava is strongly influenced by temperature. As lava cools and begins to congeal, the flow decreases in mobility and eventually halts.

A more significant factor influencing volcanic behaviour is the chemical composition of the magma. Recall from Chapter 3 that a major difference among various igneous rocks is their silica (SiO_2) content (Table 4.1). Magmas that produce mafic rocks such as basalt contain about 50 percent silica, whereas magmas that produce felsic rocks (granite and its extrusive equivalent, rhyolite) contain over 70 percent silica. The intermediate rock types—andesite and diorite—contain about 60 percent silica.

A magma's viscosity is directly related to its silica content. In general, the more silica in magma, the greater its viscosity. The flow of magma is impeded because

silica structures link together into long chains, even before crystallization begins. Consequently, because of their high silica content, rhyolitic (felsic) lavas are very viscous and tend to form comparatively short, thick flows. By contrast, mafic lavas, which contain less silica, tend to be quite fluid and have been known to travel distances of 150 kilometres or more before congealing.

Importance of Dissolved Gases

Dissolved gases tend to increase the fluidity of magma. Upon their escape from the magma, gases also provide enough force to propel molten rock from a volcanic vent.

The summits of volcanoes often begin to inflate, months or even years before an eruption occurs. This indicates that magma is migrating into a shallow reservoir located within the cone. During this phase **volatiles** (the gaseous component of magma, consisting mostly of water) tend to migrate upward and accumulate near the top of the magma chamber. Thus, the upper portion of a magma body is enriched in dissolved gases.

When an eruption starts, gas-charged magma moves from the magma chamber and rises through the volcanic conduit or vent. As the magma nears the surface, the confining pressure is greatly reduced. This reduction in pressure allows the dissolved gases to be released suddenly, just as opening a warm pop can allow carbon dioxide gas bubbles to expand and escape.

BOX 4.1

Understanding Earth
Anatomy of an Eruption

The events leading to the May 18, 1980 eruption of Mount St. Helens began about two months earlier as a series of minor Earth tremors centred beneath the awakening mountain (Figure 4.A, part A). The tremors were caused by the upward movement of magma within the mountain. The first volcanic activity took place a week later, when a small amount of ash and steam rose from the summit. Over the next several weeks, sporadic eruptions of varied intensity occurred. Prior to the main eruption, the primary concern had been the potential hazard of mudflows. These moving lobes of saturated soil and rock are created as ice and snow melt from the heat emitted from magma within the volcano.

The only warning of a potential eruption was a bulge on the volcano's north flank (Figure 4.A, part B). Careful monitoring of this dome-shaped structure indicated a very slow but steady growth rate of a few metres per day. If the growth rate of the bulge changed appreciably, an eruption might quickly follow. Unfortunately, no such variation was detected prior to the explosion. In fact, the seismic activity decreased during the two days preceding the huge blast.

Dozens of scientists were monitoring the mountain when it exploded. "Vancouver, Vancouver, this is it!" was the only warning—and last words from one scientist—that preceded the unleashing of tremendous quantities of pent-up gases. The trigger was a medium-sized earthquake. Its vibrations sent the north slope of the cone plummeting into the Toutle River, removing the overburden that had trapped the magma below (Figure 4.A, part C). With the pressure reduced, the water in the magma vapourized and expanded, causing the mountainside to rupture like an overheated steam boiler. Because the eruption originated around the bulge, several hundred metres below the summit, the initial blast was directed laterally rather than vertically. Had the full force of the eruption been upward, far less destruction would have occurred.

Mount St. Helens is one of 15 large volcanoes and innumerable smaller ones that comprise the Cascade Range, which extends from British Columbia to northern California. Eight of the largest volcanoes have been active in the past few hundred years. Of the remaining seven active volcanoes, the most likely to erupt again are Mount Baker (just south of Vancouver) and Mount Rainier in Washington, Mount Shasta and Lassen Peak in California, and Mount Hood in Oregon.

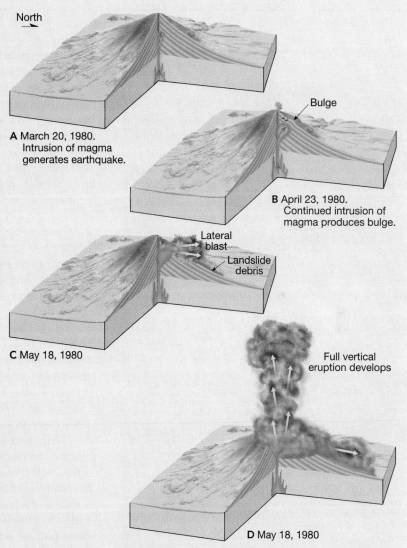

North

A March 20, 1980.
Intrusion of magma
generates earthquake.

Bulge

B April 23, 1980.
Continued intrusion of
magma produces bulge.

Lateral blast

Landslide debris

C May 18, 1980

Full vertical
eruption develops

D May 18, 1980

◆ **Figure 4.A** Idealized diagrams showing the events in the May 18, 1980, eruption of Mount St. Helens. **A.** First, a sizable earthquake recorded on Mount St. Helens indicates that renewed volcanic activity is possible. **B.** Alarming growth of a bulge on the north flank suggests increasing magma pressure below. **C.** Triggered by an earthquake, a giant landslide reduced the confining pressure on the magma body and initiated an explosive lateral blast. **D.** Within seconds a large vertical eruption sent a column of volcanic ash to an altitude of about 18 kilometres. This phase of the eruption continued for over nine hours.

TABLE 4.1   Magmas Have Different Compositions, Which Cause Their Properties to Vary

Composition	Silica Content	Viscosity	Gas Content	Tendency to Form Pyroclastics	Volcanic Landform
Mafic (Basaltic) Magma	Least (~50%)	Least	Least (1–2%)	Least	Shield Volcanoes Basalt Plateaus Cinder Cones
Intermediate (Andesitic) magma	Intermediate (~60%)	Intermediate	Intermediate (3–4%)	Intermediate	Composite Cones
Felsic (Rhyolitic) magma	Most (~70%)	Greatest	Most (4–6%)	Greatest	Volcanic Domes Pyroclastic Flows

◆ **Figure 4.4** Huge eruption column produced as Mount St. Helens erupted on July 22, 1980, two months after the most destructive event.
(Photo by David Weintraub/Photo Researchers, Inc.)

Very fluid mafic magmas allow the expanding gases to migrate upward and escape from the vent with relative ease. As they escape, the gases may propel incandescent lava hundreds of metres into the air, producing lava fountains (see Figure 4.2). Although spectacular, such fountains are mostly harmless and not generally associated with major explosive events that cause great loss of life and property.

At the other extreme, highly viscous magmas explosively expel jets of hot ash-laden gases that evolve into buoyant plumes called **eruption columns** that extend thousands of metres into the atmosphere (Figure 4.4). Prior to an explosive eruption, an extended period of *magmatic differentiation* occurs, in which iron-rich minerals crystallize and settle out, leaving the upper part of the magma enriched in silica and dissolved gases. As this volatile-rich magma moves up the volcanic vent toward the surface, these gases begin to collect as tiny bubbles. At some height in the conduit this mixture is transformed into a gas jet containing tiny bits of magma that are explosively ejected from the volcano. This type of explosive eruption was exemplified by Mount Pinatubo in the Philippines (1991) and Mount St. Helens (1980).

As magma in the upper portion of the vent is ejected, the pressure on the underlying magma drops. Thus, rather than a single "bang," volcanic eruptions are really a series of explosions. The soluble gases in a viscous magma migrate upward quite slowly. Only within the uppermost portion of the magma body does the gas content build sufficiently to trigger explosive eruptions. Thus, an explosive event is commonly followed by the quiet emission of "degassed" lavas. However, once this eruptive phase ceases, the process of gas buildup begins anew. This time lag may partially explain the sporadic eruptive patterns of volcanoes that eject viscous lavas.

To summarize, the viscosity of magma, plus the quantity of dissolved gases and the ease with which they can escape, determines the nature of a volcanic eruption. We can now begin to explain the "gentle" volcanic eruptions of hot, fluid lavas in Hawaii and the explosive and sometimes catastrophic eruptions of viscous lavas from volcanoes such as Mount St. Helens.

Materials Extruded During an Eruption

Internal Processes
↳ Igneous Activity

Volcanoes extrude lava, large volumes of gas, and pyroclastic materials (broken rock, lava "bombs," fine ash, and dust). In this section we will examine each of these materials.

Lava Flows

Because of their low silica content, hot mafic lavas are usually very fluid. They flow in thin, broad sheets or streamlike ribbons. On the island of Hawaii, such lavas have been clocked at speeds of 30 kilometres per hour down steep slopes, but flow rates of 10 to 300 metres per hour are more common. Further, mafic lavas have been known to travel distances of 150 kilometres or more before congealing. In contrast, the movement of silica-rich (rhyolitic) lava may be too slow to perceive.

When fluid mafic lavas of the Hawaiian type congeal, they commonly form a relatively smooth skin that wrinkles as the still-molten subsurface lava continues to advance (Figure 4.5A). These are known as **pahoehoe flows** (pronounced *pah-hoy-hoy*) and resemble the twisting braids in ropes.

Another common type of mafic lava, called **aa** (pronounced *ah-ah*), has a surface of rough, jagged blocks with dangerously sharp edges and spiny projections (Figure 4.5B). Active aa flows are relatively cool and thick and advance at rates from 5 to 50 metres per hour. Moreover, gases escaping from the surface produce numerous voids and sharp spines in the congealing lava. As the molten interior advances, the outer crust is broken further, giving the flow the appearance of an advancing mass of lava rubble.

Hardened lava flows commonly contain tunnels that once were channels carrying lava from the volcanic vent to the flow's leading edge. These openings develop in the interior of a flow where temperatures remain high long after the surface congeals. Under these conditions, the still-molten lava within the conduits continues its forward motion, leaving behind the cavelike tunnels called **lava tubes** (Figure 4.6). Lava tubes are important because they allow fluid lavas to advance great distances from their source.

When lava enters the ocean, or when outpourings of lava actually originate in the ocean basin, the

A

B

◆ **Figure 4.5   A.** Typical pahoehoe (ropy) lava flow, Kilauea, Hawaii. **B.** Typical slow-moving aa flow.
(Photos by J.D. Griggs, U.S. Geological Survey)

flow's outer skin quickly congeals. However, the lava is usually able to move forward by breaking through the hardened surface. This process occurs over and over, generating a lava flow composed of elongated structures resembling large bed pillows stacked atop one another. These structures, called **pillow lavas**, are useful in the reconstruction of Earth history (see Figure 18.15) because they indicate deposition in an underwater environment.

Gases

Magmas contain varying amounts of dissolved gases (*volatiles*) held in the molten rock by confining pressure,

A   B

◆ **Figure 4.6**   Lava streams that flow in confined channels often develop a solid crust and become flows within lava tubes.
A. Thurston Lava Tube, Hawaii Volcanoes National Park. **B.** View of an active lava tube as seen through the collapsed roof.
(Photo **A** by Douglas Peebles and Photo **B** by Jeffrey Judd, U.S. Geological Survey)

just as carbon dioxide is held in pop or beer. As with carbonated drinks, as soon as the pressure is reduced, the gases begin to escape.

The gaseous portion of most magmas comprises 1 to 6 percent of the total weight, with most of this in the form of water vapour. Although the percentage may be small, the actual quantity of emitted gas can exceed thousands of tonnes per day.

The composition of volcanic gases is important because they contribute significantly to the gases that make up our planet's atmosphere. Analyses of samples taken during Hawaiian eruptions indicate that the gases are about 70 percent water vapour, 15 percent carbon dioxide, 5 percent nitrogen, 5 percent sulphur dioxide, and lesser amounts of chlorine, hydrogen, and argon. Sulphur compounds are easily recognized by their pungent odour. Volcanoes are a natural source of air pollution, including sulphur dioxide, which readily combines with water to form sulphuric acid.

In addition to propelling magma from a volcano, gases play an important role in creating the narrow passageway or conduit that connects the magma chamber to the surface. First, high temperatures and the buoyant force from the magma body cracks the rock above. Then, hot blasts of high-pressure gases expand the cracks and develop a passageway to the surface. Once the passageway is completed, the hot gases armed with rock fragments erode its walls, producing a larger conduit. Because these erosive forces are concentrated on any protrusion along the pathway, the volcanic pipes that are produced have a circular shape. As the conduit enlarges, magma moves upward to the surface. Following an eruptive phase, the volcanic pipe often becomes choked with congealed magma and debris that was not thrown clear of the vent. Before the next eruption, a new surge of explosive gases may again clear the conduit.

Occasionally, eruptions emit colossal amounts of volcanic gases that rise high into the atmosphere, where they may reside for several years. Some of these eruptions may have an impact on Earth's climate, a topic we will consider later in the chapter.

Pyroclastic Materials

When mafic lava is extruded, dissolved gases escape quite freely and continually. These gases propel incandescent blobs of lava to great heights (see Figure 4.2). Some of this ejected material lands near the vent and builds a cone-shaped structure, whereas smaller particles tend to be carried great distances by the wind. By contrast, viscous (rhyolitic) magmas are highly charged with gases, and upon release they expand a thousandfold as they blow pulverized rock, lava, and glass fragments from the vent. The particles produced in both situations are called **pyroclastic materials** (*pyro* = fire, *clast* = fragment). These ejected fragments range in size from dust and sand-sized volcanic ash (less than 2 millimetres) to pieces that weigh several tonnes.

Ash and *dust* particles are produced from gas-laden viscous magma during an explosive eruption (see Figure 4.4). As magma moves up in the vent, the gases rapidly expand, generating a froth of melt. As the hot gases expand explosively, the froth is blown into very fine glassy fragments. When the hot ash falls, the glassy shards often fuse to form a rock called *welded tuff*. Sheets of this material, as well as ash deposits that later consolidate, cover vast portions of North America.

Also common are pyroclasts that range in size from the size of small beads to about the size of walnuts, termed *lapilli* ("little stones") or *cinders* (2 to 64 millimetres). Particles larger than 64 millimetres in diameter are called *blocks* when they are made of hardened lava and *bombs* when they are ejected as incandescent lava. Because bombs are semi-molten upon ejection, they often assume a streamlined shape as they hurtle through the air (Figure 4.7). Because of their size, bombs and blocks usually fall on the slopes of a cone; however, they are occasionally propelled far from the volcano by the force of escaping gases. For instance, bombs 6 metres long and weighing about 200 tons were blown

600 metres from the vent during an eruption of the Japanese volcano Asama.

So far we have distinguished various pyroclastic materials based largely on the size of the fragments. Some materials are also identified by their texture and composition. In particular, **scoria** is the name applied to vesicular (containing voids) ejecta derived from mafic magma (Figure 4.8). These black to reddish-brown fragments are generally found in the size range of lapilli. When magma with an intermediate or silica-rich composition generates vesicular ejecta, it is called **pumice** (see Figure 3.12). Pumice is usually lighter in colour and less dense than scoria. Furthermore, some pumice fragments have such a preponderance of vesicles that they may float in water for prolonged periods.

Volcanic Structures and Eruptive Styles

 Internal Processes
↳ Igneous Activity

The popular image of a volcano is that of a solitary, graceful, snowcapped cone, such as Japan's Fujiyama.

◆ **Figure 4.7** Volcanic bombs forming during an eruption of Hawaii's Kilauea volcano. Ejected lava fragments take on a streamlined shape as they sail through the air. The bomb in the insert is about 10 centimetres long. (Photo by Arthur Roy/National Audubon Society; inset photo by E. J. Tarbuck)

← 2 cm →

◆ **Figure 4.8**   Scoria is a volcanic rock that exhibits a vesicular texture. Vesicles are small holes left by escaping gas bubbles. (Photo by E. J. Tarbuck)

These picturesque, conical mountains are produced by volcanic activity that occurred intermittently over thousands, or even hundreds of thousands, of years. However, many volcanoes do not fit this image. Some volcanoes are only 30 metres high and formed during a single eruptive phase that may have lasted only a few days. Other volcanic landforms are not "volcanoes" at all.

Volcanic landforms come in a wide variety of shapes and sizes, and each structure has a unique eruptive history. Nevertheless, volcanologists have been able to classify volcanic landforms and determine their eruptive patterns. In this section we will consider the general anatomy of a volcano and look at three major volcanic types: shield volcanoes, cinder cones, and composite cones. This discussion will be followed by an overview of other significant volcanic landforms.

Anatomy of a Volcano

Volcanic activity frequently begins when a fissure (crack) develops in the crust as magma moves forcefully toward the surface. As the gas-rich magma moves up this linear fissure, its path is usually localized into a circular **conduit**, or **pipe**, that terminates at a surface opening called a **vent** (Figure 4.9). Successive eruptions of lava, pyroclastic material, or frequently a combination of both often separated by long periods of inactivity eventually build the structure we call a **volcano**.

Located at the summit of many volcanoes is a steep-walled depression called a **crater** (*crater* = a bowl). Craters are constructional features that are built upward as ejected fragments collect around the vent to form a doughnut-like structure. Some volcanoes have multiple summit craters, whereas others have very large, more or less circular depressions called **calderas**. Calderas are large collapse structures that may or may not form in association with a volcano. (We will consider the formation of various types of calderas later.)

During early stages of growth most volcanic discharges come from a central summit vent. As a volcano matures, material also tends to emerge from fissures that develop along the flanks, or base, of the volcano. Continued activity from a flank eruption may produce a small **parasitic cone** (*parasitus* = one who

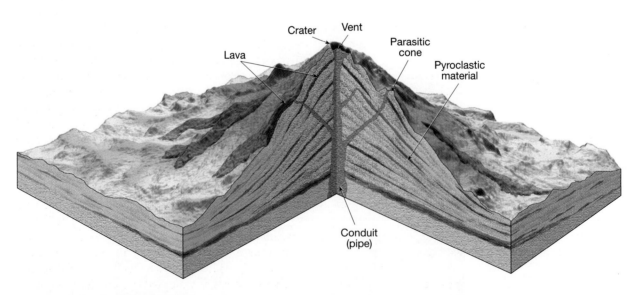

◆ **Figure 4.9**   Anatomy of a "typical" composite cone (see also Figures 4.10 and 4.14 for a comparison with a shield and cinder cone, respectively).

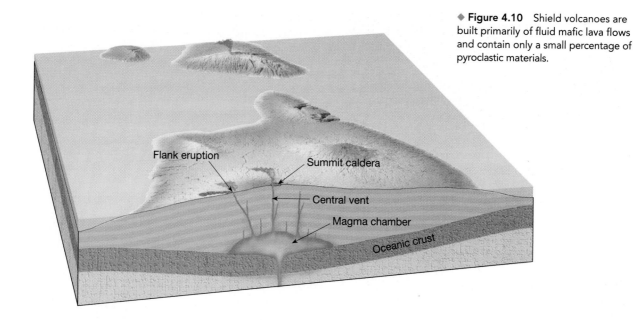

◆ **Figure 4.10** Shield volcanoes are built primarily of fluid mafic lava flows and contain only a small percentage of pyroclastic materials.

Flank eruption

Summit caldera

Central vent

Magma chamber

Oceanic crust

eats at the table of another). Mount Etna in Italy, for example, has more than 200 secondary vents, some of which have built cones. Many of these vents, however, emit only gases and are appropriately called **fumaroles** (*fumus* = smoke).

The form of a particular volcano is largely determined by the composition of the contributing magma. As you will see, fluid Hawaiian-type lavas tend to produce broad structures with gentle slopes, whereas more viscous silica-rich lavas (and some gas-rich mafic lavas) tend to generate cones with moderate to steep slopes.

Shield Volcanoes

Shield volcanoes are produced by the accumulation of fluid mafic lavas and exhibit the shape of a broad slightly domed structure that resembles a warrior's shield, with slopes typically less than 5 degrees (Figure 4.10). Most shield volcanoes have grown up from the deep ocean floor to form islands or seamounts such as the islands of the Hawaiian chain, Iceland, and the Galapagos. Some shield volcanoes, however, occur on continents. Included in this group are rather massive structures located in East Africa, like Suswa in Kenya.

Extensive study of the Hawaiian Islands confirms that each shield was built from a myriad of mafic lava flows averaging a few metres thick with less than about 1 percent pyroclastic ejecta.

Mauna Loa is one of five overlapping shield volcanoes that together comprise the Big Island of Hawaii (Figure 4.11). From its base on the floor of the Pacific Ocean to its summit, Mauna Loa is over

9 kilometres high, exceeding the height of Mount Everest. This massive pile of basaltic rock has a volume of 40,000 cubic kilometres that was extruded over a period of nearly a million years. For comparison, the volume of material composing Mauna Loa is roughly 200 times the amount composing a large composite cone such as Mt. Rainier (Figure 4.12). Most shields, however, are more modest in size. For example, the classic Icelandic shield, Skjalbreidur, rises to a height of only about 600 metres and is 10 kilometres across its base.

Despite its enormous size, Mauna Loa is not the largest known volcano in the solar system. Olympus Mons, a huge shield volcano on Mars, is 25 kilometres high and 600 kilometres wide (see Chapter 22).

Another feature common to a mature, active shield volcano is a large steep-walled caldera that occupies its summit. Calderas form when the roof of the volcano collapses as magma from the central magma reservoir migrates to the flanks, often feeding fissure eruptions. Mauna Loa's summit caldera measures 2.6 by 4.5 kilometres and has a depth that averages about 150 metres.

In their final stage of growth, the activity on mature shields is more sporadic and pyroclastic ejections are more prevalent. Further, the lavas increase in viscosity, resulting in thicker, shorter flows. These eruptions tend to steepen the slope of the summit area, which often becomes capped with clusters of cinder cones. This explains why Mauna Kea, a very mature volcano that has not erupted in historic times, has a steeper summit than Mauna Loa, which erupted as recently as 1984.

◆ **Figure 4.11** **A.** Mauna Loa is one of five shield volcanoes that together make up the island of Hawaii. **B.** The Ilgachuz Range is a large shield volcano located in western British Columbia. The irregular top of this old volcano is due to erosion.
(Photo A by Greg Vaughn. Photo B by Catherine Hickson, Geological Survey of Canada)

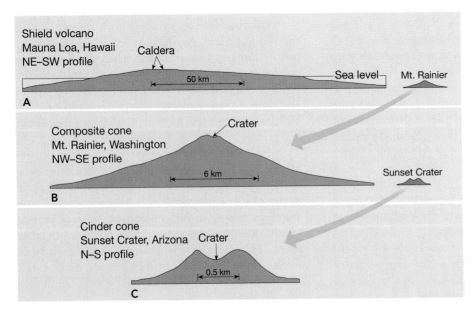

◆ **Figure 4.12** Profiles of volcanic landforms. **A.** Profile of Mauna Loa, Hawaii, the largest shield volcano in the Hawaiian chain. Note size comparison with Mt. Rainier, Washington, a large composite cone. **B.** Profile of Mt. Rainier, Washington. Note how it dwarfs a typical cinder cone. **C.** Profile of Sunset Crater, Arizona, a typical steep-sided cinder cone.

Kilauea, Hawaii: Eruption of a Shield Volcano.
Kilauea, the most active and intensely studied shield volcano in the world, is located on the island of Hawaii in the shadow of Mauna Loa. More than 50 eruptions have been witnessed here since record-keeping began in 1823. Several months before each eruptive phase, Kilauea inflates as magma gradually migrates upward and accumulates in a central chamber located a few kilometres below the summit. For up to 24 hours in advance of an eruption, swarms of small earthquakes warn of the impending activity.

Most of the activity on Kilauea during the past 50 years occurred along the flanks of the volcano in a region called the East Rift Zone. The longest and largest rift eruption ever recorded on Kilauea began in 1983 and continues to this day, with no signs of abating. The first discharge began along a 6-kilometre fissure where a 100-metre high "curtain of fire" formed as red-hot lava was ejected skyward (Figure 4.13). When the activity became localized, a cinder and spatter cone given the Hawaiian name *Puu Oo* was built. Over the next three years the general eruptive pattern consisted

of short periods (hours to days) when fountains of gas-rich lava sprayed skyward (see Figure 4.2). Each event was followed by nearly a month of inactivity.

By the summer of 1986 a new vent opened up 3 kilometres downrift. Here smooth-surfaced pahoehoe lava formed a lava lake. Occasionally the lake overflowed, but more often lava escaped through tunnels to feed pahoehoe flows that moved down the southeastern flank of the volcano toward the sea. Lava has been pouring intermittently into the ocean ever since, adding new land to the island of Hawaii.

Situated just 32 kilometres off the southern coast of Kilauea, a submarine volcano, Loihi, is also active. It, however, has another 930 metres to go before it breaks the surface of the Pacific Ocean.

Cinder Cones

As the name suggests, **cinder cones** (also called **scoria cones**) are built from ejected lava fragments that harden while in flight. These fragments range in size from fine ash to bombs but consist mostly of

◆ **Figure 4.13**   Lava extruded along the East Rift Zone, Kilauea, Hawaii. (Photo by Greg Vaughn)

pea- to walnut-size lapilli. Usually a product of relatively gas-rich mafic magma, cinder cones consist of *scoria*. Occasionally an eruption of silica-rich magma will generate a cinder cone composed of ash and pumice fragments. Although cinder cones are composed mostly of loose pyroclastic material, they sometimes extrude lava at their base.

Cinder cones have a very simple, distinctive shape determined by the slope that loose pyroclastic material maintains as it comes to rest (Figure 4.14). Because cinders have a high angle of repose (the steepest angle at which material remains stable), young cinder cones are steep-sided, having *slopes* between 30 and 40 degrees. In addition, cinder cones have large, deep craters in relation to the overall size of the structure. Although relatively symmetrical, many cinder cones are elongated, and higher on the side that was downwind during the eruptions.

Cinder cones are usually the product of a single eruptive episode that sometimes lasts only a few weeks and rarely exceeds a few years. Once this event ceases, the magma in the pipe connecting the vent to the magma chamber solidifies, and the volcano never erupts again. As a consequence of this short life span, cinder cones are small, usually between 30 metres and 300 metres and rarely exceed 700 metres in height (see Figure 4.12). Small cinder cones are common volcanic features occurring in the thousands around the globe. Canadian volcanoes with cinder cones include Mt. Edziza in British Columbia.

Parícutin: Life of a Garden-Variety Cinder Cone.
One of the very few volcanoes studied by geologists from beginning to end is the cinder cone called Parícutin, located abour 320 kilometres west of Mexico City. In 1943, its eruptive phase began in a cornfield owned by Dionisio Pulido, who witnessed the event as he prepared the field for planting.

For two weeks prior to the first eruption, numerous Earth tremors caused apprehension in the nearby village of Parícutin. Then, on February 20, sulphurous gases began billowing from a small depression that had been in the cornfield for as long as people could remember. During the night, hot, glowing rock fragments were ejected from the vent, producing a spectacular fireworks display. Explosive discharges continued, throwing hot fragments and ash occasionally as high as 6000 metres above the crater rim. Larger fragments fell near the crater, some remaining incandescent as they rolled down the slope. These built an aesthetically pleasing cone, while finer ash fell over a much larger area, burning and eventually covering the village of Parícutin. In the first day, the cone grew to 40 metres, and by the fifth day it was over 100 metres high. Within the first year, over 90 percent of the total ejecta had been discharged.

The first lava flow came from a fissure that opened just north of the cone, but after a few months, flows began to emerge from the base of the cone itself. In June 1944, a clinkery aa flow 10 metres thick moved over much of the village of San Juan Parangaricutiro, leaving only the church steeple exposed (Figure 4.15). After nine years of intermittent pyroclastic explosions and nearly continuous discharge of lava from vents at its base, the activity ceased almost as quickly as it had begun. Today Parícutin is just

◆ **Figure 4.14** Eve Cone, a cinder cone of the Mt. Edziza Volcanic Complex, British Columbia.
(Photo by Catherine Hickson, Geological Survey of Canada)

Pyroclastic material

Crater

Central vent filled with rock fragments

♦ **Figure 4.15** The village of San Juan Parangaricutiro engulfed by aa lava from Paricutin, shown in the background. Only the church towers remain. (Photo by Tad Nichols)

another one of the scores of cinder cones dotting the landscape in this region of Mexico. Like the others, it will not erupt again.

Composite Cones

Earth's most picturesque yet potentially dangerous volcanoes are **composite cones** or **stratovolcanoes** (Figure 4.16). Most are located in a relatively narrow zone that rims the Pacific Ocean, appropriately called the *Ring of Fire* (see Figure 4.33). This active zone includes a chain of continental volcanoes that are distributed along the west coast of South and North America, including the large cones of the Andes and the Cascade Range of western North America. The latter group includes Mount St. Helens and Mount Rainier in Washington state and Mount Garibaldi in British Columbia. The most active regions in the Ring of Fire are located along curved belts of volcanic islands situated adjacent to the deep ocean trenches of the northern and western Pacific. This nearly continuous chain of volcanoes stretches from the Aleutian Islands to Japan and the Philippines and ends on the North Island of New Zealand.

The classic composite cone is a large, nearly symmetrical structure composed of both lava and pyroclastic deposits. For the most part, composite cones are the product of gas-rich magma having an andesitic composition. (Composite cones may also emit various amounts of material having a mafic and/or rhyolitic

composition). Relative to shields, the silica-rich magmas typical of composite cones generate thick viscous lavas that travel short distances. In addition, composite cones may generate explosive eruptions that eject huge quantities of pyroclastic material.

The growth of a "typical" composite cone begins with both pyroclastic material and lava being emitted from a central vent. As the structure matures, lavas tend to flow from fissures that develop on the lower flanks of the cone. This activity may alternate with explosive eruptions that eject pyroclastic material from the summit crater. Sometimes both activities occur simultaneously.

A conical shape, with a steep summit area and more gradually sloping flanks, is typical of many large composite cones. This classic profile is partially a consequence of the way viscous lavas and pyroclastic ejecta contribute to the growth of the cone. Coarse fragments ejected from the summit crater tend to accumulate near their source. Due to their high angle of repose, coarse materials contribute to the steep slopes of the summit area. Finer ejecta are deposited as a thin layer over a large area, which acts to flatten the flank of the cone. In addition, during the early stages of growth, lavas tend to be more abundant and flow greater distances from the vent than do later lavas. This contributes to the cone's broad base. As the volcano matures, the short flows that come from the central vent armour and strengthen the summit area. Consequently, steep slopes exceeding 40 degrees are

◆ **Figure 4.16**   These two volcanoes, Pomerape and Parinacota, exhibit the classic shape of a composite cone. Lauca National Park, Chile.
(Photo by Michael Giannechini/Photo Researchers, Inc.)

sometimes possible. Fujiyama in Japan exhibits the classic form we expect of a composite cone, with its steep summit and gently sloping flanks.

Despite their symmetrical form, most composite cones have a complex history. Huge mounds of volcanic debris surrounding many cones provide evidence that, in the distant past, a large section of the volcano slid downslope as a massive landslide. Others develop horseshoe-shaped depressions at their summits as a result of explosive eruptions or, as occurred during the 1980 eruption of Mount St. Helens, a combination of a landslide and the eruption of magma that left a gaping void on one side of the cone. Often, so much rebuilding has occurred since these eruptions that no trace of the amphitheatre-shaped scar remains.

Living in the Shadow of a Composite Cone

More than 50 volcanoes have erupted in North America in the past 200 years (Figure 4.17). Fortunately, the most explosive of these eruptions, except for Mount St. Helens in 1980, occurred in sparsely inhabited regions of Alaska. On a global scale, numerous destructive eruptions have occurred during the past few thousand years, a few of which may have influenced the course of human civilization.

The Lost Continent of Atlantis

Anthropologists have proposed that a catastrophic eruption on the island of Santorini (also called Thera) contributed to the collapse of the advanced Minoan civilization centred around Crete in the Aegean Sea (Figure 4.18). It is suggested that this event also gave rise to the enduring legend of the lost continent of Atlantis. According to an account by the Greek philosopher Plato, an island empire named Atlantis was swallowed up by the sea in a single day and night. Although the connection between Plato's Atlantis and the Minoan civilization is somewhat tenuous, there is no doubt that a catastrophic eruption took place on Santorini about 1600 BC.

This eruption generated a tall, billowing *eruption column* composed of huge quantities of pyroclastic materials. Ash and pumice rained from this plume for days, eventually blanketing the surrounding landscape to a maximum depth of 60 metres. One nearby Minoan city, now called Akrotiri, was buried and its remains sealed until 1967, when archaeologists began investigating the area. The excavation of beautiful ceramic jars and elaborate wall paintings indicate that Akrotiri was home to a wealthy and sophisticated society.

Following the ejection of this large quantity of material, the summit of Santorini collapsed, producing a caldera 8 kilometres across. This once-majestic

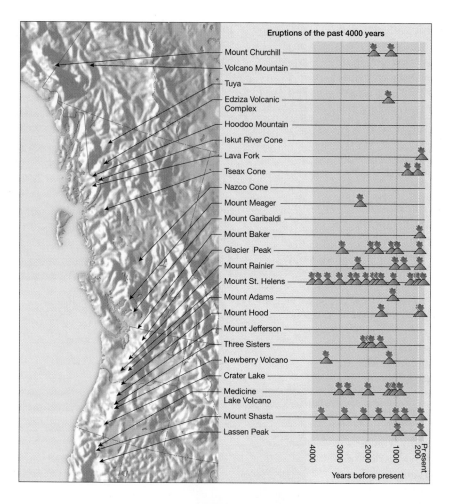

Eruptions of the past 4000 years

Mount Churchill
Volcano Mountain
Tuya
Edziza Volcanic Complex
Hoodoo Mountain
Iskut River Cone
Lava Fork
Tseax Cone
Nazco Cone
Mount Meager
Mount Garibaldi
Mount Baker
Glacier Peak
Mount Rainier
Mount St. Helens
Mount Adams
Mount Hood
Mount Jefferson
Three Sisters
Newberry Volcano
Crater Lake
Medicine Lake Volcano
Mount Shasta
Lassen Peak

4000   3000   2000   1000   Present 200

Years before present

Figure 4.17   The locations of volcanoes along the west coast of North America. Each eruption symbol in the diagram represents from one to several dozens eruptions closely spaced in time. (After U.S. Geological Survey and the Geological Survey of Canada)

Santorini

Greece   25°   Turkey

Santorini ▢

Crete

35°

Therasia

Mediterranean Sea

Phira

Thera

Aspronisi

Akrotiri

Figure 4.18   Map showing the remains of the volcanic island of Santorini after the top of the cone collapsed into the emptied magma chamber following an explosive eruption. The location of the recently excavated Minoan town of Akrotiri is shown. Volcanic eruptions over the last 500 years built the central islands. Despite the fact that another destructive eruption is likely, the city of Phira was built on the flanks of the caldera.

volcano presently consists of five small islands. The eruption and collapse of Santorini generated large sea waves (*tsunamis*) that caused widespread destruction to the coastal villages of Crete and to nearby islands to the north.

Although some scholars suggest that the eruption of Santorini contributed to the demise of the Minoan civilization, was this eruption the main cause of the dispersal of this great civilization or only one of many contributing factors? Was Santorini the island continent of Atlantis described by Plato? Whatever the answers to these questions, clearly volcanism can dramatically change the course of human events.

Eruption of Vesuvius AD 79

In addition to producing some of the most violent volcanic activity, composite cones can erupt unexpectedly. One of the best documented of these events was the AD 79 eruption of the Italian volcano Vesuvius. Prior to this eruption, Vesuvius had been dormant for centuries and had vineyards adorning its sunny slopes. On August 24, however, the tranquility ended, and in less than 24 hours the city of Pompeii (near Naples) and more than 2000 of its 20,000 residents perished.

By reconciling historical records with detailed scientific studies of the region, volcanologists have pieced together the chronology of the destruction of Pompeii. The eruption most likely began as steam discharges on the morning of August 24. By early afternoon, fine ash and pumice fragments formed a tall eruptive cloud. Shortly thereafter, debris from this cloud began to shower Pompeii, located 9 kilometres downwind of the volcano. Undoubtedly, many people fled during this early phase of the eruption. For the next several hours pumice fragments as large as 5 centimetres fell on Pompeii. One historical record states that people located more distant than Pompeii tied pillows to their heads in order to fend off the flying fragments.

The pumice fall continued for several hours, accumulating at the rate of 12 to 15 centimetres per hour. Most of the roofs in Pompeii eventually gave way. Despite the accumulation of more than 2 metres of pumice, many of the people that had not evacuated Pompeii were probably still alive the morning of August 25. Then suddenly and unexpectedly a surge of searing hot dust and gas swept rapidly down the flanks of Vesuvius, killing an estimated 2000 people who had somehow managed to survive the pumice fall. Some may have been killed by flying debris, but most died by inhaling ash-laden gases. Their remains were quickly buried by the falling ash, which rain cemented into a hard mass before their bodies decayed. The subsequent decomposition of the bodies produced cavities in the hardened ash. Nineteenth-century excavators found these cavities and created casts of the corpses by pouring plaster of Paris into the voids (Figure 4.19). Some of the plaster casts show victims trying to cover their mouths in an effort to draw their last breath.

Volcanologists now realize that several destructive flows of hot, asphyxiating ash-laden gas swept over the countryside surrounding Vesuvius. Skeletons excavated from the nearby town of Herculaneum indicate that most of its inhabitants were probably killed by these flows. Further, many of those who evacuated Pompeii probably met a similar fate. An estimated 16,000 people may have perished in this tragic and unexpected event.

Nuée Ardente: A Deadly Pyroclastic Flow

Although the destruction of Pompeii was catastrophic, **pyroclastic flows** consisting of hot gases infused with incandescent ash and larger rock fragments can be even more devastating. The most destructive of these fiery flows, called **nuée ardentes** (also referred to as *glowing avalanches*), are capable of racing down steep volcanic slopes at speeds that can approach 200 kilometres per hour (Figure 4.20).

◆ **Figure 4.19** Plaster casts of several victims of the AD 79 eruption of Mount Vesuvius that destroyed the Italian city of Pompeii.
(Photo by Leonard von Matt/Photo Researchers, Inc.)

◆ **Figure 4.20** Nuée ardente races down the slope of Mount St. Helens on August 7, 1980, at speeds in excess of 100 kilometres per hour.
(Photo by Peter W. Lipman, U.S. Geological Survey)

The ground-hugging portion of a glowing avalanche is rich in particulate matter, which is suspended by jets of buoyant gases passing upward through the flow. Some of these gases have escaped from newly erupted volcanic fragments. In addition, air that is overtaken and trapped by an advancing flow may be heated sufficiently to provide buoyancy to the particulate matter of the nuée ardente. Thus, these flows, which can include large rock fragments in addition to ash, travel downslope in a nearly frictionless environment. This helps to explain why some nuée ardente deposits are found more than 100 kilometres from their source.

Gravity causes these heavier-than-air flows to sweep downslope much like a snow avalanche. Some pyroclastic flows result when a powerful eruption blasts pyroclastic material laterally out the side of a volcano. Probably more often, nuée ardentes form from the collapse of tall eruption columns that form over a volcano during an explosive event. Once gravity overcomes the initial upward thrust provided by the escaping gases, the ejecta begin to fall. Massive amounts of incandescent blocks, ash, and pumice fragments that fall onto the summit area begin to cascade downslope under the influence of gravity. The largest fragments have been observed bouncing down the flanks of a cone, while the finer materials travel rapidly as an expanding tongue-shaped cloud.

The Destruction of St. Pierre.

In 1902 an infamous nuée ardente from Mount Pelée, a small volcano on the Caribbean island of Martinique, destroyed the port town of St. Pierre. The destruction happened in moments and was so devastating that almost all of St. Pierre's 28,000 inhabitants were killed. Only one person on the outskirts of town—a prisoner protected in a dungeon—and a few people on ships in the harbour were spared (Figure 4.21). Satis N. Coleman, in *Volcanoes, New and Old*, relates a vivid account of this event, which lasted less than five minutes.

I saw St. Pierre destroyed. The city was blotted out by one great flash of fire.... Our boat, the *Roraima*, arrived at St. Pierre early Thursday morning. For hours before entering the roadstead, we could see flames and smoke rising from Mt. Pelée.... There was a constant muffled roar. It was like the biggest oil refinery in the world burning up on the mountain top. There was a tremendous explosion about 7:45, soon after we got in. The mountain was blown to pieces. There was no warning. The side of the volcano was ripped out and there was hurled straight toward us a solid wall of flame. It sounded like a thousand cannons.... The air grew stifling hot and we were in the thick of it. Wherever the mass of fire struck the sea, the water boiled and sent up vast columns of steam.... The blast of fire from the volcano lasted only a few minutes. It shrivelled and set fire to everything it touched. Thousands of casks of rum were stored in St. Pierre, and these were exploded by the terrific heat.... Before the volcano burst the landings of St. Pierre were covered with people. After the explosion, not one living soul was seen on land.[*]

Shortly after this calamitous eruption, scientists arrived on the scene. Although St. Pierre was mantled by only a thin layer of volcanic debris, they discovered that masonry walls nearly a metre thick were knocked over like dominoes; large trees were uprooted and cannons were torn from their mounts. A further reminder of the destructive force of this nuée ardente is preserved in the ruins of the mental hospital. One of the immense steel chairs that had been used to confine alcoholic patients can be seen today, contorted, as though it were made of plastic.

Lahars: Mudflows on Active and Inactive Cones

In addition to violent eruptions, large composite cones may generate a type of mudflow referred to by its Indonesian name **lahar**. These destructive flows occur when volcanic debris becomes saturated with water and rapidly moves down steep volcanic slopes, generally following gullies and stream valleys. Some lahars may be triggered when large volumes of ice and snow melt during an eruption. Others are generated when heavy rainfall saturates weathered volcanic deposits. Thus, lahars can occur even when a volcano is *not* erupting.

When Mount St. Helens erupted in 1980, several lahars formed. These flows and accompanying flood waters raced down the valleys of the north and south forks of the Toutle River at speeds exceeding 30 kilometres per hour. Water levels in the river rose to 4 metres above flood stage, destroying or severely damaging nearly all the homes and bridges along the impacted area. Fortunately, the area was not densely populated. This was not the case in 1985, when Nevado del Ruiz, an ice-capped volcano in the Andes, erupted and generated lahars that killed nearly 20,000 people (see the section entitled "Debris Flow" in Chapter 9).

---

[*]New York: John Day, 1946, pp. 80–81.

◆ **Figure 4.21** St. Pierre as it appeared shortly after the eruption of Mount Pelée, 1902. (Reproduced from the collection of the Library of Congress)

Other Volcanic Landforms

The most obvious volcanic structure is a cone. But other distinctive and important landforms are also associated with volcanic activity.

Calderas

Calderas (*caldaria* = a cooking pot) are large collapse depressions having a more or less circular form. Their diameters exceed one kilometre, and many are tens of kilometres across. (Those less than a kilometre across are called *collapse pits*.) Most calderas are formed by one of the following processes: 1) the collapse of the summit of a large composite volcano following an explosive eruption of silica-rich pumice and ash fragments; 2) the collapse of the top of a shield volcano caused by subterranean drainage from a central magma chamber; and 3) the collapse of a large area, independent of any pre-existing volcanic structures, caused by the discharge of colossal volumes of silica-rich pumice and ash along ring fractures.

Crater Lake–Type Calderas.   Crater Lake, Oregon, is located in a caldera that has a maximum diameter of 10 kilometres and is 1175 metres deep. This caldera formed about 7000 years ago when a composite cone,

Students Sometimes Ask...

Some of the larger volcanic eruptions, like the eruption of Krakatoa, must have been impressive. What was it like?

On August 27, 1883, in what is now Indonesia, the volcanic island of Krakatoa exploded and was nearly obliterated. The sound of the explosion was heard an incredible 4800 kilometres away at Rodriguez Island in the western Indian Ocean. Dust from the explosion was propelled into the atmosphere and circled Earth on high-altitude winds. This dust produced unusual and beautiful sunsets for nearly a year.

Not many were killed directly by the explosion, because the island was uninhabited. However, the displacement of water from the energy released during the explosion was enormous. The resulting *seismic sea wave* or *tsunami* exceeded 35 metres in height. It devastated the coastal region of the Sunda Strait between the nearby islands of Sumatra and Java, drowning over 1000 villages and taking more than 36,000 lives. The energy carried by this wave reached every ocean basin and was detected by tide recording stations as far away as London and San Francisco.

later named Mount Mazama, violently extruded 50 to 70 cubic kilometres of pyroclastic material (Figure 4.22). With the loss of support, 1500 metres of the summit of this once prominent cone collapsed. After the collapse, rainwater filled the caldera (Figure 4.23). Later volcanic activity built a small cinder cone in the lake. Today this cone, called Wizard Island, provides a mute reminder of past activity.

Hawaiian–Type Calderas.   Although most calderas are produced by *collapse following an explosive eruption*, some are not. For example, Hawaii's active shield volcanoes, Mauna Loa and Kilauea, both have large calderas at their summits. Kilauea's measures 3.3 by 4.4 kilometres and is 150 metres deep. Each caldera formed by gradual subsidence of the summit as magma slowly drained laterally from the central magma chamber to a rift zone, often producing flank eruptions.

Yellowstone–Type Calderas.   Although the 1980 eruption of Mount St. Helens was spectacular, it pales by comparison to what happened 630,000 years ago in the region now occupied by Yellowstone National Park. Here approximately 1000 cubic kilometres of pyroclastic material erupted, eventually producing a caldera 70 kilometres across. This event produced showers of ash as far away as the Gulf of Mexico. Vestiges of this activity are the many hot springs and geysers in the region.

Unlike calderas associated with composite cones, these depressions are so large and poorly defined that many remained undetected until high-quality aerial, or satellite, images became available.

The formation of a large Yellowstone-type caldera begins when a silica-rich (rhyolitic) magma body is emplaced near the surface, upwarping the overlying rocks. Next, ring fractures develop in the roof, providing a pathway to the surface for the highly viscous gas-rich magma. This initiates an explosive eruption of colossal proportions, ejecting huge volumes (usually exceeding 100 cubic kilometres) of pyroclastic materials, mainly in the form of ash and pumice fragments. Typically, these materials form a pyroclastic flow that spreads across the landscape at speeds that may exceed 100 kilometres per hour, destroying most living things in its path. Finally, with the loss of support, the roof of the magma chamber collapses, generating a large caldera.

Another distinctive feature associated with most large calderas is a slow upheaval, or *resurgence*, of the floor of the caldera following an eruptive phase. Thus, these structures consist of a large, somewhat circular depression containing a central elevated region. Most large calderas exhibit a complex history. Geological evidence suggests that a magma reservoir still exists beneath Yellowstone; thus, another caldera-forming eruption is likely, but not imminent.

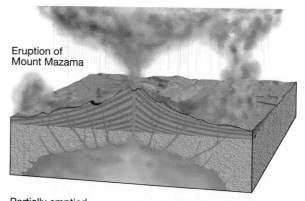

Eruption of Mount Mazama

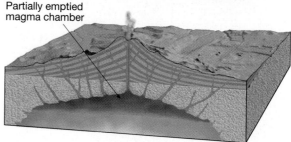

Partially emptied magma chamber

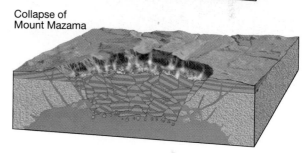

Collapse of Mount Mazama

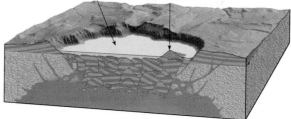

Formation of Crater Lake and Wizard Island

◆ **Figure 4.22**   Sequence of events that formed Crater Lake, Oregon. About 7000 years ago violent eruption partly emptied the magma chamber, causing the summit of former Mount Mazama to collapse. Rainfall and groundwater contributed to form Crater Lake, the deepest lake in the United States. Subsequent eruptions produced the cinder cone called Wizard Island.
(After H. Williams, *The Ancient Volcanoes of Oregon*, p. 47. Courtesy of the University of Oregon)

Calderas of the type located on the Yellowstone Plateau of northwestern Wyoming are the largest volcanic structures on Earth. Some geologists have compared their destructive force with that of the impact of a small asteroid. Fortunately, no eruption of this type has occurred in historic times.

◆ **Figure 4.23**   Crater Lake occupies a caldera about 10 kilometres in diameter. (Photo by Greg Vaughn/Tom Stack and Associates)

Fissure Eruptions and Lava Plateaus

We think of volcanic eruptions as building a cone or shield from a central vent. But by far the greatest volume of volcanic material is extruded from fractures in the crust called **fissures** (*fissura* = to split). Rather than building a cone, these long, narrow cracks may emit a low-viscosity mafic lava, blanketing a wide area.

The extensive Columbia Plateau in the northwestern United States was formed this way (Figure 4.24). Here, numerous **fissure eruptions** extruded very fluid mafic lava between 6 and 17 million years ago (Figure 4.25). Successive flows, some 50 metres thick, buried the existing landscape as they built a lava plateau nearly 1.5 kilometres thick. The fluid nature of the lava is evident, because some remained molten long enough to flow 150 kilometres from its source. Although less extensive in terms of area, basaltic lavas of similar age to those of the Columbian plateau occur in south-central British Columbia. The term **flood basalts** appropriately describes these flows.

One of the largest accumulations of flood basalt in the world is the Deccan Traps, a thick sequence of

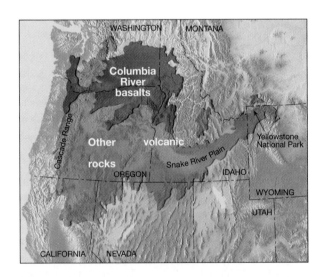

◆ **Figure 4.24**   Volcanic areas that compose the Columbia Plateau in the U.S. Pacific Northwest. The Columbia River basalts cover an area of nearly 200,000 square kilometres. Activity here began about 17 million years ago as lava began to pour out of large fissures, eventually producing a basalt plateau with an average thickness of nearly 1.5 kilometre. (After U.S. Geological Survey)

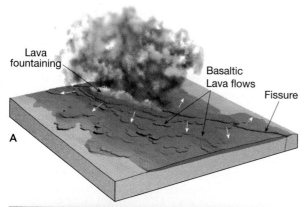

◆ **Figure 4.25** Basaltic fissure eruption. **A.** Lava fountaining from a fissure and formation of fluid lava flows called flood basalts. **B.** Photo of basalt flows near Idaho Falls. (Photo by John S. Shelton)

flat-lying basalt flows covering nearly 500,000 square kilometres of west central India. When the Deccan Traps formed about 66 million years ago, nearly 2 million cubic kilometres of lava were extruded in less than 1 million years. Another huge deposit of flood basalts, called the Ontong Java Plateau, is found on the floor of the Pacific Ocean.

Iceland, which is located astride the Mid-Atlantic Ridge, regularly experiences fissure eruptions. The largest Icelandic eruptions in historic times occurred in 1783 and were named the Laki eruptions. A rift 25 kilometres long generated over 20 separate vents, which initially extruded sulphurous gases and ash deposits that built several small cinder cones. This activity was followed by huge outpourings of very fluid mafic lava. The total volume of lava extruded by the Laki eruptions was in excess of 12 cubic kilometres.

Volcanic gases stunted grasslands and directly killed most of Iceland's livestock. The ensuing famine caused 10,000 deaths.

Lava Domes

In contrast to mafic lavas, silica-rich lavas near the felsic (rhyolitic) end of the compositional spectrum are so viscous they hardly flow. As the thick lava is "squeezed" out of the vent, it may produce a steep-sided dome-shaped mass of congealed lava called a **lava dome**.

Lava domes typify the late stages of activity of mature, chiefly andesitic composite cones (see Box 4.2). Here these rhyolitic structures form in the summit crater and as parasitic structures on the flanks of these cones following an explosive eruption of a gas-rich magma. This is exemplified by the volcanic dome that continues to "grow" from the vent that produced the 1980 eruption of Mount St. Helens (Figure 4.26). Although most lava domes form in association with pre-existing composite cones or shield volcanoes, some form independently, such as the line of rhyolitic and obsidian domes at Mono Craters, California.

Students Sometimes Ask...

If volcanoes are so dangerous, why do people live on or near them?

Realize that many who live near volcanoes did not choose the location; they were simply born there. Their ancestors may have lived in the region for generations. Historically, many have been drawn to volcanic regions because of their fertile soils. Not all volcanoes have explosive eruptions, but all active volcanoes are dangerous. Certainly, choosing to live close to an active composite cone like Mount St. Helens or Soufriére Hills has a high inherent risk. However, the time interval between successive eruptions might be several decades or more—plenty of time for generations of people to forget the last eruption and consider the volcano to be dormant and therefore safe. Other volcanoes, like Mauna Loa or those on Iceland, are continually active, so recent eruptions are fresh in the minds of local populations. Many people that choose to live near an active volcano have the belief that the *relative* risk is no higher than in other hazard-prone places. In essence, they are gambling that they will be able to live out their lives before the next major eruption.

Volcanic Pipes and Necks

Most volcanoes are fed magma through short conduits, called *pipes*, that connect a magma chamber to the surface. In rare circumstances, pipes may extend to depths exceeding 200 kilometres. When this occurs, the ultramafic magmas that migrate up these structures produce rocks that are thought to be samples of the mantle that have undergone very little alteration during their ascent. Geologists consider these unusually deep conduits to be "windows" into Earth, for they allow us to view rock normally found only at great depth.

Well-known volcanic pipes are the diamond-bearing structures of South Africa and Northwest Territories (see Chapter 21). Here, the rocks filling the pipes originated at depths of at least 150 kilometres, where pressure is high enough to generate diamonds and other high-pressure minerals. The task of transporting essentially unaltered magma (along with diamond inclusions) through 150 kilometres of solid rock is exceptional and accounts for the scarcity of natural diamonds.

Volcanoes on land are continually being lowered by weathering and erosion. Cinder cones are easily eroded, because they are composed of unconsolidated materials. However, all volcanoes eventually succumb to relentless erosion over geologic time. As erosion progresses, the rock occupying the volcanic pipe is often more resistant and may remain standing above the surrounding terrain long after most of the cone has vanished. Shiprock, New Mexico, is such a feature and is called a **volcanic neck** (Figure 4.27). This structure, higher than many skyscrapers, is but one of many such landforms that protrude conspicuously from the red desert landscapes of the American Southwest.

Intrusive Igneous Activity

Internal Processes
  ↳ Igneous Activity

Although volcanic eruptions can be among the most violent and spectacular events in nature and therefore

◆ **Figure 4.26** Following the May 1980 eruption of Mount St. Helens, a lava dome began to develop. (Photo by David Falconer/DRK Photo)

◆ **Figure 4.27**   Shiprock, New Mexico, is a volcanic neck. This structure, which stands over 420 metres high, consists of igneous rock, which crystallized in the vent of a volcano that has long since been eroded away. (Photo by Tom Bean)

worthy of detailed study, most magma is emplaced at depth. Thus, an understanding of intrusive igneous activity is as important to geologists as the study of volcanic events.

Many of the structures that result from the emplacement of igneous material at depth are called **plutons**, named for Pluto, the god of the lower world in classical mythology. Because all plutons form out of view beneath Earth's surface, they can be studied only after uplift and erosion have exposed them. The challenge lies in reconstructing the events that generated these structures millions or even hundreds of millions of years ago.

For the sake of clarity, we have separated our discussions of volcanism and plutonic activity. Keep in mind, however, that these diverse processes occur simultaneously and involve basically the same materials.

Nature of Intrusions

Intrusions occur in a great variety of sizes and shapes. Some of the most common types are illustrated in Figure 4.28. Notice that some of these structures have a tabular (tabletop) shape, whereas others are quite massive or shapeless. Also, observe that some of these bodies cut across existing structures, such as layers of sedimentary rock; others form when magma is injected between sedimentary layers. Because of these differences, intrusive igneous bodies are generally classified according to their shape as either **tabular** or **massive** and by their orientation with respect to the host rock. Intrusive rocks are said to be **discordant** (*discordare* = to disagree) if they cut across existing structures and **concordant** (*concordare* = to agree) if they form parallel to features such as sedimentary strata. As you can see in Figure 4.28A, plutons are closely associated with volcanic activity. Many of the largest intrusive bodies are the remnants of magma chambers that once fed ancient volcanoes. Large intrusive bodies of particular significance in Canada include plutons in the Cascade Range of British Columbia, the Monteregian intrusions in the Montréal region of Quebec, and the Sweetgrass Hills just south of Alberta.

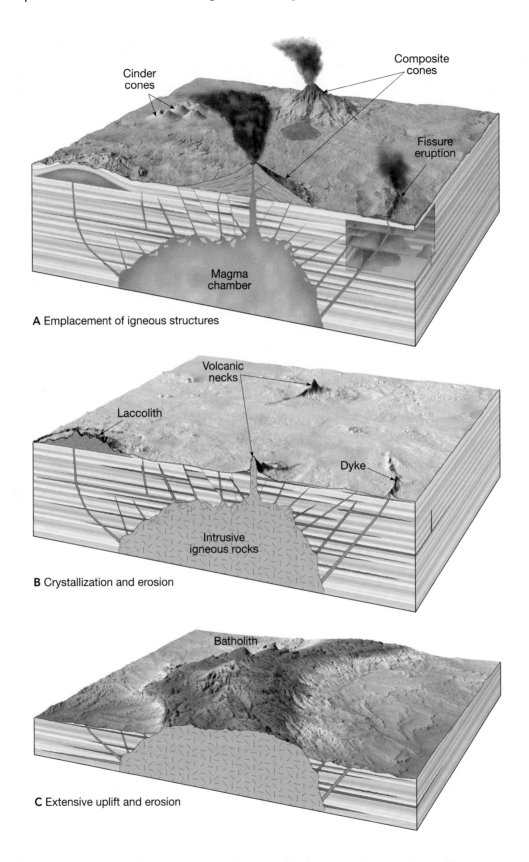

A Emplacement of igneous structures

B Crystallization and erosion

C Extensive uplift and erosion

◆ **Figure 4.28**   Illustrations showing basic igneous structures. **A.** This block diagram shows the relationship between volcanism and intrusive igneous activity. **B.** This view illustrates the basic intrusive igneous structures, some of which have been exposed by erosion long after their formation. **C.** After millions of years of uplifting and erosion, a batholith is exposed at the surface.

People and the Environment
Volcanic Crisis on Montserrat

The Caribbean's Lesser Antilles are mostly volcanic in origin and extend from near the northeast coast of South America in an arc toward Puerto Rico and the Virgin Islands (Figure 4.B). Near the beginning of the twentieth century, devastating eruptions occurred when volcanoes on Martinique (Mt. Pelée) and St. Vincent (Soufriére) killed more than 30,000 people. As the twentieth century came to a close, the Caribbean was once again a focus for volcanologists. This time their attention was on the island of Montserrat.

This small island is dominated by the Soufriére Hills volcano, which began erupting in July 1995 after thousands of years of inactivity. The volcano, like most in the Caribbean, erupts viscous lava that oozes out at the surface, forming a lava dome. Such domes have the potential to produce devastating pyroclastic flows. Because there may be little warning, such eruptions are extremely dangerous (Figure 4.C).

The activity at Soufriére Hills volcano included many large pyroclastic flows that eventually covered large parts of the island. Moreover, plumes of volcanic ash were sometimes erupted to heights of 6000 metres or more. By January 1998, many of the island's nearly 12,000 residents had been evacuated to neighbouring islands. To say the least, the erupting volcano caused serious hardship and economic distress to the people of Montserrat. On the positive side, loss of life was small.

Since the onset of eruptive activity, Soufriére Hills has become one of the most closely monitored volcanoes in the world. Almost immediately after the unexpected activity began, the Montserrat Volcano Observatory was

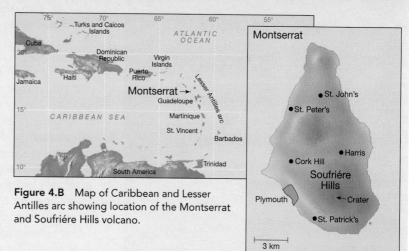

Figure 4.B Map of Caribbean and Lesser Antilles arc showing location of the Montserrat and Soufriére Hills volcano.

◆ **Figure 4.C** The August 20, 1997, eruption of Soufriére Hills volcano on the Caribbean island of Montserrat. This is one of several volcanoes in the region that owes its existence to the melting of a subducting plate. (Photo by Kevin West/Gamma Liaison)

established. Staffed by scientists from the University of the West Indies and the British Geological Survey, the mountain was wired with seismometers, tiltmeters, and gas analyzers. The valuable data being gathered may one day contribute to a reliable method of predicting volcanic eruptions.

Dykes

Dykes are tabular discordant bodies that are produced when magma is injected into fractures (Figure 4.29). The force exerted by the emplaced magma can be great enough to further separate the walls of the fracture. Once crystallized, these wall-like structures have thick-nesses ranging from less than a centimetre to more than a kilometre. The largest have lengths of hundreds of kilometres. Most dykes, however, are a few metres thick and extend laterally for no more than a few kilometres.

Dykes are often found in groups that once served as vertically oriented pathways followed by molten

◆ **Figure 4.29**   Light-coloured dykes of felsic composition cut through dark-coloured mafic volcanic rocks near Lakelse, British Columbia.
(Photo by Steve Hicock)

rock that fed ancient lava flows. The parent pluton is generally not observable. Some dykes are found radiating, like spokes on a wheel, from an eroded volcanic neck. In these situations the active ascent of magma is thought to have generated fissures in the volcanic cone out of which lava flowed.

Sills and Laccoliths

Sills and laccoliths are concordant intrusions that form when magma is intruded in a near-surface environment. They differ from one another in shape and usually in composition.

Sills.   **Sills** are tabular intrusions formed when magma is injected along sedimentary bedding surfaces (Figure 4.30). Horizontal sills are the most common, although all orientations, even vertical, are known to exist. Because of their relatively uniform thickness and large areal extent, sills are likely the product of very fluid magmas. Magmas having a low silica content are more fluid, so most sills are mafic in composition.

The emplacement of a sill requires that the overlying sedimentary rock be lifted to a height equal to the thickness of the sill. Although this is a formidable task, in shallow environments it often requires less energy than forcing the magma up the remaining distance to the surface. Consequently, sills

form only at shallow depths, where the pressure exerted by the weight of overlying rock layers is low. Although sills are intruded between layers, they can be locally discordant.

In many respects, sills closely resemble buried lava flows. Both are tabular and often exhibit columnar jointing (Figure 4.31). **Columnar joints** form as igneous rocks cool and develop shrinkage fractures that produce elongated, pillarlike columns.

When attempts are made to reconstruct the geologic history of a region, it becomes important to differentiate between sills and buried lava flows. Fortunately, under close examination these two phenomena can be readily distinguished. The upper portion of a buried lava flow usually contains voids produced by escaping gas bubbles. Further, only the rocks beneath a lava flow show evidence of metamorphic alteration. Sills, on the other hand, form when magma has been forcefully intruded between sedimentary layers. Thus, fragments of the overlying rock can be found only in sills. Lava flows, conversely, are extruded before the overlying strata are deposited. "Baked" zones in the rock above and below are trademarks of a sill and are only found on the rock below a lava flow, not on the overlying rock.

◆ **Figure 4.30**   The dark-coloured rock layer in this photograph is a sill made of gabbro. It is located in the Upper Brock River Canyon, near Paulatuk, Northwest Territories. (Reproduced with the permission of the Minister of Public Works and Government Services Canada, 2004 and courtesy of Natural Resources Canada, Geological Survey of Canada. Photo by S. Irwin.)

◆ **Figure 4.31** Columns north of Vancouver, British Columbia show the characteristic hexagonal patterns of joints in basalt. These are produced by the contraction and fracturing that result as a lava flow or sill gradually cools. (Photo by Catherine Hickson, Geological Survey of Canada)

Laccoliths.   **Laccoliths** are similar to sills because they form when magma is intruded between sedimentary layers in a near-surface environment. However, the magma that generates laccoliths is more viscous and collects as a blister-like mass that arches the overlying strata upward (see Figure 4.28B). Consequently, a laccolith can occasionally be detected because of the dome-shaped bulge it creates at the surface. Most large laccoliths are probably not much wider than a few kilometres.

Batholiths

By far the largest intrusive igneous bodies are **batholiths** (*bathos* = depth, *lithos* = stone). Most often, batholiths occur in groups that form linear structures several hundreds of kilometres long and up to 100 kilometres wide, as shown in Figure 4.32. Indirect evidence gathered from gravitational studies indicates that batholiths are also very thick, possibly extending dozens of kilometres into the crust. Based on the amount exposed by erosion, some batholiths must be at least several kilometres thick.

By definition, a plutonic body must have a surface exposure greater than 100 square kilometres to be considered a batholith. Smaller plutons of this type are termed **stocks**. Many stocks appear to be portions of batholiths that are not yet fully exposed.

Batholiths usually consist of rock types having chemical compositions toward the felsic end of the spectrum, although diorite is commonly found. Smaller batholiths can be rather simple structures composed almost entirely of one rock type. However, studies of large batholiths have shown that they consist of several distinct plutons that were intruded over a period of millions of years. The plutonic activity that created the Sierra Nevada batholith, for example, occurred nearly continuously over a 130-million-year period that ended about 80 million years ago during the Cretaceous period.

Batholiths may compose the core of mountain systems. Uplifting and erosion have removed the surrounding rock, thereby exposing the resistant igneous body. Large expanses of granitic rock also occur in the stable interiors of the continents, such as the Canadian Shield. These relatively flat exposures are the remains of ancient mountains that have long since been levelled by erosion. The exposed rocks that make up the batholiths of youthful mountain ranges, such as the Sierra Nevada, were generated near the top of a magma chamber, whereas in shield areas, the roots of former mountains and, thus, the lower portions of batholiths, are exposed.

Emplacement of Batholiths.   An interesting problem that faced geologists was trying to explain how large

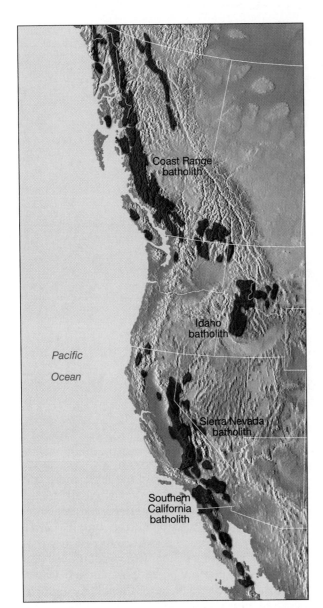

◆ **Figure 4.32** Granitic batholiths that occur along the western margin of North America. These gigantic, elongated bodies consist of numerous plutons that were emplaced during the last 150 million years of Earth history.

for itself by pushing aside the overlying rock. As the magma continues to move upward, some of the host rock that was shouldered aside will fill in the space left by the magma body as it passes.*

As a magma body nears the surface, it encounters relatively cool, brittle rock that resists deformation. Further upward movement is accomplished by a process called *stoping*. In this process, fractures that develop in the overlying host rock allow magma to rise and dislodge blocks of rock. Once incorporated into the magma body, these blocks may melt, thereby altering the composition of the magma body. Eventually the magma body will cool sufficiently so that upward movement comes to a halt. Evidence supporting the fact that magma can move through solid rock are called **xenoliths**, and are inclusions of the host rock.

Plate Tectonics and Igneous Activity

Geologists have known for decades that the global distribution of volcanism is not random. Of the more than 800 active volcanoes that have been identified, most are located along the margins of the ocean basins—most notably along the circum-Pacific belt known as the *Ring of Fire* (Figure 4.33).** This group of volcanoes consists mainly of composite cones that emit volatile-rich magma of intermediate (andesitic) composition and occasionally produce awe-inspiring eruptions.

The volcanoes comprising a second group emit very fluid mafic lavas and are confined to the deep ocean basins, including well-known examples on Hawaii and Iceland. This group also contains many active submarine volcanoes that dot the ocean floor; particularly notable are the innumerable small seamounts that form linear patterns. At these depths the pressures are so great that seawater does not boil explosively, even in contact with hot lavas. Thus, firsthand knowledge of these eruptions is limited, coming mainly from deep-diving submersibles.

granitic batholiths came to reside within only moderately deformed sedimentary and metamorphic rocks. What happened to the rock that was displaced by these igneous masses? How did the magma body make its way through several kilometres of solid rock?

We know that magma rises because it is less dense than the surrounding rock, much like a cork held at the bottom of a container of water will rise when it is released. But Earth's crust is made of solid rock. Nevertheless, at depths of several kilometres, where temperature and pressure are high, even solid rock deforms by flowing. Thus at great depths a mass of buoyant, rising magma can forcibly make room

---

*An analogous situation occurs when a can of oil-based paint is left in storage. The oily component of the paint is less dense than the pigments used for coloration; thus, oil collects into drops that slowly migrate upward, while the heavier pigments settle to the bottom.

**For our purposes, active volcanoes include those with eruptions that have been dated. At least 700 other cones exhibit geologic evidence that they have erupted within the past 10,000 years and are regarded as potentially active. Further innumerable active submarine volcanoes are hidden from view in the depths of the ocean and are not counted in these numbers.

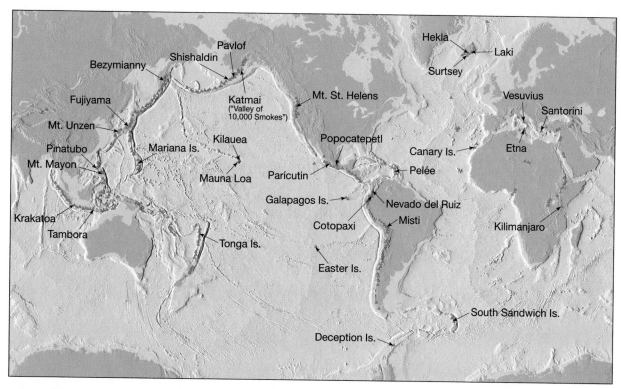

◆ **Figure 4.33**   Locations of some of Earth's major volcanoes.

A third group includes those volcanic structures that are irregularly distributed in the interiors of the continents. None are found in Australia or in the eastern two-thirds of North and South America. Africa is notable because it has many potentially active volcanoes including Mount Kilimanjaro, the highest point on the continent (5895 metres). Volcanism on continents is the most diverse, ranging from eruptions of very fluid mafic lavas, like those that generated the Columbia River basalts to explosive eruptions of silica-rich rhyolitic magma.

Until the late 1960s, geologists had no explanation for the apparently haphazard distribution of continental volcanoes, nor were they able to account for the almost continuous chain of volcanoes that circles the margin of the Pacific basin. The development of the theory of plate tectonics clarified the picture. Recall that all primary (unaltered) magma originates in the upper mantle and that the mantle is essentially solid, *not molten* rock. The basic connection between plate tectonics and volcanism is that *plate motions provide the mechanisms by which mantle rocks melt to generate magma.*

We will examine three regions of igneous activity and their relationship to plate boundaries. These active areas are located 1) along convergent plate boundaries where plates move toward each other and one sinks beneath the other; 2) along divergent plate boundaries, where plates move away from each other and new seafloor is created; and 3) areas within the plates proper that are not associated with any plate boundary. (It should be noted that volcanic activity rarely occurs along transform plate boundaries.) These three volcanic settings are depicted in Figure 4.34.

Igneous Activity at Convergent Plate Boundaries

Recall that at convergent plate boundaries, slabs of oceanic crust are bent as they descend into the mantle, generating an oceanic trench. As a slab sinks deeper into the mantle, the increase in temperature and pressure drives volatiles (mostly H_2O) from the oceanic crust. These mobile fluids migrate upward into the wedge-shaped piece of mantle located between the subducting slab and overriding plate (Figure 4.35). Once the sinking slab reaches a depth of about 100 to 150 kilometres, these water-rich fluids reduce the melting point of hot mantle rock sufficiently to trigger some melting. The partial melting of mantle rock (principally peridotite) generates magma with a mafic composition. After a sufficient quantity of magma has accumulated, it slowly migrates upward.

Volcanism at a convergent plate margin results in the development of a linear or slightly curved chain of volcanoes called a *volcanic arc*. These volcanic chains develop roughly parallel to the associated trench—at distances of 200 to 300 kilometres. Volcanic arcs can

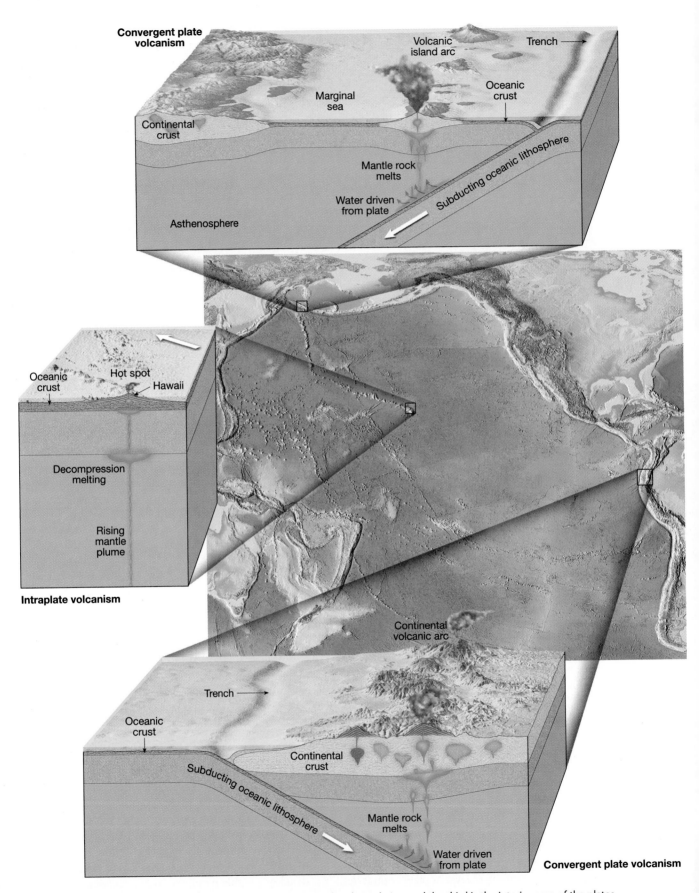

Convergent plate volcanism

Volcanic island arc

Trench

Marginal sea

Oceanic crust

Continental crust

Mantle rock melts

Subducting oceanic lithosphere

Water driven from plate

Asthenosphere

Oceanic crust

Hot spot

Hawaii

Decompression melting

Rising mantle plume

Intraplate volcanism

Continental volcanic arc

Trench

Oceanic crust

Subducting oceanic lithosphere

Continental crust

Mantle rock melts

Water driven from plate

Convergent plate volcanism

◆ **Figure 4.34**   Three zones of volcanism. Two of these zones are plate boundaries, and the third is the interior area of the plates.

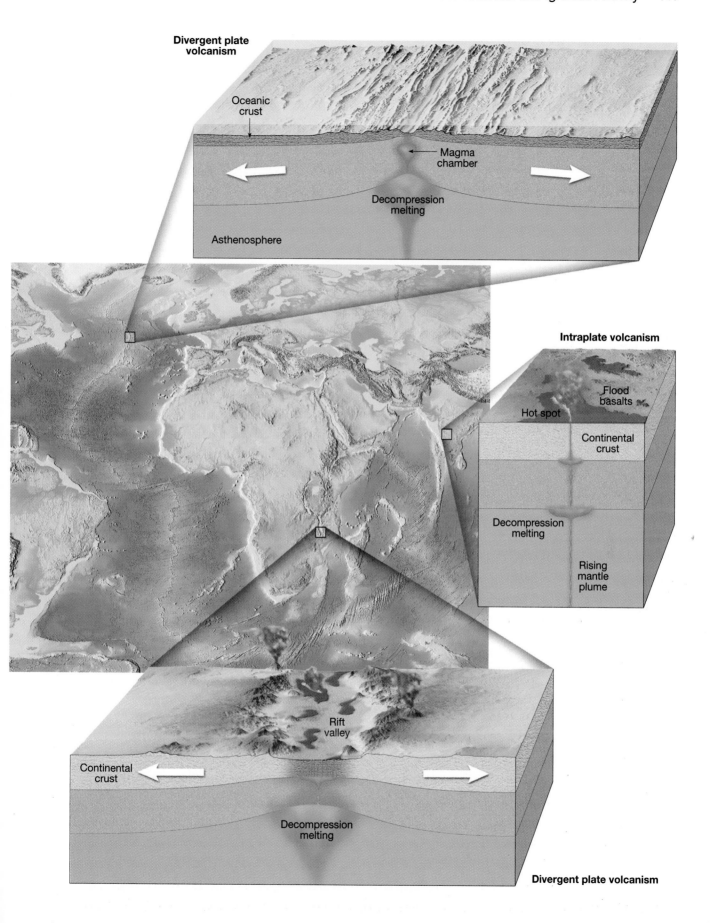

Divergent plate volcanism

Oceanic crust

Magma chamber

Decompression melting

Asthenosphere

Intraplate volcanism

Flood basalts

Hot spot

Continental crust

Decompression melting

Rising mantle plume

Rift valley

Continental crust

Decompression melting

Divergent plate volcanism

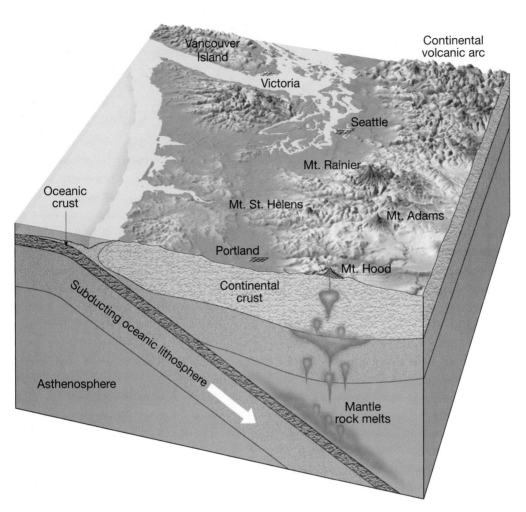

◆ **Figure 4.35**   As an oceanic plate descends into the mantle, water and other volatiles are driven from the subducting crustal rocks. These volatiles lower the melting temperature of mantle rock sufficiently to generate melt.

be constructed on oceanic, or continental, lithosphere. Those that develop within the ocean and grow large enough for their tops to rise above the surface are called **volcanic island arcs**, or simply **island arcs** (Figure 4.34). Several young volcanic island arcs of this type border the western Pacific basin, including the Aleutians, the Tongas, and the Marianas.

The early stage of island arc volcanism is typically dominated by the eruption of fluid basalts that build shieldlike structures on the ocean floor. Because this activity begins at great depth, volcanic cones must extrude a great deal of lava before their tops rise above the sea to form islands. This cone-building activity, coupled with massive mafic intrusions as well as magma that is added to the underside of the crust, thickens the arc crust through time. As a result, mature volcanic arcs are underlain by a comparatively thick crust that impedes the upward flow of the mantle-derived basalts. This in turn provides time for magmatic differentiation to occur, in which heavy iron-rich

minerals crystallize and settle out, leaving the melt enriched in silica (see Chapter 3). Consequently, as the arc matures, the magmas that reach the surface tend to erupt silica-rich andesites and even some rhyolites. In addition, magmatic differentiation tends to concentrate the available volatiles (water) into the more silica-rich components of these magmas. Because they emit viscous volatile-rich magma, island arc volcanoes typically have explosive eruptions.

Volcanism associated with convergent plate boundaries may also develop where slabs of oceanic lithosphere are subducted under continental lithosphere to produce a **continental volcanic arc** (Figure 4.35). The mechanisms that generate these mantle-derived magmas are essentially the same as those operating at island arcs. The major difference is that continental crust is much thicker and is composed of rocks having a higher silica content than oceanic crust. Hence, through the assimilation of silica-rich crustal rocks, plus extensive magmatic differentiation, a mantle-derived magma may change from a comparatively dry, fluid mafic magma to a volatile-rich, viscous andesitic or rhyolitic as it moves up through the continental crust. The volcanic chain of the Andes Mountains along the western edge of South America is perhaps the best example of a mature continental volcanic arc.

Since the Pacific basin is essentially bordered by convergent plate boundaries (and associated subduction zones), it is easy to see why the irregular belt of explosive volcanoes we call the Ring of Fire formed in this region.

Igneous Activity at Divergent Plate Boundaries

The greatest volume of magma (perhaps 60 percent of Earth's total yearly output) is produced along the oceanic ridge system in association with seafloor spreading (Figure 4.34). Here, below the ridge axis where the lithospheric plates are being continually pulled apart, the solid yet mobile mantle responds to the decrease in overburden and rises upward to fill in the rift. Recall from Chapter 3 that as rock rises, it experiences a decrease in confining pressure and undergoes melting without the addition of heat. This process, called *decompression melting*, is the most common process by which mantle rocks melt.

Partial melting of mantle rock at spreading centres produces mafic magma having a composition that is surprisingly similar to that generated along convergent plate boundaries. Because this newly formed mafic magma is less dense than the mantle rock from which it was derived, it rises faster than the mantle.

Collecting in chambers located just beneath the ridge crest, about 10 percent of this magma eventually migrates upward along fissures to erupt as flows on the ocean floor. This activity continuously adds new basaltic rock to the plate margins, temporarily welding them together, only to break again as spreading continues.

Although most spreading centres are located along the axis of an oceanic ridge, some are not. In particular, the East African Rift is a site where continental lithosphere has been ripped apart. Here, magma is generated by decompression melting in the same manner it is produced along the oceanic ridge system. Vast outpourings of fluid mafic lavas are common in this region. The East African Rift zone also contains some large composite cones, as exemplified by Mount Kilimanjaro. Like composite cones that form along convergent plate boundaries, these volcanoes form when mantle-derived basalts evolve into volatile-rich andesitic magma as they migrate up through thick silica-rich rocks of the continent.

Intraplate Igneous Activity

We know why igneous activity is initiated along plate boundaries, but why do eruptions occur in the interiors of plates? Hawaii's Kilauea is considered the world's most active volcano, yet it is situated thousands of kilometres from the nearest plate boundary in the middle of the vast Pacific plate. Other sites of **intraplate volcanism** (meaning "within the plate") include the Canary Islands, Yellowstone, and several volcanic centres in the Sahara Desert of northern Africa.

We now recognize that most intraplate volcanism occurs where a mass of hotter than normal mantle material called a **mantle plume** ascends toward the surface (Figure 4.36). Although the depth at which (at least some) mantle plumes originate is still hotly debated, many appear to form deep within Earth at the core—mantle boundary. These plumes of solid yet mobile mantle rock rise toward the surface in a manner similar to the blobs that form within a lava lamp. A mantle plume has a bulbous head that draws out a narrow stalk beneath it as it rises. Once the plume head nears the top of the mantle, decompression melting generates mafic magma that may eventually trigger volcanism at the surface. The result is a localized volcanic region a few hundred kilometres across called a **hot spot** (Figure 4.36). More than 100 hot spots have been identified, and most have persisted for millions of years. The land surface around hot spots is often elevated, showing that it is buoyed up by a plume of warm low-density material. Furthermore, by measuring the heat flow in these regions, geologists have determined that the mantle beneath hot spots must be 100–150°C hotter than normal.

The volcanic activity on the island of Hawaii, with its outpourings of mafic lava, is certainly the result of hot-spot volcanism. Where a mantle plume has persisted for long periods of time, a chain of volcanic structures may form as the overlying plate moves over it. In the Hawaiian Islands, hot-spot activity is currently centred on Kilauea. However, over the past 80 million years, the same mantle plume generated a chain of volcanic islands (and seamounts) that extend thousands of kilometres from the Big Island in a northwestward direction across the Pacific.

Mantle plumes are also thought to be responsible for the vast outpourings of mafic lava that create large basalt plateaus such as the Columbia Plateau centred in the northwestern United States, India's Deccan Plateau, and the Ontong Java Plateau in the western Pacific. The most widely accepted explanation for these eruptions, which emit extremely large volumes of mafic magma over relatively short time intervals, involves a plume with a substantial-sized head. Sometimes called *superplumes*, these structures

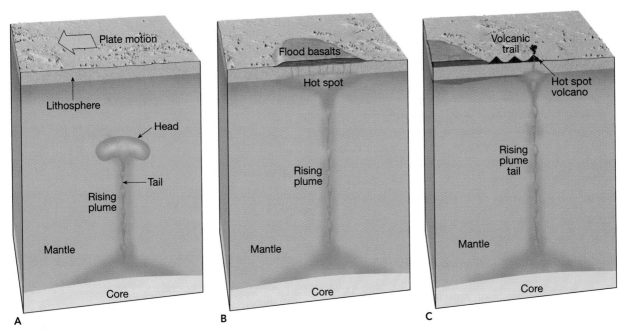

◆ **Figure 4.36** Model of a mantle plume and associated hot-spot volcanism. **A.** A rising mantle plume with large bulbous head and narrow tail. **B.** Rapid decompression melting of the head of a mantle plume produces vast outpourings of basalt. **C.** Less voluminous activity caused by the plume tail produces a linear volcanic chain on the seafloor.

may have heads that are hundreds of kilometres in diameter connected to a long, narrow tail rising from the core–mantle boundary (Figure 4.36). Upon reaching the base of the lithosphere, the temperature of the material in the plume is estimated to be 200–300°C warmer than the surrounding rock. Thus, as much as 10 to 20 percent of the mantle material making up the plume head rapidly melts. It is this melting that triggers the burst of volcanism that emits voluminous outpourings of lava to form a huge basalt plateau in a matter of a million or so years (Figure 4.36). Substantial evidence supports the idea that massive outpourings of lava associated with a superplume released large quantities of carbon dioxide into the atmosphere, which in turn may have significantly altered the climate of the Cretaceous Period (see Box 4.3). The comparatively short initial eruptive phase is followed by tens of millions of years of less voluminous activity, as the plume tail slowly rises to the surface. Thus, extending away from most large flood basalt plateaus is a chain of volcanic structures, similar to the Hawaiian chain, that terminates over an active hot spot marking the current position of the tail of the plume.

Can Volcanoes Change Earth's Climate?

One example of the interplay between different parts of the Earth system is the relationship between volcanic

activity and changes in climate. We know that changes in the composition of the atmosphere can have a significant impact on climate. Moreover, we know that volcanic eruptions can emit large quantities of gases and particles into the atmosphere, thus altering its composition (see Box 4.3). So do volcanic eruptions actually influence Earth's climate?

The idea that explosive eruptions might alter Earth's climate goes back many years. In fact, Benjamin Franklin speculated that the eruption of a large volcano in Iceland might have been responsible for the unusually cold winter of 1783–84. The basic premise is that suspended volcanic material reflects a portion of the incoming solar radiation, which in turn drops temperatures in the lowest layer of the atmosphere (this layer, called the *troposphere*, extends from Earth's surface to a height of about 10 kilometres).

Perhaps the most notable cool period linked to a volcanic event in the recent past is the "year without a summer" that followed the 1815 eruption of Indonesia's Mount Tambora. The cold temperatures of spring and summer were unprecedented and were followed by an early fall in many parts of the Northern Hemisphere, including eastern North America and Western Europe. The abnormal temperatures and shortened growing season led to crop failures, which in turn led to famine in some regions. The eruption of another Indonesian volcano, Krakatau, in 1883 was so severe that a caldera 300 metres deep is all that remains of the northern part of the volcanic island

BOX 4.3

Earth as a System
A Possible Link Between Volcanism and Climate Change in the Geologic Past

The Cretaceous Period is the last period of the Mesozoic Era, the era of *middle life* that is often called the "age of dinosaurs." It began about 144 million years ago and ended about 65 million years ago with the extinction of the dinosaurs (and many other life forms as well).

The Cretaceous climate was among the warmest in Earth's long history. Dinosaurs, which are associated with mild temperatures, ranged north of the Arctic Circle. Tropical forests existed in Greenland and Antarctica, and coral reefs grew as much as 15 degrees latitude closer to the poles than at present. Deposits of peat that would eventually form widespread coal beds accumulated at high latitudes. Sea level was as much as 200 metres higher than today, indicating that there were no polar ice sheets.

What was the cause of the unusually warm climates of the Cretaceous Period? Among the significant factors that may have contributed was an increase in the amount of carbon dioxide in the atmosphere. Carbon dioxide (CO_2) is a gas that is found naturally in the atmosphere. The importance of carbon dioxide lies in the fact that it is transparent to incoming short-wavelength solar radiation, but it is not transparent to some of the longer-wavelength, outgoing radiation emitted

---

*For more on the end of the Cretaceous, see Box 8.3, "The KT Extinction."

by Earth. A portion of the energy leaving the ground is absorbed by carbon dioxide and subsequently reemitted, part of it toward the surface, thereby keeping the air near the ground warmer than it would be without carbon dioxide. Thus, carbon dioxide is one of the gases responsible for warming the lower atmosphere. The process is called the *greenhouse effect*. Because carbon dioxide is an important heat absorber, any change in the air's carbon dioxide content could alter temperatures in the lower atmosphere.

Where did the additional CO_2 come from that contributed to the Cretaceous warming? Many geologists suggest that the probable source was volcanic activity. Carbon dioxide is one of the gases emitted during volcanism, and there is now considerable geologic evidence that the Middle Cretaceous was a time when there was an unusually high rate of volcanic activity. Several huge oceanic lava plateaus were produced on the floor of the western Pacific during this span. These vast features were associated with hot spots that may have been the product of very large mantle plumes that geologists call *superplumes* (see Figure 4.36). Massive outpourings of lava over millions of years would have been accompanied by the release of huge quantities of CO_2, which in turn would have enhanced the atmospheric greenhouse effect. *Thus, the warmth that*

characterized the Cretaceous may have had its origins deep in Earth's mantle.

There were other probable consequences of this extraordinarily warm period that are linked to volcanic activity. For example, the high global temperatures and enriched atmospheric CO_2 in the Cretaceous led to increases in the quantity and types of phytoplankton (tiny, mostly microscopic plants such as algae) and other life forms in the ocean. This expansion in marine life is reflected in the widespread chalk deposits associated with the Cretaceous Period (see Figure 6.11). Chalk consists of the calcite-rich hard parts of microscopic marine organisms. Oil and gas originate from the alteration of biological remains (chiefly phytoplankton). Some of the world's most important oil and gas fields occur in marine sediments of the Cretaceous age, a consequence of the greater abundance of marine life during this warm time.

This list of possible consequences linked to the extraordinary period of volcanism during the Cretaceous Period is far from complete, yet it serves to illustrate the interrelationships among parts of the Earth system. Materials and processes that at first might seem to be completely unrelated turn out to be linked. Here you have seen how processes originating deep in Earth's interior are connected directly or indirectly to the atmosphere, the oceans, and the biosphere.

---

that once stood 600 metres above sea level. It is estimated that ash ejected from the volcano remained in the atmosphere for up to 2 years.

When Mount St. Helens erupted in 1980, there was almost immediate speculation: Can an eruption such as this change our climate? There is no doubt that the large quantity of volcanic ash emitted by the explosive eruption had local and regional effects for a short period (Figure 4.37). Still, studies indicated that any longer-term lowering of hemispheric temperatures was negligible. The cooling was so slight, probably less than 0.1°C, that it could not be distinguished from other natural temperature fluctuations.

Two years of monitoring and studies following the 1982 eruption of Mexico's El Chichón volcano indicated that its cooling effect on global mean temperature was greater than that of Mount St. Helens on the order of 0.3 to 0.5°C. The eruption of El Chichón was less explosive than the Mount St. Helens blast, so why did it have a greater impact on global temperatures? The reason is that the material emitted by Mount St. Helens was largely fine ash that settled out in a relatively short time. El Chichón, on the other hand, emitted far greater quantities of sulphur dioxide gas (an estimated 40 times more) than Mount St. Helens. This gas combines with water vapour to produce a dense cloud of tiny sulphuric-acid particles in the stratosphere (the atmospheric layer between about 10 and 50 kilometres above the surface). The particles, called *aerosols*, may take up to several years to settle out completely. They lower the troposphere's mean temperature because they reflect solar radiation back to space.

The Philippines volcano, Mount Pinatubo, erupted explosively in June 1991, injecting 23 million to 27 million tonnes of sulphur dioxide into the stratosphere. The event provided scientists with an opportunity to study the climatic impact of a major explosive volcanic eruption using NASA's spaceborne Earth Radiation Budget Experiment. During the next year, the haze of tiny aerosols lowered global temperatures by 0.5°C.

We now understand that volcanic clouds that remain in the stratosphere for a year or more can be composed largely of sulphuric-acid droplets and not solely of volcanic ash, as was once thought. Thus, the volume of ash emitted during a typical explosive event is *not* necessarily an accurate criterion for predicting the global atmospheric effects of an eruption.

Even more difficult to assess than the climatic effects of recent volcanic eruptions are those of immense eruptions that took place in the more distant past. The possible effects of the "supereruption" that took place in Toba, Indonesia about 73,500 years ago are particularly intriguing. Evidence supporting this explosive event includes a caldera measuring approximately 100 kilometres in diameter and an ash layer with a thickness of 15 centimetres at distances over 1000 kilometres away from the caldera. Based on current estimates, a huge volume of volcanic ash, roughly equivalent to 2800 km^3 of solid rock, was injected up to 30 kilometres into the atmosphere. Computer models of atmospheric response to the eruption suggest that the dust-sized ash and aerosols were of sufficient volume to drive tropical temperatures to the freezing point for days to weeks and to produce abnormally cool temperatures for as long as a decade. Such temperatures would have been detrimental to any humans, and it has been speculated that humans may have even teetered on the edge of extinction. While more research is required to confirm the details of this scenario, it is apparent that the climatic effects of the Toba event were probably much more profound than for any of the volcanic eruptions witnessed in historic times.

◆ **Figure 4.37** When Mount St. Helens erupted on May 18, 1980, huge quantities of volcanic ash were blown into the atmosphere. This satellite image was taken less than eight hours after the eruption. The ash cloud had already spread as far as western Montana. Volcanic ash has little long-term impact on global climate because it settles quickly from the air. A more significant factor affecting climate is the quantity of sulphur dioxide gas emitted during an eruption. (Photo courtesy of the National Environmental Satellite Service)

Chapter Summary

- The primary factors that determine the nature of volcanic eruptions include the magma's *composition*, its *temperature*, and the *amount of dissolved gases* it contains. As lava cools, it begins to congeal and, as *viscosity* increases, its mobility decreases. The *viscosity of magma is directly related to its silica content*. Rhyolitic (felsic) lava, with its high silica content (over 70 percent), is very viscous and forms short, thick flows. Mafic lava, with a lower silica content (about 50 percent), is more fluid and may travel a long distance before congealing. Dissolved gases tend to increase the fluidity of magma and, as they expand, provide the force that propels molten rock from the vent of a volcano.

- The materials associated with a volcanic eruption include (1) *lava flows* (*pahoehoe* flows, which resemble twisted braids; and *aa* flows, consisting of rough, jagged blocks; both form from mafic lavas); (2) *gases* (primarily *water vapour*); and (3) *pyroclastic material* (pulverized rock and lava fragments blown from the volcano's vent, which include *ash, pumice, lapilli, cinders, blocks,* and *bombs.*)

- Successive eruptions of lava from a central vent result in a mountainous accumulation of material known as a *volcano*. Located at the summit of many volcanoes is a steep-walled depression called a *crater*. *Shield cones* are broad, slightly domed volcanoes built primarily of fluid, mafic lava. *Cinder cones* have steep slopes composed of pyroclastic material. *Composite cones*, or *stratovolcanoes*, are large, nearly symmetrical structures built of interbedded lavas and pyroclastic deposits. Composite cones produce some of the most violent volcanic activity. Often associated with a violent eruption is a *nuée ardente*, a fiery cloud of hot gases infused with incandescent ash that races down steep volcanic slopes. Large composite cones may also generate a type of mudflow known as a *lahar*.

- Most volcanoes are fed by *conduits* or *pipes*. As erosion progresses, the rock occupying the pipe is often more resistant and may remain standing above the surrounding terrain as a *volcanic neck*. The summits of some volcanoes have large, nearly circular depressions called *calderas* that result from collapse following an explosive eruption. Calderas also form on shield volcanoes by subterranean drainage from a central magma chamber, and the largest calderas form by the discharge of colossal volumes of silica-rich pumice along ring fractures. Although volcanic eruptions from a central vent are the most familiar, by far the largest amounts of volcanic material are extruded from cracks in the crust called *fissures*. The term *flood basalts* describes fluid, waterlike, mafic lava flows that cover extensive regions. When silica-rich magma is extruded, *pyroclastic flows* consisting largely of ash and pumice fragments usually result.

- Intrusive igneous bodies are classified according to their *shape* and by their *orientation with respect to the host rock*, generally sedimentary rock. The two general shapes are *tabular* (sheetlike) and *massive*. Intrusive igneous bodies that cut across existing sedimentary beds are said to be *discordant*; those that form parallel to existing sedimentary beds are *concordant*.

- *Dykes* are tabular, discordant igneous bodies produced when magma is injected into fractures that cut across rock layers. Tabular, concordant bodies, called *sills*, form when magma is injected along the bedding surfaces of sedimentary rocks. In many respects, sills closely resemble buried lava flows. *Laccoliths* are similar to sills but form from less fluid magma that collects as a lens-shaped mass that arches the overlying strata upward. *Batholiths*, the largest intrusive igneous bodies with surface exposures of more than 100 square kilometres, frequently make up the cores of mountains.

- *Most active volcanoes are associated with plate boundaries*. Active areas of volcanism are found along mid-ocean ridges where seafloor spreading is occurring (*divergent plate boundaries*), in the vicinity of ocean trenches where one plate is being subducted beneath another (*convergent plate boundaries*), and in the interiors of plates themselves (intraplate volcanism). Rising plumes of hot mantle rock are the source of most intraplate volcanism.

- Explosive volcanic eruptions are regarded as an explanation for some aspects of Earth's climatic variability. The basic premise is that suspended volcanic material will filter out a portion of the incoming solar radiation, which in turn will reduce air temperatures in the lower atmosphere.

Review Questions

1. What event triggered the May 18, 1980, eruption of Mount St. Helens? (See Box 4.1.)

2. List three factors that determine the nature of a volcanic eruption. What role does each play?

3. Why is a volcano fed by highly viscous magma likely to be a greater threat than a volcano supplied with very fluid magma?

4. Describe pahoehoe and aa lava.

5. List the main gases released during a volcanic eruption. Why are gases important in eruptions?

6. How do volcanic bombs differ from blocks of pyroclastic debris?

7. What is scoria? How is scoria different from pumice?

8. Compare a volcanic crater to a caldera.

9. Compare and contrast the three main types of volcanoes (size, composition, shape, and eruptive style).

10. Name a prominent volcano for each of the three types.

11. Explain why in Canada volcanoes are only found in British Columbia and Yukon Territory.

12. Contrast the destruction of Pompeii with the destruction of St. Pierre (time frame, volcanic material, and nature of destruction).

13. Describe the formation of Crater Lake. Compare it to the caldera found on shield volcanoes, such as Kilauea.

14. What are the largest volcanic structures on Earth?

15. What is Shiprock, New Mexico, and how did it form?

16. How do the eruptions that created the Columbia Plateau differ from eruptions that create volcanic peaks?

17. Where are fissure eruptions most common?

18. Extensive pyroclastic flow deposits are most often associated with which volcanic structures?

19. Describe each of the four intrusive features discussed in the text (dyke, sill, laccolith, and batholith).

20. Why might a laccolith be detected at Earth's surface before being exposed by erosion?

21. What is the largest of all intrusive igneous bodies? Is it tabular or massive? Concordant or discordant?

22. Describe how batholiths are emplaced.

23. Volcanism at divergent plate boundaries is associated with which rock type? What causes rocks to melt in these regions?

24. What is the Ring of Fire?

25. What type of plate boundary is associated with the Ring of Fire?

26. Are volcanoes in the Ring of Fire generally described as quiescent or violent? Name a volcano that would support your answer.

27. Describe the situation that generates magma along convergent plate boundaries.

28. What is the source of magma for intraplate volcanism?

29. What is meant by hot-spot volcanism?

30. How do geologists identify hot spots other than by the presence of volcanism?

31. The Hawaiian Islands and Yellowstone are associated with which of the three zones of volcanism? Cascade Range? Flood basalt plateaus?

32. What component released by a volcanic eruption is thought to have a short-term effect on climate? What component may have a long-term effect? (See Box 4.3.)

Key Terms

aa flow (p. 93)
batholith (p. 115)
caldera (p. 96)
cinder cone (p. 99)
columnar joints (p. 114)
composite cone (p. 101)
concordant (p. 111)
conduit (p. 96)
continental volcanic
    arc (p. 120)
crater (p. 96)
dyke (p. 113)
discordant (p. 111)

eruption column (p. 92)
fissure (p. 108)
fissure eruption (p. 108)
flood basalt (p. 108)
fumarole (p. 97)
hot spot (p. 121)
intraplate volcanism
    (p. 121)
island arc (p. 120)
laccolith (p. 115)
lahar (p. 105)
lava dome (p. 109)
lava tube (p. 93)

mantle plume (p. 121)
massive (p. 111)
nuée ardente (p. 104)
pahoehoe flow (p. 93)
parasitic cone (p. 96)
pillow lava (p. 93)
pipe (p. 96)
pumice (p. 95)
pyroclastic flow (p. 104)
pyroclastic material
    (p. 94)
scoria (p. 95)
scoria cone (p. 99)

shield volcano (p. 97)
sill (p. 114)
stock (p. 115)
stratovolcano (p. 101)
tabular (p. 111)
vent (p. 96)
viscosity (p. 90)
volcanic island arc (p.
    120)
volcanic neck (p. 110)
volcano (p. 96)
volatiles (p. 90)
xenolith (p. 116)

Web Resources

The *Earth* Web site uses the resources and flexibility of the Internet to aid in your study of the topics in this chapter. Written and developed by geology instructors, this site will help improve your understanding of geology. Visit **http://www.pearson.ca/tarbuck** and click on the cover of the text to find:

■ Online review quizzes.
■ Web-based critical thinking and writing exercises.
■ Links to chapter-specific Web resources.
■ Internet-wide key-term searches.

http://www.pearson.ca/tarbuck

Chapter 5

Weathering and Soil

The spectacular badland topography of southern Alberta is a testament to the processes of weathering and erosion.
(Photo by Mike Schering)

129

Earth's surface is constantly changing. Rock is disintegrated and decomposed, moved to lower elevations by gravity, and carried away by water, wind, or ice. In this manner, Earth's physical landscape is sculptured. This chapter focuses on the first step of this never-ending process—weathering. What causes solid rock to crumble, and why does the type and rate of weathering vary from place to place? Soil, an important product of the weathering process and a vital resource, is also examined.

Earth's External Processes

Weathering, mass wasting, and erosion are called **external processes** because they occur at or near Earth's surface and are powered by energy from the Sun. External processes are a basic part of the rock cycle because they are responsible for transforming solid rock into sediment.

To the casual observer, the face of Earth may appear to be unchanging. In fact, a mere 200 years ago most people fervently believed that mountains, lakes, and deserts were permanent features of a young Earth, no more than a few thousand years old. Today we know that Earth is at least 4.5 billion years old and that mountains eventually wear down, lakes fill with sediment or are drained by streams, and deserts come and go with changes in climate.

Earth is a dynamic body. Some parts of Earth's surface are gradually elevated by mountain building and volcanic activity. These **internal processes** derive their energy from Earth's interior. Meanwhile, opposing external processes are continually breaking rock apart and moving the debris to lower elevations (Figure 5.1). The latter processes include:

1. **Weathering**—the physical breakdown (disintegration) and chemical alteration (decomposition) of rock at or near Earth's surface.
2. **Mass wasting**—the transfer of rock and soil downslope under the influence of gravity.
3. **Erosion**—the physical removal of material by mobile agents such as water, wind, or ice.

In this chapter we will focus on rock weathering and the products generated by this activity. However, weathering cannot be easily separated from mass wasting and erosion because as weathering breaks rocks apart, erosion and mass wasting remove the rock debris. This transport of material by erosion and mass wasting promotes further disintegration and decomposition of the rock.

Weathering

Weathering goes on all around us, but it seems so slow and subtle that it is easy to underestimate its impor-

◆ **Figure 5.1**   This is an interesting example of the differential weathering of cross-bedded sandstone in Writing-on-Stone Provincial Park, Davis Coulee, Alberta. When weathering accentuates differences in rocks, spectacular landforms are sometimes created. As the rock gradually disintegrates and decomposes, mass wasting and erosion remove the products of weathering.
(Photo by Rudi Meyer)

tance. However, weathering is an essential part of the rock cycle and thus a key process in the Earth system.

All materials are subject to weathering. Consider, for example, the fabricated product concrete. A newly poured concrete sidewalk has a smooth, fresh look. However, a few years later the same sidewalk will appear chipped, cracked, and rough, with pebbles exposed at the surface. Roots of nearby trees may heave and buckle the concrete as well. The same natural processes that break apart a concrete sidewalk also break down rock.

Weathering occurs when rock is mechanically fragmented (disintegrated) and/or chemically altered (decomposed). **Mechanical weathering** is accomplished by physical forces that break rock into smaller and smaller pieces without changing the rock's mineral composition. **Chemical weathering** involves a chemical transformation of rock into one or more new compounds. These two concepts can be illustrated with a piece of paper. As an analogy, paper can be disintegrated by tearing it into smaller and smaller pieces, whereas decomposition occurs when the paper is burned.

Why does rock weather? Simply, weathering is the response of Earth materials to a changing environment. For instance, after millions of years of uplift and erosion, the rocks overlying a large, intrusive igneous body may be removed, exposing it at the surface. This mass of crystalline rock—formed deep below ground where temperatures and pressures are high—is now subjected to a very different and comparatively hostile surface environment. In response, this rock mass will gradually change. This transformation of rock is what we call weathering.

In the following sections, we will discuss the various modes of mechanical and chemical weathering. Although we consider these two processes separately, keep in mind that they usually work simultaneously in nature.

Mechanical Weathering

Earth Materials
↳ Sedimentary Rocks

When a rock undergoes *mechanical weathering*, it is broken into smaller and smaller pieces, each retaining the characteristics of the original material. The end result is many small pieces from a single large one. Figure 5.2 shows that breaking a rock into smaller pieces increases the surface area available for chemical attack. An analogous situation occurs when sugar is added to tea. In this situation, a cube of sugar will dissolve much slower than an equal volume of sugar granules because the cube has much less surface area available for dissolution. Hence, by breaking rocks into smaller pieces, mechanical weathering increases the amount of surface area available for chemical weathering.

In nature, four important physical processes lead to the fragmentation of rock: frost wedging, expansion resulting from unloading, thermal expansion,

and biological activity. In addition, although the work of erosional agents such as wind, glacial ice, and running water is usually considered separately from mechanical weathering, it is nevertheless important. As these mobile agents move rock debris, they relentlessly disintegrate these materials.

Frost Wedging

Repeated freezing and thawing are important processes of mechanical weathering. Liquid water has the unique property of expanding about 9 percent upon freezing, because water molecules in the regular crystalline structure of ice are farther apart than they are in liquid water near the freezing point. As a result, water freezing in a confined space exerts tremendous outward pressure on the walls of its container. To verify this, consider an unopened pop can placed in the freezer. As the water in the pop freezes, the container can burst.

In nature, water works its way into cracks in rock and, upon freezing, expands and enlarges these openings. After many freeze–thaw cycles, the rock is broken into angular fragments. This process is appropriately called **frost wedging** (Figure 5.3). Frost wedging is most pronounced in mountainous regions where a daily freeze–thaw cycle often exists. Here, sections of rock are wedged loose and may tumble into large piles called **talus slopes** that often form at the base of steep rock outcrops (Figure 5.3).

Frost wedging also causes great destruction to highways in Canada, particularly in the early spring when the freeze–thaw cycle is well established. Roadways acquire numerous potholes and are occasionally heaved and buckled by this destructive force.

Unloading

When large masses of igneous rock, particularly granite, are exposed by erosion, concentric slabs begin to break loose. The process generating these onion-like layers is called **sheeting**. It is thought that this occurs, at

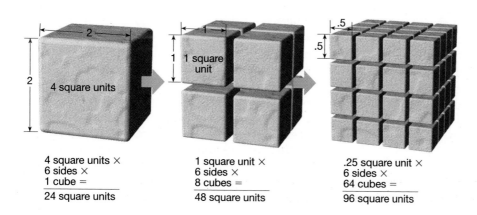

4 square units ×
6 sides ×
1 cube =
_____
24 square units

1 square unit ×
6 sides ×
8 cubes =
_____
48 square units

.25 square unit ×
6 sides ×
64 cubes =
_____
96 square units

◆ **Figure 5.2** Chemical weathering can occur only to those portions of a rock that are exposed to the elements. Mechanical weathering breaks rock into smaller and smaller pieces, thereby increasing the surface area available for chemical attack.

◆ **Figure 5.3** Frost wedging. As water freezes, it expands, exerting a force great enough to break rock. When frost wedging occurs in a setting such as this, the broken rock fragments fall to the base of the cliff and create a cone-shaped accumulation known as talus. Inset: This boulder, a glacial erratic in the polar desert conditions of the Northwest Territories, was probably split by frost wedging. (Main photo by Tom & Susan Bean, Inc.; Inset photo by C. Jefferson. Reproduced with the permission of the Minister of Public Works and Government Services Canada, 2004 and Courtesy of Natural Resources Canada, Geological Survey of Canada.)

least in part, because of the great reduction in pressure when the overlying rock is eroded away, a process called *unloading*. Accompanying this unloading, the outer layers expand more than the rock below and thus separate from the rock body (Figure 5.4). Continued weathering eventually causes the slabs to separate and spall off (flake away), creating **exfoliation domes** (*ex* = off, *folium* = leaf). Excellent examples of exfoliation domes are Stone Mountain, Georgia; and Half Dome and Liberty Cap in Yosemite National Park (Figure 5.5).

Deep underground mining provides us with another example of how rocks behave once the confining pressure is removed. Large rock slabs have been known to explode off the walls of newly cut mine tunnels because of the abruptly reduced pressure. Evidence of this type, plus the fact that fracturing occurs parallel to the floor of a quarry when large blocks of rock are removed, strongly supports the process of unloading as the cause of sheeting.

Although many fractures are created by expansion, others are produced by contraction during the crystallization of magma, and still others by tectonic forces during mountain building. Fractures produced by these activities generally form a definite pattern and are called *joints* (Figure 5.6). Joints allow water to penetrate deeply and start the process of weathering long before the rock is exposed.

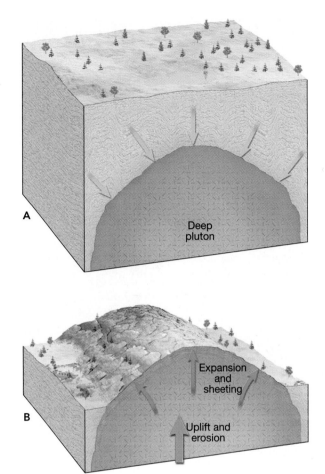

◆ **Figure 5.4**   Sheeting is caused by the expansion of crystalline rock as erosion removes the overlying material. When the deeply buried pluton in **A** is exposed at the surface following uplift and erosion in **B**, the igneous mass fractures into thin slabs. The photo of Olmstead Point in Yosemite National Park, California, illustrates the onion-like layers created by sheeting. (Jeff Gnass Photography/Corbis/Stock Market)

Thermal Expansion

The daily cycle of temperature may weaken rocks, particularly in hot deserts where daily variations may exceed 30°C. Heating a rock causes expansion, and cooling causes contraction. Repeated swelling and shrinking of minerals with different expansion rates should logically exert some stress on the rock's outer shell.

Although this process was once thought to be of major importance in the disintegration of rock, laboratory experiments have not substantiated this. In one test, unweathered rocks were heated to temperatures much higher than those normally experienced on Earth's surface and then cooled. This procedure was repeated many times to simulate hundreds of years of weathering, but the rocks showed little apparent change.

Nevertheless, pebbles in desert areas do show evidence of shattering, which may have been caused by temperature changes (Figure 5.7). A proposed solution to this dilemma suggests that rocks must first be weakened by chemical weathering before they can be broken down by thermal activity. Further, this process may be aided by the rapid cooling of a desert rainstorm. Additional data are needed before a definite conclusion can be reached as to the impact of temperature variation on rock disintegration.

Biological Activity

Weathering is also accomplished by the activities of organisms, including plants, burrowing animals, and humans. Plant roots in search of nutrients and water grow into fractures (Figure 5.8). Burrowing animals further break down rock by moving fresh material to the surface, where physical and chemical processes can more effectively attack it. Decaying organisms also produce acids that contribute to chemical weathering. Where rock has been blasted in search of minerals or for road construction, the impact of humans is particularly noticeable.

TABLE 5.2   **Generalized descriptions of Canadian soils according to the Canadian Soil Classification System**

Soil Order	Distinguishing Features of Soil Profile	Environments of Soil Formation
Regosolic	Very poor development of A and B horizons.	Areas where mass wasting or repeated influx of parent material prevents significant maturation of soil profile.
Chernozemic	Dark, brown to black-coloured (organic-rich) A horizon grading downward into B horizon.	Cool subarid to subhumid grasslands and forest-grassland transition zones.
Brunisolic	Brown-coloured B horizon.	Young forested areas affected by relatively dry climatic conditions and long winters.
Gleysolic	B horizon often grey or mottled with grey and red.	Areas affected by frequent flooding or poor drainage where iron reduction is prevalent.
Luvisolic	Strongly leached A horizon and a clay-rich B horizon.	Areas hosting deciduous or mixed forest with moderate to cool climate, and calcareous parent material.
Podzolic	Poorly decomposed organic layer, strongly leached A horizon, clay- and aluminum/iron-rich B horizon.	Cool, moist, temperate areas dominated by coniferous forest and strongly affected by leaching.
Solonetzic	High salt content at or near soil surface, iron-stained B horizons, salt-rich C horizon.	Semiarid climates where evaporation is extreme. Vegetation is dominated by grass or mixed grass-forest cover.
Organic	Very high organic content and low mineral content.	Cool, wet areas where organic matter accumulation is high.
Cryosolic	Layer of permafrost within one metre of the soil surface. Disruption of soil horizons by repeated freezing and thawing.	Tundra landscape dominated by moss and lichens.
Vertisolic	Clay-rich, with deep cracks and other disruptive features due to repeated shrinking and swelling.	Semi-arid regions of Canadian prairies underlain by parent material rich in swelling clays. Distribution highly localized.

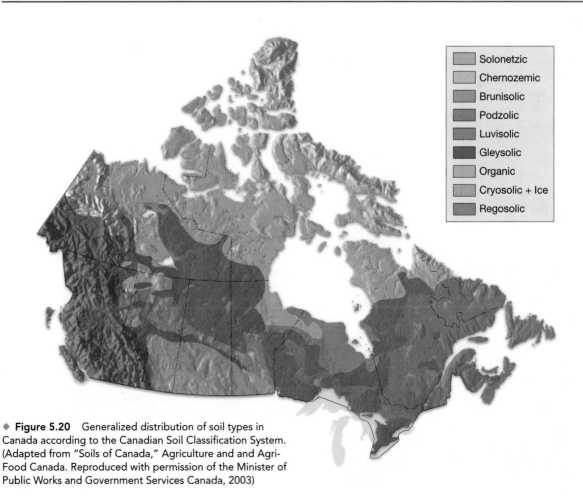

◆ **Figure 5.20** Generalized distribution of soil types in Canada according to the Canadian Soil Classification System. (Adapted from "Soils of Canada," Agriculture and and Agri-Food Canada. Reproduced with permission of the Minister of Public Works and Government Services Canada, 2003)

soils. Such features include traces and remains of plant roots, cracks produced by repeated wetting and drying, horizons showing the reduction or oxidation of iron, and beds of caliche. Specific combinations of such features preserved in any given paleosol can provide information on water content, drainage capacity, fertility, and other attributes of the land area in which the original soil formed. Paleosols have proven particularly useful in studies of glacial-interglacial cycles of the last ice age, serving as indicators of fluctuations in temperature and rates of precipitation. In order to be preserved in the geologic record, a soil must reside in an area where it can be buried. Thus, paleosols developed in lowland or coastal areas are most likely to be preserved (Figure 5.21).

Land plants were not present prior to the Paleozoic to contribute to the formation of soils. However, weathered, soil-like horizons containing oxidized iron have been noted in very old rocks extending as far back in time as the Proterozoic. The oldest of such horizons, estimated to be nearly 2 billion years old, are of great geologic significance because they indicate that oxygen levels had reached sufficient concentrations in the atmosphere by that time to allow iron to rust on land.

Soil Erosion

Soils are just a tiny fraction of all Earth materials, yet they are a vital resource. Because soils are necessary for the growth of rooted plants, they are the very foundation of the human life-support system. While human ingenuity can increase the agricultural productivity of soils through fertilization and irrigation, soils can also be damaged or destroyed by careless activities. Despite their basic role in providing food, fibre, and other basic materials, soils are among our most abused resources.

Perhaps this neglect and indifference has occurred because a substantial amount of soil seems to remain even where soil erosion is serious. Nevertheless, although the loss of fertile topsoil may not be obvious to the untrained eye, it is a growing problem as human activities expand and disturb more and more of Earth's surface.

How Soil Is Eroded

Soil erosion is part of the constant recycling of Earth materials that we call the *rock cycle*. Once soil forms, erosional forces, especially water and wind, move soil components from one place to another. Every time it rains, raindrops strike the land with surprising force. Each drop acts like a tiny bomb, blasting moveable soil particles out of their positions in the soil mass.

◆ **Figure 5.21**   A paleosol preserved in a Cretaceous coastal plain deposit of British Columbia. In this case, soil development occurred in mudstone with alternating organic-rich (dark-coloured) and organic-poor (light-coloured) layers. The drab colour of the paleosol indicates that iron in the soil was not oxidized, probably due to poor soil drainage on a coastal plain. The thin, dark-coloured lines are fossil plant roots that penetrate through the sedimentary layers of the parent material of the soil. The small increments of the scale bar represent centimetres. (Photo by Guy Plint)

Then, water flowing across the surface carries away the dislodged soil particles. Because the soil is moved by thin sheets of water, this process is termed *sheet erosion*.

After flowing as a thin, unconfined sheet for a relatively short distance, threads of current typically develop, and tiny channels called *rills* begin to form. When normal farm cultivation cannot eliminate the channels, we know the rills have grown large enough to be called *gullies* (Figure 5.22). Although most dislodged soil particles move only a short distance during each rainfall, substantial quantities eventually leave the fields and make their way downslope to a stream. Once in the stream channel, these soil particles, which can now be called *sediment*, are transported downstream and eventually deposited.

BOX 5.2

People and the Environment
Laterites and the Clearing of the Rainforest

Laterites are thick red soils that form in the wet tropics and subtropics where chemical weathering is extreme. Because lush tropical rainforests have laterite soils, we might assume the soils are fertile, and suitable for agriculture. Paradoxically, laterites are among the poorest soils for farming. How can this be?

Because laterites develop under rainforest conditions of high temperature and heavy rainfall, they are severely leached. Leaching destroys fertility because most plant nutrients are removed by the large volume of downward-percolating water. Therefore, even though the vegetation may be dense and luxuriant, the soil itself contains few available nutrients.

Most nutrients that support the rainforest are locked up in the trees themselves. As vegetation dies and decomposes, the roots of the rainforest trees quickly absorb the nutrients before they are leached from the soil. The nutrients are continuously recycled as trees die and decompose.

When rainforests are cleared to provide land for farming or to harvest the timber, most of the nutrients are removed as well (Figure 5.B). What remains is a soil that contains little to nourish planted crops.

The clearing of rainforests not only removes plant nutrients but also accelerates erosion. When vegetation is present, its roots anchor the soil, and its leaves and branches provide a canopy that protects the ground by deflecting the full force of the frequent heavy rains.

The removal of vegetation also exposes the ground to strong direct sunlight. When baked by the Sun, laterites harden to a bricklike consistency and become practically impenetrable to water and crop roots. In only a few years, lateritic soils in a freshly cleared area may be rendered useless for cultivation.

The term *laterite* is derived from the Latin word *latere*, meaning "brick," and was first applied to the use of this

◆ **Figure 5.B** Clearing the tropical rainforest in West Kalimantan (Borneo), Indonesia. The thick lateric soil is highly leached. (Photo by Wayne Lawler/Photo Researchers, Inc.)

◆ **Figure 5.C** This ancient temple at Angkor Wat, Cambodia, was built of bricks made of laterite. (Photo by R. Ian Lloyd/The Stock Market)

material for brick making in India and Cambodia. Labourers simply excavated the soil, shaped it, and allowed it to harden in the Sun. Ancient but still well-preserved structures built of laterite remain standing today in the wet tropics (Figure 5.C). Such structures have withstood centuries of weathering because all of the original soluble materials were already removed from the soil by chemical weathering. Laterites are therefore virtually insoluble and very stable.

◆ **Figure 5.22** Gully erosion in poorly protected soil, southern Colombia.
(Photo by Carl Purcell/Photo Researchers, Inc.)

Rates of Erosion

We know that soil erosion is the ultimate fate of practically all soils. In the past, erosion occurred at slower rates than it does today because more of the land surface was covered and protected by trees, shrubs, grasses, and other plants. However, human activities such as farming, logging, and construction, which remove or disrupt the natural vegetation, have greatly accelerated the rate of soil erosion. Without the stabilizing effect of plants, the soil is more easily swept away by the wind or carried downslope by sheet wash.

Natural rates of soil erosion vary greatly from one place to another and depend on soil characteristics as well as such factors as climate, slope, and type of vegetation. Over a broad area, erosion caused by surface runoff may be estimated by determining the sediment loads of streams that drain the region. When studies of this kind were made on a global scale, they indicated that prior to the appearance of humans, sediment transport by rivers to the ocean amounted to just over 9 billion tonnes per year. By contrast, the amount of material currently transported to the sea by rivers is about 24 billion tonnes per year, or more than two and a half times the earlier rate.

It is more difficult to measure the loss of soil due to wind erosion. Soil removal by wind is generally much less significant than by water erosion except during periods of prolonged drought. When dry conditions prevail, strong winds can remove large quantities of soil from unprotected fields. Such was the case in the 1930s in the prairie lands that came to be called the Dust Bowl (see Box 5.3).

In many regions the rate of soil erosion is significantly greater than the rate of soil formation. This means that a renewable resource has become non-renewable in these places. At present, it is estimated that topsoil is eroding faster than it forms on more than one-third of the world's croplands. The result is lower productivity, poorer crop quality, reduced agricultural income, and an uncertain future.

Sedimentation and Chemical Pollution

Another problem related to excessive soil erosion involves the deposition of sediment. Each year in North America hundreds of millions of tons of eroded soil are deposited in lakes, reservoirs, and streams. The detrimental impact of this process can be significant. For example, as more and more sediment is deposited in a reservoir, the capacity of the reservoir is diminished, limiting its usefulness for flood control, water supply, and/or hydroelectric power generation. In addition, sedimentation in streams and other waterways can restrict navigation and lead to costly dredging operations.

In some cases soil particles are contaminated with pesticides used in farming. When these chemicals are introduced into a lake or reservoir, the quality of the water supply is threatened and aquatic organisms may be endangered. In addition to pesticides, nutrients found naturally in soils as well as those added by agricultural fertilizers make their way into streams and lakes, where they stimulate the growth of plants. Over a period of time, excess nutrients accelerate the process by which plant growth, and subsequent decay, leads to the depletion of oxygen and death of the lake.

The availability of good soils is critical if the world's rapidly growing population is to be fed. On every continent unnecessary soil loss occurs as a result of a lack of conservation measures. Although it is a recognized fact that soil erosion can never be completely eliminated, soil conservation programs can substantially reduce its severity. Windbreaks (rows of trees), terracing, and plowing along the contours of hills are effective measures, as are special tillage practices and crop rotation. Innovative experiments on the use of waste material in soil remediation also hold great promise for the sustainability of soils (see Box 5.4).

BOX 5.3

People and the Environment
Return to the Dust Bowl?

During a span of dry years in the 1930s, large dust storms plagued the Great Plains of the U.S.A. Because of the size and severity of these storms, the region came to be called the Dust Bowl, and the time period, the Dirty Thirties. The heart of the dustbowl was nearly 40 million hectares in the panhandles of Texas and Oklahoma and adjacent parts of Colorado, New Mexico, and Kansas.

Meanwhile, in Canada, low rainfall rates in the early thirties caused severe shortages of surface water and groundwater. Farmers desperately planted seed in the dry fields only to see it blown away along with the topsoil. Many Canadian farmers fell deeply into debt and moved to cities to look for new jobs. Between the late 1920s and early 1930s, the average income of prairie families dropped over 90 percent.

What caused the Dust Bowl? Clearly, the fact that portions of the Great Plains experience some of North America's strongest winds is important. However, it was the expansion of agriculture that set the stage for the disastrous period of soil erosion. Thanks to mechanization, cultivation expanded nearly tenfold between the 1870s and 1930. When precipitation was adequate, the soil remained in place. However, the prolonged drought conditions of the 1930s rendered the unprotected fields prone to wind erosion. It was

◆ **Figure 5.D**   Dust storm near Edmonton Alberta in the summer of 2000. (Photo by L. Peleshok)

only in late 1930s when precipitation rates increased that soil erosion was again reduced.

Soil conservation practices have greatly improved since the Dust Bowl era. In particular, farmers minimize tilling and leave a layer of stubble on the ground to reduce soil erosion. Still, the severity of summer droughts in the past couple of decades serves as an eerie reminder of the Dust Bowl years (Figure 5.D). 2002 was reportedly the driest year in over a century for parts of Alberta and Saskatchewan, significantly reducing hay and grain productivity. The hay shortage forced many cattle ranchers to sell much of their herd. Forest fires, grasshopper epidemics, and dust storms were commonplace. Many fear that global warming is contributing to the severe conditions recently witnessed in the prairies, signalling a possible return to Dust Bowl conditions in the new millennium.

BOX 5.4

People and the Environment
"Waste" Materials As Possible Resources For Land Regeneration

Brian Hart*

"Land degradation" is a concept that reflects the needs of humans and is typically a result of misuse due to human-based activities (e.g., agricultural practices, overgrazing by livestock, and deforestation). In fact, there is really no such thing as "degraded" land. Even for land classified as "severely degraded" (beyond the point of regeneration—approximately 10 percent of the total degraded land worldwide), the only real constraints on regeneration are the amount and kind of financial inputs/sources that can be availed.

As population increases and soil conditions worsen, the perception of the amount of money that should be used to rejuvenate degraded soil will probably increase. This scenario is one of the major challenges facing the world today and new technologies are drastically needed to meet the challenge. In a country like India, the health of soils and their ability to produce vegetative matter is a key issue for quality of life, especially when 75 percent of the population is rural and the vast majority of daily living requirements come from the land.

Inputs for increasing soil productivity have historically been confined to chemical fertilizers and water management. However, recent work has shown that the use of fertilizers can have negative effects on soil pH, microorganisms, groundwater, and surface water. Alternative measures to enhance the biotic function of soil by using various types of organic and inorganic additives are becoming increasingly more appealing and, since many of these materials are considered "waste," they are also becoming more available.

Common by-products of human activities, such as composted sewage sludge (SS) and coal combustion ash (fly ash: FA) can be used as additives.

◆ **Figure 5.E**   Kandai plantation site, Chattisgarth, India: a view of the area before the plantation was established. The soil is a highly weathered laterite, rich in Fe and Al (hence the red colour), very low in organic material, low in pH, low in nutrient availability, and poor in productivity. Soil temperatures in the summer months reach 60° C. (Photo by Brian Hart)

Combined, they contain large amounts of organic matter along with nitrogen, phosphorus, potassium, calcium, magnesium, and numerous beneficial trace elements. They can also act as conditioners to improve soil properties such as aeration, bulk density, microbial activity, and water retention. This makes them favourable for use as amendments in poor quality soils.

The use of these materials does, however, have the potential to create environmental problems. Both materials can contain high concentrations of potentially toxic trace metals. These can accumulate in the soil, be mobilized into groundwater, taken up by plants, or transported to drinking water supplies, eventually entering the food chain. Therefore, with the potential for environmental damage, the importance of rigorous and detailed analyses of the raw materials (SS and FA) and continual

monitoring of the amended sites cannot be overemphasized.

What follows is a brief summary of an ongoing project, conducted by scientists at the University of Western Ontario, that seeks to demonstrate the technology of mixing FA with SS as a soil amendment to regenerate degraded lands in India. Amended sites were used to establish tree plantations and carry out edible crop trials. Growth rates, biomass production, and soil and vegetation analyses were conducted regularly over a 4-year period to assess the impact of the amendment.

Tree Plantations.   Approximately 400 hectares of wasteland was brought under mixed plantation with more than 100 different tree species in various agricultural zones of India (Figure 5.E).

For the majority of the species planted at different sites, plant growth has proven successful (Figure 5.F). The four-year-old standing tree volume of *Gmelina arborea* (gamhar) shows 1.6 times more wood yield than that of the traditional methods and is estimated to rise to around 2 times in a six-year time span. The average survival rate of plants at all the sites on amended lateritic soil is around 85 percent.

Soil and vegetation analyses reveal that there was no build-up of any heavy metal in the soil at any of the sites after four years of amendment application. The concentrations of most of the potentially toxic elements were within the normal ranges. No accumulation of any trace metal (beyond their respective threshold values) was found in vegetative tissues of plants on amended soils.

Edible Crop Trials. In addition to the 400 hectares of plantation, 3 hectares of agricultural trials were established to test the suitability of the technology on edible crops like peanuts. Applications improved soil quality, resulting in crop growth responses better than, or similar to, traditional fertilizer inputs. Metal loading in soils was not observed after amendment application and metal content in crops and crop residues did not exceed recommended guidelines. Based on the trials, crop responses vary according to amendment composition, soil types, climatic conditions, and individual crops. For example alkaline soils responded much better to high proportions of sludge in the mix; sludges tend to have a high content of organic acids and hence a low pH. However, for neutral or slightly acidic soils, high proportions of sludge in the mix reduces peanut production. Presumably, lower pH reduces the ability of soil microorganisms to promote the formation of root nodules and so nut production falls off.

For the most part, the results from the project were positive; the production of biomass in vegetation grown in mixtures of sludge and ash was equal to or greater than vegetation grown in traditionally fertilized soils. Furthermore, potential toxicity arising from metal accumulation after amendment was not observed. However, reduced nut production shows that the technology is not without its drawbacks. It is important to recognize that soils (degraded or not) are extremely variable and therefore amendments should be tailored for the soil to which they are applied.

---

*Brian Hart is a research scientist in the Department of Earth Sciences, The University of Western Ontario

◆ **Figure 5.F** The Kandai plantation 2.5 years after establishment. These trees (in treatments of soil amended with sludge and ash) are on average 1.5 x greater in girth than the trees planted with traditional amendments. Toxicity was not detected in the trees, nor was metal accumulation detected in the soil after amendment.
(Photo by Brian Hart)

Chapter Summary

- External processes include (1) *weathering*—the disintegration and decomposition of rock at or near Earth's surface; (2) *mass wasting*—the transfer of rock material downslope under the influence of gravity; and (3) *erosion*—the removal of material by a mobile agent, usually water, wind, or ice. They are called *external processes* because they occur at or near Earth's surface and are powered by energy from the Sun. By contrast, *internal processes*, such as volcanism and mountain building, derive their energy from Earth's interior.

- *Mechanical weathering* is the physical breaking up of rock into smaller pieces. Rocks can be broken into smaller fragments by *frost wedging* (where water works its way into cracks or voids in rock and, upon freezing, expands and enlarges the openings), *unloading* (expansion and breaking due to a great reduction in pressure when the overlying rock is eroded away), *thermal expansion* (weakening of rock as the result of expansion and contraction as it heats and cools), and *biological activity* (by humans, burrowing animals, plant roots, etc.).

- *Chemical weathering* alters a rock's chemistry, changing it into different substances. Water is by far the most important agent of chemical weathering. *Dissolution* occurs when water-soluble minerals such as halite become dissolved in water. Oxygen dissolved in water will *oxidize* iron-rich minerals. When carbon dioxide (CO_2) is dissolved in water, it forms *carbonic acid*, which accelerates the decomposition of silicate minerals by *hydrolysis*. The chemical weathering of silicate minerals frequently produces (1) soluble products containing sodium, calcium, potassium, and magnesium ions, and silica in solution; (2) insoluble iron oxides; and (3) clay minerals.

- The rate at which rock weathers depends on such factors as (1) *particle size*—small pieces generally weather faster than large pieces; (2) *mineral makeup*—calcite readily dissolves in mildly acidic solutions, and silicate minerals that form first from magma are least resistant to chemical weathering; and (3) *climatic factors*, particularly temperature and moisture. Frequently, rocks exposed at Earth's surface do not weather at the same rate. This *differential weathering* of rocks is influenced by such factors as mineral makeup and degree of jointing.

- *Soil* is a combination of mineral and organic matter, water, and air—the portion of the *regolith* (the layer of rock and mineral fragments produced by weathering) that supports the growth of plants. About half of the total volume of a good-quality soil is a mixture of disintegrated and decomposed rock (mineral matter) and *humus* (the decayed remains of animal and plant life); the remaining half consists of pore spaces, where air and water circulate. The most important factors that control soil formation are *parent material, time, climate, plants and animals*, and *slope*.

- Soil-forming processes operate from the surface downward and produce zones or layers in the soil that are called *horizons*. From the surface downward, the soil horizons are respectively designated as *O* (largely organic matter), *A* (largely mineral matter), *E* (where the fine soil components and soluble materials have been removed by *eluviation* and *leaching*), *B* (or *subsoil*, often referred to as the *zone of accumulation*), and *C* (partially altered parent material). Together the *O* and *A* horizons make up what is commonly called the *topsoil*.

- Although there are hundreds of soil types and subtypes worldwide, the three generic types are (1) *pedalfer*—characterized by an accumulation of iron oxides and aluminum-rich clays in the *B* horizon; (2) *pedocal*—characterized by an accumulation of calcium carbonate; and (3) *laterite*—deep soils that develop in the hot, wet tropics and are poor for growing crops because they are highly leached.

- Soil erosion is a natural process; it is part of the constant recycling of Earth materials that we call the rock cycle. Once in a stream channel, soil particles are transported downstream and eventually deposited. *Rates of soil erosion* vary from one place to another and depend on the soil's characteristics as well as such factors as climate, slope, and type of vegetation.

Review Questions

1. Describe the role of external processes in the rock cycle.

2. If two identical rocks were weathered, one mechanically and the other chemically, how would the products of weathering for the two rocks differ?

3. In what type of environment is frost wedging most effective?

4. Describe the processes of sheeting and spheroidal weathering. How are they different and how are they similar?

5. How does mechanical weathering add to the effectiveness of chemical weathering?

6. Granite and basalt are exposed at the surface in a hot, wet region.
   (a) Which type of weathering will predominate?
   (b) Which of these rocks will weather most rapidly? Why?

7. Heat speeds up a chemical reaction. Why then does chemical weathering proceed slowly in a hot desert?

8. How is carbonic acid (H_2CO_3) formed in nature? What results when this acid reacts with potassium feldspar?

9. List some possible environmental effects of acid precipitation (see Box 5.1).

10. What is the difference between soil and regolith?

11. What factors might cause different soils to develop from the same parent material, or similar soils to form from different parent materials?

12. Which of the controls of soil formation is most important? Explain.

13. How can slope affect the development of soil? What is meant by the term *slope orientation*?

14. List the characteristics associated with each of the horizons in a well-developed soil profile. Which of the horizons constitute the solum? Under what circumstances do soils lack horizons?

15. Distinguish between pedalfers and pedocals.

16. What soil type is associated with tropical rainforests? As this soil is associated with luxuriant natural vegetation, is it also excellent for growing crops? Briefly explain.

17. List three detrimental effects of soil erosion other than the loss of topsoil from croplands.

18. Briefly describe the conditions that led to the Dust Bowl of the 1930s (see Box 5.3).

Key Terms

caliche (p. 146)
chemical weathering (p. 130)
differential weathering (p. 141)
dissolution (p. 135)
eluviation (p. 145)
erosion (p. 130)
exfoliation dome (p. 132)

external process (p. 130)
frost wedging (p. 131)
hardpan (p. 146)
horizon (p. 144)
humus (p. 142)
hydrolysis (p. 136)
internal process (p. 130)
laterite (p. 146)
leaching (p. 145)

mass wasting (p. 130)
mechanical weathering (p. 130)
oxidation (p. 136)
parent material (p. 142)
pedalfer (p. 146)
pedocal (p. 146)
regolith (p. 142)

sheeting (p. 131)
soil (p. 142)
soil profile (p. 144)
solum (p. 146)
spheroidal weathering (p. 138)
talus slope (p. 131)
weathering (p. 130)

Web Resources

 The *Earth* Web site uses the resources and flexibility of the Internet to aid in your study of the topics in this chapter. Written and developed by geology instructors, this site will help improve your understanding of geology. Visit **http://www.pearson.ca/tarbuck** and click on the cover of *Earth* to find:

■ Online review quizzes.
■ Web-based critical thinking and writing exercises.
■ Links to chapter-specific Web resources.
■ Internet-wide key-term searches.

http://www.pearson. ca/tarbuck

Chapter 6
Sedimentary Rocks

Cambrian sedimentary rocks exposed below Popes Peak near Lake Louise, Alberta (Photo by Christopher Collom)

Chapter 5 gave you the background needed to understand the origin of sedimentary rocks. Recall that weathering of existing rocks begins the process. Next, erosional agents such as running water, wind, waves, and ice remove the products of weathering and carry them to a new location where they are deposited. Usually the particles are broken down further during the transport phase. Following deposition, this material, which is now called **sediment**, becomes lithified. In most cases, the sediment is lithified into solid sedimentary rock by the processes of *compaction* and *cementation*.

What Is a Sedimentary Rock?

 Earth Materials
    ↳ Sedimentary Rocks

The products of mechanical and chemical weathering constitute the raw materials for sedimentary rocks. The word *sedimentary* indicates the nature of these rocks, for it is derived from the Latin *sedimentum*, which means "to settle," a reference to solid material settling out of a fluid (water or air). Most, but not all, sediment is deposited in this fashion. Weathered debris is constantly being swept from bedrock, carried away, and eventually deposited in lakes, river valleys, seas, and countless other places. The particles in a desert sand dune, the mud on the floor of a swamp, the gravel in a streambed, and even household dust are examples of this never-ending process. Because the weathering of bedrock and the transport and deposition of the weathering products are continuous, sediment is found almost everywhere. As piles of sediment accumulate, the materials near the bottom are compacted. Over long periods, these sediments become cemented together by mineral matter deposited in the pores between particles, forming solid rock.

Geologists estimate that sedimentary rocks account for only about 5 percent (by volume) of Earth's outer 16 kilometres. However, the importance of this group of rocks is far greater than this percentage would imply. If we were to sample the rocks exposed at the surface, we would find that the great majority are sedimentary. Indeed, about 75 percent of all rock outcrops on the continents are sedimentary (Figure 6.1). Therefore, we may think of sedimentary rocks as comprising a relatively thin and somewhat discontinuous layer in the uppermost portion of the crust. This fact is readily understood when we consider that sediment accumulates at the surface.

Because sediments are deposited at Earth's surface, the rock layers that they eventually form contain evidence of past events that occurred at the surface. By their very nature, sedimentary rocks contain within them indications of past environments in which their particles were deposited and, in some cases, clues to the mechanisms involved in their transport. Furthermore, it is sedimentary rocks that contain fossils, which are vital to the relative dating and correlation of rock units. Thus, it is largely from this group of rocks that geologists must reconstruct the details of Earth history.

Finally, it should be mentioned that many sedimentary rocks are very important economically. Coal, which is burned to provide a significant portion of energy in North America, is classified as a sedimentary rock. Our other major energy sources, petroleum and natural gas, are associated with sedimentary rocks. So are major sources of iron, aluminum, manganese, and fertilizer, plus numerous materials essential to the construction industry.

Turning Sediment into Sedimentary Rock: Diagenesis and Lithification

A great deal of change can occur to sediment from the time it is deposited until it becomes a sedimentary rock and is subsequently subjected to the temperatures and pressures that convert it to metamorphic rock. The term **diagenesis** (*dia* = change; *genesis* = origin) is a collective term for all of the chemical, physical, and biological changes that take place after sediments are deposited and during and after lithification but prior to metamorphism.

Burial promotes diagenesis because as sediments are buried, they are subjected to increasingly higher temperatures and pressures. Diagenesis occurs within the upper few kilometres of Earth's crust at temperatures that are generally less than 150° to 200°C. Beyond this somewhat arbitrary threshold, metamorphism is said to occur.

One example of diagenetic change is *recrystallization*, the development of more stable minerals from less stable ones. It is illustrated by the mineral aragonite, the less stable form of calcium carbonate ($CaCO_3$). Aragonite is secreted by many marine organisms to form shells and other hard parts, such as the skeletal structures produced by modern corals. In some environments, large quantities of these solid materials accumulate as sediment. As burial takes place, aragonite recrystallizes to the more stable form of calcium carbonate, calcite (the main constituent in the sedimentary rock limestone).

Lithification comprises the processes by which unconsolidated sediments are transformed into solid sedimentary rocks (*lithos* = stone; *fic* = making). Basic lithification processes include compaction and cementation.

◆ **Figure 6.1**   These eroded sedimentary rocks near Cheltenham, Ontario are very colourful. Sedimentary rocks are exposed at the surface more than igneous and metamorphic rocks. Because they contain fossils and other clues about our geologic past, sedimentary rocks are important in the study of Earth history. (Photo by C. Tsujita)

As sediment accumulates, the weight of overlying material compresses the deeper sediments. This process is called **compaction**. The deeper a sediment is buried, the more it is compacted, and the firmer it becomes. As the grains are pressed closer and closer together, there is considerable reduction in pore space (the open space between particles). For example, when clays are buried beneath several thousand metres of material, the volume of clay may be reduced by as much as 40 percent. As pore space decreases, much of the water that was trapped in the sediments is driven out. Because sands and other coarse sediments are less compressible, compaction is most significant as a lithification process in fine-grained sedimentary rocks.

Cementation is the most important process by which sediments are converted to sedimentary rock. It is a chemical diagenetic change that involves the precipitation of minerals among the individual sediment grains. The cementing materials are carried in solution by water percolating through the open spaces between particles. Through time, the cement precipitates onto the sediment grains, fills the open spaces, and joins the particles. Just as the amount of pore space is reduced during compaction, the addition of cement into a sedimentary deposit reduces its porosity as well.

Calcite, silica, and iron oxide are the most common cements. It is often a relatively simple matter to identify the cementing material. Calcite cement will effervesce with dilute hydrochloric acid. Silica is the hardest cement and thus produces the hardest sedimentary rocks. An orange or dark red colour in a sedimentary rock means that iron oxide is present.

Most sedimentary rocks are lithified by means of compaction and cementation. However, some initially form as solid masses of intergrown crystals rather than beginning as accumulations of separate particles that later become solid. Other crystalline sedimentary rocks do not begin that way but are transformed into masses of interlocking crystals sometime after the sediment is deposited.

For example, with time and burial, loose sediment consisting of delicate calcareous skeletal debris may be recrystallized into a relatively dense crystalline limestone. Because crystals grow until they fill all the available space, pore spaces are frequently lacking in crystalline sedimentary rocks. Unless the rocks later develop joints and fractures, they will be relatively impermeable to fluids like water and oil.

Types of Sedimentary Rocks

 Earth Materials
      ↳ Sedimentary Rocks

Sediment has two principal sources. First, sediment may be an accumulation of material that originates and is transported as solid particles derived from both mechanical and chemical weathering of pre-existing rocks. Deposits of this type are termed *detrital*, and the sedimentary rocks that they form are called **detrital sedimentary rocks**. The second major source of sediment is soluble material produced largely by chemical weathering. When these dissolved substances are precipitated by either inorganic or organic processes, the material is known as chemical sediment and the rocks formed from it are called **chemical sedimentary rocks**.

We will now look at each type of sedimentary rock, and some examples of each.

Detrital Sedimentary Rocks

 Earth Materials
    ↳ Sedimentary Rocks

Though a wide variety of minerals and rock fragments may be found in detrital rocks, clay minerals and quartz are the chief constituents of most sedimentary rocks in this category. Recall from Chapter 5 that clay minerals are the most abundant product of the chemical weathering of silicate minerals, especially the feldspars. Clays are fine-grained minerals with sheetlike crystalline structures similar to the micas. The other common mineral, quartz, is abundant because it is extremely durable and very resistant to chemical weathering. Thus, when igneous rocks such as granite are attacked by weathering processes, individual quartz grains are freed.

Other common minerals in detrital rocks are feldspars and micas. Because chemical weathering rapidly transforms these minerals into new substances, their presence in sedimentary rocks indicates that erosion and deposition were fast enough to preserve some of the primary minerals from the source rock before they could be decomposed.

Particle size is the primary basis for distinguishing among various detrital sedimentary rocks. Table 6.1 presents the size categories for particles making up detrital rocks. Particle size is not only a convenient method of dividing detrital rocks; the sizes of the component grains also provide useful information about environments of deposition. Currents of water or air sort the particles by size; the stronger the cur-

Students Sometimes Ask...

According to Table 6.1, clay is a term used to indicate a microscopic particle size. I thought clays were a group of platy silicate minerals. Which one is right?
Both are. In the context of detrital particle size, the term *clay* refers only to those grains less than 1/256 millimetre, thus being microscopic in size. It does not indicate that these particles are of a particular composition. However, the term *clay* is also used to denote a specific composition: namely, a group of silicate minerals related to the micas. Although most of these clay minerals are of clay size, not all clay-size sediment consists of clay minerals!

rent, the larger the particle size carried. Gravels, for example, are moved by swiftly flowing rivers as well as by landslides and glaciers. Less energy is required to transport sand; thus, it is common to such features as windblown dunes and some river deposits and beaches. Very little energy is needed to transport clay, so it settles very slowly. Accumulation of these tiny particles is generally associated with the quiet water of a lake, lagoon, swamp, or certain marine environments.

Common detrital sedimentary rocks, in order of increasing particle size, are mudrocks, sandstone, and conglomerate or breccia. We will now look at each type and how it forms.

Shale and Other Mudrocks

Mudrocks, the group of sedimentary rocks that include shale, mudstone, and siltstone, consist of clay- to silt-sized particles (Figure 6.2). These fine-grained detrital rocks account for well over half of all sedimentary rocks. The particles in these rocks are so small that they cannot be readily identified without great magnification and for this reason make mudrocks more difficult to study and analyze than most other sedimentary rocks.

Much of what can be learned is based on particle size. The tiny grains in mudrocks indicate that deposition occurs as the result of gradual settling from relatively quiet, nonturbulent currents. Such environments include lakes, river floodplains, lagoons, and portions of the deep-ocean basins. Even in these "quiet" environments, there is usually enough turbulence to keep clay-size particles suspended almost indefinitely. Consequently, much of the clay is deposited only after the individual particles coalesce to form larger aggregates.

Sometimes the chemical composition of the rock provides additional information. One example is black shale, which is black because it contains abundant organic matter (carbon). When such a rock is found, it strongly implies that deposition occurred in

TABLE 6.1   Particle size classification for detrital rocks

Size Range (millimetres)	Particle Name	Common Sediment Name	Detrital Rock
>256	Boulder		
64–256	Cobble	Gravel	Conglomerate or breccia
4–64	Pebble		
2–4	Granule		
1/16–2	Sand	Sand	Sandstone
1/256–1/16	Silt	Mud	Mudrocks (Shale, mudstone, or siltstone)
<1/256	Clay		

10   20   30   40   50   60
Scale in millimetres

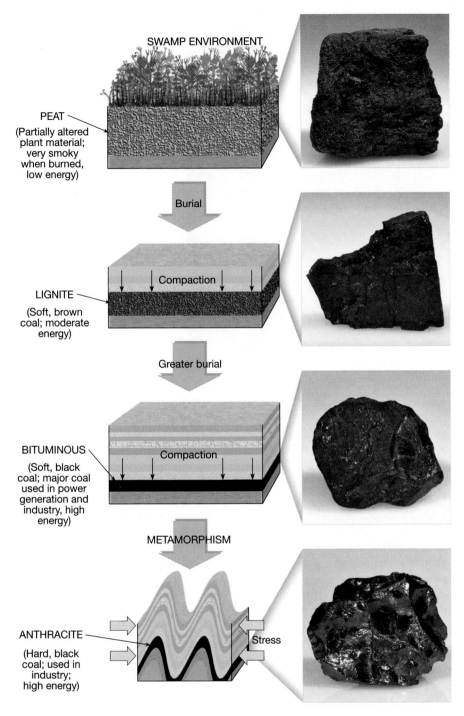

SWAMP ENVIRONMENT

PEAT
(Partially altered
plant material;
very smoky
when burned,
low energy)

Burial

Compaction

LIGNITE
(Soft, brown
coal; moderate
energy)

Greater burial

Compaction

BITUMINOUS
(Soft, black
coal; major coal
used in power
generation and
industry, high
energy)

METAMORPHISM

ANTHRACITE
(Hard, black
coal; used in
industry;
high energy)

Stress

◆ **Figure 6.14** Successive stages in the formation of coal. (Photos by E. J. Tarbuck)

to *lignite*, a soft brown coal. Lignite is actively mined in Saskatchewan, but also occurs in Alberta and Manitoba. Burial increases the temperature of sediments as well as the pressure on them.

The higher temperatures bring about chemical reactions within the plant materials and yield water and organic gases (volatiles). As the load increases from more sediment on top of the developing coal, the water and volatiles are pressed out and the pro-

portion of *fixed carbon* (the remaining solid combustible material) increases. The greater the carbon content, the greater the coal's energy ranking as a fuel. During burial, the coal also becomes increasingly compact. For example, deeper burial transforms lignite into a harder, more compacted black rock called *bituminous* coal. Compared to the peat from which it formed, a bed of bituminous coal may be only one-tenth as thick. Canada is a major pro-

ducer of bituminous coal, the largest deposits occurring in the Maritimes, Alberta, and British Columbia.

Lignite and bituminous coals are sedimentary rocks. However, when sedimentary layers are subjected to the folding and deformation associated with mountain building, the heat and pressure cause a further loss of volatiles and water, thus increasing the concentration of fixed carbon. This metamorphoses bituminous coal into *anthracite*, a very hard, shiny black *metamorphic* rock. Accordingly, it is no accident that the only Canadian occurrences of anthracite are in the mountainous regions of northern British Columbia and the Yukon. Although anthracite is a clean-burning fuel, only a relatively small amount is mined. Anthracite is not widespread and is more difficult and expensive to extract than the relatively flat-lying layers of bituminous coal.

Coal is a major energy resource. Its role as a fuel and some of the problems associated with burning coal are discussed in Chapter 21.

Classification of Sedimentary Rocks

 Earth Materials
↳ Rock Review

The classification scheme in Table 6.2 divides sedimentary rocks into two major groups: detrital and chemical. Further, we can see that the main criterion for subdividing the detrital rocks is particle size, whereas the primary basis for distinguishing among different rocks in the chemical group is their mineral composition.

As is the case with many (perhaps most) classifications of natural phenomena, the categories presented in Table 6.2 are more rigid than the actual state of nature. In reality, many of the sedimentary rocks classified into the chemical group also contain at least small quantities of detrital sediment. Many limestones, for example, contain varying amounts of mud or sand, giving them a "sandy" or "shaley" quality. Conversely, because practically all detrital rocks are cemented with material that was originally dissolved in water, they too are far from being "pure."

As was the case with the igneous rocks examined in Chapter 3, *texture* is a part of sedimentary rock classification. There are two major textures used in the classification of sedimentary rocks: clastic and nonclastic. The term **clastic** is taken from a Greek word meaning "broken." Rocks that display a clastic texture consist of discrete fragments and particles that are cemented and compacted together. Although cement is present in the spaces between particles, these openings are rarely filled completely. Table 6.2 shows that *all* detrital rocks have a clastic texture. The table also shows that some chemical sedimentary rocks exhibit this texture, too. For example, coquina, the limestone composed of shells and shell fragments, is obviously as clastic as a conglomerate or sandstone. The same applies for some varieties of oolitic limestone.

Some chemical sedimentary rocks have a **nonclastic** texture in which the minerals form a pattern of interlocking crystals. The crystals may be microscopically small or large enough to be visible without

TABLE 6.2   Classification of Sedimentary Rocks

DETRITAL ROCKS			
Texture	**Sediment Name and Particle Size**	**Comments**	**Rock Name**
Clastic	Gravel (>2 mm)	Rounded rock fragments Angular rock fragments	Conglomerate Breccia
Clastic	Sand (1/16–2 mm)	Quartz predominates Quartz with considerable feldspar Dark colour; quartz with considerable feldspar, clay, and rocky fragments	Quartz sandstone Arkose Greywacke
Clastic	Mud (<1/16 mm)	Splits into thin layers Breaks into clumps or blocks	Shale Mudstone

CHEMICAL ROCKS			
Group	**Texture**	**Composition**	**Rock Name**
Inorganic	Clastic or nonclastic Nonclastic Nonclastic Nonclastic Nonclastic	Calcite, $CaCo_3$ Dolomite, $CaMg(CO_3)_2$ Microcrystalline quartz, SiO_2 Halite, $NaCl$ Gypsum, $CaSO4 \cdot 2H_2O$	Limestone Dolostone Chert Rock salt Rock gypsum
Biochemical	Clastic or nonclastic Nonclastic Nonclastic	Calcite, $CaCO_3$ Microcrystalline quartz, SiO_2 Altered plant remains	Limestone Chert Coal

magnification. Common examples of rocks with non-clastic textures are those deposited when seawater evaporates (Figure 6.16). The materials that make up many other nonclastic rocks may actually have originated as detrital deposits. In these instances, the particles probably consisted of shell fragments and other hard parts rich in calcium carbonate or silica. The clastic nature of the grains was subsequently obliterated or obscured because the particles recrystallized when they were consolidated into limestone or chert.

Nonclastic rocks consist of intergrown crystals, and some may resemble igneous rocks, which are

A

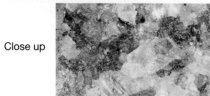

Close up

B

◆ **Figure 6.16   A.** Rock salt, the substance that made Goderich, Ontario famous, is said to have a nonclastic texture because it is composed of intergrown crystals. **B.** A similar crystalline texture is seen in this sample of potash from Saskatchewan.
(Photo A by E.J. Tarbuck; Photo B by C. Tsujita)

also crystalline. The two rock types are usually easy to distinguish because the minerals contained in nonclastic sedimentary rocks are quite unlike those found in most igneous rocks. For example, rock salt, gypsum, and some forms of limestone consist of intergrown crystals, but the minerals within these rocks (halite, gypsum, and calcite) are seldom associated with igneous rocks.

Sedimentary Environments

 Earth Materials
     ↳ Sedimentary Rocks

Sedimentary rocks are important in the interpretation of Earth history. By understanding the conditions under which sedimentary rocks form, geologists can often deduce the history of a rock, including information about the origin of its component particles, the method and length of sediment transport, and the nature of the place where the grains eventually came to rest; that is, the environment of deposition.

An **environment of deposition** or **sedimentary environment** is simply a geographic setting where sediment is accumulating. Each site is characterized by a particular combination of geologic processes and environmental conditions. Some sediments, such as the chemical sediments that precipitate in water bodies, are solely the product of their sedimentary environment. That is, their component minerals originated and were deposited in the same place. Other sediments originate far from the site where they accumulate. These materials are transported great distances from their source by some combination of gravity, water, wind, and ice.

At any given time the geographic setting and environmental conditions of a sedimentary environment determine the nature of the sediments that accumulate. Therefore, geologists carefully study the sediments in present-day depositional environments because the features they find can also be observed in ancient sedimentary rocks.

By applying a thorough knowledge of present-day conditions, geologists attempt to reconstruct the ancient environments and geographical relationships of an area at the time a particular set of sedimentary layers were deposited. Such analyses often lead to the creation of maps, which depict the geographic distribution of land and sea, mountains and river valleys, deserts and glaciers, and other environments of deposition.

Types of Sedimentary Environments

Sedimentary environments are commonly placed into one of three categories: continental, marine, or transitional (shoreline). Each category includes many specific subenvironments. Figure 6.17 is an idealized diagram illustrating a number of important sedimentary environments associated with each category. Realize that this is just a sampling of the great diversity of depositional environments. The remainder of this section provides a brief overview of each category. Later, Chapters 10 through 14, as well as portions of Chapter 18, will examine many of these environments in greater detail. Each is an area where sediment accumulates and where organisms live and die. Each produces a characteristic sedimentary rock or assemblage that reflects prevailing conditions.

Continental Environments.   Continental environments are dominated by the erosion and deposition associated with streams. In some cold regions, moving masses of glacial ice replace running water as the dominant process. In arid regions (as well as some coastal settings) wind takes on greater importance. Clearly, the nature of the sediments deposited in continental environments is strongly influenced by climate.

Streams are the dominant agent of landscape alteration, eroding more land and transporting and depositing more sediment than any other process. In addition to channel deposits, large quantities of sediment are dropped when floodwaters periodically inundate broad, flat valley floors (called *floodplains*). Where rapid streams emerge from a mountainous area onto a flatter surface, a distinctive cone-shaped accumulation of sediment known as an *alluvial fan* forms.

In frigid high-latitude or high-altitude settings, glaciers pick up and transport huge volumes of sediment. Materials deposited directly from ice are typically poorly sorted mixtures of particles that range in size from clay to boulders. Water from melting glaciers transports and redeposits some of the glacial sediment, creating stratified, sorted accumulations.

The work of wind and its resulting deposits are referred to as *eolian*, after Aeolus, the Greek god of wind. Unlike glacial deposits, eolian sediments are well sorted. Wind can lift fine dust high into the atmosphere and transport it great distances. Where winds are strong and the surface is not anchored by vegetation, sand is transported closer to the ground, where it accumulates in *dunes*. Deserts and coasts are common sites for this type of deposition.

In addition to being areas where dunes sometimes develop, desert basins are the sites where shallow *playa lakes* occasionally form following heavy rains or periods of snowmelt in adjacent mountains. They rapidly dry up, sometimes leaving behind evaporites and other characteristic deposits. In humid regions lakes are more enduring features, and their quiet waters are excellent sediment traps. Small deltas, beaches, and bars form along the lakeshore, with finer sediments coming to rest on the lake floor.

Marine Environments.   Marine depositional environments are divided according to depth. The *shallow marine* environment reaches to depths of about 200 metres and extends from the shore to the outer edge of the continental shelf. The *deep marine* environment lies seaward of the continental shelf in waters deeper than 200 metres.

The shallow marine environment borders all of the world's continents. Its width varies greatly, from practically nonexistent in some places to broad expanses extending as far as 1500 kilometres in other locations. On the average this zone is about 80 kilometres wide. The kind of sediment deposited here depends on several factors, including distance from shore, the elevation of the adjacent land area, water depth, water temperature, and climate.

Due to the ongoing erosion of the adjacent continent, the shallow marine environment receives huge quantities of land-derived sediment. Where the influx of such sediment is small and the seas are relatively warm, carbonate-rich muds may be the predominant sediment. Most of this material consists of the skeletal debris of carbonate-secreting organisms mixed with inorganic precipitates. Coral reefs are also associated with warm, shallow marine environments. In hot regions where the sea occupies a basin with restricted circulation, evaporation triggers the precipitation of soluble materials and the formation of marine evaporite deposits.

Deep marine environments include all the floors of the deep ocean. Far from landmasses, tiny particles from many sources remain adrift for long time spans. Gradually these small grains "rain" down on the ocean floor, where they accumulate very slowly as *biogenic ooze*. Significant exceptions are thick deposits of relatively coarse sediment that occur at the base of the continental slope. These materials are moved down from the continental shelf by turbidity currents—dense gravity-driven masses of sediment and water—and are deposited as turbidites.

Transitional Environments.   In transition zones characterized by low gradients and/or quiet water conditions, fine-grained sediment may be alternately deposited under shallow sheets of water and exposed to air, producing mud-dominated *tidal flats*. At the other extreme, where the shoreline is characterized by energetic wave and current action, sediments delivered to the sea by streams may be sorted and deposited to produce a sand or gravel *beach*. Further reworking and redistribution of sand by waves and currents along a shoreline can produce linear sand bodies such as *spits*, *bars*, and *barrier islands*. Finally, a sheltered, calm-water area called a *lagoon* may develop behind an offshore bar or reef, within which fine-grained sediments may be deposited.

Deltas are among the most significant deposits associated with transitional environments. The complex accumulations of sediment build outward into the sea when rivers experience an abrupt loss of velocity and drop their load of detritus.

Sedimentary Facies

When a series of sedimentary layers is studied, we can see the changes in environmental conditions that occurred at a particular place with the passage of time. Changes in past environments may also be seen when a single unit of sedimentary rock is traced laterally. This is true because at any one time many different depositional environments can exist over a broad area. For example, when sand is accumulating in a beach environment, finer muds are often being deposited in quieter offshore waters. Still farther out, perhaps in a zone where biological activity is high and land-derived sediments are scarce, the deposits consist largely of the calcareous remains of small organisms. In this example, different sediments are accumulating adjacent to one another at the same time. Each unit possesses a distinctive set of characteristics reflecting the conditions in a particular environment. To describe such sets of sediments, the term **sedimentary facies** is used. When a sedimentary unit is examined in cross-section from one end to the other, each facies grades laterally into another that formed at the same time but which exhibits different characteristics (Figure 6.18). Lateral shifting of facies through time produces vertical facies patterns that reflect the central relationship of one facies to another (see Box 6.2).

Sedimentary Structures

 Earth Materials
↳ Sedimentary Rocks

In addition to variations in grain size, mineral composition, and texture, sediments exhibit a variety of structures. Some, such as graded beds, are created when sediments are accumulating and are a reflection of the transporting medium. Others, such as *mud cracks*, form after the materials have been deposited and result from processes occurring in the environment. When present, sedimentary structures provide additional information that can be useful in the interpretation of Earth history.

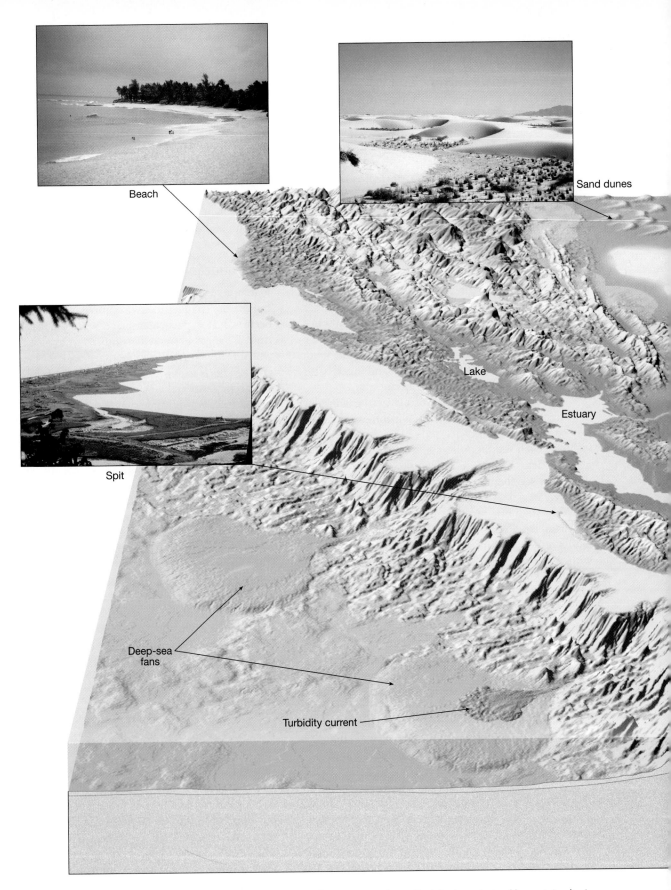

Labels within image: Beach, Sand dunes, Spit, Lake, Estuary, Deep-sea fans, Turbidity current

◆ **Figure 6.17** Sedimentary environments are those places where sediment accumulates. Each is characterized by certain physical, chemical, and biological conditions. Because each sediment contains clues about the environment in which it was deposited, sedimentary rocks are important in the interpretation of Earth history. A number of important continental, transitional, and marine sedimentary environments are represented in this idealized diagram.
(Photos by E. J. Tarbuck, except alluvial fan, by Martin G. Miller)

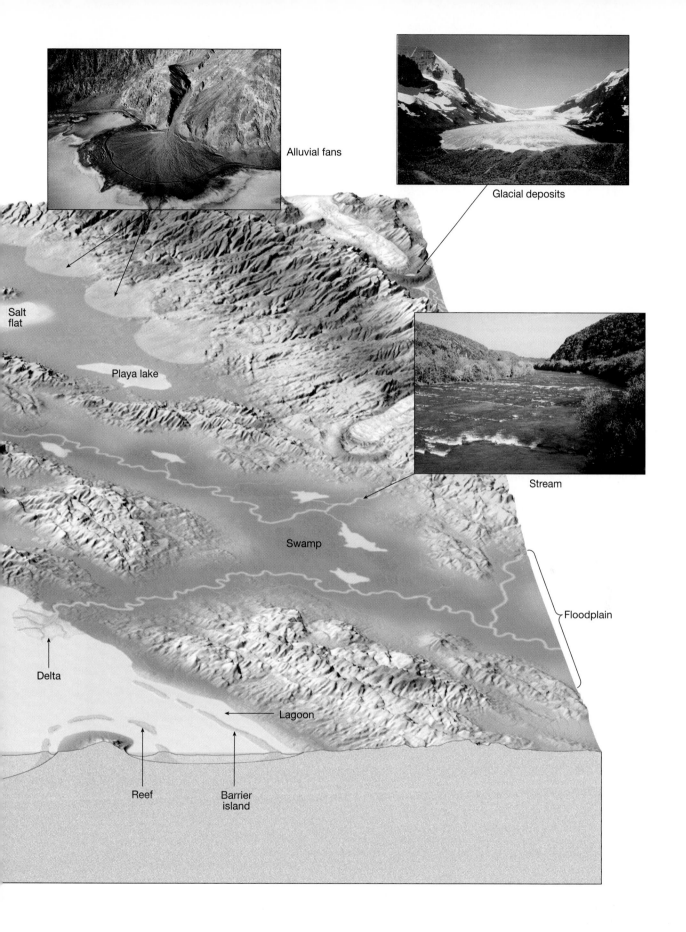

Alluvial fans

Glacial deposits

Salt flat

Playa lake

Stream

Swamp

Floodplain

Delta

Lagoon

Reef

Barrier island

177

BOX 6.2

Understanding Earth
A Closer Look At Facies

W. Glen E. Caldwell*

The derivation of the term facies is Latin and French and literally means "face, form, aspect, or condition." American Geological Institute defines a facies as "The aspect, appearance, and characteristics of a rock unit, usually reflecting the conditions of its origin, especially as differentiating the unit from adjacent or associated units." In geology, the term has been used to designate a wide variety of entities. Among the plethora of facies recognized in geology are sedimentary facies, biofacies, lithofacies, igneous facies, metamorphic facies, environmental facies, and mineral facies.

The use and misuse of the term facies in sedimentary geology calls for an historical account of how the term was originally applied to sedimentary rocks. Upon accepting the uniformitarian concepts of Hutton, geologists became increasingly aware of the role of environment in the accumulation of sediments. A significant step in looking beyond the simple "layer cake" model

of sediment deposition was made in 1838 by French geologist Constant Prevost. Prevost described the effect of environment on accumulating sediment and the sediment's contained fossils and pointed out that, within limits, similar rocks and fossil assemblages may indicate similarity in environmental setting rather than equivalence in age.

At about the same time, Swiss geologist Armanz Gressly, studying Mesozoic rocks of the Jura Mountains, found that he could not adequately describe the stratigraphy of the region solely in terms of vertical successions of rock units. He recognized significant lateral differences in rock characteristics within individual stratigraphic units. To designate the lateral changes in sedimentary rock type and fossil content of each unit, he coined the term facies.

While Gressly's definition of facies referred to lateral changes in rock units of approximately the same age, he indicated that rock of the same characteristics may appear in vertical succession without regard to stratigraphic

boundaries. Many geologists misinterpreted this to mean that facies could refer to vertical as well as horizontal differences in rock characteristics. Among these geologists was Johannes Walther, a German who, in 1893, introduced his Law of Correlation of Facies, known today as **Walther's Law**. This principle states that, within a given sedimentary cycle, the same succession of facies that occurs laterally is also present as components of the vertical succession.

Walther's Law remains a ruling principle in modern stratigraphic studies and is central to the explanation and prediction of the patterns of sedimentary deposits in time and space (Figure 6B). It should be kept in mind, however, that the use of facies in the context of Walther's Law is much different from Gressly's original definition.

―――――

*W. Glen E. Caldwell is a Professor Emeritus, Department of Earth Sciences, The University of Western Ontario

Gressly vs. Walther concepts of facies

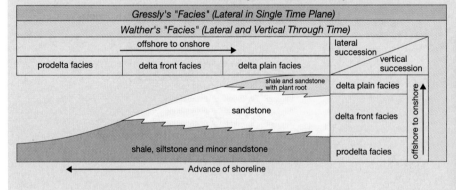

♦ **Figure 6.B**   Distribution of rocks representing depositional environments in space and time resulting from the advance of a shoreline. Gressly originally noted facies as defining lateral differences in depositional environment at a given time. Walther noted that differences in depositional environments could be tracked both laterally and vertically (through time). The outcrop on the right shows the vertical sequence of sedimentary facies (in the sense of Walther) in the Smoky River Valley of Alberta produced by the advance of a delta as depicted in the diagram on the left. (Photo by Guy Plint)

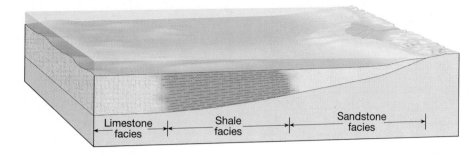

◆ **Figure 6.18** When a single sedimentary layer is traced laterally, we may find that it is made up of several different rock types. This can occur because many sedimentary environments can exist at the same time over a broad area. The term *facies* is used to describe such sets of sedimentary rocks. Each facies grades laterally into another that formed at the same time but in a different environment.

Sedimentary rocks form as layer upon layer of sediment accumulates in various depositional environments. These layers, called **strata**, or **beds**, are probably *the single most common and characteristic feature of sedimentary rocks*. Each stratum is unique. It may be a coarse sandstone, a fossil-rich limestone, a black shale, and so on. When you look at Figure 6.19 or look back through this chapter at the title page photo, Figure 6.3 (page 162), and Figure 6.12B (p. 169), you see many such layers, each different from the others. The variations in texture, composition, and thickness reflect the different conditions under which each layer was deposited.

The thickness of beds ranges from microscopically thin to tens of metres thick. Separating the strata are **bedding planes**, flat surfaces along which rocks tend to separate or break. Changes in the grain size or in the composition of the sediment being deposited can create bedding planes. Pauses in deposition can also lead to layering because chances are slight that newly deposited material will be exactly the same as

previously deposited sediment. Generally, each bedding plane marks the end of one episode of sedimentation and the beginning of another.

Because sediments usually accumulate as particles that settle from a fluid, most strata are originally deposited as horizontal layers. There are circumstances, however, when sediments do not accumulate in horizontal beds. Sometimes when a bed of sedimentary rock is examined, we see layers within it that are inclined to the horizontal. When this occurs, it is called **cross-bedding** and is most characteristic of sand dunes, river deltas, and certain stream channel deposits (Figure 6.20).

Graded beds represent another special type of bedding. In this case the particles within a single sedimentary layer gradually change from coarse at the bottom to fine at the top (Figure 6.21). Graded beds are most characteristic of rapid deposition from water containing sediment of varying sizes. When a current experiences a rapid energy loss, the largest particles settle first, followed by successively smaller grains. The

◆ **Figure 6.19** These Cretaceous strata exposed near Drumheller, Alberta, illustrate the characteristic layering of sedimentary rocks. The dark-coloured beds are composed of organic-rich mudstone whereas the light-coloured beds are composed of sandstone. (Photo by S. Hicock)

A

B

◆ **Figure 6.20**   **A.** The cut-away section of this sand dune shows cross-bedding. **B.** The cross-bedding of this sandstone indicates it was once a sand dune. (Photo A by John S. Shelton; Photo B by David Muench Photography, Inc.)

deposition of a graded bed is most often associated with a turbidity current, a mass of sediment-choked water that is denser than clear water and that moves downslope along the bottom of a lake or ocean.* Graded beds may also be produced when sediment particles thrown into suspension during a storm settle out after the storm passes.

As geologists examine sedimentary rocks, much can be deduced. A conglomerate, for example, may indicate a high-energy environment, such as a surf zone or rushing stream, where only coarse materials settle out and finer particles are kept suspended (Figure 6.22). If the rock is arkose, it may signify a dry climate where little chemical alteration of feldspar is possible. Carbonaceous shale is a sign of a low-energy, organic-rich environment, such as a swamp or lagoon.

Knowledge of how sedimentary structures form in modern sedimentary deposits provides additional clues to the environments represented by sedimentary rocks. For example, wet mud is observed to shrink upon exposure to air, forming **mud cracks** (Figure 6.23). Ripple marks are other such features. **Ripple marks** are small waves of sand that develop on the surface of a sediment layer by the action of moving water or air (Figure 6.24). The ridges form at right angles to the direction of motion. If the ripple marks were formed by air or water moving in essentially one direction, their form will be asymmetrical. These *current ripple marks* will have steeper sides in the downcurrent direction and more gradual slopes on the upcurrent side. Ripple marks produced by a stream flowing across a sandy channel or by wind blowing over a sand dune are two common examples of current ripples. When present in solid rock, they may be used to determine the direction of movement of ancient wind or water currents. Other ripple marks have a symmetrical form. These features, called *oscillation ripple marks*, result from the back-and-forth movement of surface waves in a shallow nearshore environment.

The significance of storms in the deposition of sediment in the sea has become apparent in the past couple of decades. Storms are capable of eroding, transporting, and depositing large volumes of sediment in a very short time, on the order of hours to days. This contrasts greatly with the small volumes of sediment that are generally observed during calm water periods. A common type of sedimentary structure found in sandy storm deposits is **hummocky cross stratification** (HCS), characterized by low-angle laminations that define wave-like undulations up to one metre long in cross-section (Figure 6.25A).

---

*More on these currents and graded beds may be found in the section on "Submarine Canyons and Turbidity Currents" in Chapter 18.

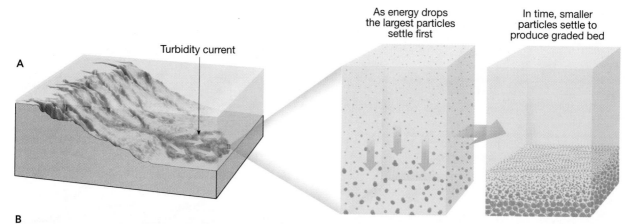

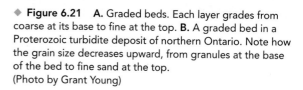

As energy drops the largest particles settle first

In time, smaller particles settle to produce graded bed

A

Turbidity current

◆ **Figure 6.21**   **A.** Graded beds. Each layer grades from coarse at its base to fine at the top. **B.** A graded bed in a Proterozoic turbidite deposit of northern Ontario. Note how the grain size decreases upward, from granules at the base of the bed to fine sand at the top.
(Photo by Grant Young)

B

This type of structure is believed to be formed by large storm waves in the shallow marine environment. Erosion and sorting of sediment on the seafloor by storms can also form concentrations of pebbles, shells, and other coarse-grained debris on areas of the seafloor that are otherwise dominated by fine-grained sediment (Figure 6.25B).

◆ **Figure 6.22**   In a turbulent stream channel, only large particles settle out. Finer sediments remain suspended and continue their downstream journey. This energetic river is in the Rocky Mountains of British Columbia.
(Photo by Mike Schering)

A

B

◆ **Figure 6.23   A.** Mud cracks developed on a modern riverbank. This example is from the Thames River, London, Ontario. **B.** Mud cracks preserved in mudstone. (Photo A courtesy of R.W. Hodder; Photo B by Gary Yeowell/Stone)

A

B

◆ **Figure 6.24   A.** Ripple marks in sand of a river channel **B.** Ripple marks preserved in Proterozoic sandstone. (Photo A by Galen Rowell / Mountain Light Photography Inc.; Photo B by Guy Plint)

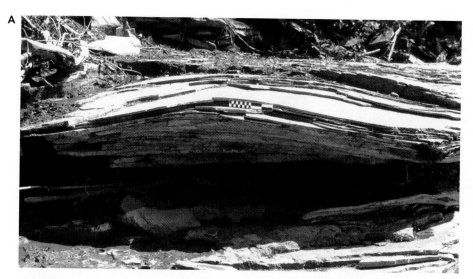

◆ **Figure 6.25** **A.** Hummocky cross stratification, as seen in this Cretaceous sandstone, is a sedimentary structure believed to reflect wave and current action generated by storms. Smallest increments of scale bar are in centimetres. **B.** Storms can also concentrate coarse-grained material such as shell debris while washing away much of the mud they were once buried in. Note that the lowermost shell-rich horizon in this photograph marks the base of a limestone bed containing hummocky cross stratification. (Photo A by Guy Plint; Photo B by Jisuo Jin)

Fossils: Evidence of Past Life

Fossils, the remains or traces of prehistoric life, are important inclusions in sediment and sedimentary rocks. Fossils can provide information on sedimentary environments, such as water salinity, temperature, turbidity, depth, and sediment softness/firmness that may not otherwise be obvious in the physical characteristics of the host rocks. Further, fossils are important time indicators and play a key role in correlating rocks of similar ages among different areas of the world. Fossils can be loosely classified into two categories, *body fossils* which preserve evidence of the tissues of an organism, and *trace fossils* which preserve evidence of an organism's activities.

Only a tiny fraction of the organisms that lived during the geologic past have left evidence of their tissues as body fossils. Normally, the remains of an animal or plant are totally destroyed. Under what circumstances are they preserved?

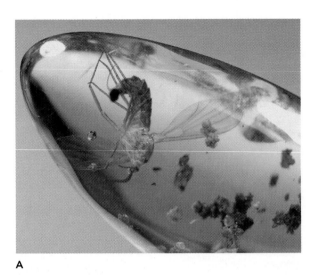

A

B

◆ **Figure 6.26   A.** The soft tissues of this insect are beautifully preserved in amber. **B.** Imprints of spindle- and frond-shaped, soft bodied Ediacaran organisms from Mistaken Point, Newfoundland.
(Photo A by Breck P. Kent; Photo B by Guy Narbonne)

The preservation of any organic remains requires their isolation from scavengers and microbes that consume and disturb dead tissue and from the physical and chemical agents that assist in the breakdown of mineralized tissues. This means that fossil preservation is more likely to occur when the remains are rapidly entombed, or better yet, when rapid entombment is accompanied by the suppression of scavenging and decomposition. Such conditions were apparently met in the soft-tissue preservation of organisms in amber (fossil resin; Figure 6.26A), tar, and permafrost. However, these cases are most exceptional. More commonly, the remains of dead organisms are buried in sediment where scavengers and microbes degrade the remains to some degree.

Very rarely, the combination of rapid burial of remains and severely low oxygen levels that slow down the decomposition process allow soft tissues to be preserved. These remarkable circumstances were responsible for the soft tissue preservation observed in the famous Cambrian strata of the Burgess Shale in Yoho National Park, British Columbia (see Box 6.3). It is believed that the Burgess Shale organisms were rapidly buried by sediment slumps at the foot of a large, submarine escarpment, a deep-water environment site with little or no oxygen to support decomposition. The Burgess Shale fossils are preserved as a result of both *carbonization*, the preservation of tissue as a carbon film upon the loss of volatile tissue components, and precipitation of clay or mica minerals on the tissue surfaces. The fossil assemblage of the Burgess Shale is extremely important because it provides a fairly faithful snapshot of the communities (dominated by soft-bodied organisms) that

existed shortly after the initial explosive diversification of life at the beginning of the Paleozoic Era.

Soft tissue can also be represented in the fossil record by *imprints* despite not being preserved itself (Figure 6.26B). This type of preservation is typical in Precambrian fossils of the oldest complex multicellular animals known (collectively known as Ediacaran fauna), all of which lacked hard skeletons. Were it not for this type of preservation, the existence of complex life prior to the Paleozoic would be known only from tracks and trails of a select few organisms.

Hard parts (e.g., shells, bones, and teeth) have a much higher preservation potential than soft tissue due to their physical and chemical durability. Therefore, these parts represent the vast majority of body fossils in the geologic record. If the hard parts are physically and chemically stable in their host sediments, they may be preserved unaltered from their original state. For example, shells composed of calcite or silica are commonly preserved without significant changes in their composition. However, hard parts composed of less stable materials (e.g., calcium carbonate in the form of aragonite) may recrystallize or be completely dissolved away. Even hard parts have varying degrees of preservational quality.

The preservation of hard parts can follow a number of pathways, depending on the processes they are subjected to both before and after burial. In some cases, fossilization merely involves the filling of pores in the hard parts by minerals that are precipitated from sediment porewaters. This simple addition of minerals to the original remains is called *petrifaction* and is a common mode of preservation observed in highly porous remains such as bone and wood (Figure 6.27).

◆ **Figure 6.27**   Petrified dinosaur bone from southern Alberta. Pores that were once occupied by blood vessels have been infilled with siderite (iron carbonate) that precipitated from the porewaters of the surrounding sediment. (Photo by Denis Tetreault)

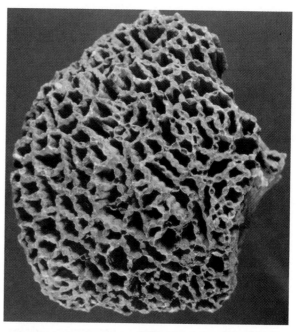

◆ **Figure 6.28**   Fossil coral replaced by silica (Manitoulin Island, Ontario). Note that the calcite matrix has dissolved away, leaving only the silicified (silica-replaced) fossil. (Photo by Denis Tetreault)

Alternatively, the original minerals comprising the hard parts may undergo *replacement*, a process that involves the substitution of original material by a different substance. The process of replacement is poorly understood, but appears to operate at a molecular level and can faithfully preserve even the most delicate structures. It is not uncommon to find originally calcareous fossil remains replaced by silica (Figure 6.28) or pyrite. Cases in which both hard and soft tissues have been replaced by phosphate are also known.

Body fossils can also be preserved by virtue of the moulding properties of their surrounding sediments (Figure 6.29). For example, a buried bivalved organism with its valves clamped shut will leave an *external mould* in the sediment surrounding it. If soft sediment is able to fill the empty space between the valves, it will form an *internal mould*, preserving internal features of the shell such as muscle scars. If the shell completely dissolves away, an internal mould, an external mould, and an empty space where the

A   B

◆ **Figure 6.29** **A.** Internal moulds of brachiopod shells, surrounded by empty spaces where the shells were dissolved. Internal moulds represent moulds of the interior surface of a fossilized object. Each of these particular shells contained a vertical plate that supported a feeding apparatus. This has been dissolved away to leave a v-shaped slit in each specimen. **B.** An external mould of a bivalve (clam) shell, showing external ribs and growth lines that adorned the outer surface of the shell. (Photos by Denis Tetreault)

original shell existed is observed. If the empty space is not filled and is later filled by minerals, a *cast*, or replica of the external form of the original shell will be produced within the external mould (Figure 6.30).

Trace fossils do not represent the actual remains of organisms, but rather, record their activities. Such traces include footprints (Figure 6.31A), burrows, borings, trails, fossil excrement (Figure 6.31B), and gastroliths ("gizzard stones" of dinosaurs). As in body fossils, the preservation of trace fossils in sediment requires isolation from agents of disturbance that would otherwise obscure the evidence left by the organism. Burrows are commonly preserved because they automatically reside within the sediment and are therefore protected from surface processes. The trace in question is either formed within the sediment or, if produced on the surface, is rapidly buried to avoid being obscured by potential agents of disturbance.

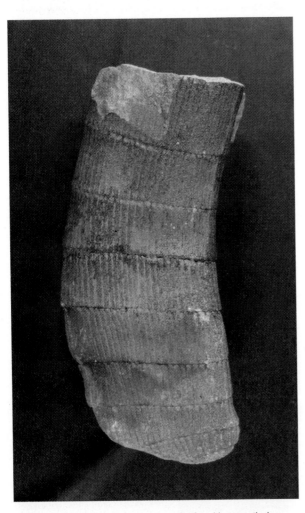

◆ **Figure 6.30**   Cast of the stem of a fossil horsetail plant from sandstone of Joggins, Nova Scotia. The space once occupied by wood was filled in with sediment and produced a replica of the original exterior surface of the stem. (Photo by Denis Tetreault)

◆ **Figure 6.31**   **A.** Footprint of an ancient amphibian from Nova Scotia. **B.** Coprolite (fossil excrement), from Cretaceous rocks of Saskatchewan, possibly produced by a small dinosaur. The coprolite is approximately 15 centimetres long. (Photo by Denis Tetreault)

BOX 6.3

Canadian Profile
The Burgess Shale: Yoho National Park, British Columbia

Matt Devereux*

The fossil record can be rich with the remains of durable skeletal materials such as shell, bone, teeth, and wood. Only in extremely rare circumstances are the softer tissues like skin, muscle, and gut ever preserved, and consequently, there are very few localities in the world where soft tissue remains are found. The best known of these is the Middle Cambrian (515-million-year-old) Burgess Shale of Yoho National Park in British Columbia.

A prominent American scientist by the name of Charles Doolittle Walcott discovered the Burgess Shale in 1909. In seven summer collecting seasons between 1909 and 1924, he managed to excavate over 60,000 specimens representing over 100 different animals. These fossils now reside at the Smithsonian Institution in Washington, D.C. More recent excavations, led by Desmond Collins of the Royal Ontario Museum, were just as successful, and many new species have been identified since Walcott's time.

Additional fossils have come from a few hundred cubic metres of shale excavated from a mountain ridge within walking distance of the town of Field, British Columbia. The shales were originally deposited as fine mud along the base of an ancient underwater cliff near the continental margin of ancient North America. This cliff has been traced over a distance of some 20 kilometres south of Walcott's quarry, along which other fossil localities have been identified. Though new and important discoveries have been made, none yet rival the original discovery of the Burgess Shale.

The incredibly well-preserved fossils give us a glimpse of life as it was shortly after the "big bang" of animal evolution known as the Cambrian Explosion. Some of animals were probably the ancestors of modern sea life, but a significant proportion are so strange that they continue to baffle scientists almost 100 years after their initial discovery. For example, one animal called *Opabinia* had five eyes on the top of its head and a long snout ending in a claw. Another animal, *Wiwaxia*, looks like a computer mouse armed with blade-like spines. Many animals reveal delicate appendages and others have even been found with their last meal still visible in their gut cavities (Figure 6.C). More remarkable still, components of some of the gut remains are still identifiable.

---

*Matt Devereux studied the Burgess Shale fauna as part of his M.Sc. thesis and is currently the Science Outreach Coordinator at The University of Western Ontario

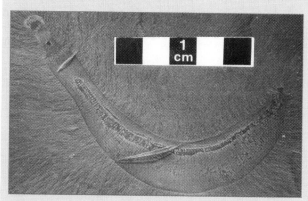

◆ **Figure 6.C**   An example of exceptional preservation of soft tissues in the Middle Cambrian Burgess Shale animal *Ottoia*. The animal's gut, containing partly digested food, is represented by the dark-coloured, U-shaped band that runs down the length of the animal's body. (Photo by Matt Devereux)

Chapter Summary

- *Sedimentary rock* consists of *sediment* that in most cases has been *lithified* into solid rock by the processes of *compaction* and *cementation*. Sediment has two principal sources: (1) as *detrital material*, which originates and is transported as solid particles from both mechanical and chemical weathering, which, when lithified, forms *detrital sedimentary rocks*; and (2) from soluble material produced largely by *chemical weathering*, which, when precipitated, forms *chemical sedimentary rocks*.

- *Diagenesis* refers to all of the physical, chemical, and biological changes that occur after sediments are deposited and during and after the time they are turned into sedimentary rock. Burial promotes diagenesis. Diagenesis includes lithification.

- *Lithification* refers to the processes by which unconsolidated sediments are transformed into solid sedimentary rock. Most sedimentary rocks are lithified by means of *compaction* and/or *cementation*. Compaction occurs when the weight of overlying materials compresses the deeper sediments. Cementation, the most important process by which sediments are converted to sedimentary rocks, occurs when soluble cementing materials, such as *calcite*, *silica*, and *iron oxide*, are precipitated onto sediment grains, fill open spaces, and join the particles. Although most sedimentary rocks are lithified by compaction or cementation, certain chemical rocks, such as the evaporites, initially form as solid masses of intergrown crystals.

- *Particle size* is the primary basis for distinguishing among various detrital sedimentary rocks. The size of the particles in a detrital rock indicates the energy of the medium that transported them. For example, gravels are moved by swiftly flowing rivers, whereas less energy is required to transport sand. Common detrital sedimentary rocks include *mudrocks* (silt- and clay-size particles), *sandstone*, and *conglomerate* (rounded gravel-size particles) or *breccia* (angular gravel-size particles).

- Precipitation of chemical sediments occurs in two ways: (1) by *inorganic processes*, such as evaporation and chemical activity; or by (2) *organic processes* of water-dwelling organisms that produce sediments of *biochemical origin*. *Limestone*, the most abundant chemical sedimentary rock, consists of the mineral calcite ($CaCO_3$) and forms either by inorganic means or as the result of biochemical processes. Inorganic limestones include *travertine*, which is commonly seen in caves, and *oolitic limestone*, consisting of small spherical grains of calcium carbonate. Other common chemical sedimentary rocks include *dolostone* (composed of the calcium-magnesium carbonate mineral dolomite), *chert* (made of microcrystalline quartz), *evaporites* (such as rock salt and rock gypsum), and *coal* (lignite and bituminous).

- Sedimentary rocks can be divided into two main groups: *detrital* and *chemical*. All detrital rocks have a *clastic texture*, which consists of discrete fragments and particles that are cemented and compacted together. The main criterion for subdividing the detrital rocks is particle size. Common detrital rocks include *conglomerate*, *sandstone*, and *mudrocks*. The primary basis for distinguishing among different rocks in the chemical group is their mineral composition. Some chemical rocks, such as those deposited when seawater evaporates, have a *nonclastic texture* in which the minerals form a pattern of interlocking crystals. However, in reality, many of the sedimentary rocks classified into the chemical group also contain at least small quantities of detrital sediment. Common chemical rocks include *limestone*, *rock gypsum*, and *coal* (e.g., lignite and bituminous).

- Sedimentary environments are those places where sediment accumulates. They are grouped into continental, marine, and transitional (shoreline) environments. Each is characterized by certain physical, chemical, and biological conditions. Because sediment contains clues about the environment in which it was deposited, sedimentary rocks are important in the interpretation of Earth's history.

- Sedimentary rocks are particularly important in interpreting Earth's history because, as layer upon layer of sediment accumulates, each records the nature of the environment at the time the sediment was deposited. These layers, called *strata*, or *beds*, are probably the single most characteristic feature of sedimentary rocks. Other features found in some sedimentary rocks, such as *ripple marks*, *mud cracks*, *cross-bedding*, and *fossils*, also provide clues to past environments.

Review Questions

1. How does the volume of sedimentary rocks in Earth's crust compare with the volume of igneous rocks in the crust? Are sedimentary rocks evenly distributed throughout the crust?

2. What is diagenesis? Give an example.

3. Compaction is most important as a lithification process with which sediment grain size?

4. List three common cements for sedimentary rocks. How might each be identified?

5. What minerals are most common in detrital sedimentary rocks? Why are these minerals so abundant?

6. What is the primary basis for distinguishing among various detrital sedimentary rocks?

7. Why does shale usually crumble easily?

8. How are the degree of sorting and the amount of rounding related to the transportation of sand grains?

9. Distinguish between conglomerate and breccia.

10. Distinguish between the two categories of chemical sedimentary rocks.

11. What are evaporite deposits? Name a rock that is an evaporite.

12. When a body of seawater evaporates, minerals precipitate in a certain order. What determines this order?

13. Each of the following statements describes one or more characteristics of a particular sedimentary rock. For each statement, name the sedimentary rock that is being described.
    (a) An evaporite used to make plaster.
    (b) A fine-grained detrital rock that exhibits *fissility*.
    (c) Dark-coloured sandstone containing angular rock particles as well as clay, quartz, and feldspar.
    (d) The most abundant chemical sedimentary rock.
    (e) A dark-coloured hard rock made of microcrystalline quartz.
    (f) A variety of limestone composed of small spherical grains.

14. How is coal different from other biochemical sedimentary rocks?

15. What is the primary basis for distinguishing among different chemical sedimentary rocks?

16. Distinguish between clastic and nonclastic textures. What type of texture is common to all detrital sedimentary rocks?

17. Some nonclastic sedimentary rocks closely resemble igneous rocks. How might the two be distinguished easily?

18. List three categories of sedimentary environments. Provide one or more examples for each category.

19. What is probably the single most characteristic feature of sedimentary rocks?

20. Distinguish between cross-bedding and graded bedding.

21. How do current ripple marks differ from oscillation ripple marks?

22. List two conditions that favour the preservation of organisms as fossils.

23. What type of fossilization is indicated by each of the following statements for body fossils? Which one is an example of a trace fossil?
    (a) A leaf preserved as a thin carbon film.
    (b) Small internal cavities and pores of a log filled with mineral matter.
    (c) Fossil dung.
    (d) This is created when a mould is filled with mineral matter.

Key Terms

bedding plane (p. 179)
beds (strata) (p. 179)
biochemical (p. 165)
cementation (p. 159)
chemical sedimentary rock (p. 159)
clastic (p. 172)
compaction (p. 159)

cross-bedding (p. 179)
detrital sedimentary rock (p. 159)
diagenesis (p. 158)
environment of deposition (p. 174)
evaporite deposit (p. 169)
fissility (p. 161)

fossil (p. 183)
graded bed (p. 179)
hummocky cross stratification (p. 180)
lithification (p. 158)
mud crack (p. 180)
nonclastic (p. 173)
ripple mark (p. 180)

salt flat (p. 170)
sediment (p. 158)
sedimentary environment (p. 174)
sedimentary facies (p. 175)
sorting (p. 161)
strata (beds) (p. 179)
Walther's Law (p. 178)

Web Resources

 The *Earth* Web site uses the resources and flexibility of the Internet to aid in your study of the topics in this chapter. Written and developed by geology instructors, this site will help improve your understanding of geology. Visit **http://www.pearson.ca/ tarbuck** and click on the cover of *Earth* to find:

■ Online review quizzes.
■ Web-based critical thinking and writing exercises.
■ Links to chapter-specific Web resources.
■ Internet-wide key-term searches.

http://www.pearson.ca/tarbuck

Metamorphism and Metamorphic Rocks

Deformed Precambrian metamorphic rocks north of Barrie, Ontario. (Photo by Mike Schering)

As we have seen in the previous chapters, igneous and sedimentary processes are relatively straightforward. Igneous rocks result from the intrusion or extrusion of magma generated within the Earth, whereas sedimentary rocks result from the deposition of particles derived from pre-existing rock or by the precipitation of minerals from a solution. Metamorphism, on the other hand, involves some very complex physical and chemical processes that can affect pre-existing rocks as the result of heating, changes in pressure, the flow of fluids deep below Earth's surface or a combination of these factors.

The conditions and time scales involved in metamorphism almost defy the imagination. To begin to appreciate the dynamics of metamorphism one need only consider the belt of metamorphic rocks that, in Canada stretches from the northern Georgian Bay area of Ontario, through the northern St. Lawrence Region of Quebec, and into Newfoundland and Labrador. Around one billion years ago, these rocks lay several kilometres below the surface and were subjected to temperatures in excess of 800°C. This is because these very rocks occupied the roots of a mountain chain that may have risen as high as the present-day Himalayas. Incredibly, the intense deformation (Figure 7.1) and radical mineral changes in these rocks occurred entirely while the rock was solid.

As metamorphism occurs under extreme conditions, and almost always deep below the Earth's surface, we cannot directly observe active metamorphic processes in nature. Because of this, our understanding of metamorphic processes is largely built

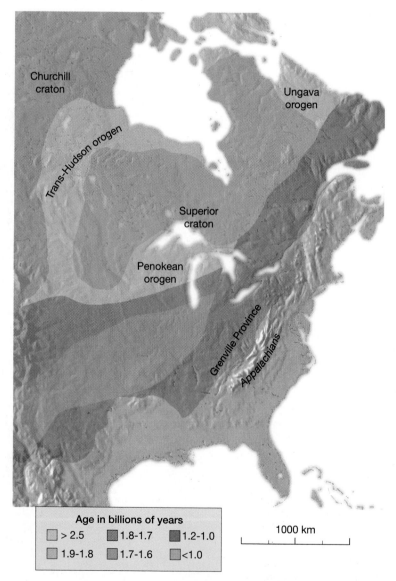

Churchill craton

Ungava orogen

Trans-Hudson orogen

Superior craton

Penokean orogen

Grenville Province

Appalachians

Age in billions of years

> 2.5   1.8-1.7   1.2-1.0
1.9-1.8   1.7-1.6   <1.0

1000 km

A

B

◆ **Figure 7.1   A.** Metamorphic rocks of the Grenville Province were formed in the roots of a Precambrian mountain chain that formed 1 billion years ago and may have risen as high as the Himalayas. Most of these metamorphic rocks are now covered by a thick pile of younger sedimentary rocks. **B.** Folded Precambrian metamorphic rocks of the Grenville Province at Parry Sound, Ontario.
(Photo by W.R. Church)

on observations of ancient metamorphic rocks and experiments conducted in special high temperature/high-pressure laboratories.

Metamorphism

The term **metamorphism**, meaning to "change form," collectively refers to the changes that take place *when a rock is subjected to temperatures and/or pressures markedly different from those in which it originally formed.* These changes happen in a solid state, although hot fluids migrating through the rock may play a key role in the process. Physical and chemical characteristics of the rock change in order to establish equilibrium with the new temperature/pressure conditions it encounters.

The type and degree of change in the transformation of a pre-existing **parent rock** to a metamorphic rock depends on the type and intensity of individual or combined metamorphic processes involved. On one hand, lightly metamorphosed slate may be difficult to distinguish from its shale parent rock. On the other hand, metamorphism can be so severe that all of the distinguishing characteristics of the parent rock, including its original mineral content, become unrecognizable (Figure 7.2).

Controlling Factors in Metamorphism

The overall chemical composition of a metamorphic rock largely reflects that of its parent rock, but the *texture* and *specific mineral makeup* of the rock depend on the particular physical processes involved in the metamorphic transformation. The main factors controlling metamorphism are heat, pressure, and chemically active fluids. In this section, we will examine the significance of these fundamental factors.

Composition of the Parent Rock

All metamorphic rocks ultimately originate from pre-existing parent rocks. Accordingly, the composition of a parent rock greatly determines the characteristics of the resultant metamorphic rock. Even though elements may be redistributed between minerals, and some elements or volatiles such as water may leak in or out during metamorphism, the chemical composition of a metamorphic rock is usually inherited largely from its parent. For example, the metamorphism of a mafic igneous rock will result in the formation of a metamorphic rock that is relatively rich in iron- and magnesium-bearing minerals, whereas the metamorphism of a felsic igneous rock will result in a metamorphic rock that is depleted in such minerals but rich in feldspar and quartz. Likewise, clay minerals contained

in shale can recrystallize to form micas, and the calcite in marble is obviously derived from a limestone parent. It is only in extreme cases of element loss or gain due to fluid migration that a metamorphic rock becomes substantially different in composition from its parent rock.

Heat as a Metamorphic Agent

Heat is one of the most important agents of metamorphism. Not only does heat drive the basic chemical reactions that transform minerals into new forms, but it also influences the mobility and reactivity of chemically active fluids. The heating of parent rocks can occur in two ways. First, heat given off by an intrusive igneous body can locally bake the rocks surrounding it. Second, rocks can be cooked by Earth's internal heat if they are brought to great depths below the surface.

◆ **Figure 7.2** Deformed metamorphic rocks exposed in southern Greenland.
(Photo by Carl Ozyer)

When considering the effects of temperature on rocks, it is important to note that even though minerals are solid and might appear static to us, the individual atoms contained in a mineral can migrate between different sites within its crystal structure. With an increase in temperature, atoms become more energetic and move around more freely within the crystal structures of rock-forming minerals. As a result, the atoms can achieve configurations that are stable under higher temperature conditions. Changes in atomic configuration can be achieved by simple recrystallization or the formation of new minerals. If a metamorphic environment were a truly closed system, isolated from any possible interactions with fluids, the overall chemical composition of a metamorphic rock would remain identical to that of its parent rock. In such a case, no atoms would be added or lost, but rather, merely redistributed within the rock. In nature, however, many metamorphic rocks have chemical compositions that are probably similar, but not identical, to their parent rocks. This is because some addition or subtraction of components can occur through small leaks in the system.

It should also be kept in mind that different minerals are stable at different temperatures and therefore do not all change at the same time. For example, at temperatures between about 150° and 200°C, clay minerals become unstable and start changing into micas. On the other hand, minerals such as quartz remain stable at these temperatures and can remain so in environments with much higher temperatures.

Pressure (Stress) as a Metamorphic Agent

Whereas the primary role of heat in metamorphism is to change the distribution of chemical components within a rock, pressure is primarily responsible for changing the physical characteristics of the rock. Pressure does, however, also influence the temperature at which certain minerals are stable and therefore plays a role in dictating the distribution of elements within a metamorphic rock. When discussing the deformation of rocks, the term "stress" is sometimes substituted for "pressure."

Confining Pressure (Uniform Stress) **Confining pressure**, also called **uniform stress**, involves the squeezing of a rock in all directions, "confining" the material into the smallest possible volume with no change in shape. Water pressure is a familiar form of confining pressure that increases with water depth under the weight of the overlying water mass. When you dive into deep water, you can feel the increasing water pressure in your ears, but your body is certainly not pressed flat. This is because the pressure is not only transmitted through the water from above,

but is applied from all directions. In the same way, a rock below the Earth's surface experiences uniform pressure from all directions under the weight of the overlying rock column (Figure 7.3A). Accordingly, confining pressure increases with increasing depth of burial. At great depths, confining pressure can force mineral grains to crowd together and crystal arrangements to become more compact. Because no internal deformation is involved, these simple changes form metamorphic rocks that are harder and denser, and contain new minerals, but are otherwise similar in appearance to their parent rocks (Figure 7.4A).

Directed Pressure (Differential Stress) **Directed pressure**, also called **differential stress**, refers to the unequal application of pressure, resulting in the distortion of a body (Figure 7.3B). Within the Earth, directed pressure is largely associated with the horizontal shortening and stretching of the Earth's crust associated with plate tectonic activity.

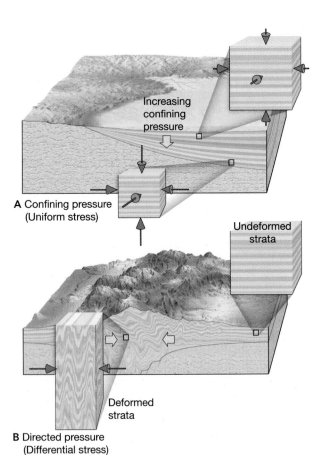

◆ **Figure 7.3** The effects of pressure on the texture of metamorphic rocks. **A.** If solely affected by confining pressure (uniform stress), metamorphic rocks retain features of their parent rock such as bedding. **B.** Directed pressure (differential stress) causes rocks to crumple in the plane of least stress.

The Upper Limit of Regional Metamorphism

In the most extreme environments in regional metamorphism, rocks begin to partially melt, and pass into a transition zone between metamorphism and the igneous realm. In such environments, part of the rock remains solid and can be considered as part of the metamorphic process, but part of it is melted to form magma. Rocks called **migmatites** represent this somewhat ambiguous state and include both metamorphic and igneous components. They are generally included in metamorphic rocks because they represent the very last possible step a metamorphic rock can take in its passage through conditions of progressively higher temperature and pressure. Recall from our discussion of igneous rocks that different minerals melt at different temperatures. Light-coloured silicates such as quartz and potassium feldspar have the lowest melting temperatures and begin to melt first, whereas the mafic silicates, such as amphibole and biotite, remain solid. When the partially melted rock cools, the light bands have an igneous appearance, whereas the dark bands consist of solid metamorphic components (Figure 7.20). The igneous bands often form tortuous folds and sometimes contain angular fragments of the dark components.

Regional Metamorphism and Nonfoliated Rocks

The fact that many rocks produced by regional metamorphism are foliated does not mean that all of them show this texture. This is due to the fact that some parent rocks lack the necessary elements to produce the platy or elongate mineral grains that define foliation in metamorphic rocks. Minerals such as quartz

◆ **Figure 7.20** Migmatite. The lightest-coloured layers are igneous rock composed of quartz and feldspar, while the darker layers have a metamorphic origin.
(Photo by Harlan H. Roepke)

and calcite occur in various shapes and therefore do not as readily define a foliation as minerals such as micas do. As in contact metamorphism, regional metamorphism transforms quartz sandstone to quartzite and limestone to marble. As a result of the lack of platy and elongate minerals, pure quartzites (Figure 7.14A) and marbles (Figure 7.14B) produced by regional metamorphism may look identical to those produced by contact metamorphism, at least to the unaided eye. However, in some cases impurities, such as clay in the host rocks, may produce some faint banding in regionally metamorphosed rocks that are otherwise homogeneous in appearance.

Subduction Zone Metamorphism

Subduction zone metamorphism is technically a form of regional metamorphism, but is discussed here in a separate category because of the special conditions involved. Metamorphism in the region of a subducted slab is characterized by very high pressures but relatively low temperatures (Figure 7.9). Rocks lying on both sides of the plate boundary are affected. Because a subducted oceanic slab is cold and is a poor heat conductor, rocks on the top of the subducted slab and on the underside of the overlying plate encounter a large increase in pressure at a rapid rate but remain relatively cool. Under these conditions, a peculiar, sodium-rich, blue-coloured amphibole called glaucophane forms. Due to the preferred orientation of elongate glaucophane grains, the rock exhibits foliation and is called blueschist (Figure 7.21A). Continued subduction drags rocks into the mantle, where the blueschist is further transformed into a rock called eclogite, characterized by pyroxene and garnet (Figure 7.21B).

Hydrothermal Metamorphism

In some cases, metamorphism primarily involves the interaction of minerals within a rock with hydrothermal (*hydra* = water, *therm* = heat) fluids. This form of metamorphism, appropriately called **hydrothermal metamorphism,** is often closely associated with igneous activity, since that is what provides the heat required to circulate chemically active fluids. Thus, hydrothermal metamorphism often occurs simultaneously with contact metamorphism in regions where large plutons are emplaced.

Hydrothermal metamorphism occurs on a much larger scale at spreading ridges, where hot, mineral-rich fluids circulate through cracks in newly formed oceanic crust and cause the alteration of minerals in basalt. As in cases where hydrothermal metamorphism is associated with contact metamorphism, the

A Blueschist

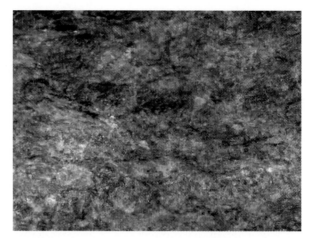

B Eclogite

◆ **Figure 7.21**   Rocks produced by subduction zone metamorphism. **A.** Blueschist, representing low temperature, high pressure conditions (note blue-coloured amphibole called glaucophane). **B.** Eclogite, representing high temperature, high-pressure conditions of the mantle. Note pink grains of garnet and green grains of pyroxene. (Photos by C. Tsujita)

heating of rock is made possible by rising magma. The primary difference, however, is that this special form of hydrothermal metamorphism, called **seafloor metamorphism**, involves the interaction of rocks with circulated seawater rather than with chemically active fluids derived directly from the magma.

When hot seawater percolates through the young, hot oceanic crust, it chemically reacts with the anhydrous minerals of basalt. At relatively shallow depths, and hence relatively low temperatures, ferromagnesian minerals such as olivine and pyroxene are converted into hydrated silicates like serpentine and talc. At greater depths (and higher temperatures), index minerals similar to those found in regionally metamorphosed rocks are associated with the gabbro. Unlike regionally metamorphosed rocks, however, these rocks are not foliated because they are not subjected to high differential pressures.

At the same time, precipitation of metallic sulphide minerals in chimneys on the seafloor (black smokers) and in veins just below the seafloor can produce large deposits of valuable minerals. These are discussed further in Chapter 21.

Metamorphic Facies and Plate Tectonics

A **metamorphic facies** is defined by a distinctive assemblage of minerals that may be used to designate a particular set of metamorphic environmental conditions recorded in the rocks (Figure 7.22). One advantage of this scheme is that rocks produced by different types of metamorphism can be interpreted within a single, unified context.

You will notice that the names of metamorphic facies refer to metamorphic rocks derived specifically from a basaltic parent rock. This is because Pennti Eskola, the geologist who proposed the metamorphic facies scheme in 1915, concentrated on the metamorphism of basalts, and his basic terminology, although now slightly modified, has stuck ever since. The names of Eskola's facies serve as convenient labels for a particular combination of temperature and pressure values. In other words, even if a non-basaltic parent rock produces different indicator minerals than basalt under a given set of metamorphic conditions, the names of Eskola's facies are used for the purpose of denoting the temperature and pressure ranges embodied by the metamorphic rock.

Comparison of Figure 7.22 and Figure 7.9 illustrates the usefulness of the metamorphic facies concept in the interpretation of metamorphic environments. Figure 7.23 shows how facies fit into the context of plate tectonics. The *hornfels facies*, not surprisingly, is associated with the high temperatures and low pressures involved in contact metamorphism. The *zeolite facies* reflects very low-grade metamorphism involving low heat and pressure and is generally associated with burial metamorphism and the very lowest grade of regional metamorphism. The increasing temperatures and pressures associated with regional metamorphism are recorded by the *greenschist-amphibolite-granulite facies* sequence. Subduction zone metamorphism is characterized by the *blueschist facies* representing relatively low temperatures but high pressures. Finally, rocks of the *eclogite facies* are diagnostic of very high temperatures and pressures encountered by a deeply subducted slab of oceanic lithosphere.

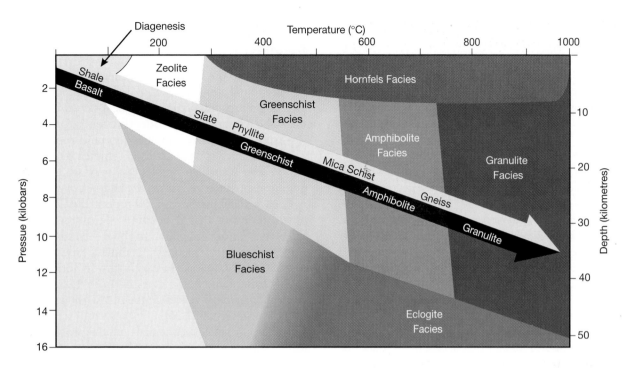

◆ **Figure 7.22** Metamorphic facies and corresponding temperature and pressure conditions. Note the equivalent metamorphic rocks produced from regional metamorphism of shale and basalt parent rocks.

Other Types of Metamorphism

Cataclastic Metamorphism

Cataclastic metamorphism is caused by extreme directed pressures (shear) and occurs locally along large faults and fault systems. In such high-stress environments, rocks and their constituent mineral grains are either flattened into new and unusual shapes or destroyed entirely. Near the surface, where most rocks exhibit brittle behaviour, rocks in fault zones become fractured and crushed into angular fragments, forming fault breccia (Figure 7.24). At great depths, where rocks deform by ductile flow, rock fragments and minerals in fault zones tend to become elongated and give the rock a lineated appearance. The resulting rocks are called mylonites (*mylo* = mill, *ite* = a stone; Figure 7.24B).

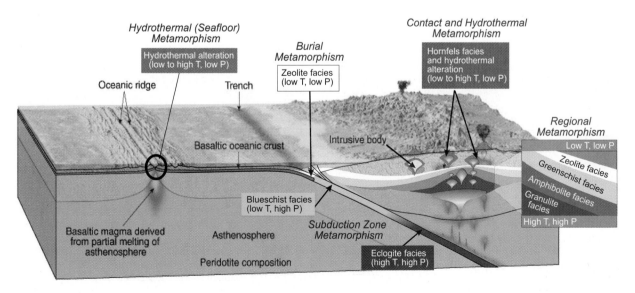

◆ **Figure 7.23** The association of metamorphic facies with plate tectonic environments.

This type of metamorphism is generally restricted in scale, but in some instances may be quite dramatic. For example, displacements along California's San Andreas fault have created a zone of fault breccia and related rock types over 1000 kilometres long and up to 3 kilometres wide.

Impact (Shock) Metamorphism

Impact (shock) metamorphism is a rather rare but distinctive type of metamorphism that occurs when a rock is suddenly and almost instantaneously subjected to intense heat and pressure. Not surprisingly, this type of metamorphism is associated with impacts of large extraterrestrial bodies such as asteroids and comets (Box 7.1).

Upon impact, the energy of the rapidly falling meteorite is transformed into heat and shock waves that pass through the surrounding rocks. The rock is pulverized, shattered, and sometimes melted. Solids formed from the molten material cannot, of course, be considered metamorphic products. Metamorphic minerals formed under these extreme conditions include coesite (a very dense form of quartz) and minute diamonds (produced from carbon in the rock or originally from the impacting body itself). These high-pressure minerals indicate that the pressures and temperatures generated during meteorite impacts are at least as great as those that exist in the mantle. Shock metamorphism also affects the structure of rocks. Distinctive features associated with shock metamorphism are *shock lamellae* and **shatter cones** (see Box 7.1).

Ancient Metamorphic Environments

In addition to the linear belts of metamorphic rocks that are found along the axes of most mountain belts, even larger expanses of metamorphic rocks exist within the stable continental interiors (Figure 7.25). These relatively flat expanses of metamorphic rocks and associated igneous rocks are called **shields**. One such region, the Canadian Shield, has very little topographic expression and forms the bedrock over much of Canada. Radiometric dating of rocks in the Canadian Shield indicates that it is composed of rocks that range in age from 1.8 billion to about 4 billion years. Rocks of the Acasta Gneiss Complex, exposed just east of Great Slave Lake in the Northwest Territories, have yielded radiometric dates of up to 4.03 billion years and currently represent the oldest known rocks in the world (Figure 7.26).

Because shields are ancient and their rock structure is similar to that found in the cores of younger, deformed mountainous terrains, they are assumed to be remnants of much earlier periods of mountain building (Figure 7.25). This evidence strongly supports the generally accepted view that Earth has been a dynamic planet throughout most of its history. Studies of these vast areas of metamorphism in the context of the plate tectonic model have given geologists new insights into the problem of discerning just how the continents came to exist. We will consider this topic further in Chapter 20.

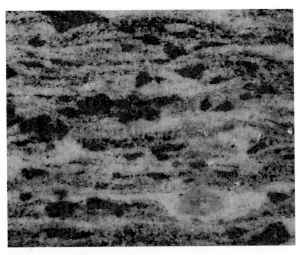

A   B

◆ **Figure 7.24   A.** Fault breccia consisting of large angular fragments. This outcrop, located in Titus Canyon, Death Valley, California, was produced along a fault zone. **B.** Mylonite, characterized by "milled" and deformed mineral grains, is another rock type produced by cataclastic metamorphism. This mylonite was produced from intense shearing of a granitic parent rock.
(Photo A by A.P. Trujillo/APT Photos; Photo B by C. Tsujita)

Understanding Earth
Impact Metamorphism

BOX 7.1

It is now clear that comets and asteroids have collided with Earth far more frequently than previously thought, based on over one hundred giant impact structures called astroblemes (*astro* = star, *blema* = wound from a projectile) that have been found worldwide. The Barringer crater in Arizona bears the familiar bowl-like shape that we generally associated with impact craters (Figure 7.A). Many astroblemes also exist in Canada, and include the Manicouagan crater in Québec, the Sudbury Basin in Ontario, and the Eagle Butte structure in Alberta.

When high-velocity projectiles (comets, asteroids) impact Earth's surface, pressures reach millions of atmospheres and temperatures exceed 2000°C momentarily. The result is pulverized, shattered, and melted rock. Where impact craters are relatively fresh, shock-melted ejecta and rock fragments ring the impact site. Although most material is deposited close to its source, some ejecta can travel great distances. One example is tektites (*tektos* = molten), beads of silica-rich glass up to a few centimetres in size resulting from rapid quenching of blobs of the molten ejecta. These can fall hundreds of kilometres away from the impact site.

Shock metamorphism produces some distinctive deformation features. At the microscopic end of the scale are shock lamellae, linear features developed due to small scale, partial melting. Shock lamellae are recognized relatively easily in minerals such as quartz that do not have planar cleavage or fracture planes that the shock lamellae could otherwise be confused with. On the large scale of deformation features are shatter cones. These are nested, cone-like structures that emanate from the point of impact and can be compared to the small-scale patterns produced in glass when a rock damages the windshield of your car. The presence of these structures in rocks surrounding the Sudbury Basin provides evidence suggesting that Sudbury Basin represents the site of a large meteorite impact that occurred during the Precambrian (Figure 7.B). Because the Sudbury Basin has been deformed, the orientations of shatter cones vary from place to place, but it is assumed that before this deformation, the pointed ends of all the shatter cones were consistently oriented toward the site of impact.

◆ **Figure 7.A** Barringer Crater, located west of Winslow, Arizona. (Photo by Michael Collier)

◆ **Figure 7.B** Upward-pointing shatter cones made in Precambrian greywacke near Sudbury, Ontario. Inset: close-up view of a single shatter cone. (Photos by Grant Young)

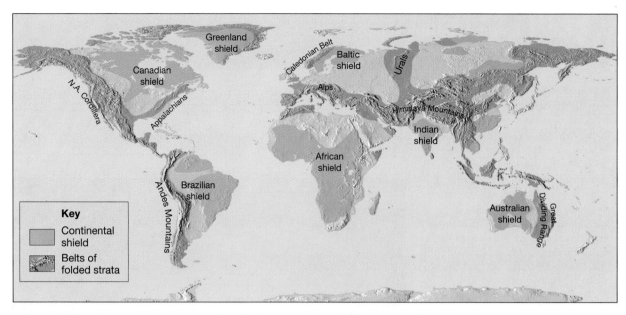

◆ **Figure 7.25**   Occurrences of metamorphic rocks. The continental shields of the world are composed largely of metamorphic rocks and associated igneous intrusions. In addition, the deformed portions of many mountain belts are also metamorphic. The remaining area shown in this map is the stable continental interior, which generally consists of undeformed sedimentary beds that overlie metamorphic and igneous basement rocks.

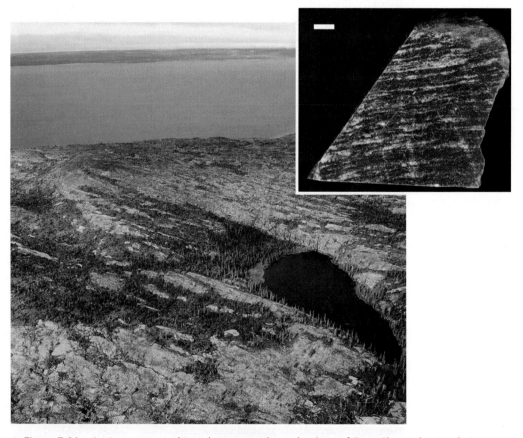

◆ **Figure 7.26**   Ancient metamorphic and igneous rocks on the shore of Great Slave Lake, Northwest Territories. These rocks, which are part of the Canadian Shield, are among the most ancient on Earth. Inset: A sample of the Acasta Gneiss, determined by radiometric dating to be 4.03 billion years old and currently representing the oldest known rock in the world (Scale bar: 1 centimetre). (Main photo by Tom and Susan Bean, Inc.; Inset photo by C. Tsujita)

Chapter Summary

- *Metamorphism* is the transformation of one rock type into another. *Metamorphic* rocks form from pre-existing rocks (either igneous, sedimentary, or other metamorphic rocks) that have been altered by the agents of metamorphism, which include *heat, pressure (stress)*, and *chemically active fluids*. During metamorphism the material essentially remains solid. The changes that occur in metamorphosed rocks are textural as well as mineralogical.

- The three agents of metamorphism are *heat, pressure (stress)*, and *chemically active fluids*. The mineral makeup of the parent rock determines, to a large extent, the degree to which each metamorphic agent will cause change. Heat is the most important agent because it provides the energy to drive chemical reactions that result in the recrystallization of minerals. Pressure, like temperature, also increases with depth. When subjected to *confining pressure (uniform stress)*, minerals may recrystallize into more compact forms. During mountain-building, rocks are subjected to *directed pressure (differential stress)*, which tends to shorten them in the direction pressure is applied from and lengthens them in the direction perpendicular to that force. At depth, rocks are warm and ductile, which accounts for their ability to deform by flowing when subjected to differential stresses. Chemically active fluids, usually dominated by water, also enhance the metamorphic process by acting as catalysts in chemical reactions, by transporting dissolved substances and, in some cases, by becoming incorporated into the chemical makeup of certain minerals.

- The *grade of metamorphism* is reflected in the *texture and mineralogy* of metamorphic rocks. In terms of mineralogy, metamorphic grade is indicated by the presence of certain *index minerals*, which reflect the temperature/pressure conditions experienced by the rock during metamorphism. In terms of texture, rocks affected by temperature alone retain a nonfoliated texture regardless of grade, whereas those affected by both temperature and differential stress show a general increase in the degree of preferred orientation among platy/elongate minerals.

- The *preferred orientation* of platy and elongate minerals defines the *foliation* of rocks affected by regional metamorphism. Foliation develops as platy or elongated minerals are rotated into parallel alignment, are plastically deformed into flattened grains that exhibit a planar alignment, or are recrystallized to form new grains with a preferred orientation.

- Foliation is particularly well-displayed in metamorphic rocks produced from the metamorphism of shale. *Rock cleavage* is a type of foliation in which rocks split cleanly into thin slabs along surfaces where platy minerals are aligned. *Schistosity* is a type of foliation defined by the parallel alignment of medium- to-coarse-grained platy minerals. During high-grade metamorphism, ion migrations can cause minerals to segregate into distinct layers or bands. Metamorphic rocks with a banded texture are called *gneiss*. Metamorphism of basalt produces rocks that are similarly foliated but contain different minerals. The regional metamorphism of basalt produces, in order of increasing grade, *greenschist, amphibolite*, and *granulite*.

- Metamorphic rocks composed of only one mineral forming equidimensional crystals often appear *nonfoliated*, regardless of the specific pressure conditions involved in their formation. Both *Marble* (metamorphosed limestone) and *quartzite* are examples of such rocks. *Hornfels* is a nonfoliated rock that is diagnostic of contact metamorphism.

- A *metamorphic facies* is defined by a distinctive assemblage of minerals and is used to designate a particular set of metamorphic environmental conditions recorded in the rocks (Figure 7.22). The metamorphic facies concept provides a convenient context within which many different types of metamorphism can be interpreted in terms of temperature and pressure conditions.

- Metamorphism most often occurs in two settings: (1) when rock is in contact with magma, *contact metamorphism* occurs, (2) during continental convergence and mountain-building, where extensive areas of rock undergo *regional metamorphism* and *subduction zone metamorphism*. However, metamorphism can also occur during the simple burial of rocks *(burial metamorphism)*, by the interaction of rocks with hot water and its dissolved chemical components *(hydrothermal metamorphism)*, by the milling of brittle rocks along faults *(cataclastic metamorphism)*, and by meteorite impact *(impact metamorphism)*.

Review Questions

1. What is metamorphism? What are the agents that change rocks?

2. Why is heat such an important agent of metamorphism?

3. How is confining pressure (uniform stress) different than directed pressure (differential stress)?

4. What role do chemically active fluids play in metamorphism?

5. How does the parent rock affect the metamorphic process?

6. What is foliation? Distinguish between schistosity and gneissic textures.

7. Briefly describe the three mechanisms by which minerals develop a preferred orientation.

8. List some changes that might occur to a rock in response to metamorphic processes.

9. Slate and phyllite resemble each other. How might you distinguish one from the other?

10. Each of the following statements describes one or more characteristics of a particular metamorphic rock. For each statement, name the metamorphic rock that is being described.
    (a) calcite-rich and nonfoliated
    (b) loosely coherent rock composed of broken fragments that formed along a fault zone
    (c) represents a grade of metamorphism between slate and schist
    (d) very fine-grained and foliated; excellent rock cleavage
    (e) foliated and composed predominantly of platy minerals
    (f) composed of alternating bands of light and dark silicate minerals
    (g) hard, nonfoliated rock resulting from contact metamorphism.

11. Distinguish between contact metamorphism and regional metamorphism. Which creates the greatest quantity of metamorphic rock?

12. Where does most hydrothermal metamorphism occur?

13. Describe burial metamorphism.

14. How do geologists use index minerals?

15. Briefly describe the textural changes that occur in the transformation of slate to phyllite to schist and then to gneiss.

16. How are gneisses and migmatites related?

17. With which type of plate boundary is regional metamorphism associated?

18. Why do the cores of Earth's major mountain chains contain metamorphic rocks?

19. Explain the concept of metamorphic facies. Why are metamorphic facies useful in the interpretation of metamorphic conditions rocks have experienced?

20. What are shields? How are these relatively flat areas related to mountains?

Key Terms

aureole (p. 198)
burial metamorphism (p. 198)
cataclastic metamorphism (p. 209)
chemically active fluids (p. 195)
confining pressure (uniform stress) (p. 194)
contact metamorphism (p. 198)

directed pressure (differential stress) (p. 194)
foliation (p. 195)
gneissic texture (p. 205)
grade (p. 197)
hydrothermal metamorphism (p. 207)
impact (shock) metamorphism (p. 210)
index mineral (p. 197)

metamorphic facies (p. 208)
metamorphism (p. 193)
migmatite (p. 207)
nonfoliated (p. 197)
parent rock (p. 193)
porphyroblastic texture (p. 199)
regional metamorphism (p. 201)

rock cleavage (p. 203)
schistosity (p. 204)
seafloor metamorphism (p. 208)
shatter cones (p. 210)
shield (p. 210)
subduction zone metamorphism (p. 207)

Web Resources

 The *Earth* Web site uses the resources and flexibility of the Internet to aid in your study of the topics in this chapter. Written and developed by geology instructors, this site will help improve your understanding of geology. Visit **http://www.pearson.ca/tarbuck** and click on the cover of the text to find:

■ Online review quizzes.

■ Web-based critical thinking and writing exercises.

■ Links to chapter-specific Web resources.

■ Internet-wide key-term searches.

http://www.pearson.ca/tarbuck

Chapter 8
Geologic Time

Carbonate rocks at Chute Vauréal, Anticosti Island, Québec, preserve a record of geological processes that operated hundreds of millions of years ago, during the Ordovician Period. (Photo by Jisuo Jin)

In the late eighteenth century, James Hutton recognized the immensity of Earth history and the importance of time as a component in all geological processes. In the nineteenth century, Sir Charles Lyell and others effectively demonstrated that Earth had experienced many episodes of mountain building and erosion, which must have required great spans of geologic time. Although these pioneering scientists understood that Earth was very old, they had no way of knowing its true age. Was it tens of millions, hundreds of millions, or even billions of years old? A geologic time scale was developed that showed the sequence of events based on relative dating principles. What are these principles? What part do fossils play? With the discovery of radioactivity and radiometric dating techniques, geologists now can assign fairly accurate dates to many of the events in Earth history. What is radioactivity? Why is it a good "clock" for dating the geologic past?

Geology Needs a Time Scale

Like the pages in a long and complicated history book, rocks record the geological events (Figure 8.1) and changing life forms of the past. The book, however, is not complete. Many pages, especially in the early chapters, are missing. Others are tattered, torn, or smudged. Yet enough of the book remains to allow much of the story to be deciphered, much like the rock record.

Interpreting Earth history is a prime goal of the science of geology. Like a modern-day sleuth, the geologist must interpret the clues found preserved in the rocks. By studying rocks, especially sedimentary rocks, and the features they contain, geologists can unravel the complexities of the past.

Geologic events by themselves, however, have little meaning until they are put into a time perspective. Studying history, whether it is that of humans or dinosaurs, requires a calendar. Among geology's major contributions to science is the *geologic time scale* and the discovery that Earth history is exceedingly long.

Relative Dating—Key Principles

Geologic Time
↳ Relative Dating

The geologists who developed the geologic time scale revolutionized the way people think about time and perceive our planet. They learned that Earth is much older than anyone had previously imagined and that its surface and interior have been changed over and over again by the same geologic processes that operate today.

During the late 1800s and early 1900s, attempts were made to determine Earth's age. Although some of the methods appeared promising at the time, none of these early efforts proved to be reliable. What these scientists were seeking was a **numerical date**. Such dates specify the actual number of years that have passed since an event occurred. Today our understanding of radioactivity allows us to accurately determine numerical dates for rocks that represent important events in Earth's distant past. We will study radioactivity later in this chapter. Prior to the discovery of radioactivity, geologists had no reliable method of numerical dating and had to rely solely on relative dating.

Relative dating means that rocks are placed in their proper *sequence of formation*—which ones formed first, second, third, and so on. Relative dating cannot tell us how long ago something took place, only that it followed one event and preceded another. The relative dating techniques that were developed are valuable and still widely used. Numerical dating methods did not replace these techniques; they simply supplemented them. To establish a relative time scale, a few basic principles or rules had to be discovered and applied. Although they may seem obvious to us today, they were major breakthroughs in thinking at the time, and their discovery was an important scientific achievement.

◆ **Figure 8.1**   The history of this large boulder illustrates the immensity of geologic time. The boulder is composed of metamorphic rock that once lay several kilometres below the Earth's surface in the roots of a great Precambrian mountain chain. Uplift and erosion over hundreds of millions of years exposed the rock at the surface. During the last ice age, the boulder was dislodged from the bedrock surface and transported to a site near its present location. It now lies on this rocky beach, near Port Severn, Ontario, where it is being slowly broken down by physical and chemical weathering. (Photo by Mike Schering)

Law of Superposition

Nicolaus Steno, a Danish anatomist, geologist, and priest (1638–1686), is credited with being the first to recognize a sequence of historical events in an outcrop of sedimentary rock layers. Working in the mountains of western Italy, Steno applied a very simple rule that has come to be the most basic principle of relative dating—**the law of superposition** (*super* = above, *positum* = to place). The law simply states that in an undeformed sequence of sedimentary rocks, each bed is older than the one above and younger than the one below. Although it may seem obvious that a rock layer could not be deposited with nothing beneath it for support, it was not until 1669 that Steno clearly stated this principle.

This rule also applies to other surface-deposited materials, such as lava flows and beds of ash from volcanic eruptions. Applying the law of superposition to the beds exposed in the upper portion of the Grand Canyon (Figure 8.2), we can easily place the layers in their proper order. Among those that are pictured, the sedimentary rocks in the Supai Group are the oldest, followed in order by the Hermit Shale, Coconino Sandstone, Toroweap Formation, and Kaibab Limestone.

Principle of Original Horizontality

Steno is also credited with recognizing the importance of another basic principle, called the **principle of original horizontality**. Simply stated, it means that layers of sediment are generally deposited in a horizontal position. Thus, if we observe rock layers that are flat, it means they have not been disturbed and still have their *original* horizontality. The layers in the Chute Vauréal illustrate this in the chapter-opening photo, as do those

A

B

◆ **Figure 8.2**   Applying the law of superposition to these layers exposed in the upper portion of the Grand Canyon, the Supai Group is the oldest and the Kaibab Limestone is the youngest. (Photo by E. J. Tarbuck)

of the Grand Canyon (Figure 8.2). But if the layers are folded or inclined at a steep angle, they must have been moved into that position by crustal disturbances some time after their deposition (Figure 8.3).

Principle of Cross-Cutting Relationships

When a fault cuts through other rocks, or when magma intrudes and crystallizes, we can assume that the fault or intrusion is younger than the rocks affected. For example, in Figure 8.4, the faults and dykes clearly must have occurred after the sedimentary layers were deposited.

This is the **principle of cross-cutting relationships**. By applying the cross-cutting principle, you can see that fault A occurred *after* the sandstone layer was deposited because it "broke" the layer. Likewise, fault A occurred *before* the conglomerate was laid down, because that layer is unbroken.

We can also state that dyke B and its associated sill are older than dyke A because dyke A cuts the sill. In the same manner, we know that the batholith was emplaced after movement occurred along fault B but before dyke B was formed. This is true because the batholith cuts across fault B while dyke B cuts across the batholith.

Inclusions

Sometimes inclusions can aid the relative dating process. **Inclusions** (*includere* = to enclose) are fragments of one rock unit that have been enclosed within another. The basic principle is logical and straightforward. The rock mass adjacent to the one containing the inclusions must have been there first in order to provide the rock fragments. Therefore, the rock mass containing inclusions is the younger of the two. Figure 8.5 provides an example. Here the inclusions of intrusive

◆ **Figure 8.3** Most layers of sediment are deposited in a nearly horizontal position. Thus, when we see rock layers that are inclined, we can assume that they must have been moved into that position by crustal disturbances after their deposition. Hartland Quay, Devon, England. (Photo by Tom Bean/DRK Photo)

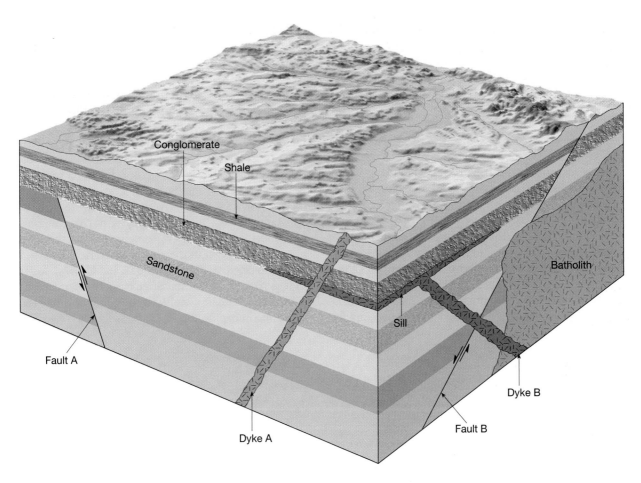

◆ **Figure 8.4**   Cross-cutting relationships represent one principle used in relative dating. An intrusive rock body is younger than the rocks it intrudes. A fault is younger than the rock layers it cuts.

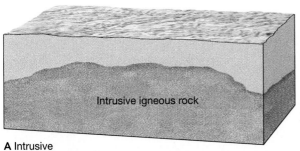

A Intrusive
igneous rock

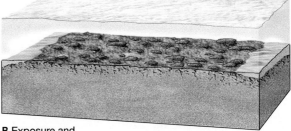

B Exposure and
weathering of intrusive igneous rock

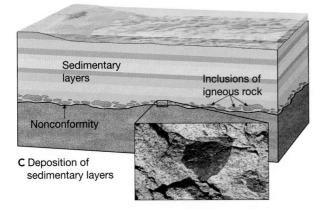

C Deposition of
sedimentary layers

◆ **Figure 8.5**   These three diagrams illustrate one way that inclusions can form, and a type of unconformity termed a *nonconformity*. In the third diagram we know the igneous rock must be older because pieces of it are included in the overlying sedimentary bed. When older intrusive igneous rocks are overlain by younger sedimentary layers, a nonconformity is said to exist. The photo shows an inclusion of dark igneous rock in a lighter-coloured and younger host rock. (Photo by Tom Bean)

igneous rock in the adjacent sedimentary layer indicate that the sedimentary layer was deposited on top of a weathered igneous mass rather than being intruded from below by magma that later crystallized.

Unconformities

When we observe layers of rock that have been deposited essentially without interruption, we call them **conformable**. Particular sites exhibit conformable beds representing certain spans of geologic time. However, no place on Earth has a complete set of conformable strata.

Throughout Earth history, the deposition of sediment has been interrupted over and over again. All such breaks in the rock record are termed unconformities. An **unconformity** represents a long period during which deposition ceased, erosion removed previously formed rocks, and then deposition resumed. In each case, uplift and erosion are followed by subsidence and renewed sedimentation. Unconformities are important features because they represent significant geologic events in Earth history. Moreover, their recognition helps us identify what intervals of time are not represented by strata and thus are missing from the geologic record.

The rocks exposed in the Grand Canyon of the Colorado River represent a tremendous span of geologic history. It is a wonderful place to take a trip through time. The canyon's colourful strata record a long history of sedimentation in a variety of environments—advancing seas, rivers, and deltas, tidal flats, and sand dunes. But the record is not continuous. Unconformities represent vast amounts of time that have not been recorded in the canyon's layers. Figure 8.6 is a geologic cross-section of the Grand Canyon. Refer to it as you read about the three basic types of unconformities: angular unconformities, disconformities, and nonconformities.

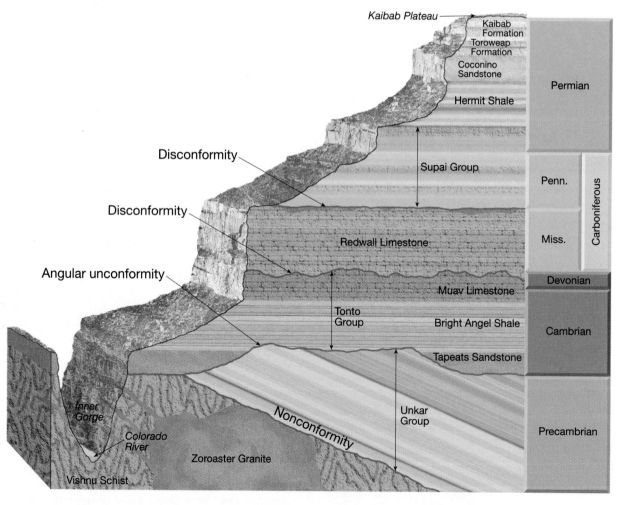

◆ **Figure 8.6** This cross-section through the Grand Canyon illustrates the three basic types of unconformities. An angular unconformity can be seen between the tilted Precambrian Unkar Group and the Cambrian Tapeats Sandstone. Two disconformities are marked, above and below the Redwall Limestone. A nonconformity occurs between the igneous and metamorphic rocks of the Inner Gorge and the sedimentary strata of the Unkar Group.

Chitinozoans are tiny (0.05-0.4 millimetres) flask-shaped fossils of unknown biological affinity that are important Paleozoic index fossils (Early Ordovician to Late Devonian).

Swimmers

Conodonts are microscopic, (0.2-5 millimetres) phosphatic, tooth-like objects that were elements of the feeding apparatus of small, eel-like animals that were active swimmers. The conodont animal is tentatively classified as a primitive, extinct chordate. Conodonts are important index fossils for the Paleozoic and Early Mesozoic (Triassic).

Graptolites are extinct colonial animals that formed long, saw-blade-shaped, organic-walled skeletons and are classified in the group of animals called hemichordates. Forms that were passive floaters are useful in the relative dating of Paleozoic strata, particularly those of Ordovician to Silurian age.

Ammonoids are extinct molluscs related to squids and octopuses and were probably active swimmers. Their calcareous shells are important index fossils for the late Paleozoic and the Mesozoic.

Throughout much of the Phanerozoic Eon, an important factor that restricted the geographic distribution of many bottom-dwelling organisms was the stagnation of bottom waters in deep water areas of sedimentary basins; at times, this prevented much of the seafloor from being colonized by organisms. During times of adverse seafloor conditions, the remains of most bottom-dwelling organisms tend to have a patchy distribution on a global scale and are therefore undesirable for correlation. Even so, there are a few organisms that had an unusually high tolerance of low-oxygen conditions on the seafloor and are used as index fossils for regional correlations. Some of these are described below.

Bottom dwellers

Trilobites are extinct arthropods, the group of organisms that include present day crabs, shrimps, lobsters, insects, spiders, scorpions, and barnacles. Their calcium carbonate skeletons are used as index fossils primarily in Cambrian to Devonian strata.

Bivalves (known less formally as clams) are molluscs that construct a bivalved shell of calcium carbonate. A group of clams informally known as "flat clams" were extremely abundant and very tolerant of hostile seafloor conditions during the Mesozoic. Various forms of flatclams are important index fossils of the Mesozoic. Scallop-like flatclams called monotids are important in Triassic strata, whereas oyster-like forms called inoceramids serve as index fossils in Cretaceous strata.

---

happens, the potassium-argon clock is reset and dating the sample will give only the time of thermal resetting, not the true age of the rock. For other radiometric clocks, a loss of daughter atoms can occur if the rock has been subjected to weathering or leaching. To avoid such a problem, one simple safeguard is to use only fresh, unweathered material and not samples that may have been chemically altered.

Dating with Carbon-14

To date very recent events, carbon-14 is used. Carbon-14 is the radioactive isotope of carbon. The process is often called **radiocarbon dating**. Because the half-life of carbon-14 is only 5730 years, it can be used for dating events from the historic past as well as those from very recent geologic history (see Box 8.3). In some cases carbon-14 can be used to date events as far back as 75,000 years.

Carbon-14 is continuously produced in the upper atmosphere as a consequence of cosmic-ray bombardment. Cosmic rays (high-energy nuclear particles) shatter the nuclei of gas atoms, releasing neutrons. Some of the neutrons are absorbed by nitrogen atoms (atomic number 7, mass number 14), causing each nucleus to emit a proton. As a result, the atomic number decreases by 1 (to 6), and a different element, carbon-14, is created (Figure 8.14A). This isotope of carbon quickly becomes incorporated into carbon dioxide, which

Students Sometimes Ask...

If parent/daughter ratios are not always reliable, how can meaningful radiometric dates for rocks be obtained? One common precaution against sources of error is the use of cross checks. Often this simply involves subjecting a sample to two different radiometric methods. If the two dates agree, the likelihood is high that the date is reliable. If, on other hand, there is an appreciable difference between the two dates, other cross checks must be employed (such as the use of fossils or correlation with other, well-dated marker beds) to determine which date—if either—is correct.

circulates in the atmosphere and is absorbed by living matter. As a result, all organisms contain a small amount of carbon-14, including yourself.

As long as an organism is alive, the decaying radiocarbon is continually replaced, and the proportions of carbon-14 and carbon-12 remain constant. Carbon-12 is the stable and most common isotope of carbon. However, when any plant or animal dies, the amount of carbon-14 gradually decreases as it decays to nitrogen-14 by beta emission (Figure 8.14B). By comparing the proportions of carbon-14 and carbon-12 in a sample, radiocarbon dates can be determined.

BOX 8.3

Understanding Earth
Using Tree Rings to Date and Study the Recent Past

If you look at the top of a tree stump or at the end of a log, you will see that it is composed of a series of concentric rings. Each of these *tree rings* becomes larger in diameter outward from the centre (Figure 8.E). Every year in temperate regions trees add a layer of new wood under the bark. Characteristics of each tree ring, such as size and density, reflect the environmental conditions (especially climate) that prevailed during the year when the ring formed. Favourable growth conditions produce a wide ring; unfavourable ones produce a narrow ring. Trees growing at the same time in the same region show similar tree-ring patterns.

Because a single growth ring is usually added each year, the age of the tree when it was cut can be determined by counting the rings. If the year of cutting is known, the age of the tree and the year in which each ring formed can be determined by counting back from the outside ring.*

This procedure can be used to determine the dates of recent geologic events, for example, the minimum number of years since a new land surface was created by a landslide or a flood. The dating and study of annual rings in trees is called *dendrochronology*.

To make the most effective use of tree rings, extended patterns known as

*Scientists are not limited to working with trees that have been cut down. Small, nondestructive core samples can be taken from living trees.

◆ **Figure 8.E**   Each year a growing tree produces a layer of new cells beneath the bark. If the tree is felled and the trunk examined (or if a core is taken, to avoid cutting the tree), each year's growth can be seen as a ring. Because the amount of growth (thickness of a ring) depends upon precipitation and temperature, tree rings are useful records of past climates. (Photo by Stephen J. Krasemann/DRK Photo)

ring chronologies are established. They are produced by comparing the patterns of rings among trees in an area. If the same pattern can be identified in two samples, one of which has been dated, the second sample can be dated from the first by matching the ring pattern common to both. This technique, called *cross dating*, is illustrated in Figure 8.F. Tree-ring chronologies extending back for thousands of years have been established for some regions. To date a timber sample of unknown age, its ring pattern is matched against the reference chronology.

Tree-ring chronologies are unique archives of environmental history and have important applications in such disciplines as climate, geology, ecology, and archaeology. For example, tree rings are used to reconstruct climate variations within a region for spans of thousands of years prior to human historical records. Knowledge of such long-term variations is of great value in making judgments regarding the recent record of climate change.

In summary, dendrochronology provides useful numerical dates for events in the historic and recent prehistoric past. Moreover, because tree rings are a storehouse of data, they are a valuable tool in the reconstruction of past environments.

Although carbon-14 is only useful in dating the last small fraction of geologic time, it has become a very valuable tool for anthropologists, archaeologists, and historians, as well as for geologists who study very recent Earth history. In fact, the development of radiocarbon dating was considered so important that the chemist who discovered this application, Willard F. Libby, received a Nobel Prize in 1960.

Importance of Radiometric Dating

Bear in mind that although the basic principle of radiometric dating is simple, the actual procedure is quite complex. The analysis that determines the quantities of parent and daughter must be painstakingly precise. In addition, some radioactive materials do not decay directly into the stable daughter product, as was the case with our hypothetical example, a fact that may further complicate the analysis. In the case of uranium-238, there are 13 intermediate unstable daughter products formed before the fourteenth and last daughter product, the stable isotope lead-206, is produced (see Figure 8.12).

Radiometric dating methods have produced literally thousands of dates for events in Earth history. Rocks from several localities have been dated at

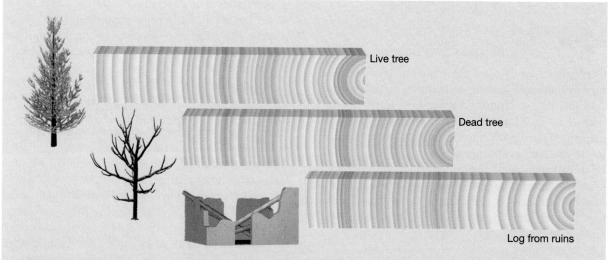

◆ **Figure 8.F** Cross dating is a basic principle in dendrochronology. Here it was used to date an archaeological site by correlating tree-ring patterns for wood from trees of three different ages. First, a tree-ring chronology for the area is established using cores extracted from living trees. This chronology is extended further back in time by matching overlapping patterns from older, dead trees. Finally, cores taken from beams inside the ruin are dated using the chronology established from the other two sites.

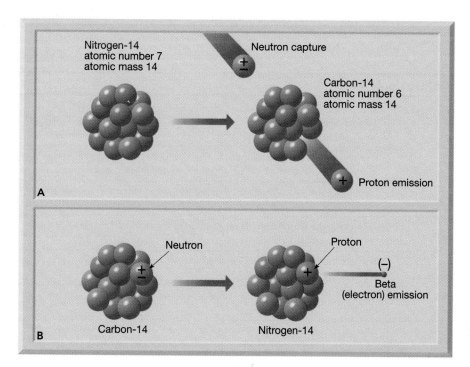

◆ **Figure 8.14** **A.** Production and **B.** decay of carbon-14. These sketches represent the nuclei of the respective atoms.

more than 3 billion years, and geologists realize that still-older rocks exist. For example, a granite from South Africa has been dated at 3.2 billion years and contains inclusions of quartzite. (Remember that inclusions are older than the rock containing them.) Quartzite, a metamorphic rock, was originally the sedimentary rock sandstone. Sandstone, in turn, is the product of the lithification of sediments produced by the weathering of existing rocks. Thus, we have a positive indication that even older rocks existed.

Radiometric dating has vindicated the ideas of Hutton, Darwin, and others, who over 150 years ago inferred that geologic time must be immense. Indeed, radiometric dating has proved that there has been enough time for the processes we observe to have accomplished tremendous tasks.

The Geologic Time Scale

Geologic Time
↳ Geologic Time Scale

Geologists have divided the whole of geologic history into units of varying magnitude. Together, they comprise the **geologic time scale** of Earth history (Figure 8.15). The major units of the time scale were delineated during the nineteenth century, principally by workers in Western Europe and Great Britain. Because radiometric dating was unavailable at that time, the entire time scale was created using methods of relative dating. It was only in the twentieth century that radiometric methods permitted numerical dates to be added.

Structure of the Time Scale

The geologic time scale subdivides the 4.5-billion-year history of Earth into many different units and provides a meaningful time frame within which the events of the geologic past are arranged. As shown in Figure 8.15, **eons** represent the greatest expanses of time. The eon that began about 540 million years ago

is the **Phanerozoic**, a term derived from Greek words meaning *visible life*. It is an appropriate description because the rocks and deposits of the Phanerozoic eon contain abundant fossils that document major evolutionary trends.

Another glance at the time scale reveals that the Phanerozoic eon is divided into **eras**. The three eras within the Phanerozoic are the **Paleozoic** (*paleo* = ancient, *zoe* = life), the **Mesozoic** (*meso* = middle, *zoe* = life), and the **Cenozoic** (*ceno* = recent, *zoe* = life). As the names imply, the eras are bounded by profound worldwide changes in life forms (see Box 8.4).

Each era is subdivided into time units known as **periods**. The Paleozoic has six, the Mesozoic three, and the Cenozoic three. Each of these dozen periods is characterized by a somewhat less profound change in life forms as compared with the eras. The eras and periods of the Phanerozoic, with brief explanations of each, are shown in Table 8.2.

Finally, each of the 12 periods is divided into still smaller units called **epochs**. As you can see in Figure 8.15, seven epochs have been named for the periods of the Cenozoic. The epochs of other periods usually are simply termed *early, middle,* and *late.*

TABLE 8.2   Major Divisions of Geologic Time

Era	Period	Description
Cenozoic Era (Age of Recent Life)	Quaternary Period	The several geologic eras were originally named Primary, Secondary, Tertiary, and Quaternary. The first two names are no longer used; Tertiary and Quaternary have been retained but used as period designations.
	Tertiary Period	
Mesozoic Era (Age of Middle Life)	Cretaceous Period	Derived from Greek word for chalk (Kreta) and first applied to extensive deposits that form white cliffs along the English Channel (see Figure 6.11).
	Jurassic Period	Named for the Jura Mountains, located between France and Switzerland, where rocks of this age were first studied.
	Triassic Period	Taken from the word "trias" in recognition of the threefold character of these rocks in Europe.
Paleozoic Era (Age of Ancient Life)	Permian Period	Named after the province of Perm, Russia, where these rocks were first studied.
	Pennsylvanian Period*	Named for the state of Pennsylvania where these rocks have produced much coal.
	Mississippian Period*	Named for the Mississippi River Valley where these rocks are well exposed.
	Devonian Period	Named after Devonshire County, England, where these rocks were first studied.
	Silurian Period	Named after Celtic tribes, the Silures and the Ordovices, that lived in Wales during the Roman Conquest.
	Ordovician Period	
	Cambrian Period	Taken from Roman name for Wales (Cambria), where rocks containing the earliest evidence of complex forms of life were first studied.
Precambrian		The time between the birth of the planet and the appearance of complex forms of life. About 88 percent of Earth's estimated 4.5 billion years fall into this span.

Source: U.S. Geological Survey
*Outside of the United States, the Mississippian and Pennsylvanian periods are considered subperiods and are combined into the Carboniferous Period.

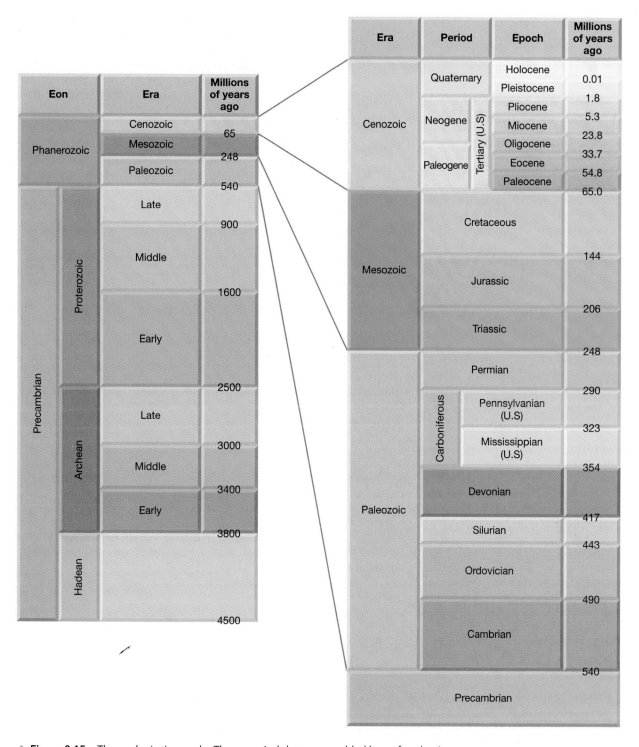

◆ **Figure 8.15** The geologic time scale. The numerical dates were added long after the time scale had been established using relative dating techniques.
(Data from Geological Society of America)

Precambrian Time

Notice that the detail of the geologic time scale does not begin until about 540 million years ago, the date for the beginning of the Cambrian Period. The nearly four billion years prior to the Cambrian are divided into three eons, the **Hadean** (*Hades* = mythological subterranean world of departed spirits), the **Archean** (*archaios* = ancient), and the **Proterozoic** (*proteros* = before, *zoe* = life). It is also common for this vast expanse of time

Earth as a System
Demise of the Dinosaurs

BOX 8.4

The boundaries between divisions on the geologic time scale represent times of significant geological and/or biological charge. Of special interest is the boundary between the Mesozoic Era ("middle life") and Cenozoic Era ("recent life"), about 65 million years ago. Around this time, more than half of all plant and animal species died out in a *mass extinction*. This boundary marks the end of the era in which dinosaurs and other reptiles dominated the landscape and the beginning of the era when mammals become very important (Figure 8.G). Because the last period of the Mesozoic is the Cretaceous (abbreviated K to avoid confusion with other "C" periods), and the first period of the Cenozoic is the Tertiary (abbreviated T), the time of this mass extinction is called the *Cretaceous-Tertiary* or *KT boundary*.

The extinction of the dinosaurs is generally attributed to this group's inability to adapt to some radical change in environmental conditions. What event could have triggered the rapid extinction of the dinosaurs—one of the most successful groups of land animals ever to have lived?

The most strongly supported hypothesis proposes that about 65 million years ago a large meteorite approximately 10 kilometres in diameter collided with Earth (Figure 8.H). The speed at impact was about 70,000 kilometres per hour! The force of the impact vaporized the meteorite and trillions of tonnes of Earth's crust. Huge quantities of dust and other metamorphosed debris were blasted high into the atmosphere. For months the encircling dust cloud greatly restricted the sunlight reaching Earth's surface. With insufficient sunlight for photosynthesis, delicate food chains most likely would have collapsed. Large plant-eating dinosaurs would have been affected more adversely than smaller life forms because of the tremendous volume of vegetation they consumed. Next would have come the demise of large carnivores. When the sunlight returned, more than half of the

◆ **Figure 8.G** Dinosaurs dominated the Mesozoic landscape until their extinction at the close of the Cretaceous period. These dinosaur bones are exposed in Dinosaur National Monument, Utah.
(Photo by Tom & Susan Bean, Inc.)

species on Earth, including numerous marine organisms, had become extinct.

This period of mass extinction appears to have affected all land animals larger than dogs. Supporters of the catastrophic-event scenario suggest that small, ratlike mammals were able to survive a breakdown in their food chains lasting perhaps several months. Large dinosaurs, they contend, could not survive this event. The loss of these large reptiles opened a habitat for the small mammals that remained. This new habitat, along with evolutionary forces, led to the development of the large mammals that occupy our modern world.

What evidence points to such a catastrophic collision 65 million years ago? First, a thin layer of sediment nearly 1 centimetre thick has been discovered at the KT boundary, worldwide. This sediment contains a high level of the element *iridium*, rare in Earth's crust but found in high proportions in stony meteorites. Could this layer be the scattered remains of the meteorite that was responsible for the environmental changes that led to the demise of many reptile groups?

Despite its growing support, some scientists disagree with the impact hypothesis. They suggest instead that huge volcanic eruptions led to the breakdown in the food chain. To sup-

port their hypothesis, they cite enormous outpourings of lavas in the Deccan Plateau of northern India about 65 million years ago.

Whatever caused the KT extinction, we now have a greater appreciation of the role of catastrophic events in shaping the history of our planet and the life that occupies it. Could a catastrophic event having similar results occur today? This possibility may explain why an event that occurred 65 million years ago has captured the interest of so many.

◆ **Figure 8.H** Chicxulub crater is a giant impact crater that formed about 65 million years ago and has since been filled with sediments. About 180 kilometres in diameter, Chicxulub crater is regarded by some researchers as the impact site of the meteorite that resulted in the demise of the dinosaurs.

to simply be referred to as the **Precambrian**. Although it represents about 88 percent of Earth history, the Precambrian is not divided into nearly as many smaller time units as the Phanerozoic eon.

Why is the huge expanse of Precambrian time not divided into numerous eras, periods, and epochs? The reason is that Precambrian history is not known in great enough detail. The quantity of information geologists have deciphered about Earth's past is somewhat analogous to the detail of human history. The farther back we go, the less that is known. Certainly more data and information exist about the past 10 years than for the first decade of the twentieth century; the events of the nineteenth century have been documented much better than the events of the first century AD; and so on. So it is with Earth history. The more recent past has the freshest, least disturbed, and more observable record. The further back in time the geologist goes, the more fragmented the record and clues become. There are other reasons to explain our lack of a detailed time scale for this vast segment of Earth history:

1. The first abundant fossil evidence does not appear in the geologic record until the beginning of the Cambrian period. Prior to the Cambrian, simple life forms such as algae, bacteria, fungi, and worms predominated. All of these organisms lack hard parts, an important condition favouring preservation. For this reason, there is only a meagre Precambrian fossil record. Many exposures of Precambrian rocks have been studied in some detail, but correlation is often difficult when fossils are lacking.

2. Because Precambrian rocks are very old, most have been subjected to a great many changes. Much of the Precambrian rock record is composed of highly distorted metamorphic rocks. This makes the interpretation of past environments difficult, because many of the clues present in the original sedimentary rocks have been destroyed.

Radiometric dating has provided a partial solution to the troublesome task of dating and correlating Precambrian rocks. But untangling the complex Precambrian record still remains a daunting task.

Difficulties in Dating the Geologic Time Scale

Although reasonably accurate numerical dates have been worked out for the periods of the geologic time scale (Figure 8.15), the task is not without difficulty.

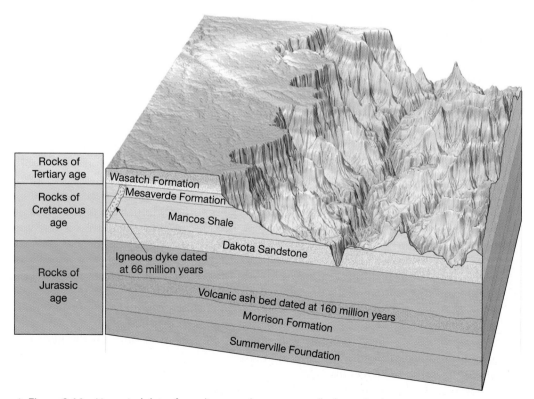

◆ **Figure 8.16** Numerical dates for sedimentary layers are usually determined by examining their relationship to igneous rocks. (After U.S. Geological Survey)

The primary difficulty in assigning numerical dates to units of time is the fact that not all rocks can be dated by radiometric methods. Recall that for a radiometric date to be useful, all the minerals in the rock must have formed at about the same time. For this reason, radioactive isotopes can be used to determine when minerals in an igneous rock crystallized and when pressure and heat created new minerals in a metamorphic rock.

However, samples of sedimentary rock can only rarely be dated directly by radiometric means. Although a detrital sedimentary rock may include particles that contain radioactive isotopes, the rock's age cannot be accurately determined because the grains composing the rock are not the same age as the rock in which they occur. Rather, the sediments have been weathered from rocks of diverse ages.

Radiometric dates obtained from metamorphic rocks may also be difficult to interpret, because the age of a particular mineral in a metamorphic rock does not necessarily represent the time when the rock initially formed. Instead, the date might indicate any one of a number of subsequent metamorphic phases in which the "clock" may have been reset.

If samples of sedimentary rocks rarely yield reliable radiometric ages, how can numerical dates be assigned to sedimentary layers? Usually the geologist must relate the strata to datable igneous masses, as in Figure 8.16.

This is one example of literally thousands that illustrate how datable materials are used to bracket the various episodes in Earth history within specific time periods. It shows the necessity of combining laboratory dating methods with field observations of rocks.

Chapter Summary

- The two types of dates used by geologists to interpret Earth history are (1) *relative dates*, which put events in their *proper sequence of formation*, and (2) *numerical dates*, which pinpoint the *time in years* when an event occurred.

- Relative dates can be established using the *law of superposition* (in an underformed sequence of sedimentary rocks or surface-deposited igneous rocks, each bed is older than the one above, and younger than the one below), *principle of original horizontality* (most layers are deposited in a horizontal position), *principle of cross-cutting relationships* (when a fault or intrusion cuts through another rock, the fault or intrusion is younger than the rocks cut through), and *inclusions* (the rock mass containing the inclusion is younger than the rock that provided the inclusion).

- *Unconformities* are gaps in the rock record. Each represents a long period during which deposition ceased, erosion removed previously formed rocks, and then deposition resumed. The three basic types of unconformities are *angular unconformities* (tilted or folded sedimentary rocks that are overlain by younger, more flat-lying strata), *disconformities* (the strata on either side of the unconformity are essentially parallel), and *nonconformities* (where a break separates older metamorphic or intrusive igneous rocks from younger sedimentary strata).

- *Correlation*, the matching up of two or more geologic phenomena in different areas, is used to develop a geologic time scale that applies to the whole Earth.

- Fossils are used to *correlate* sedimentary rocks that are from different regions by using the rocks' distinctive fossil content and applying the *principle of fossil succession*. It is based on the work of *William Smith* in the late 1700s and states that fossil organisms succeed one another in a definite and determinable order, and therefore any time period can be recognized by its fossil content. The use of *index fossils*, those that are widespread geographically and are limited to a short span of geologic time, provides an important method for matching rocks of the same age.

- Each atom has a nucleus containing *protons* (positively charged particles) and *neutrons* (neutral particles). Orbiting the nucleus are negatively charged *electrons*. The *atomic number* of an atom is the number of protons in the nucleus. The *mass number* is the number of protons plus the number of neutrons in an atom's nucleus. *Isotopes* are variants of the same atom, but with a different number of neutrons and hence a different mass number.

- *Radioactivity* is the spontaneous breaking apart (decay) of certain unstable atomic nuclei. Three common types of radioactive decay are

(1) emission of *alpha particles* from the nucleus, (2) emission of *beta particles* from the nucleus, and (3) *capture of electrons* by the nucleus.

- An unstable *radioactive isotope*, called the *parent*, will decay and form stable *daughter products*. The length of time for half of the nuclei of a radioactive isotope to decay is called the *half-life* of the isotope. If the half-life of the isotope is known, and the parent/daughter ratio can be measured, the age of a sample can be calculated. An accurate radiometric date can only be obtained if the mineral containing the radioactive isotope has remained in a closed system during the entire period since its formation.

- The *geologic time scale* divides Earth's history into units of varying magnitude. It is commonly presented in chart form, with the oldest time and event at the bottom and the youngest at the top. The principal subdivisions of the geologic time scale, called *eons*, include the *Hadean*, *Archean*, *Proterozoic* (together, these three eons are commonly referred to as the *Precambrian*), and, beginning about 540 million years ago, the *Phanerozoic*. The Phanerozoic (meaning "visible life") eon is divided into the following *eras: Paleozoic* ("ancient life"), *Mesozoic* ("middle life"), and *Cenozoic* ("recent life").

- A significant problem in assigning numerical dates is that *not all rocks can be radiometrically dated*. A sedimentary rock may contain particles of many ages that have been weathered from different rocks that formed at various times. One way geologists assign numerical dates to sedimentary rocks is to relate them to datable igneous masses, such as volcanic ash beds.

Review Questions

1. Distinguish between numerical and relative dating.

2. What is the law of superposition? How are crosscutting relationships used in relative dating?

3. Refer to Figure 8.4 (p. 221) and answer the following questions:
   (a) Is fault A older or younger than the sandstone layer?
   (b) Is dike A older or younger than the sandstone layer?
   (c) Was the conglomerate deposited before or after fault A?
   (d) Was the conglomerate deposited before or after fault B?
   (e) Which fault is older, A or B?
   (f) Is dike A older or younger than the batholith?

4. When you observe an outcrop of steeply inclined sedimentary layers, what principle allows you to assume that the beds were tilted after they were deposited?

5. A mass of granite is in contact with a layer of sandstone. Using a principle described in this chapter, explain how you might determine whether the sandstone was deposited on top of the granite, or whether the granite was intruded from below after the sandstone was deposited?

6. Distinguish among angular unconformity, disconformity, and nonconformity.

7. What is meant by the term *correlation?*

8. Describe William Smith's important contribution to the science of geology.

9. Why are fossils such useful tools in correlation?

10. Figure 8.17 (p. 242) is a block diagram of a hypothetical area in Canada. Place the lettered features in the proper sequence, from oldest to youngest. Identify an angular unconformity and a nonconformity.

11. If a radioactive isotope of thorium (atomic number 90, mass number 232) emits 6 alpha particles and 4 beta particles during the course of radioactive decay, what are the atomic number and mass number of the stable daughter product?

12. Why is radiometric dating the most reliable method of dating the geologic past?

13. A hypothetical radioactive isotope has a half-life of 10,000 years. If the ratio of radioactive parent to stable daughter product is 1:3, how old is the rock containing the radioactive material?

14. Briefly describe why tree rings might be helpful in studying the geologic past (see Box 8.3).

15. In order to provide a reliable radiometric date, a mineral must remain a closed system from the time of its formation until the present. Why is this true?

16. What precautions are taken to ensure reliable radiometric dates?

17. To make calculations easier, let us round the age of Earth to 5 billion years.
    (a) What fraction of geologic time is represented by recorded history (assume 5000 years for the length of recorded history)?
    (b) The first abundant fossil evidence does not appear until the beginning of the Cambrian period (540 million years ago). What percent of geologic time is represented by abundant fossil evidence?

18. What subdivisions make up the geologic time scale?
19. Explain the lack of a detailed time scale for the vast span known as the Precambrian.
20. Briefly describe the difficulties in assigning numerical dates to layers of sedimentary rock.

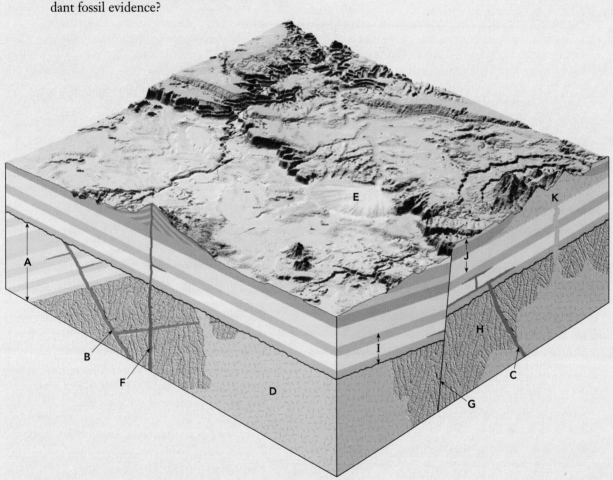

◆ **Figure 8.17** Use this block diagram in conjunction with Review Question 10.

Key Terms

angular unconformity (p. 223)
Archean Eon (p. 237)
Cenozoic Era (p. 236)
conformable (p. 222)
correlation (p. 224)
cross-cutting relationships, principle of (p. 220)
disconformity (p. 223)

eon (p. 236)
epoch (p. 236)
era (p. 236)
fossil succession, principle of (p. 226)
geologic time scale (p. 236)
Hadean Eon (p. 237)
half-life (p. 230)
inclusions (p. 220)

index fossil (p. 226)
Mesozoic Era (p. 236)
nonconformity (p. 223)
numerical date (p. 218)
original horizontality, principle of (p. 219)
Paleozoic Era (p. 236)
period (p. 236)
Phanerozoic Eon (p. 236)
Precambrian (p. 239)

Proterozoic Eon (p. 237)
radioactivity (p. 228)
radiocarbon dating (p. 233)
radiometric dating (p. 229)
relative dating (p. 218)
superposition, law of (p. 219)
unconformity (p. 222)

Web Resources

 The *Earth* Web site uses the resources and flexibility of the Internet to aid in your study of the topics in this chapter. Written and developed by geology instructors, this site will help improve your understanding of geology. Visit **http://www.pearson.ca/tarbuck** and click on the cover of the text to find:

- Online review quizzes.
- Web-based critical thinking and writing exercises.
- Links to chapter-specific Web resources.
- Internet-wide key-term searches.

http://www.pearson.ca/tarbuck

Chapter 9
Mass Wasting: The Work of Gravity

This rock slide occurred in January 1997 on Highway 140 near the Arch Rock entrance to Yosemite National Park, California. (Roger J. Wyan/AP/Wide World Photos)

Earth's surface is never perfectly flat, but instead consists of slopes of different lengths and gradients. Slopes can be mantled with soil and vegetation or consist of barren rock and rubble. Taken together, slopes are important elements in our physical landscape. Although most slopes may appear to be stable and unchanging, the force of gravity ultimately causes material to move downslope. At one extreme, the movement may be gradual and practically imperceptible. At the other extreme, it may consist of a roaring debris flow or a thundering rock avalanche. When these hazardous natural processes lead to loss of life and property, they become natural disasters.

A Landslide Disaster in Peru

We periodically hear news reports relating the terrifying and often grim details of landslides. On May 31, 1970, one such event occurred when a gigantic rock avalanche buried more than 20,000 people in Yungay and Ranrahirca, Peru (Figure 9.1). There was little warning of the impending disaster; it began and ended in just a matter of minutes. The avalanche started about 14 kilometres from Yungay, near the summit of the 6700-metre Nevado Huascaran, the loftiest peak in the Peruvian Andes. Triggered by the ground motion from a strong offshore earthquake, a huge mass of rock and ice broke free from the precipitous north face of the mountain. After plunging nearly a kilometre, the material pulverized on impact

and immediately began rushing down the mountainside, made fluid by trapped air and melted ice.

The falling debris ripped loose millions of tonnes of additional debris as it roared downhill. Hurricane-speed winds generated as compressed air that escaped from beneath the avalanche mass created thunderlike noise and stripped nearby hillsides of vegetation. Although the material followed a previously eroded gorge, a portion of the debris jumped a 200- to 300-metre bedrock ridge that had protected Yungay from similar events in the past and buried the entire city. After destroying another town in its path, Ranrahirca, the mass of debris finally reached the bottom of the valley. There, its momentum carried it across the Rio Santa and tens of metres up the opposite valley wall.

This was not the first such disaster in the region and will probably not be the last. Just eight years earlier a less spectacular, but still devastating, rock avalanche took the lives of an estimated 3500 people on the heavily populated valley floor at the base of the mountain. Fortunately, mass movements such as these are infrequent and only occasionally affect large numbers of people.

Mass Wasting and Landform Development

Landslides are spectacular examples of a basic geologic process called mass wasting. **Mass wasting** refers to the downslope movement of rock, regolith, and soil

A

B

◆ **Figure 9.1**   This Peruvian valley was devastated by a rock avalanche that was triggered by an offshore earthquake in May 1970. **A**. Before. **B**. After the rock avalanche.
(Photos courtesy of Iris Lozier)

under the direct influence of gravity. It is distinct from the erosional processes examined in subsequent chapters because mass wasting does not require a transporting medium such as water, wind, or glacial ice.

The Role of Mass Wasting

Landforms develop as products of weathering are removed from the places where they originate. Once weathering weakens and breaks rock apart, mass wasting carries the debris downslope, where a stream, acting as a conveyor belt, usually carries it away. Although there may be many intermediate stops along the way, the sediment is eventually transported to the sea.

The combined effects of mass wasting and running water produce stream valleys, the most common and conspicuous of Earth's landforms. If streams alone were responsible for creating the valleys in which they flow, the valleys would be very narrow. The fact that most river valleys are much wider than they are deep indicates the significance of mass-wasting processes in supplying material to streams. This is illustrated by the Grand Canyon (Figure 9.2). The walls of the canyon extend far from the Colorado River, owing to the transfer of weathered debris downslope to the river and its

tributaries by mass wasting processes. In this manner, streams and mass wasting combine to sculpt the surface along with groundwater, waves, and wind.

Slopes Change Through Time

Clearly, if mass wasting is to occur, there must be slopes that rock, soil, and regolith can move down. It is Earth's mountain-building and volcanic processes that produce these slopes, sporadically changing the elevations of landmasses and the ocean floor. If dynamic internal processes did not continually produce regions of higher elevations, the transfer of debris to lower elevations would gradually slow and eventually cease.

Most extreme mass-wasting events occur in areas of rugged, geologically young mountains. Newly formed mountains are rapidly eroded by rivers and glaciers, producing steep and unstable slopes. It is in such settings that massive destructive landslides, such as the Yungay disaster, occur. As mountain-building subsides, mass wasting and erosional processes lower the land. Over time, steep and rugged mountains are worn down to form gentler, more subdued terrain with smaller, less dramatic downslope movements.

◆ **Figure 9.2** The walls of the Grand Canyon extend far from the channel of the Colorado River. This results primarily from the transfer of weathered debris downslope to the river and its tributaries by mass wasting processes. (Photo by Tom and Susan Bean, Inc.)

Controls and Triggers of Mass Wasting

Gravity is the controlling force of mass wasting, but several factors play an important role in promoting downslope movements. Long before a landslide occurs, various processes work to weaken slope material, gradually making it more and more susceptible to failure. During this span, the slope becomes less stable. Eventually, the strength of the slope is weakened to the point that something **triggers**, or initiates, downslope movement. Remember that the trigger is not the sole cause of the mass-wasting event, but just the last of many causes. Among the common factors that trigger mass-wasting processes are saturation of material with water, oversteepening of slopes, removal of anchoring vegetation, and ground vibrations from earthquakes.

The Role of Water

Mass wasting is sometimes triggered when heavy rains or periods of snowmelt saturate surface materials. This was the case in October 1998 when torrential downpours associated with Hurricane Mitch triggered devastating mudflows in Central America (Figure 9.3).

When the pores in sediment become filled with water, the cohesion among particles is destroyed, allowing them to slide past one another with relative ease. For example, when sand is slightly moist, it sticks together quite well. However, if enough water is added to fill the openings between the grains, the sand oozes out in all directions. Thus, saturation reduces the internal strength of materials, which are then easily set in motion by the force of gravity. When clay is wetted, it becomes very slick—another example of the "lubricating" effect of water. Water also adds considerable weight to a mass of material. The added weight in itself may be enough to cause the material to slide or flow downslope.

Oversteepened Slopes

Oversteepening of slopes can also trigger mass movements. There are many natural settings where oversteepening takes place. A stream undercutting a valley wall and waves pounding against the base of a cliff are but two familiar examples. Furthermore, through their activities, people often create oversteepened and unstable slopes that become prime sites for mass wasting (see Box 9.1).

Unconsolidated, granular particles (sand-size or coarser) assume a stable slope called the **angle of repose** (*repose* = to be at rest). This is the steepest angle at which material remains stable. Depending on the size and shape of the particles, the angle varies from 25 to 40 degrees. The larger, more angular particles maintain the steepest slopes. If the angle is increased, the rock debris will adjust by moving downslope.

Oversteepening is not just important because it triggers movements of unconsolidated granular materials. Oversteepening also produces unstable slopes and mass movements in cohesive soils, regolith, and bedrock. The response will not be immediate, as with loose, granular material, but sooner or later one or more mass-wasting processes will eliminate the oversteepening and restore stability to the slope.

Removal of Vegetation

Plants contribute to the stability of slopes because their root systems bind soil and regolith together. Where plants are lacking, mass wasting is enhanced, especially if slopes are steep and water is plentiful. When anchoring vegetation is removed by forest fires or by people (for timber, farming, or development), surface materials frequently move downslope.

In July 1994 a severe wildfire swept Storm King Mountain west of Glenwood Springs, Colorado, denuding the slopes of vegetation. Two months later heavy rains resulted in numerous debris flows, inundating a major highway with tons of rock, mud, and burned trees. Since extensive wildfires occurred in the summer of 2000, similar types of mass wasting threaten highways and other development near fire-ravaged hillsides throughout the Western U.S. (Figure 9.4).

◆ **Figure 9.3** Heavy rains from Hurricane Mitch in the fall of 1998 triggered devastating mudflows in Central America. Water plays an important role in many mass-wasting processes. (Photo by Noel Quidu/Liaison Agency, Inc.)

BOX 9.1

People and the Environment
The Vaiont Dam Disaster

A massive rock avalanche in Peru is described at the beginning of this chapter. As with most occurrences of mass wasting, this tragic episode was triggered by a natural event—in this case, an earthquake. However, disasters also result from the mass movement of surface material triggered by the actions of humans.

In 1960 a large dam, almost 265 metres tall, was built across Vaiont Canyon in the Italian Alps. It was engineered without good geological input, and the result was a disaster only three years later.

The bedrock in Vaiont Canyon slanted steeply downward toward the lake impounded behind the dam. The bedrock was weak, highly fractured limestone strata with beds of clay and numerous solution cavities. As the reservoir filled behind the completed dam, rocks became saturated and the clays became swollen and more plastic. The rising water reduced the internal friction that had kept the rock in place.

Measurements made shortly after the reservoir was filled hinted at the problem, because they indicated that a portion of the mountain was slowly creeping downhill at the rate of 1 centimetre per week. In September 1963 the rate increased to 1 centimetre per day, then 10–20 centimetres per day, and eventually as much as 80 centimetres on the day of the disaster.

Finally, the mountainside let loose. In just an instant, 240 million cubic

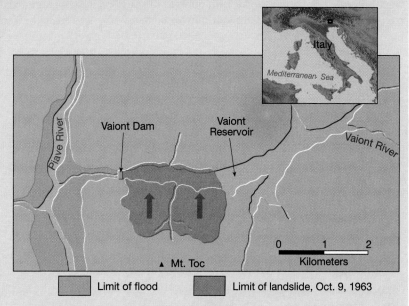

metres of rock and rubble slid down the mountainside and filled nearly 2 kilometres of the gorge to heights of 150 meters above the reservoir level (Figure 9.A). This pushed the water over the dam in a wave more than 90 meters high. More than 1.5 kilometres downstream, the wall of water was still 70 meters high, destroying everything in its path.

The entire event lasted less than seven minutes, yet it claimed an esti-

mated 2600 lives. Although this is known as the worst dam disaster in history, the Vaiont Dam itself remained intact. And while the catastrophe was triggered by human interference with the Vaiont River, the slide eventually would have occurred on its own; however, the effects would not have been nearly as tragic.

◆ **Figure 9.A**   Sketch map of the Vaiont River area showing the limits of the landslide, the portion of the reservoir that was filled with debris, and the extent of flooding downstream.
(After G.A. Kiersch, "Vaiont Reservoir Disaster," *Civil Engineering* 34 (1964): 32–39.)

An unusual example illustrating the anchoring effect of plants occurred several decades ago on steep slopes near Menton, France. Farmers replaced olive trees, which have deep roots, with a more profitable but shallow-rooted crop of carnations. When the slope became unstable and failed, the landslide took 11 lives.

Earthquakes as Triggers

Conditions that favour mass wasting may exist in an area for a long time before movement occurs. An additional factor is sometimes necessary to trigger the

movement. Among the more important and dramatic triggers are earthquakes. An earthquake and its aftershocks can dislodge enormous volumes of rock and unconsolidated material. The event in the Peruvian Andes described at the beginning of the chapter is one tragic example.

Landslides Triggered by the Northridge Earthquake
In January 1994 an earthquake struck the Los Angeles region of southern California. Named for its epicentre in the town of Northridge, the 6.7-magnitude event produced estimated losses of $20 billion. Some

of relatively thin, isolated masses. Farther north, the area and thickness gradually increase to the point where the permafrost is essentially continuous and its thickness may approach or even exceed 500 metres. In the discontinuous zone, land-use planning is frequently more difficult than in the continuous zone farther north because the occurrences of permafrost are patchy and difficult to predict.

When people disturb the surface, such as by removing the insulating vegetation mat or by building roads and buildings, the delicate thermal balance is disturbed, and the permafrost can thaw (Figure 9.C). Thawing produces unstable ground that may slide, slump, subside, and undergo severe frost heaving.

As Figure 9.D illustrates, when a heated structure is built directly on permafrost that contains a high proportion of ice, thawing creates soggy material into which a building can sink. One solution is to place buildings and other structures on piles, like stilts. Such piles allow subfreezing air to circulate between the floor of the building and the soil and thereby keep the ground frozen.

When oil was discovered on Alaska's North Slope, many people were concerned about the building of a pipeline linking the oil fields at Prudhoe Bay to the ice-free port of Valdez 1300 kilo-metres to the south. There was serious concern that such a massive project might damage the sensitive permafrost environment. Many also worried about possible oil spills.

Because oil must be heated to about 60°C to flow properly, special engineering procedures had to be developed to isolate this heat from the permafrost. Methods included insulating the pipe, elevating portions of the pipeline above ground level, and even placing cooling devices in the ground to keep it frozen. The Alaskan pipeline is clearly one of the most complex and costly projects ever built in the Arctic tundra. Detailed studies and careful engineering helped minimize adverse effects resulting from the disturbance of frozen ground.

◆ **Figure 9.C** When a rail line was built across this permafrost landscape in Alaska, the ground subsided. (Photo by Lynn A. Yehle, U.S. Geological Survey)

◆ **Figure 9.D** These buildings in Dawson, Yukon, are tilted because they rest directly on permafrost that is melting from the heat of the buildings. (Photo by Steve Hicock)

A   B

◆ **Figure 9.18** A. Solifluction lobes northeast of Fairbanks, Alaska. Solifluction occurs in permafrost regions when the active layer thaws in summer. B Closeup of a gravelly solifluction lobe (above shovel) in mainland Nunavut. (Photo by Steve Hicock)

chain as well as along the continental shelf and slope of North America (see Box 18.1). In fact, many submarine landslides, mostly in the form of slumps and debris avalanches, appear to be much larger than any similar mass-wasting events that occur on land.

Among the most spectacular underwater landslides are those that occur on the flanks of submarine volcanoes (called *seamounts*) and volcanic islands such as Hawaii. One of the largest yet mapped, known as the Nuuanu debris avalanche, is on the northeastern side of Oahu. It extends for nearly 25 kilometres across the ocean floor, then rises up a 300-metre slope at its terminus, indicating that it must have had great power and momentum. Giant blocks many kilometres across were transported by this giant landslide. It is probable that when such large and rapid events occur, they produce giant sea waves called tsunami that race across the Pacific.*

The massive submarine slides discovered on the flanks of the Hawaiian Islands are almost certainly related to the movement of magma while a volcano is active. As huge quantities of lava are added to the seaward margin of a volcano, the buildup of

material eventually triggers a great landslide. In the Hawaiian chain, it appears that this process of growth and collapse is repeated at intervals of 100,000 to 200,000 years while the volcano is active.

Along the continental margins of North America, large slumps and debris-flow scars mark the continental slope. These processes are triggered by the rapid buildup of unstable sediments or by such forces as storm waves and earthquakes (see Box 18.1). Submarine mass wasting is especially active near deltas, the massive deposits of sediment at the mouths of rivers. Here, as great loads of water-saturated clay and organic-rich sediments accumulate, they become unstable and readily flow down even gentle slopes. Some of these movements have been forceful enough to damage large offshore drilling platforms.

Mass wasting appears to be an integral part of the growth of passive continental margins. Sediments supplied to the continental shelf by rivers move across the shelf to the upper continental slope. From here, slumps, slides, and debris flows move sediment down to the continental rise and sometimes beyond.

---

*For more on these destructive waves, see the section on tsunamis in Chapter 16.

Chapter Summary

- *Mass wasting* refers to the downslope movement of rock, regolith, and soil under the direct influence of gravity. In the evolution of most landforms, mass wasting is the step that follows weathering. The combined effects of mass wasting and erosion by running water produce stream valleys.

- *Gravity is the controlling force of mass wasting.* Other factors that influence or trigger downslope movements are saturation of the material with water, oversteepening of slopes beyond the *angle of repose*, removal of vegetation, and ground shaking by earthquakes.

- The various processes included under the name of mass wasting are divided and described on the basis of (1) the type of material involved (debris, mud, earth, or rock); (2) the type of motion (fall, slide, or flow); and (3) the rate of movement (rapid or slow).

- The more rapid forms of mass wasting include *slump*, the downward sliding of a mass of rock or unconsolidated material moving as a unit along a curved surface; *rockslide*, blocks of bedrock breaking loose and sliding downslope; *debris flow*, a relatively rapid flow of soil and regolith containing a large amount of water; and *earthflow*, an unconfined flow of saturated clay-rich soil that most often occurs on a hillside in a humid area following heavy precipitation or snowmelt.

- The slowest forms of mass wasting include *creep*, the gradual downhill movement of soil and regolith; and *solifluction*, the gradual flow of a saturated surface layer that is underlain by an impermeable zone. Common sites for solifluction are regions underlain by *permafrost* (permanently frozen ground associated with tundra and ice-cap climates).

- Mass wasting is not confined to land; it also occurs underwater. Many *submarine landslides*, mostly slumps and debris avalanches, are much larger than those that occur on land.

Review Questions

1. Describe how mass-wasting processes contribute to the development of stream valleys.

2. How did the building of a dam contribute to the Vaiont Canyon disaster? Was the disaster avoidable? (See Box 9.1, p. 249).

3. What is the controlling force of mass wasting?

4. How does water affect mass-wasting processes?

5. Describe the significance of the angle of repose.

6. How might the removal of vegetation by fire or logging promote mass wasting?

7. How are earthquakes linked to landslides?

8. Distinguish among fall, slide, and flow.

9. Why can rock avalanches move at such great speeds?

10. Both slump and rockslide move by sliding. In what ways do these processes differ?

11. What factors led to the massive rockslide at Frank, Alberta?

12. Compare and contrast mudflow and earthflow.

13. Describe the mass wasting that occurred at Mount St. Helens during its active period in 1980 and at Nevado del Ruiz in 1985.

14. Because creep is an imperceptibly slow process, what evidence might indicate that this phenomenon is affecting a slope?

15. What is permafrost? What portion of Earth's land surface is affected? (See Box 9.2, p. 260.)

16. During what season does solifluction occur in permafrost regions?

Key Terms

angle of repose (p. 248)
creep (p. 259)
debris flow (p. 256)
debris slide (p. 254)

earthflow (p. 258)
fall (p. 251)
flow (p. 252)
lahar (p. 257)

mass wasting (p. 246)
permafrost (p. 260)
rock avalanche (p. 252)
rockslide (p. 254)

slide (p. 252)
slump (p. 253)
solifluction (p. 259)
triggers (p. 248)

Web Resources

 The *Earth* Web site uses the resources and flexibility of the Internet to aid in your study of the topics in this chapter. Written and developed by geology instructors, this site will help improve your understanding of geology. Visit **http://www.pearson.ca/tarbuck** and click on the cover of the text to find:

■ Online review quizzes.

■ Web-based critical thinking and writing exercises.

■ Links to chapter-specific Web resources.

■ Internet-wide key-term searches.

http://www.pearson.ca/tarbuck

Chapter 10
Running Water

The St. Mary River meanders through the prairie landscape of southern Alberta. The bank on the left is being eroded, while sediment is being deposited on the opposite bank. (Photo by C. Tsujita)

Rivers are very important to people. We use them as highways for moving goods, as sources of water for irrigation, and as an energy source. Their fertile floodplains have been cultivated since the dawn of civilization. When viewed as part of the Earth system, rivers and streams represent a basic link in the constant cycling of the planet's water. Moreover, running water is the dominant agent of landscape alteration, eroding more terrain and transporting more sediment than any other process. Because so many people live near rivers, floods are among the most destructive of all geologic hazards. Despite huge investments in levees and dams, rivers cannot always be controlled (Figure 10.1).

Earth as a System: The Hydrologic Cycle

External Processes
↳ Hydrologic Cycle

All the rivers run into the sea; yet the sea is not full; unto the place from whence the rivers come, thither they return again.
(Ecclesiastes 1:7)

As the perceptive writer of Ecclesiastes indicated, water is continually on the move, from the ocean to the land and back again in an endless cycle. Water is just about everywhere on Earth—in the oceans, glaciers, rivers, lakes, the air, soil, and in living tissue. All of these "reservoirs" constitute Earth's hydrosphere. In all, the water content of the hydrosphere is about 1.36 billion cubic kilometres.

The vast bulk of it, about 97.2 percent, is stored in the global oceans (Figure 10.2). Ice sheets and glaciers account for another 2.15 percent, leaving only 0.65 percent to be divided among lakes, streams, subsurface water, and the atmosphere (Figure 10.2). Although the percentages of Earth's water found in each of the latter sources is just a small fraction of the total inventory, the absolute quantities are great.

The water found in each of the reservoirs depicted in Figure 10.2 does not remain in these places indefinitely. Water can readily change from one state of matter (solid, liquid, or gas) to another at the temperatures and pressures that occur at Earth's surface. Therefore, water is constantly moving among the hydrosphere, the atmosphere, the solid Earth, and the biosphere. This unending circulation of Earth's water supply is called the **hydrologic cycle**. The cycle shows us many critical interrelationships among different parts of the Earth system.

The hydrologic cycle is a gigantic worldwide system powered by energy from the Sun in which the atmosphere provides the vital link between the oceans and continents (Figure 10.3). Water evaporates into the atmosphere from the ocean and to a much lesser extent from the continents. Winds transport this moisture-laden air, often for great distances, until conditions cause the moisture to condense into clouds and precipitation to fall. The precipitation that falls into the ocean has completed its cycle and is ready to begin another. The water that falls on land, however, must make its way back to the ocean.

◆ **Figure 10.1**   Most Canadians have witnessed stream flooding associated with the melting of snow during the spring. This photo shows a typical spring flood in the Thames River valley in London, Ontario.
(Photo courtesy of R.W. Hodder)

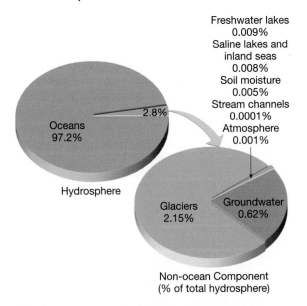

Freshwater lakes 0.009%
Saline lakes and inland seas 0.008%
Soil moisture 0.005%
Stream channels 0.0001%
Atmosphere 0.001%

Oceans 97.2%
2.8%
Hydrosphere

Glaciers 2.15%
Groundwater 0.62%

Non-ocean Component (% of total hydrosphere)

◆ **Figure 10.2**   Distribution of Earth's water.

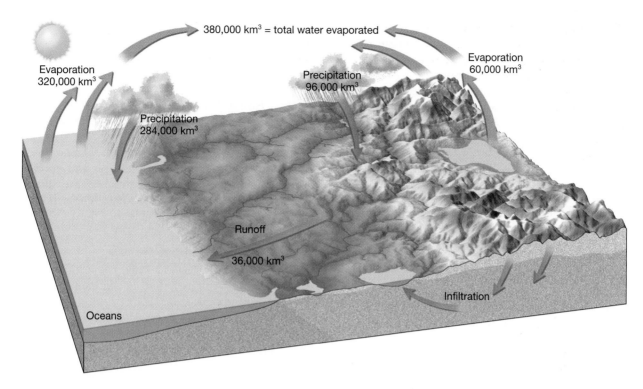

♦ **Figure 10.3** Earth's water balance. Each year, solar energy evaporates about 320,000 cubic kilometres of water from the oceans, while evaporation from the land (including lakes and streams) contributes 60,000 cubic kilometres of water. Of this total of 380,000 cubic kilometres of water, about 284,000 cubic kilometres fall back to the ocean, and the remaining 96,000 cubic kilometres fall on the land surface. Of that 96,000 cubic kilometres, only 60,000 cubic kilometres of water evaporate from the land, leaving 36,000 cubic kilometres of water to erode the land during the journey back to the oceans.

What happens to precipitation once it has fallen on the land? A portion of the water soaks into the ground (called **infiltration**) moving downward, then laterally, and finally seeping into lakes, streams, or directly into the ocean. When the rate of rainfall is greater than the land's ability to absorb it, the additional water flows over the surface into lakes and streams, a process called **runoff**. Much of the water that infiltrates or runs off eventually finds its way back to the atmosphere because of evaporation from the soil, lakes, and streams. Also, some of the water that infiltrates the ground is absorbed by plants, which later release it into the atmosphere. This process is called **transpiration** (*trans* = across, *spiro* = to breathe).

Each year a field of crops may transpire an amount of water equivalent to a layer 60 centimetres deep over the entire field. The same area of trees may pump twice this amount into the atmosphere. Because we cannot clearly distinguish between the amount of water that is evaporated and the amount that is transpired by plants, the term **evapotranspiration** is often used for the combined effect.

When precipitation falls in very cold areas—at high elevations or high latitudes—the water may not immediately soak in, run off, or evaporate. Instead, it may become part of a snowfield, or ultimately, a glacier. In this way, glaciers store large quantities of water on land. If present-day glaciers were to melt and release their stored water, sea level would rise by several tens of metres worldwide and submerge many heavily populated coastal areas. As we shall see in Chapter 12, over the past two million years huge ice sheets have formed and melted on several occasions, each time changing the balance of the hydrologic cycle.

Figure 10.3 also shows Earth's overall *water balance*, or the volume of water that passes through each part of the cycle annually. The amount of water vapour in the air is just a tiny fraction of Earth's total water supply. But the absolute quantities that are cycled through the atmosphere over a one-year period are immense—some 380,000 cubic kilometres. Estimates show that over North America almost six times as much water is carried by moving currents of air as is transported by all the continent's rivers.

It is important to know that the hydrologic cycle is balanced. Because the total water vapour in the atmosphere remains about the same, average annual precipitation over Earth must be equal to the quantity of water evaporated. However, for all of the continents taken together, precipitation exceeds evaporation.

Conversely, over the oceans, evaporation exceeds precipitation. Since the level of the world ocean is not dropping, the system must be in balance.

The erosional work accomplished by the 36,000 cubic kilometres of water that flows annually from the land to the ocean is enormous. Arthur Bloom effectively described it as follows:

> The average continental height is 823 metres above sea level.... If we assume that the 36,000 cubic kilometres of annual runoff flow downhill an average of 823 metres, the potential mechanical power of the system can be calculated. Potentially, the runoff from all lands would continuously generate almost 9×10^9 kW. If all this power were used to erode the land, it would be comparable to having... one horse-drawn scraper or scoop at work on each 3-acre piece of land, day and night, year round. Of course, a large part of the potential energy of runoff is wasted as frictional heat by turbulent flow and splashing of water.*

Although only a small percentage of the energy of running water is used to erode the surface, running water nevertheless is *the single most important agent sculpting Earth's land surface.*

To summarize, the hydrologic cycle represents the continuous movement of water from the oceans to the atmosphere, from the atmosphere to the land, and from the land back to the sea. The wearing down of Earth's land surface is largely attributable to the last of these steps and is the primary focus of the remainder of this chapter.

Running Water

Although we have always depended to a great extent on running water, its source eluded us for centuries. Not until the sixteenth century did we realize that streams were supplied by surface runoff and underground water, which ultimately had their sources as rain and snow.

Runoff initially flows in broad, thin sheets across the ground, appropriately termed **sheet flow**. The amount of water that runs off in this manner rather than sinking into the ground depends upon the **infiltration capacity** of the soil. Infiltration capacity is controlled by many factors, including: (1) the intensity and duration of the rainfall, (2) the prior wetted condition of the soil, (3) the soil texture, (4) the slope of the land, and (5) the nature of the vegetative cover. When the soil becomes saturated, sheet

_____

Geomorphology: A Systematic Analysis of Late Cenozoic Landforms (Englewood Cliffs, N.J.: Prentice Hall, 1978), p. 97.

---

flow commences as a layer only a few millimetres thick. After flowing as a thin, unconfined sheet for only a short distance, threads of current typically develop and tiny channels called **rills** begin to form and carry the water to a stream.

The remainder of this chapter will concentrate on that part of the hydrologic cycle in which the water moves in stream channels. The discussion will deal primarily with streams in humid regions. Streams are also important in arid landscapes, but we will examine that in Chapter 13, "Deserts and Winds."

Streamflow

 External Processes
↳ Running Water

Water may flow in one of two ways, either as **laminar flow** or **turbulent flow**. When the movement is laminar, the water particles flow in straight-line paths that are parallel to the channel. By contrast, when the flow is turbulent, the water moves in a confused and erratic fashion that is often characterized by swirling, whirlpool-like eddies (Figure 10.4).

The stream's velocity is a primary factor that determines whether the flow is laminar or turbulent. Laminar flow is possible only when water is moving very slowly through a smooth channel. If the velocity increases or the channel becomes rough, laminar flow changes to turbulent flow. The movement of water in streams is usually fast enough that flow is turbulent. The multidirectional movement of turbulent flow is very effective both in eroding a stream's channel and in keeping sediment suspended within the water so that it can be transported downstream.

Flowing water makes its way to the sea under the influence of gravity. Some sluggish streams flow at less than 1 kilometre per hour, whereas a few rapid ones may exceed 30 kilometres per hour. Velocities are determined at gauging stations where measurements

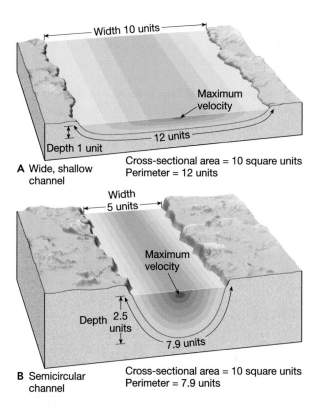

◆ **Figure 10.4** Most streamflow is turbulent, although it is usually not as rough as that experienced by this rafter on the Colorado River.
(Photo by Tom & Susan Bean, Inc.)

are taken at several locations across the stream channel and then averaged. This is done because the rate of water movement is not uniform within a stream channel. When the channel is straight, the highest velocities occur in the centre just below the surface (Figure 10.5). It is here that friction is least. Minimum velocities occur along the sides and bottom (bed) of the channel where friction is always greatest. When a stream channel is crooked or curved, the fastest flow is not in the centre. Rather, the zone of maximum velocity shifts toward the outside of each bend. As we shall see later, this shift plays an important part in eroding the stream's channel on that side.

The ability of a stream to erode and transport material is directly related to its velocity. Even slight variations in velocity can lead to significant changes in the load of sediment that water can transport. Several factors determine the velocity of a stream and therefore control the amount of erosional work a stream may accomplish. These factors include (1) the gradient, (2) the shape, size, and roughness of the channel, and (3) the discharge.

Gradient and Channel Characteristics

Certainly one of the most obvious factors controlling stream velocity is the **gradient**, or slope, of a stream channel. Gradient is typically expressed as the vertical drop of a stream over a fixed distance. Gradients vary considerably from one stream to another as well as along the course of a given stream.

C Gauging station

◆ **Figure 10.5** Influence of channel shape on velocity. **A.** The stream in this wide, shallow channel moves more slowly than does water in the semicircular channel because of greater frictional drag. **B.** The cross-sectional area of this semicircular channel is the same as the one in part A, but it has less water in contact with its channel and therefore less frictional drag. Thus, water will flow more rapidly in channel B, all other factors being equal. **C.** Continuous records of stage and discharge are collected by the U.S. Geological Survey at more than 7000 gauging stations in the United States. Average velocities are determined by using measurements from several spots across the stream. This station is on the Rio Grande south of Taos, New Mexico.
(Photo by E. J. Tarbuck)

Portions of the lower Mississippi River, for example, have very low gradients of 10 centimetres per kilometre and less. By way of contrast, some steep

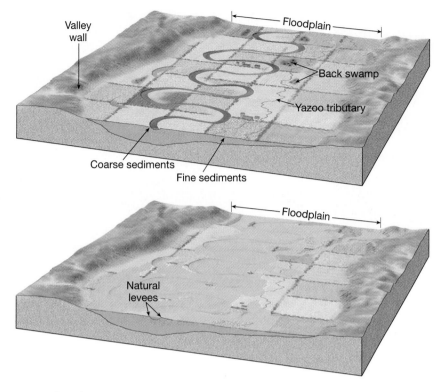

Valley wall

Floodplain

Back swamp

Yazoo tributary

Coarse sediments

Fine sediments

Floodplain

Natural levees

◆ **Figure 10.16** Natural levees are gently sloping structures that are created by repeated floods. Because the ground next to the stream channel is higher than the adjacent floodplain, back swamps and yazoo tributaries may develop.

stream has deep, interwoven, multiple channels separated by stable islands that are often heavily vegetated (Figure 10.17A). In other words, anastomosed streams are produced by the actual cutting of multiple channels rather than diversion around buildups of sediment. Also, anastomosed streams tend to be dominated by fine grained-suspended load as opposed to coarse-grained bedload.

Streams of anastomosed form are the laziest of rivers, tending to occur in areas of extremely low gradient and therefore possessing little energy to carve their banks. Like meandering stream systems, anastomosed systems tend to have broad, muddy floodplains and levees along their channel banks.

The relatively small amounts of coarse-grained bedload in anastomosed systems are ultimately deposited in channels. However, unlike the channels of meandering streams that migrate laterally by eroding their outside bank and depositing sediment on point bars, those of anastomosed streams tend to be filled with sediment in a vertical direction with little or no lateral accretion. In the geologic record, channel deposits of anastomosed streams can be recognized as lens-shaped bodies of horizontally bedded sandstones or conglomerates encased in muddy (floodplain) sediments (Figure 10.17B).

Alluvial Fan and Delta Systems

Two of the most common landforms composed of alluvium are alluvial fans and deltas. They are sometimes similar in shape and are deposited for essentially the same reason: an abrupt loss of competence in a stream. The key distinction between them is that alluvial fans are deposited on land whereas deltas are deposited in a body of water. In addition, alluvial fans can be quite steep, but deltas are relatively flat, barely protruding above the level surface of the ocean or lake in which they formed.

Alluvial Fans   **Alluvial fans** typically develop where a high-gradient stream leaves a narrow valley in mountainous terrain and comes out suddenly onto a broad, flat plain or valley floor (Figure 10.18). Alluvial fans form in response to the abrupt drop in gradient combined with the change from a narrow channel of a mountain stream to less confined channels at the base of the mountains. The sudden drop in velocity causes the stream to dump its load of sediment quickly in a distinctive cone- or fan-shaped accumulation. As illustrated by Figure 10.18, the surface of the fan slopes outward in a broad arc from an apex at the mouth of the steep valley. Usually, coarse material is dropped near the apex of the fan, while finer material is carried toward the base of the deposit. As we learned in Chapter 9, steep canyons in dry regions are prime locations for debris flows. Therefore, it should be expected that many alluvial fans in arid areas have debris-flow deposits interbedded with the alluvium.

Deltas   In contrast to an alluvial fan, a **delta** forms when a stream enters an ocean or a lake. Figure 10.19A

A

B

◆ **Figure 10.17   A.** The upper Columbia River in southeastern British Columbia is a good example of an anastomosed stream. Note the multiple active channels and intervening islands with wetlands. The light-coloured sandy deposit on the left bank of the main channel is a crevasse splay, produced when the stream flow locally breached the channel bank. **B.** A sand-filled channel (lightest coloured) of an ancient anastomosed river (Cretaceous Dunvegan Formation, Pine River area, British Columbia), encased in muddy floodplain deposits.
(Photo A by Derald Smith; Photo B by Guy Plint)

depicts the structure of a simple delta that might form in the relatively quiet waters of a lake. As the stream's forward motion is checked upon entering

◆ **Figure 10.18**   Alluvial fans develop where the gradient of a stream changes abruptly from steep to flat. Such a situation exists in Death Valley, California, where streams emerge from the mountains into a flat basin. As a result, Death Valley has many large alluvial fans.
(Photo by Michael Collier)

the lake, the dying current deposits its load of sediments. These deposits occur in three types of beds. *Foreset beds* are composed of coarser particles that drop almost immediately upon entering the lake to form layers that slope downcurrent on the *delta front.* The foreset beds are usually covered by thin, horizontal *topset beds* that are deposited on the delta plain during flood stage. The finer silts and clays settle out some distance from the mouth in nearly horizontal layers called *bottomset beds* on an area called the *prodelta.*

As the delta grows outward, the stream's gradient continually lessens. This circumstance eventually causes the channel to become choked with sediment from the slowing water. As a consequence, the river seeks a shorter, higher-gradient route to base level, as illustrated in Figure 10.19B. This illustration shows the main channel dividing into several smaller ones, called **distributaries**. Most deltas are characterized by these shifting channels that act in the opposite way to tributaries. Rather than carrying water into the main channel, distributaries carry water away from the main channel in varying paths to base level. After numerous shifts of the channel, the simple delta may grow into the triangular shape of the Greek letter delta (Δ), for which it was named. Note, however, that many deltas do not exhibit this shape. Differences in the configurations of shorelines, and variations in the nature and strength of wave activity, result in many different shapes (Figure 10.20).

Although deltas that form in the ocean generally exhibit the same basic form as the simple lake-deposited feature just described, most large marine deltas are far more complex and have foreset beds that are inclined at a much lower angle than those depicted in Figure 10.19A. Indeed, many of the world's great rivers have created massive deltas, each with its own peculiarities and none as simple as the one illustrated in Figure 10.19B.

The Mississippi Delta.

Many large rivers have deltas that extend over thousands of square kilometres. The delta of the Mississippi River is one such feature. It resulted from the accumulation of huge quantities of sediment derived from the vast region drained by the river and its tributaries. Today New Orleans rests where there was ocean less than 5000 years ago. Figure 10.21 shows that portion of the Mississippi delta which has been built over the past 5000 to 6000 years. As the figure illustrates, the delta is actually a series of seven coalescing subdeltas. Each was formed when the river left its existing channel to find a shorter, more direct path to the Gulf of Mexico. The individual subdeltas interfinger and partially cover one another to produce a very complex structure. It is also apparent from Figure 10.21 that after each portion was abandoned, coastal erosion modified the features. The present subdelta, called a *bird-foot* delta because of the configuration of its distributaries, has been built by the Mississippi in the last 500 years.

At present this active bird-foot delta has extended about as far as natural forces will allow. In fact, for many years the river has been struggling to cut through a narrow neck of land and shift its course to that of the Atchafalaya River (see inset in Figure 10.21). If this were to happen, the Mississippi would abandon its lowermost 500-kilometre path in favour of the Atchafalaya's much shorter 225-kilometre route. From the early 1940s until the 1950s, an increasing portion of the Mississippi's discharge was diverted to this new path, indicating that the river was ready to shift and begin building a new subdelta.

To prevent such an event and to keep the Mississippi following its present course, a damlike structure was erected at the site where the channel was trying to break through. Floods in the early 1970s weakened the control structure, and the river again threatened to shift until a massive auxiliary dam was completed

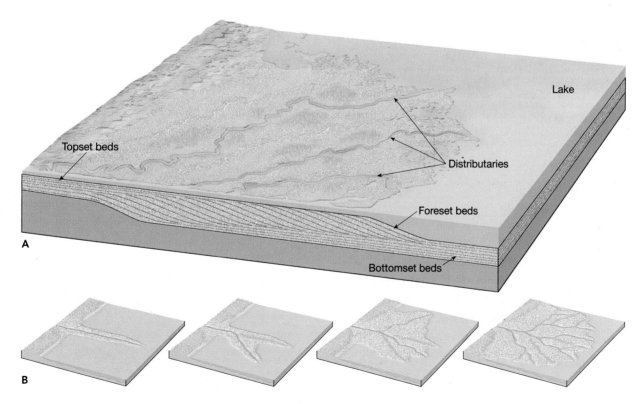

◆ **Figure 10.19**   **A.** Structure of a simple delta that forms in the relatively quiet waters of a lake. **B.** Growth of a simple delta. As a stream extends its channel, the gradient is reduced. Frequently, during flood stage the river is diverted to a higher-gradient route, forming a new distributary. Old abandoned distributaries are gradually invaded by aquatic vegetation and fill with sediment.
(After Ward's Natural Science Establishment, Inc., Rochester, N.Y.)

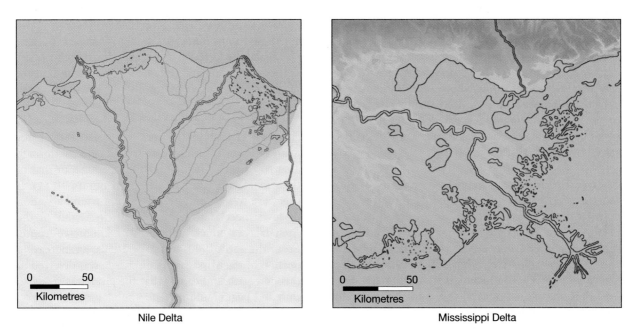

◆ **Figure 10.20**   The shapes of deltas vary and depend on such factors as a river's sediment load and the strength and nature of shoreline processes. The triangular shape of the Nile delta was the basis for naming this feature. The present Mississippi delta is called a *bird-foot delta*.

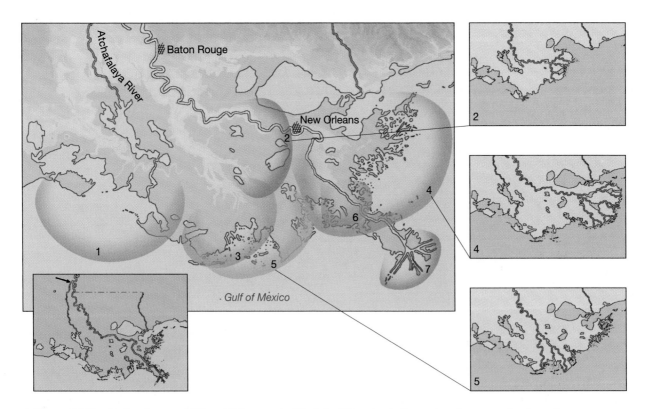

◆ **Figure 10.21**   During the past 5000 to 6000 years, the Mississippi River has built a series of seven coalescing subdeltas. The numbers indicate the order in which the subdeltas were deposited. The present bird-foot delta (number 7) represents the activity of the past 500 years. Without ongoing human efforts, the present course will shift and follow the path of the Atchafalaya River. The inset on left shows the point where the Mississippi may someday break through (arrow) and the shorter path it would take to the Gulf of Mexico. (After C.R. Kolb and J.R. Van Lopik)

in the mid-1980s. For the time being, at least, the inevitable has been avoided, and the Mississippi River will continue to flow past Baton Rouge and New Orleans on its way to the Gulf of Mexico.

Stream Valleys

 External Processes
↳ Running Water

Valleys are the most common landforms on Earth's surface. In fact, they exist in such large numbers that they have never been counted except in limited areas used for study. Prior to the turn of the nineteenth century, it was generally believed that valleys were created by catastrophic events that pulled the crust apart and created troughs in which streams could flow. Today, however, we know that with a few exceptions, streams create their own valleys.

One of the first clear statements of this fact was made by English geologist John Playfair in 1802. In his well-known work, *Illustrations of the Huttonian Theory of the Earth*, Playfair stated the principle that has come to be called **Playfair's law**:

> Every river appears to consist of a main trunk, fed from a variety of branches, each running in a valley proportioned to its size, and all of them together forming a system of valleys, communicating with one another, and having such a nice adjustment of their declivities, that none of them join the principal valley, either on too high or too low a level; a circumstance that would be indefinitely improbable, if each of these valleys were not the work of the stream that flows in it.

Not only were Playfair's observations essentially correct but they were written in a style that is seldom achieved in scientific prose.

Stream valleys can be divided into two general types: narrow V-shaped valleys and wide valleys with flat floors. These exist as the ideal forms, with many gradations in between.

Narrow Valleys

In some arid regions, where downcutting is rapid and weathering is slow, and in places where rock is particularly resistant, narrow valleys may have nearly vertical walls (see Figure 10.10, p. 275). However, most valleys, even those that are narrow at the base, are much broader at the top than the width of the channel at the bottom. This would not be the case if the only agent responsible for eroding valleys were the streams flowing through them.

Students Sometimes Ask...

I know that all rivers carry sediment. Do all rivers have deltas?

Surprisingly, no. Streams that transport large loads of sediment may lack deltas at their mouths because ocean waves and powerful currents quickly redistribute the material as soon as it is deposited (the Colombia River in the Pacific Northwest of the U.S. is one such example). In other cases, rivers do not carry sufficient quantities of sediment to build a delta. The St. Lawrence River, for example, has little opportunity to pick up much sediment between Lake Ontario and its mouth in the Gulf of St. Lawrence.

The sides of most valleys are shaped primarily as the result of weathering, sheet flow, and mass wasting. Measurements and calculations have shown that stream channel erosion during periods of increased discharge can account for only a portion of the additional sediment transported by a stream. Therefore, much of the increased load must be delivered to the stream by sheet flow and mass wasting.

A narrow V-shaped valley (Figure 10.22) indicates that the primary work of the stream has been downcutting toward base level. The most prominent features in such a valley are **rapids** and **waterfalls**. Both occur where the stream profile drops rapidly, a situation usually caused by variations in the erodibility of the bedrock into which the stream channel is cutting. A resistant bed produces rapids by acting as a temporary base level upstream while allowing downcutting to continue downstream. Once erosion has eliminated the resistant rock, the stream profile smoothes out again.

◆ **Figure 10.22** V-shaped valley of part of the Fraser River, British Columbia, produced by the downcutting of the river, combined with mass wasting of the valley walls. (Photo by Steve Hicock)

◆ **Figure 10.23**   The larger Canadian Horseshoe Falls at Niagara Falls. The river plunges over the falls and erodes the shale beneath the more resistant Lockport Dolostone. As a section of dolostone is undercut, it loses support and breaks off. (Photo by Alec Pytlowany. © 2004 Masterfile Corporation)

Waterfalls are places where the stream makes a vertical drop. One type of waterfall is exemplified by Niagara Falls (Figure 10.23). Here, the falls are supported by a resistant bed of dolostone that is underlain by a less resistant shale. As the water plunges over the lip of the falls, it erodes the less resistant shale, undermining a section of dolostone, which eventually breaks off. In this manner the waterfall retains its vertical cliff while slowly but continually retreating upstream. Over the past 12,000 years, Niagara Falls has retreated more than 11 kilometres upstream. The course of the river itself has also changed substantially since pre-glacial times (Box 10.2).

Wide Valleys

Once a stream has cut its channel closer to base level, it approaches a graded condition, and downward erosion becomes less dominant. At this point more of the stream's energy is directed from side to side. The reason for this change is not fully understood, but the reduced gradient probably is an important factor.

Nevertheless, it does occur, and the result is a widening of the valley as the river erodes first one bank and then the other (Figures 10.24 and 10.25). In this manner the flat valley floor, or floodplain, is produced. This is an appropriate name because the river is confined to its channel except during flood stage, when it overflows its banks and inundates the floodplain.

When a river erodes laterally and creates a floodplain as just described, it is called an *erosional floodplain*. However, floodplains can be depositional as well. *Depositional floodplains* are produced by a major fluctuation in conditions, such as a change in base level. The floodplain in California's Yosemite Valley is one such feature; it was produced when a glacier gouged the former stream valley about 300 metres deeper than it had been. After the glacial ice melted, the stream readjusted itself to its former base level by refilling the valley with alluvium. Similarly, few of Canada's major rivers fit their valleys because they occupy ancient spillways of glacial meltwater streams that had higher discharge rates than those of the present rivers.

BOX 10.2

Canadian Profile
The History of Niagara Falls

Rick Cheel and Keith Tinkler*

Niagara Falls, on the Niagara River that is shared between Ontario and the United States, is certainly one of the best-known natural features in Canada. It was so well known to the native North Americans that Jacques Cartier was told about the great falls almost 80 years before the first European had the opportunity to witness its majesty. The name "Niagara" is derived from the Iroquois name "Onguiaahra," meaning "the Strait."

Waters from Lakes Superior, Michigan, Huron and Erie all flow through the Niagara River where it passes over the falls. The current "natural" discharge of the river averages 5760 cubic metres per second (almost 500 million cubic metres per day). This discharge can vary due to a number of natural causes: ice dams at the mouth of the river can stop the discharge altogether, whereas winds that blow from west to east across Lake Erie can raise the water levels at the River's mouth and have produced a discharge as high as 9760 cubic metres per second. Today the discharge over the falls is controlled so that some of the flow is re-routed through turbines to produce hydro-electricity on both the Canadian and U.S. sides of the river. To facilitate power production while maintaining a significant discharge over the falls for tourists, only 50 percent of the discharge of the river is routed through turbines during summer daytime hours whereas 75 percent of the discharge is routed around the falls during summer evenings and over the winter months.

Niagara Falls exists because the Niagara River flows over the Niagara Escarpment. The action of the water flowing over the falls erodes the underlying bedrock so that its location recedes upstream over time, cutting the Niagara Gorge below the Falls. About 12,500 years ago the river fell over the brow of the escarpment (Figure 10.C) but it has eroded its gorge, and the position of the Falls, southward to its current position some 10 kilometres upstream. Prior to the onset of power generation (in 1907)

Niagara Falls was eroding southward at a rate of 1.5 metres per year. However, due to the reduction of discharge and other measures to stop erosion of the bedrock, Niagara Falls is receding at a rate of only about 0.1 metres per year.

The evolution of the modern Falls has been complex due to events associated with the last glaciation of North America. Prior to glaciation, the Niagara River followed a course that cut a gorge that extended to the northwest of the modern river, passing beyond the Escarpment near the village of St. Davids (Figure 10.C). This old, now buried, gorge intersects the modern

gorge at The Whirlpool (Figure 10.C); the location of the pre-glacial Falls is shown in Figure 10.C. As the area was covered by the thick sheet of sediment-laden continental ice the old gorge became plugged with glacial sediment. When the glaciers receded from the area, the river bypassed the old gorge and flowed to the edge of the Niagara Escarpment, near Queenston, Ontario, where the modern gorge intersects the Escarpment. Immediately following the retreat of the glaciers, discharge through the river was comparable to that today; much of that discharge was derived from glacier ice melting as it

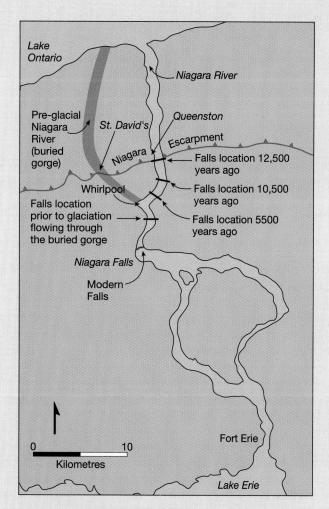

◆ **Figure 10.C**   A map of the modern Niagara River and the buried Niagara Gorge showing past locations of the falls as it evolved since prior to the last glaciation.

retreated northwards. The Falls receded from the brow of the Escarpment, cutting its gorge and moving the falls to a location approximately 2.2 kilometres to the south by 10,500 years ago. At that time the discharge through the river diminished dramatically due to events that took place to the north.

As the continental glaciers retreated to the region north of Sudbury, Ontario, the isostatically depressed land surface became the site of a major river known as the Nipissing Spillway (Figure 10.D). The Spillway captured all of the waters flowing from Lakes Superior, Huron, and Michigan and carried them to the Ottawa River. Being fed only by Lake Erie, flow through the Niagara River was greatly reduced and over the next 5000 years the Falls receded by only 1.5 kilometres (Figure 10.C). By 5500 years ago, isostatic rebound in the North Bay area was sufficient to stop the flow through the Nipissing Spillway, allowing water from all of the Great Lakes upstream of the Falls to flow through the Niagara River. With the increase in discharge, the rate of erosion of the bedrock increased significantly and the Falls progressed towards the Whirlpool. Once the Falls had receded to the Whirlpool, it encountered the fill of glacial sediment that had plugged its earlier gorge.

The resistance to erosion of this sediment was much less than that of the bedrock that it had been eroding, so that the rate of erosion, and the southward recession of the falls, was greatly speeded up. Within as little as two weeks, the Falls had receded to its preglacial position, where erosion was once again of bedrock and progressed at a much slower rate until it reached its present position (Figure 10.C).

*Rick Cheel and Keith Tinkler are professors in Department of Earth Sciences, Brock University.

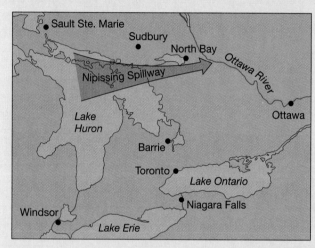

◆ **Figure 10.D**   The route of the Nipissing Spillway which captured flow from Lakes Superior, Huron, and Michigan between 10,500 and 5,500 years ago.

Owing to the slope of the channel, erosion is more effective on the downstream side of a meander. Therefore, in addition to growing laterally, the bends also gradually migrate down the valley. Sometimes the downstream migration of a meander is slowed when it reaches a more resistant portion of the floodplain. This allows the next meander upstream to "catch up." Gradually the neck of land between the meanders is narrowed. When they get close enough, the river may erode through the narrow neck of land to the next loop (Figure 10.26). The new, shorter channel segment is called a **cutoff**, and because of its shape, the abandoned bend is called an **oxbow lake** (Figure 10.27). Over a period of time, the oxbow lake fills with sediment to create a **meander scar**.

The process of meander cutoff formation has the effect of shortening the river and was described humorously by Mark Twain in *Life on the Mississippi*.

In the space of one hundred and seventy-six years the lower Mississippi has shortened itself two hundred and forty-two miles. This is an average of a trifle over one mile and a third per year. Therefore, any calm person, who is not blind or idiotic, can see that in the Old Oolitic Silurian Period, just a million years ago next November, the Lower Mississippi River was upwards of one million three hundred thousand miles long, and stuck out over the Gulf of Mexico like a fishing rod. And, by the same token, any person can see that seven hundred and forty-two years from now the Lower Mississippi will be only a mile and three quarters long, and Cairo and New Orleans will have joined their streets together, and be plodding comfortably along under a single mayor and a mutual board of aldermen. One gets such wholesale returns of conjecture out of such a trifling investment of fact.

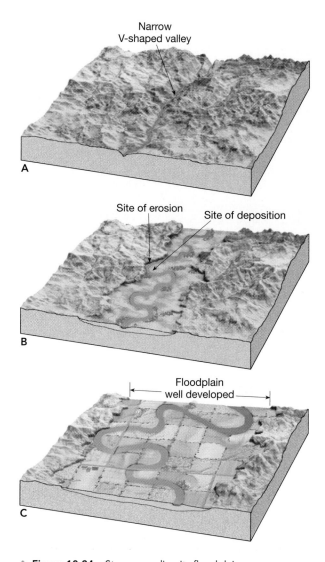

Figure 10.24 Stream eroding its floodplain.

Although the data used by Mark Twain may be reasonably accurate, he intentionally forgot to include the fact that the Mississippi also created many new meanders, thus lengthening its course by a similar amount. In fact, with the growth of its delta, the Mississippi is actually getting longer, not shorter.

Incised Meanders and Stream Terraces

We usually expect a stream with a highly meandering course to be on a floodplain in a wide valley. However, certain rivers exhibit meandering channels that flow in steep, narrow valleys. Such meanders are called **incised** (*incisum* = to cut into) **meanders** (Figure 10.28). How do such features form?

Originally the meanders probably developed on the floodplain of a stream that was near base level. Then, a change in base level caused the stream to begin downcutting. One of two events could have occurred. Either base level dropped or the land upon which the river was flowing was uplifted.

An example of the first circumstance happened during the last Ice Age when large quantities of water were withdrawn from the ocean and locked up in glaciers on land (see Chapter 12). The result was that sea level (ultimate base level) dropped, causing rivers flowing into the ocean to begin to downcut. Of course, this activity ceased at the close of the last Ice Age when the glaciers melted and the ocean rose to its former level.

Figure 10.25 Erosion of a cut bank along the Newaukum River, Washington. **A.** January 1965. **B.** March 1965. (Photos by P.A. Glancy, U.S. Geological Survey)

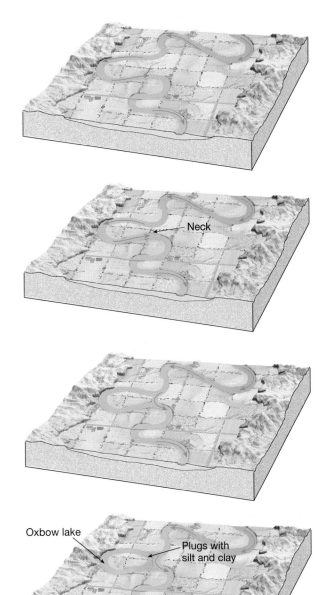

◆ **Figure 10.27**   Oxbow lakes occupy abandoned meanders. As they fill with sediment, oxbow lakes gradually become swampy meander scars. Aerial view of a meandering river north of Sault Sainte Marie, Ontario, showing a cutoff and oxbow lake.
(Photo courtesy of R.W. Hodder)

◆ **Figure 10.26**   Formation of a cutoff and oxbow lake.

Drainage Networks

External Processes
↳ Running Water

A stream is just a small component of a larger system. Each system consists of a **drainage basin**, the land area that contributes water to the stream. The drainage basin of one stream is separated from another by an imaginary line called a **divide** (Figures 10.30 and 10.31). Divides range in size from a ridge separating two small gullies to continental divides, which delineate enormous drainage basins. The Mississippi River has the largest drainage basin in North America. Extending between the Rocky Mountains in the West and the Appalachian Mountains in the East, the Mississippi River and its tributaries collect water from more than 3.2 million square kilometres of the continent.

Drainage Patterns

All drainage systems are made up of an interconnected network of streams that together form particular patterns. The nature of a drainage pattern can vary greatly from one type of terrain to another, primarily in response to the kinds of rock on which the streams developed or the structural pattern of faults and folds.

The most commonly encountered drainage pattern is the **dendritic pattern** (Figure 10.32A). This pattern is characterized by irregular branching of tributary streams that resembles the branching pattern of a deciduous tree. In fact, the word *dendritic* means "treelike." The dendritic pattern forms where

Regional uplift of the land, the second cause for incised meanders, is exemplified by the Colorado Plateau in the southwestern United States. Here, as the plateau was gradually uplifted, numerous meandering rivers adjusted to being higher above base level by downcutting (Figure 10.28).

After a river has adjusted to a relative drop in base level by downcutting, it may once again produce a floodplain at a level below the old one. The remnants of a former floodplain are sometimes present in the form of flat surfaces called **terraces** (Figure 10.29).

A

B

◆ **Figure 10.28   A.** This high-altitude image shows incised meanders of the Delores River in western Colorado. (Courtesy of USDA-ASCS) **B.** A close-up view of incised meanders of the Colorado River in Canyonlands National Park, Utah. In both places, meandering streams began downcutting because of the uplift of the Colorado Plateau. (Photo by Michael Collier)

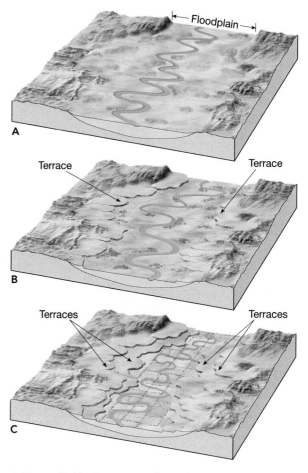

A

B

C

◆ **Figure 10.29**   Terraces can form when a stream downcuts through previously deposited alluvium. This may occur in response to a lowering of base level or as a result of regional uplift.

underlying bedrock is relatively uniform, such as flat-lying sedimentary strata or massive igneous rocks. Because the underlying material is essentially uniform in its resistance to erosion, it does not control the pattern of streamflow. Rather, the pattern is determined chiefly by the direction of the slope of land.

When streams diverge from a central area, like spokes from the hub of a wheel, the pattern is said to be **radial** (Figure 10.32B). This pattern typically develops on isolated volcanic cones and domal uplifts.

Figure 10.32C illustrates a **rectangular pattern**, with many right-angle bends. This pattern develops when the bedrock is crisscrossed by a series of joints and faults. Because these structures are eroded more easily than unbroken rock, their geometric pattern guides the directions of streams as they carve their valleys.

Figure 10.32D illustrates a **trellis drainage pattern**, a rectangular pattern in which tributary streams are nearly parallel to one another and have the appearance of a garden trellis. This pattern forms in areas underlain by alternating bands of resistant and less

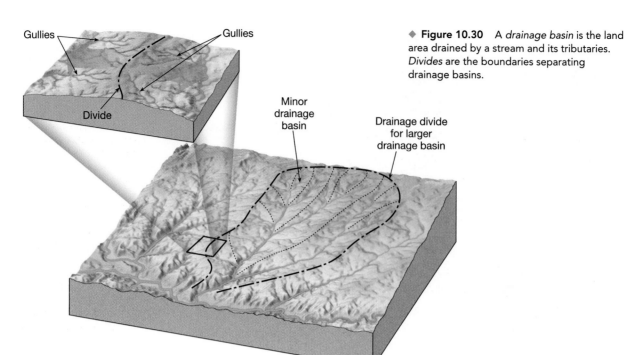

◆ **Figure 10.30**   A *drainage basin* is the land area drained by a stream and its tributaries. *Divides* are the boundaries separating drainage basins.

resistant rock and is particularly well displayed in the folded Appalachian Mountains, where both weak and strong strata crop out in nearly parallel belts.

Headward Erosion and Stream Piracy

We have seen that a stream can lengthen its course by building a delta at its mouth. A stream also lengthens its course by **headward erosion**; that is, by extending the head of its valley upslope. As sheet flow converges and becomes concentrated at the head of a stream channel, its velocity, and hence its power to erode, increases. The result can be vigorous erosion at the head of the valley. Thus, through headward erosion, the valley becomes extended into previously undissected terrain (Figure 10.33).

Headward erosion by streams plays a major role in the dissection of upland areas. In addition, an understanding of this process helps explain changes that take place in drainage patterns. One cause for changes that occur in the pattern of streams is **stream piracy**, the diversion of the drainage of one stream because of the headward erosion of another stream. Piracy can occur, for example, if a stream on one side of a divide has a steeper gradient than a stream on the other side. Because the stream with the steeper gradient has more energy, it can extend its valley headward, eventually breaking down the divide and capturing part or all of the drainage of the slower stream. In Figure 10.34 the flow of stream *A* was captured when a more swiftly flowing tributary of stream *B* breached the divide at its head and diverted stream *A*.

Stream piracy also explains the existence of narrow, steep-sided gorges that have no active streams running through them. These abandoned water gaps (called *wind gaps*) form when the stream that cut the notch has its course changed by a pirate stream. In Figure 10.34, one water gap that had been created by stream *A* became a wind gap as a result of stream piracy.

Formation of a Water Gap

Sometimes to understand fully the pattern of streams in an area, we must understand the history of the streams. For example, in many places a river valley can be seen cutting through a ridge or mountain that lies across its path. The steep-walled notch followed by the river through the structure is called a *water gap* (Figure 10.35).

Why does a stream cut across such a structure and not flow around it? One possibility is that the stream existed before the ridge or mountain was formed. In this situation, the stream, called an *antecedent stream*, would have to keep pace downcutting while the uplift progressed. That is, the stream would maintain its course as folding or faulting raised an area of the crust across the path of the stream.

A second possibility is that the stream was *superposed*, or let down, upon the structure (Figure 10.35). This can occur when a ridge or mountain is buried beneath layers of relatively flat-lying sediments or sedimentary strata. Streams originating on this cover would establish their courses without regard to the structures below. Then, as the valley was deepened

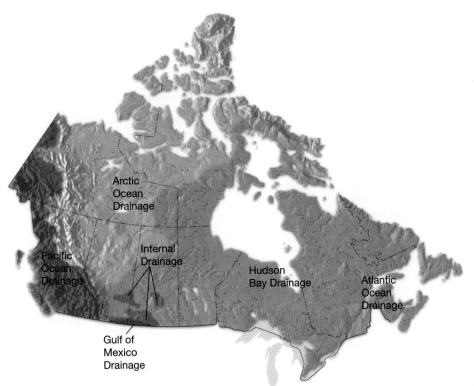

◆ **Figure 10.31** All major river systems in Canada, with the exception of a few in the prairie provinces, ultimately drain into the oceans surrounding North America. Note that streams in the southernmost areas of Alberta and Saskatchewan feed into the Mississippi drainage system that empties into the Gulf of Mexico over 6000 kilometres to the southeast. The boundaries between the drainage basins are called divides. (Map modified with permission from image courtesy of Ministry of Public Works and Government Services Canada, 2004 and Courtesy of Natural Resources Canada, Geological Survey of Canada)

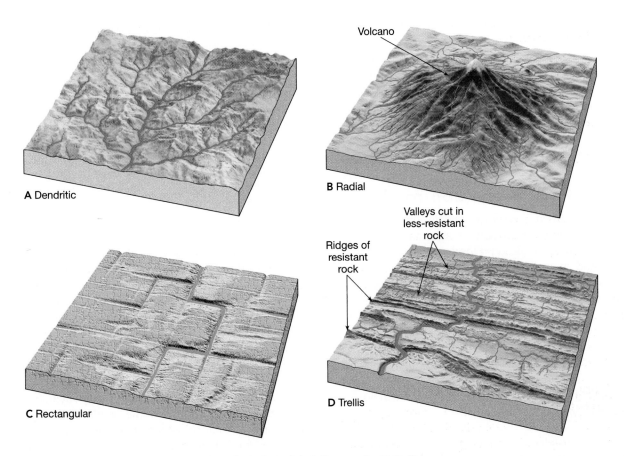

◆ **Figure 10.32**   Drainage patterns. **A**. Dendritic. **B**. Radial. **C**. Rectangular. **D**. Trellis.

◆ **Figure 10.33** By headward erosion, valleys extend into previously undissected terrain. The San Rafael River is shown above its confluence with the Green River in Utah. (Photo by Michael Collier)

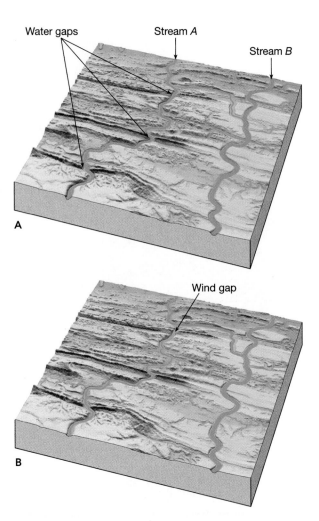

◆ **Figure 10.34** Stream piracy and the formation of wind gaps. A tributary of stream B erodes headward until it eventually captures and diverts stream A. A water gap through which stream A flowed is abandoned because of the piracy. As a result, this feature is now a wind gap. In this valley and ridge-type setting, the softer rocks in the valleys are eroded more easily than the resistant ridges. Consequently, as the valleys are lowered, the ridges and wind gaps become elevated relative to the valleys.

and the structure was encountered, the river would continue to cut its valley into it. The folded Appalachians provide good examples. Major rivers of the eastern U.S., such as the Potomac and the Susquehanna, cut across the folded strata on their way to the Atlantic.

Floods and Flood Control

When the discharge of a stream becomes so great that it exceeds the capacity of its channel, it overflows its banks as a flood. **Floods** are the most common and most destructive of all geologic hazards. They are, nevertheless, simply part of the *natural* behaviour of streams.

Causes and Types of Floods

Floods can be the result of several naturally-occurring and human-induced factors. Among the common types of floods are regional floods, flash floods, ice-jam floods, and dam-failure floods.

Regional Floods Some regional floods are seasonal. Rapid melting of snow in spring and/or heavy spring

rains often overwhelm a river. The extensive 1997 flood along the Red River is a notable example of an event triggered by rapid snowmelt (see Box 10.1). Early spring floods are sometimes made worse if the ground is frozen. This reduces infiltration into the soil, thereby increasing runoff. Extended wet periods any time of the year can create saturated soils, after which any additional rain runs off into streams until capacities are exceeded. Regional floods are often caused by slow-moving storm systems, including decaying hurricanes. For example, southern Quebec was subjected to torrential rain between July 19th and 21st, 1996 when a storm system stalled over the region. The Saguenay area was particularly hard hit,

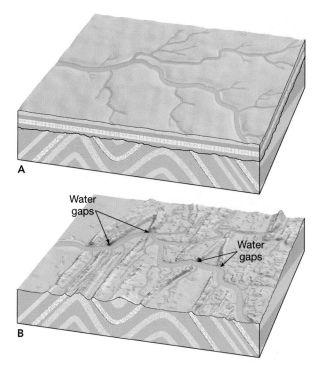

◆ **Figure 10.35** Development of a superposed stream.
A. The river establishes its course on relatively uniform strata.
B. It then encounters and cuts through the underlying structure.
C. Harpers Ferry gap at the confluence of the Shenandoah and Potomac rivers near the West Virginia-Maryland border. Water gaps such as this one are common in parts of the Appalachians.
(Photo by John S. Shelton)

in water levels and can have a devastating flow velocity. Several factors influence flash flooding. Among them are rainfall intensity and duration, surface conditions, and topography. Urban areas are susceptible to flash floods because a high percentage of the surface area is composed of impervious roofs, streets, and parking lots, where runoff is very rapid. Mountainous areas are susceptible because steep slopes can funnel runoff into narrow canyons.

Ice-Jam Floods   Frozen rivers are susceptible to ice-jam floods. As the level of a stream rises, it will break up the ice and create ice flows that can pile up on channel obstructions. Such an ice jam creates a dam across the channel. Water upstream from the ice dam can rise rapidly and overflow the channel banks. When the ice dam fails, the water stored behind the dam is released, causing a flash flood downstream.

Dam-Failure Floods   Human interference with a stream system can cause floods. A prime example is the failure of a dam or an artificial levee. Dams and artificial levees are built for flood protection. They are designed to contain floods of a certain magnitude. If a larger flood occurs, the dam or levee is over-topped. If the dam or levee fails or is washed out, the water behind it is released to become a flash flood.

◆ **Figure 10.36**   Downstream along Rivière des Ha!Ha!, a tributary of the Saguenay River, several properties were covered by up to 2 metres of sediment during the flood of 1996. The sediment was derived from upstream areas that experienced severe erosion during the flood.
(Reproduced with the permission of the Minister of Public Works and Government Services Canada, 2004 and Courtesy of Natural Resources Canada Geological Survey of Canada)

enduring well over 200 millimetres of rain in 50 hours. Flooding of the Saguenay River and its tributaries during caused over 700 million dollars in damage to the area, largely due to widespread erosion in some parts of the drainage basin and significant sediment deposition in others (Figure 10.36).

Flash Floods   A flash flood can occur with little warning and can be deadly because it produces a rapid rise

Flood Control

Several strategies have been devised to eliminate or reduce the catastrophic effects of floods. Engineering efforts include the construction of artificial levees, the building of flood-control dams, and river channellization.

Artificial Levees   *Artificial levees* are earthen mounds built on the banks of a river to increase the volume of water the channel can hold. These most common of stream-containment structures have been used since ancient times and continue to be used today.

Artificial levees are usually easy to distinguish from natural levees because their slopes are much steeper. When a river is confined by levees during periods of high water, it frequently deposits material in its channel as the discharge diminishes. This is sediment that otherwise would have been dropped on the floodplain. Thus, each time there is a high flow, deposits are left on the riverbed, and the bottom of the channel is built up. With the buildup of the bed, less water is required to overflow the original levee. As a result, the height of the levee may have to be raised periodically to protect the floodplain. Moreover, many artificial levees are not built to withstand periods of extreme flooding. For example, levee failures were numerous in the U.S. Midwest during the summer of 1993, when the upper Mississippi and many of its tributaries experienced record floods (Figure 10.37).

Flood-Control Dams   *Flood-control dams* are built to store floodwater and then let it out slowly. This lowers the flood crest by spreading it out over a longer time span. Many dams have significant nonflood-related functions such as providing water for irrigated agriculture and for hydroelectric power generation. Many reservoirs are also major regional recreational facilities.

Although dams may reduce flooding and provide other benefits, building these structures also has significant costs and consequences. For example, reservoirs created by dams may cover fertile farmland, useful forests, historic sites, and scenic valleys. Of course, dams trap sediment. Therefore, deltas and floodplains downstream erode because they are no longer replenished with silt during floods. Large dams can also cause significant ecological damage to river environments that took thousands of years to establish.

Building a dam is not a permanent solution to flooding. Sedimentation behind a dam means that the volume of its reservoir will gradually diminish, reducing the effectiveness of this flood-control measure.

Channellization   *Channellization* involves altering a stream channel in order to speed the flow of water to prevent it from reaching flood height. This may simply

◆ **Figure 10.37**   Water rushes through a break in an artificial levee in Monroe County, Illinois. During the record-breaking 1993 Midwest floods, many artificial levees could not withstand the force of the floodwaters. Sections of many weakened structures were overtopped or simply collapsed. (Photo by James A. Finley/AP/Wide World Photos)

involve clearing a channel of obstructions or dredging a channel to make it wider and deeper.

A more radical alteration involves straightening a channel by creating *artificial cutoffs*. The idea is that by shortening the stream, the gradient and the velocity are increased. By increasing velocity, the larger discharge associated with flooding can be dispersed more rapidly. However, because the river's tendency toward meandering still exists, preventing the river from returning to its previous course is difficult.

Students Sometimes Ask...

Sometimes when there is a major flood, it is described as a 100-year flood. What does that mean?
The phrase "100-year flood" is misleading because it leads people to believe that such an event happens only once every 100 years. The truth is that an uncommonly big flood can happen *any* year. The phrase "100-year flood" is really a statistical designation, indicating that there is a *1-in-100-chance* that a flood this size will happen during any year. Perhaps a better term would be the "1-in-100-chance flood."

Many flood designations are re-evaluated and changed over time as more data are collected or when a river basin is altered in a way that affects the flow of water. Dams and urban development are examples of some human influences in a basin that affect floods.

A Nonstructural Approach   All of the flood-control measures described so far have involved structural solutions aimed at "controlling" a river. These solutions are expensive and often give people residing on the floodplain a false sense of security.

Today many scientists and engineers advocate a nonstructural approach to flood control. They suggest that an alternative to artificial levees, dams, and channellization is sound floodplain management. By identifying high-risk areas, appropriate zoning regulations can be implemented to minimize development and promote more appropriate land use.

Chapter Summary

- The *hydrologic cycle* describes the continuous interchange of water among the oceans, atmosphere, and continents. Powered by energy from the Sun, it is a global system in which the atmosphere provides the link between the oceans and continents. The processes involved in the hydrologic cycle include *precipitation*, *evaporation*, *infiltration* (the movement of water into rocks or soil through cracks and pore spaces), *runoff* (water that flows over the land), and *transpiration* (the release of water vapour to the atmosphere by plants). *Running water is the single most important agent sculpting Earth's land surface.*

- The amount of water running off the land rather than sinking into the ground depends upon the *infiltration capacity* of the soil. Initially, runoff flows as broad, thin sheets across the ground, appropriately termed *sheet flow*. After a short distance, threads of current typically develop, and tiny channels called *rills* form.

- The factors that determine a stream's *velocity* are *gradient* (slope of the stream channel), *cross-sectional shape*, *size*, and *roughness* of the channel, and the stream's *discharge* (amount of water passing a given point per unit of time, frequently measured in cubic meters or cubic feet per second). Most often, the gradient and roughness of a stream decrease downstream, while width, depth, discharge, and velocity increase.

- The two general types of *base level* (the lowest point to which a stream may erode its channel) are (1) *ultimate base level* (sea level) and (2) *temporary*, or *local*, *base level*. Any change in base level will cause a stream to adjust and establish a new balance. Lowering base level will cause a stream to erode, whereas raising base level results in deposition of material in the channel.

- Streams transport their load of sediment in solution (*dissolved load*), in suspension (*suspended load*), and along the bottom of the channel (*bed load*). Much of the dissolved load is contributed by groundwater. Most streams carry the greatest part of the load in suspension. The bed load moves only intermittently and usually represents the smallest portion of a stream's load.

- A stream's ability to transport solid particles is described using two criteria: *capacity* (the maximum load of solid particles a stream can carry) and *competence* (the maximum particle size a stream can transport). Competence increases as the square of stream velocity, so if velocity doubles, water's force increases fourfold.

- Streams deposit sediment when velocity slows and competence is reduced. This results in *sorting*, the process by which like-sized particles are deposited together. Stream deposits are called *alluvium* and may occur as channel deposits called *bars*, as floodplain deposits, which include *natural levees*, and as *deltas* or *alluvial fans* at the mouths of streams.

- Although many gradations exist, the two general types of stream valleys are (1) *narrow V-shaped valleys* and (2) *wide valleys with flat floors*. Because the dominant activity is downcutting toward base level, narrow valleys often contain *waterfalls* and *rapids*. When a stream has cut its channel closer to base level, its energy is directed from side to side, and erosion produces a flat valley floor, or *floodplain*. Streams that flow upon floodplains often move in sweeping bends called *meanders*. Widespread meandering may result in shorter channel segments, called *cutoffs*, and/or abandoned bends, called *oxbow lakes*.

- The land area that contributes water to a stream is called a *drainage basin*. Drainage basins are separated by an imaginary line called a *divide*. Common *drainage patterns* (the form of a network of streams) produced by a main channel and its tributaries include (1) *dendritic*, (2) *radial*, (3) *rectangular*, and (4) *trellis*.

- *Headward erosion* lengthens a stream course by extending the head of its valley upslope. This process can lead to *stream piracy* (the diversion of the drainage of one stream by another). Former water gaps called *wind gaps* can result from stream piracy.
- Floods are triggered by heavy rains and/or snowmelt. Sometimes human interference can worsen or even cause floods. Flood-control measures include the building of artificial levees and dams, as well as channellization, which could involve creating artificial cutoffs. Many scientists and engineers advocate a nonstructural approach to flood control that involves more appropriate land use.

Review Questions

1. Describe the movement of water through the hydrologic cycle. Once precipitation has fallen on land, what paths are available to it?

2. Over the oceans, evaporation exceeds precipitation. Why does sea level not drop?

3. List several factors that influence infiltration capacity.

4. "Water in streams moves primarily in laminar flow." Briefly explain whether this statement is true or false.

5. A stream originates at 2000 metres above sea level and travels 250 kilometres to the ocean. What is its average gradient in metres per kilometre?

6. Suppose that the stream mentioned in Question 5 developed extensive meanders so that its course was lengthened to 500 kilometres. Calculate this new gradient. How does meandering affect gradient?

7. When the discharge of a stream increases, what happens to the stream's velocity?

8. What typically happens to channel width, channel depth, velocity, and discharge from the point where a stream begins to the point where it ends? Briefly explain why these changes take place.

9. Define *base level*. Name the main river in your area. For what streams does it act a base level?

10. Why do most streams have low gradients near their mouths?

11. Describe three ways in which a stream may erode its channel. Which one of these is responsible for creating potholes?

12. If you were to collect a jar of water from a stream, what part of the load would settle to the bottom of the jar? What portion would remain in the water? What part of a stream's load would probably not be present in your sample?

13. What is settling velocity? What factors influence settling velocity?

14. Distinguish between capacity and competence.

15. What factors promote the frequent flooding of broad areas along the Red River? (See Box 10.1, p. 277)

16. Describe a situation that might cause a stream channel to become braided.

17. Briefly describe the formation of a natural levee. How is this feature related to backswamps and yazoo tributaries?

18. In what way is a delta similar to an alluvial fan? In what way are they different?

19. Why does a river flowing across a delta eventually change course?

20. Each of the following statements refers to a particular drainage pattern. Identify the pattern.
    (a) Streams diverging from a central high area such as a dome.
    (b) Branching, "treelike" pattern.
    (c) A pattern that develops when bedrock is crisscrossed by joints and faults.

21. Describe how a water gap might form.

22. Contrast regional floods and flash floods. Which type is deadliest?

23. List and briefly describe three basic flood-control strategies. What are some drawbacks of each?

Key Terms

alluvial fan (p. 281)
alluvium (p. 278)
anastomosed stream (p. 280)
back swamp (p. 280)
bar (p. 278)
base level (p. 272)
bed load (p. 274)
braided stream (p. 278)
capacity (p. 276)
competence (p. 276)
cut bank (p. 279)
cutoff (p. 288)
delta (p. 282)
dendritic pattern (p. 290)
discharge (p. 270)
dissolved load (p. 274)
distributary (p. 282)
divide (p. 290)

drainage basin (p. 290)
evapotranspiration (p. 267)
flood (p. 294)
floodplain (p. 278)
graded stream (p. 272)
gradient (p. 269)
head (headwaters) (p. 271)
headward erosion (p. 292)
hydrologic cycle (p. 266)
incised meander (p. 289)
infiltration (p. 267)
infiltration capacity (p. 268)
laminar flow (p. 268)
lateral accretion surface (p. 280)

local (temporary) base level (p. 272)
longitudinal profile (p. 270)
meander (p. 279)
meander scar (p. 288)
mouth (p. 271)
natural levee (p. 280)
oxbow lake (p. 288)
Playfair's law (p. 285)
point bar (p. 278)
pothole (p. 273)
radial pattern (p. 291)
rapids (p. 285)
rectangular pattern (p. 291)
rills (p. 268)
runoff (p. 267)

saltation (p. 276)
settling velocity (p. 275)
sheet flow (p. 268)
sorting (p. 278)
stream piracy (p. 292)
suspended load (p. 274)
temporary (local) base level (p. 272)
terrace (p. 290)
transpiration (p. 267)
trellis drainage pattern (p. 291)
turbulent flow (p. 268)
ultimate base level (p. 275)
waterfall (p. 285)
yazoo tributary (p. 280)

Web Resources

 The *Earth* Web site uses the resources and flexibility of the Internet to aid in your study of the topics in this chapter. Written and developed by geology instructors, this site will help improve your understanding of geology. Visit **http://www.pearson.ca/tarbuck** and click on the cover of the text to find:

■ Online review quizzes.

■ Web-based critical thinking and writing exercises.

■ Links to chapter-specific Web resources.

■ Internet-wide key-term searches.

http://www.pearson.ca/tarbuck

Chapter 11
Groundwater

Wintertime scene at Midway Geyser Basin
in Wyoming's Yellowstone National Park.
(Photo by Mark Muench/David Muench
Photography, Inc.)

Worldwide, wells and springs provide water for cities, crops, livestock, and industry. Groundwater is the drinking water for more than 50 percent of the population, is 40 percent of the water used for irrigation, and provides more than 25 percent of industry's needs. In some areas, however, overuse of this basic resource has resulted in water shortage, streamflow depletion, land subsidence, contamination by saltwater, increased pumping cost, and groundwater pollution.

Importance of Underground Water

 External Processes
↳ Groundwater

Groundwater is one of our most important and widely available resources, yet people's perceptions of the subsurface environment from which it comes are often unclear and incorrect. The reason is that the groundwater environment is largely hidden from view except in caves and mines, and the impressions people gain from these subsurface openings are misleading. Observations on the land surface give an impression that Earth is "solid." This view remains when we enter a cave and see water flowing in a channel that appears to have been cut into solid rock.

Because of such observations, many people believe that groundwater occurs only in underground "rivers." In reality, most of the subsurface environment is not "solid" at all. It includes countless tiny *pore spaces* between grains of soil and sediment, plus narrow joints and fractures in bedrock. Together, these spaces add up to an immense volume. It is in these small openings that groundwater collects and moves.

Considering the entire hydrosphere, or all of Earth's water, only about six-tenths of 1 percent occurs underground. Nevertheless, this small percentage, stored in the rocks and sediments beneath Earth's surface, is a vast quantity. When the oceans are excluded and only sources of freshwater are considered, the significance of groundwater is more obvious.

Table 11.1 contains estimates of the distribution of fresh water in the hydrosphere. Clearly the largest volume occurs as glacial ice. Second in rank is groundwater, with slightly more than 14 percent of the total. However, when ice is excluded and just liquid water is considered, more than 94 percent of all fresh water is groundwater. Without question, *groundwater represents the largest reservoir of fresh water that is readily available to humans*. Its value in terms of economics and human well-being is incalculable.

Geologically, groundwater is important as an erosional agent. The dissolving action of groundwater slowly removes soluble rock such as limestone, allowing surface depressions known as *sinkholes* to form as well as creating subterranean caverns (Figure 11.1). Groundwater also moderates streamflow. Much of the water that flows in rivers is not direct runoff from rain and snowmelt. Rather, a large percentage of precipitation soaks in and then moves slowly underground to feed stream channels. Groundwater is thus a form of storage that sustains streams during periods when rain does not fall. The information in Table 11.1 reinforces this point. Here we see that the rate of exchange for groundwater is 280 years. This figure represents the amount of time required to replace the water now stored underground. By contrast, the rate of water exchange for rivers is just slightly more than 11 days: If the groundwater supply to a river were cut off and no rain fell, the river would run dry in just over 11 days. Thus, when we see water flowing in a river during a dry period, it represents rain that fell at some earlier time and was stored underground.

Distribution of Underground Water

 External Processes
↳ Groundwater

When rain falls, some of the water runs off, some evaporates, and the remainder infiltrates the ground. This last path is the primary source of practically all underground water. The amount of water that takes each of these paths, however, varies greatly both in time and space. Influential factors include steepness of slope, nature of surface material, intensity of rainfall, and type and amount of vegetation. Heavy rains falling on steep slopes underlain by impervious materials will obviously result in a high percentage of the water running off. Conversely, if rain falls steadily and gently upon more gradual slopes composed of materials that are easily penetrated by the water, a much larger percentage of water soaks into the ground.

Some of the water that soaks in does not travel far, because it is held by molecular attraction as a surface film on soil particles. This near-surface zone is called the **belt of soil moisture**. It is crisscrossed by roots, voids left by decayed roots, and animal and worm burrows that enhance the infiltration of rainwater into the soil. Soil water is used by plants in life functions and transpiration. Some water also evaporates directly back into the atmosphere.

Water that is not held as soil moisture percolates downward until it reaches a zone where all of the open spaces in sediment and rock are completely filled with water (Figure 11.2). This is the **zone of saturation**. Water within it is called **groundwater**.

TABLE 11.1   Fresh Water of the Hydrosphere

Parts of the Hydrosphere	Volume of Fresh Water (km³)	Share of Total Volume of Fresh Water (percent)	Rate of Water Exchange
Ice sheets and glaciers	24,000,000	84.945	8000 years
Groundwater	4,000,000	14.158	280 years
Lakes and reservoirs	155,000	0.549	7 years
Soil moisture	83,000	0.294	1 year
Water vapor in the atmosphere	14,000	0.049	9.9 days
River water	1,200	0.004	11.3 days
Total	28,253,200	100.000	

Source: U.S. Geological Survey Water Supply Paper 2220, 1987

◆ **Figure 11.1** A view of the interior of Lehman Caves in Great Basin National Park, Nevada. The dissolving action of groundwater created the caverns. Later, groundwater deposited the limestone decorations.
(Photo by David Muench/ David Muench Photography, Inc.)

The upper limit of this zone is known as the **water table**. Extending upward from the water table is the **capillary fringe** (*capillus* = hair). Here groundwater is held by surface tension in tiny passages between grains of soil or sediment. The area above the water table that includes the capillary fringe and the belt of soil moisture is called the **zone of aeration**. Although a considerable amount of water can be present in the zone of aeration, this water cannot be pumped by wells because it clings too tightly to rock and soil particles. By contrast, below the water table, the water pressure is great enough to allow water to enter wells, thus permitting groundwater to be withdrawn for use. We will examine wells more closely later in the chapter.

The Water Table

External Processes
↳ Groundwater

The water table, the upper limit of the zone of saturation, is a very significant feature of the groundwater system. The water table level is important in predicting the productivity of wells, explaining the changes in the flow of springs and streams, and accounting for fluctuations in the levels of lakes.

Variations in the Water Table

The depth of the water table is highly variable and can range from zero, at the surface, to hundreds of metres in some places. An important characteristic of the water table is that its configuration varies seasonally and yearly because the addition of water to the groundwater system is closely related to the quantity, distribution, and timing of precipitation. Except where the water table is at the surface, we cannot observe it directly. Nevertheless, its elevation can be mapped and studied in detail where wells are numerous because the water level in wells coincides with the water table (Figure 11.3). Such maps reveal that the water table is rarely level, as we might expect a table to be. Instead, its shape is usually a subdued replica of the surface topography, reaching its highest elevations beneath hills and then descending toward valleys (Figure 11.2). Wetlands (swamps) occur where the water table is right at the surface. Lakes and streams generally occupy areas low enough that the water table is above the land surface.

Several factors contribute to the irregular surface of the water table. One important influence is the

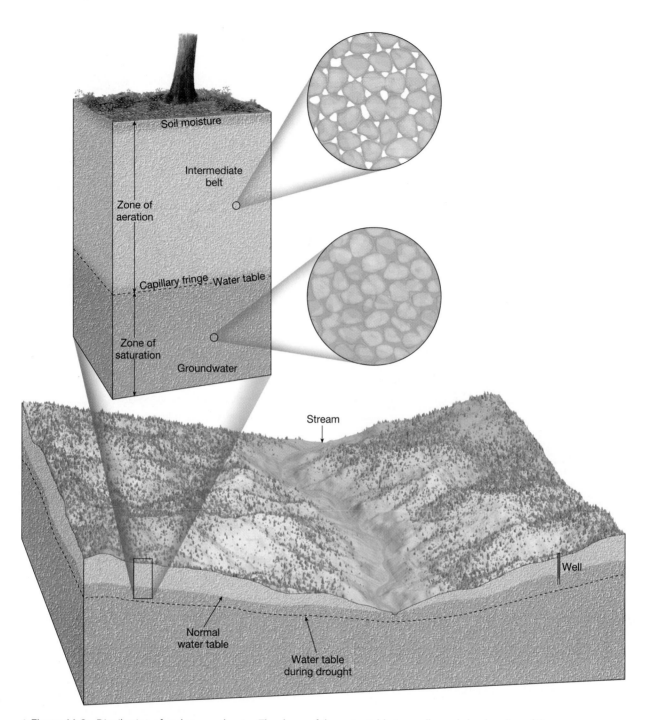

♦ **Figure 11.2** Distribution of underground water. The shape of the water table is usually a subdued replica of the surface topography. During periods of drought, the water table falls, reducing streamflow and drying up some wells.

fact that groundwater moves very slowly and at varying rates under different conditions. Because of this, water tends to "pile up" beneath high areas between stream valleys. If rainfall were to cease completely, these water-table "hills" would slowly subside and gradually approach the level of the valleys. However, new supplies of rainwater are usually added frequently enough to prevent this. Nevertheless, in times of extended drought, the water table may drop enough to dry up

shallow wells (Figure 11.2). Other causes for the uneven water table are variations in rainfall and permeability from place to place.

Interaction Between Groundwater and Streams

The interaction between the groundwater system and streams is a basic link in the hydrologic cycle. It can take place in one of three ways. Streams may gain water from the inflow of groundwater through the streambed. Such streams are called **gaining streams** (Figure 11.4A). For this to occur, the elevation of the water table must be higher than the level of the surface of the stream. Streams may lose water to the groundwater system by outflow through the streambed. The term **losing stream** is applied to this situation (Figure 11.4B). When this happens, the elevation of the water table must be lower than the surface of the stream. The third possibility is a combination of the first two—a stream gains in some sections and loses in others.

Losing streams can be connected to the groundwater system by a continuous saturated zone or they can be disconnected from the groundwater system by an unsaturated zone. Compare parts B and C in Figure 11.4. When the stream is disconnected, the water table

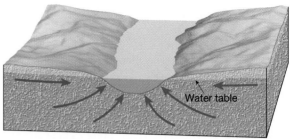

A Gaining stream

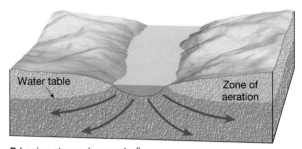

B Losing stream (connected)

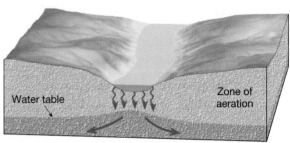

C Losing stream (disconnected)

◆ **Figure 11.4** Interaction between the groundwater system and streams. **A.** Gaining streams receive water from the groundwater system. **B.** Losing streams lose water to the groundwater system. **C.** When losing streams are separated from the groundwater system by the zone of aeration, a bulge may form in the water table.
(After U.S. Geological Survey)

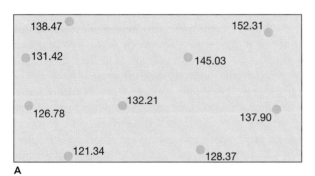

A

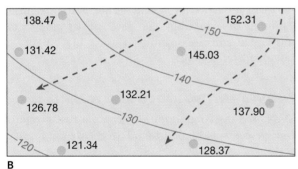

B

EXPLANATION

⬤   Location of well and altitude of water table above sea level, in metres

⌐120⌐   Water table contour shows altitude of water table, contour interval 10 metres

◄ - - - - Groundwater flow line

◆ **Figure 11.3** Preparing a map of the water table. The water level in wells coincides with the water table. **A.** First, the locations of wells and the elevation of the water table above sea level are plotted on a map. **B.** These data points are used to guide the drawing of water-table contour lines at regular intervals. On this sample map the interval is 10 metres. Groundwater flow lines can be added to show water movement in the upper portion of the zone of saturation. Groundwater tends to move approximately perpendicular to the contours and down the slope of the water table.

may have a discernible bulge beneath the stream if the rate of water movement through the streambed and zone of aeration is greater than the rate of groundwater movement away from the bulge.

In some settings, a stream might always be a gaining stream or always be a losing stream. However, in many situations flow direction can vary a great deal along a stream; some sections receive groundwater and other sections lose water to the groundwater system. Moreover, the direction of flow can change over a short time span as the result of storms adding water near the streambank or when temporary flood peaks move down the channel.

Groundwater contributes to streams in most geologic and climatic settings. Even where streams are primarily losing water to the groundwater system,

water table, as existed at Well 5, and heavy rainfall or ponded conditions, can result in the rapid downward migration of contaminants to bedrock aquifers. Significant rainfall did occur in Walkerton from May 8 to 12 and probably caused the contamination from the farm manure to enter the aquifer and move rapidly to Well 5. In the case of Well 5 it was estimated the vertical migration of surface contaminants could occur in as little as one day.[3] The incoming contamination overwhelmed the chlorine being added at the well and sufficient chlorine residuals were not maintained. This was aggravated by the fact that the operators of the Walkerton water system routinely used less chlorine than was required. Had continuous chlorine residual been maintained and turbidity monitors been installed at Well 5, the outbreak could have been prevented.

A key objective of the hydrogeological investigations that followed the May 2000 contamination was to access the degree of groundwater–surface water interaction resulting from the pumping of Well 5. In particular the scope of the work "was to investigate, using hydrogeological tools and principles, how surface contaminants such as farm runoff could have accessed the bedrock aquifer supplying the Walkerton municipal wells in a relatively short period of time[3] and with little apparent attenuation." Near Well 5 were two groundwater springs and areas of ponded surface water or small wetlands.

During a pump test on Well 5, one ponded area that was approximately 10 centimetres deep at the start of the test completely dried up within 30 minutes, while the water level in a deeper ponded area farther away saw its water level fall at least 27 centimetres over the remainder of the pump test. In addition, the flow of both groundwater springs, located within 30 metres of the well, was also substantially reduced during the pump test. Within one hour turbid water was seen entering the well. Turbid water commonly indicates the presence of surface water and/or poorly completed wells. Turbidity directly reduces the ability of water chlorination to kill bacteria. These observations illustrated a direct connection between the shallow aquifer supplying Well 5 and the ponded surface water and springs. At a later date, tracer materials (sodium chloride and sodium fluorescein, a green fluorescent dye) were introduced near one of the springs. These tracer materials were detected in the water from Well 5 within 60 minutes (electrical conductivity from sodium chloride) and within 77 minutes (sodium fluorescein) of their introduction, again confirming a direct surface water connection at Well 5 through the spring. Groundwater samples taken during the pump test of Well 5 indicated populations of both total coliform and *E. coli* organisms in the groundwater. Furthermore, DNA typing of the animals and the manure on the farm revealed that *E. coli O157:H7* strains on the farm matched the human outbreak strains predominating in Walkerton in May 2000. However, O'Connor was clear in his report that the owner of the farm cannot be faulted for the outbreak, as he engaged in acceptable farm practices.

Another key avenue of surface water–groundwater connection and shallow groundwater to deep groundwater connection are other nearby wells that have improperly grouted or corroded casings or abandoned wells that were not plugged in accordance with provincial guidelines. These wells can provide rapid routes for ground water flow, bypassing the natural aquifer materials that are often able to filter out or attenuate viruses and bacteria. Near Well 5 an unsealed test well, located only 6 metres away, would have provided a direct pathway for surface water or shallow ground water to access the aquifer.

A key recommendation of the O'Connor reports [1,2] was the establishment of groundwater protection zones within the capture areas of wells. In the case of Well 5, farm fields, which are subject to fertilization with manure and application of farm chemicals, were located within a few metres of the well, and fuel storage tanks were located within a few hundred metres. While Ontario had diverse policies and objectives on water quality, it did not have standards for protecting water sources such that these water quality objectives could be achieved. The establishment of wellhead protection zones based on aquifer vulnerability and appropriate watershed-based land use guidelines, as is currently the case in other jurisdictions (e.g., the United States), should greatly reduce the risk of harmful substances reaching groundwater supply wells at unacceptable concentrations.

3. Report on Hydrogeological Assessment, Bacteriological Impacts, Walkerton Town Wells, Municipality of Brockton, County of Bruce, Ontario, Golder Associates Ltd., London, Ontario, Canada, September 2000, 001-3105-1

Robert A. Schincariol is a professor in the Department of Earth Sciences at University of Western Ontario.

Other sources and types of contamination also threaten groundwater supplies (Figure 11.21). These include widely used substances such as highway salt, fertilizers that are spread across the land surface, and pesticides. In addition, a wide array of chemicals and industrial materials may leak from pipelines, storage tanks, landfills, and holding ponds. Some of these pollutants are classified as *hazardous*, meaning that they are flammable, corrosive, explosive, or toxic. In land disposal, potential contaminants are heaped onto mounds or spread directly over the ground. As rainwater oozes through the refuse, it may dissolve a variety of organic and inorganic materials. If the leached material reaches the water table, it will mix with the groundwater and contaminate the supply. Similar problems may result from leakage of shallow excavations, called holding ponds, into which a variety of liquid wastes are disposed.

Because groundwater movement is usually slow, polluted water can go undetected for a long time. In

A   B

◆ **Figure 11.21**   Sometimes agricultural chemicals **A**. and materials leached from landfills **B**. find their way into the groundwater. These are two of the potential sources of groundwater contamination. (Photo A by Roy Morsch/Corbis/The Stock Market; Photo B by F. Rossotto/Corbis/The Stock Market)

fact, most contamination is discovered only after drinking water has been affected and people become ill. By this time, the volume of polluted water may be very large, and even if the source of contamination is removed immediately, the problem is not solved. Although the sources of groundwater contamination are numerous, there are relatively few solutions.

Once the source of the problem has been identified and eliminated, the most common practice is simply to abandon the water supply and allow the pollutants to be flushed away gradually. This is the least costly and easiest solution, but the aquifer must remain unused for many years. To accelerate this process, polluted water is sometimes pumped out and treated. Following removal of the tainted water, the aquifer is allowed to recharge naturally, or in some cases the treated water or other fresh water is pumped back in. This process is costly, time-consuming, and it may be risky because there is no way to be certain that all of the contamination has been removed. Clearly, the most effective solution to groundwater contamination is prevention.

The Geologic Work of Groundwater

Groundwater dissolves rock. This fact is key to understanding how caverns and sinkholes form. Because soluble rocks, especially limestone, underlie millions of square kilometres of Earth's surface, it is here that

the groundwater carries on its important role as an erosional agent. Limestone is nearly insoluble in pure water but is quite easily dissolved by water containing small quantities of carbonic acid, and most groundwater contains this acid. It forms because rainwater readily dissolves carbon dioxide from the air and from decaying plants. Therefore, when groundwater comes in contact with limestone, the carbonic acid reacts with the calcite (calcium carbonate) in the rocks to form calcium bicarbonate, a soluble material that is then carried away in solution.

Students Sometimes Ask...

Is carbonic acid the only acid that creates limestone caverns?

No. It appears as though sulphuric acid (H_2SO_4) creates some caves. One example is Lechuguilla Cave in the Guadalupe Mountains near Carlsbad, New Mexico. Here solutions under pressure containing hydrogen sulphide (H_2S) derived from deep petroleum-rich sediments migrated upward through rock fractures. When these solutions mixed with groundwater containing oxygen, they formed sulphuric acid that then dissolved the limestone. Lechuguilla Cave is one of the deepest known caves in the United States, with a vertical range of 478 metres, and is also one of the largest in the country, with 170 kilometres of passages.

Caverns

The most spectacular results of groundwater's erosional handiwork are limestone **caverns**. In the United States alone, about 17,000 caves have been discovered and new ones are being found every year. Although most are relatively small, some have spectacular dimensions. Mammoth Cave in Kentucky and Carlsbad Caverns in southeastern New Mexico are famous examples. The Mammoth Cave system is the most extensive in the world, with more than 540 kilometres of interconnected passages. The dimensions at Carlsbad Caverns are impressive in a different way. Here we find the largest and perhaps most spectacular single chamber. The Big Room at Carlsbad Caverns has an area equivalent to 14 football fields and enough height to accommodate the Parliament Buildings in Ottawa.

Most caverns are created at or just below the water table in the zone of saturation. Here acidic groundwater follows lines of weakness in the rock, such as joints and bedding planes. As time passes, the dissolving process slowly creates cavities and gradually enlarges them into caverns. Material that is dissolved by the groundwater is eventually discharged into streams and carried to the ocean.

In many caves, development has occurred at several levels, with the current cavern-forming activity occurring at the lowest elevation. This situation reflects the close relationship between the formation of major subterranean passages and the river valleys into which they drain. As streams cut their valleys deeper, the water table drops as the elevation of the river drops. Consequently, during periods when surface streams are rapidly downcutting, surrounding groundwater levels drop rapidly and cave passages are abandoned by the water while the passages are still relatively small in cross-sectional area. Conversely, when the entrenchment of streams is slow or negligible, there is time for large cave passages to form.

Certainly the features that arouse the greatest curiosity for most cavern visitors are the stone formations that give some caverns a wonderland appearance. These are not erosional features, like the cavern itself, but depositional features created by the seemingly endless dripping of water over great spans of time. The calcium carbonate that is left behind produces the limestone we call travertine. These cave deposits, however, are also commonly called *dripstone*, an obvious reference to their mode of origin. Carbonate and salt minerals can also precipitate from water that forms pools on the cavern floor. Although the formation of caverns takes place in the zone of saturation, the deposition of calcium carbonate is not possible until the caverns are above the water table in the zone of aeration. As soon as the chamber is filled with air, the stage is set for the decoration phase of cavern-building to begin.

The various cavern features produced by mineral precipitation are collectively called **speleothems** (*spelaion* = cave, *them* = put), no two of which are exactly alike (Figure 11.22). Perhaps the most familiar speleothems are **stalactites** (*stalaktos* = trickling). These icicle-like pendants hang from the ceiling of the cavern and form where water seeps through cracks above. When the water reaches air in the cave, some of the dissolved carbon dioxide escapes from the drop, and calcite precipitates. Deposition occurs as a ring around the edge of the water drop. As drop after drop follows, each leaves an infinitesimal trace of calcite behind, and a hollow limestone tube is created. Water then moves through the tube, remains suspended momentarily at the end, contributes a tiny ring of calcite, and falls to the cavern floor. The stalactite just described is appropriately called a *soda straw* (Figure 11.23). Often the hollow tube of the soda straw becomes plugged or its supply of water increases. In either case, the water is forced to flow and hence deposit along the outside of the tube. As deposition continues, the stalactite takes on the more common conical shape.

Speleothems that form on the floor of a cavern and reach upward toward the ceiling are called **stalagmites** (*stalagmos* = dropping). The water supplying the calcite for stalagmite growth falls from the ceiling and splatters over the surface. As a result, stalagmites do not have a central tube and are usually more massive in appearance and rounded on their upper ends than stalactites. Given enough time, a downward-growing stalactite and an upward-growing stalagmite may join to form a *column*.

Karst Topography

Many areas of the world have landscapes that to a large extent have been shaped by the dissolving power of groundwater. Such areas are said to exhibit **karst topography**, named for the Kras Plateau in Slovenia (formerly a part of Yugoslavia) located along the northeastern shore of the Adriatic Sea, where such topography is strikingly developed. In the United States, karst landscapes occur in many areas that are underlain by limestone, including portions of Kentucky, Tennessee, Alabama, southern Indiana, and central and northern Florida. Generally, arid and semi-arid areas are too dry to develop karst topography. When solution features exist in such regions, they are likely to be remnants of a time when rainier conditions prevailed.

Karst areas typically have irregular terrain punctuated with many depressions, called **sinkholes** or

◆ **Figure 11.22**   Speleothems are of many types, including stalactites, stalagmites, and columns. Big Room, Carlsbad Caverns National Park. The large stalagmite is called the Totem Pole. (Photo by David Muench/David Muench Photography)

sinks. In the limestone areas of Florida, Kentucky, and southern Indiana, there are literally tens of thousands of these depressions, varying in depth from just a metre or two to a maximum of more than 50 metres (Figure 11.24A).

Sinkholes commonly form in two ways. Some develop gradually over many years without any physical disturbance to the rock. In these situations the limestone immediately below the soil is dissolved by downward-seeping rainwater that is freshly charged with carbon dioxide. Over time the bedrock surface is lowered and the fractures into which the water seeps are enlarged. As the fractures grow in size, soil subsides into the widening voids, from which it is removed by groundwater flowing in the passages below. These depressions are usually shallow and have gentle slopes.

By contrast, sinkholes can also form abruptly and without warning when the roof of a cavern collapses under its own weight. Typically, the depressions created in this manner are steep-sided and deep. In Canada, sinkholes are common in some parts of the Maritime Provinces that are underlain by easily dissolved evaporite and carbonate bedrock (Figure 11.24B).

In addition to a surface pockmarked by sinkholes, karst regions characteristically show a striking lack of surface drainage (streams). Following a rainfall, the runoff is quickly funnelled belowground through sinks. It then flows through caverns until it finally reaches the water table. Where streams do exist at the surface, their paths are usually short. The names of such streams often give a clue to their fate. In the Mammoth Cave area of Kentucky, for example, there

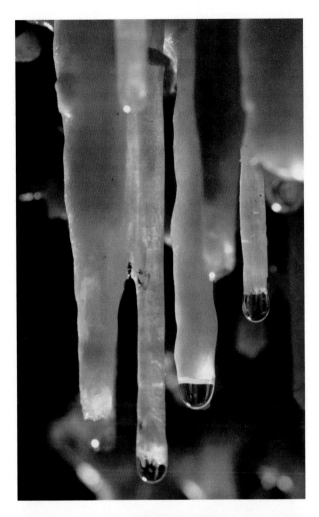

is Sinking Creek, Little Sinking Creek, and Sinking Branch. Some sinkholes become plugged with clay and debris, creating small lakes or ponds. The development of a karst landscape is depicted in Figure 11.25.

Some regions of karst development exhibit landscapes that look very different from the sinkhole-studded terrain depicted in Figure 11.25. One striking example is an extensive region in southern China that is described as exhibiting *tower karst*. As the image in Figure 11.26 shows, the term *tower* is appropriate because the landscape consists of a maze of isolated steep-sided hills that rise abruptly from the ground. Each is riddled with interconnected caves and passageways. This type of karst topography forms in wet tropical and subtropical regions having thick beds of highly jointed limestone. Here groundwater has dissolved large volumes of limestone, leaving only these residual towers. Karst development is more rapid in tropical climates due to the abundant rainfall and the greater availability of carbon dioxide from the decay of lush tropical vegetation. The extra carbon dioxide in the soil means there is more carbonic acid for dissolving limestone. Other tropical areas of advanced karst development include portions of Puerto Rico, western Cuba, and northern Vietnam.

◆ **Figure 11.23** "Live" soda straw stalactites. Lehman Caves, Great Basin National Park, Nevada. (Photo by Tom & Susan Bean, Inc.)

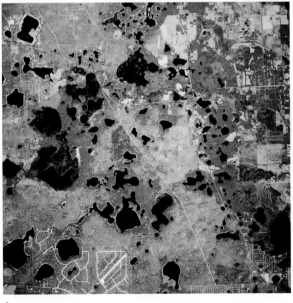

A

B

◆ **Figure 11.24** **A**. This high-altitude infrared image shows an area of Karst topography in central Florida. The numerous lakes occupy sinkholes. **B**. Sinkhole in the Codroy Valley, western Newfoundland. This sinkhole resulted from the collapse of a cave in gypsum bedrock. (Photo A courtesy of USDA-ASCS; Photo B by Trevor Bell)

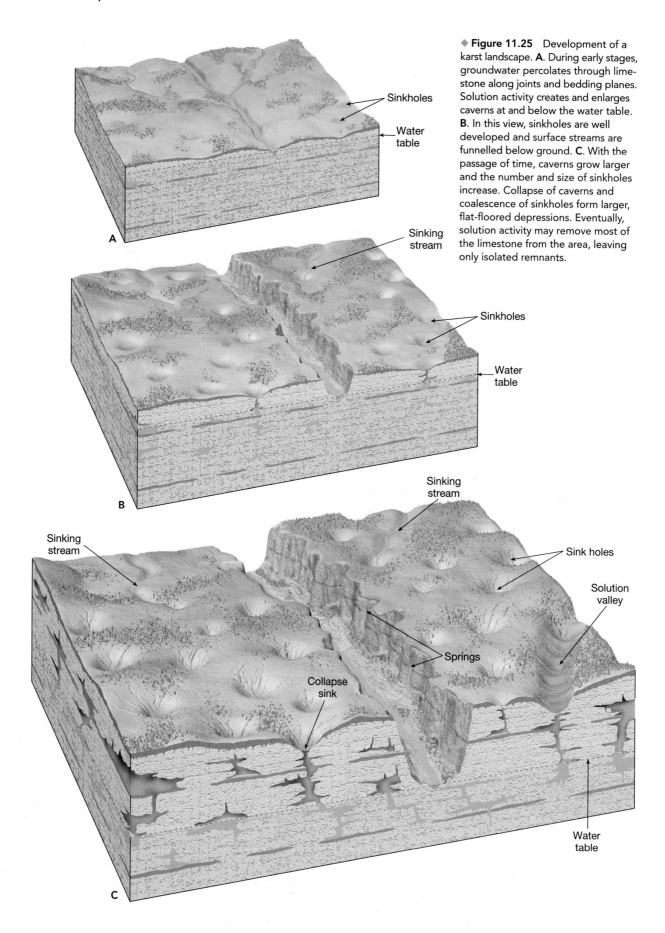

◆ **Figure 11.25** Development of a karst landscape. **A**. During early stages, groundwater percolates through limestone along joints and bedding planes. Solution activity creates and enlarges caverns at and below the water table. **B**. In this view, sinkholes are well developed and surface streams are funnelled below ground. **C**. With the passage of time, caverns grow larger and the number and size of sinkholes increase. Collapse of caverns and coalescence of sinkholes form larger, flat-floored depressions. Eventually, solution activity may remove most of the limestone from the area, leaving only isolated remnants.

Sinkholes

Water table

A

Sinking stream

Sinkholes

Water table

B

Sinking stream

Sinking stream

Sink holes

Solution valley

Springs

Collapse sink

Water table

C

◆ **Figure 11.26** One of the best-known and most distinctive regions of tower karst development is the Guilin District of southeastern China. (Photo by Jose Fuste Raga/Corbis/ The Stock Market)

Chapter Summary

- As a resource, *groundwater* represents the largest reservoir of fresh water that is readily available to humans. Geologically, the dissolving action of groundwater produces *caves* and *sinkholes*. Groundwater is also an equalizer of streamflow.

- Groundwater is water that completely fills the pore spaces in sediment and rock in the subsurface *zone of saturation*. The upper limit of this zone is the *water table*. The *zone of aeration* is above the water table where the soil, sediment, and rock are not saturated.

- The interaction between streams and groundwater takes place in one of three ways: streams gain water from the inflow of groundwater (*gaining stream*); they lose water through the streambed to the groundwater system (*losing stream*); or they do both, gaining in some sections and losing in others.

- Materials with very small pore spaces (such as clay) hinder or prevent groundwater movement and are called *aquitards*. *Aquifers* consist of materials with larger pore spaces (such as sand) that are permeable and transmit groundwater freely.

- Groundwater moves in looping curves that are a compromise between the downward pull of gravity and the tendency of water to move toward areas of reduced pressure.

- *Springs* occur whenever the water table intersects the land surface and a natural flow of groundwater results. *Wells*, openings bored into the zone of saturation, withdraw groundwater and create roughly conical depressions in the water table known as *cones of depression*. *Artesian wells* occur when water rises above the level at which it was initially encountered.

- When groundwater circulates at great depths, it becomes heated. If it rises, the water may emerge as a *hot spring*. *Geysers* occur when groundwater is heated in underground chambers, expands, and some water quickly changes to steam, causing the geyser to erupt. The source of heat for most hot springs and geysers is hot igneous rock.

- Some of the current environmental problems involving groundwater include (1) *overuse* by intense irrigation, (2) *land subsidence* caused by groundwater withdrawal, (3) *saltwater contamination*, and (4) contamination by pollutants.

- Most *caverns* form in limestone at or below the water table when acidic groundwater dissolves rock along lines of weakness, such as joints and bedding planes. The various *dripstone* features found in caverns are collectively called *speleothems*. Landscapes that to a large extent have been shaped by the dissolving power of groundwater exhibit *karst topography*, an irregular terrain punctuated with many depressions, called *sinkholes* or *sinks*.

Review Questions

1. What percentage of fresh water is groundwater? If glacial ice is excluded and only liquid fresh water is considered, about what percentage is groundwater?

2. Geologically, groundwater is important as an erosional agent. Name another significant geological role for groundwater.

3. Compare and contrast the zones of aeration and saturation. Which of these zones contains groundwater?

4. Although we usually think of tables as being flat, the water table generally is not. Explain.

5. Contrast a gaining stream and a losing stream.

6. Distinguish between porosity and permeability.

7. What is the difference between an aquitard and an aquifer?

8. Under what circumstances can a material have a high porosity but not be a good aquifer?

9. As illustrated in Figure 11.5, groundwater moves in looping curves. What factors cause the water to follow such paths?

10. Briefly describe the important contribution to our understanding of groundwater movement made by Henry Darcy.

11. When an aquitard is situated above the main water table, a localized saturated zone may be created. What term is applied to such a situation?

12. What is the source of heat for most hot springs and geysers? How is this reflected in the distribution of these features?

13. Two neighbours each dig a well. Although both wells penetrate to the same depth, one neighbour is successful and the other is not. Describe a circumstance that might explain what happened.

14. What is meant by the term *artesian*?

15. In order for artesian wells to exist, two conditions must be present. List these conditions.

16. How was the Walkerton tragedy influenced by the geology of the area surrounding Walkerton, Ontario? (See Box 11.2)

17. What problem is associated with the pumping of groundwater for irrigation?

18. Briefly explain what happened in the San Joaquin Valley as the result of excessive groundwater withdrawal (see Box 11.1).

19. In a particular coastal area, the water table is 4 metres above sea level. Approximately how far below sea level does the fresh water reach?

20. Why does the rate of natural groundwater recharge decrease as urban areas develop?

21. Which aquifer would be most effective in purifying polluted groundwater: coarse gravel, sand, or cavernous limestone?

22. What is meant when a groundwater pollutant is classified as hazardous?

23. Name two common speleothems and distinguish between them.

24. Areas whose landscapes largely reflect the erosional work of groundwater are said to exhibit what kind of topography?

25. Describe two ways in which sinkholes are created.

Key Terms

aquifer (p. 306)
aquitard (p. 306)
artesian (p. 311)
belt of soil moisture
  (p. 302)
capillary fringe (p. 303)
cavern (p. 321)
cone of depression
  (p. 311)
Darcy's law (p. 307)

drawdown (p. 311)
flowing artesian well
  (p. 312)
gaining stream (p. 305)
geyser (p. 308)
groundwater (p. 302)
head (p. 307)
hot spring (p. 308)
hydraulic gradient
  (p. 307)

karst topography (p. 321)
losing stream (p. 305)
nonflowing artesian well
  (p. 312)
perched water table
  (p. 308)
permeability (p. 306)
porosity (p. 306)
sinkhole (sink) (p. 321)
speleothem (p. 321)

spring (p. 307)
stalactite (p. 321)
stalagmite (p. 321)
water table (p. 303)
well (p. 310)
zone of aeration (p. 303)
zone of saturation (p. 302)

Web Resources

The *Earth* Web site uses the resources and flexibility of the Internet to aid in your study of the topics in this chapter. Written and developed by geology instructors, this site will help improve your understanding of geology. Visit **http://www.pearson.ca/tarbuck** and click on the cover of the text to find:

■ Online review quizzes.
■ Web-based critical thinking and writing exercises.
■ Links to chapter-specific Web resources.
■ Internet-wide key-term searches.

http://www.pearson.ca/tarbuck

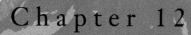

Chapter 12

Glaciers and Glaciation

34

Alsek Glacier, St. Elias Mountains, Glacier Bay
National Park, Alaska.
(Photo copyright © by Carr Clifton.
All rights reserved.)

◆ **Figure 12.1**   Moraine Lake, Banff National Park, Alberta.
(Photo by Christopher Collom)

Today glaciers cover nearly 10 percent of Earth's land surface; however, in the recent geologic past ice sheets were three times as extensive, covering vast areas with ice thousands of metres thick. Many regions still bear the mark of these glaciers (Figure 12.1). The basic character of such diverse places as the Rocky Mountains, Canadian Shield, and Gros Morne National Park was fashioned by now-vanished masses of glacial ice. Moreover, the Great Lakes and the fjords of Norway and British Columbia all owe their existence to glaciers. Glaciers, of course, are not just a phenomenon of the geologic past. As you will see, they are still sculpting the landscape and depositing debris in many regions today.

Glaciers: A Part of Two Basic Cycles

Glaciers are a part of two fundamental cycles in the Earth system—the hydrologic cycle and the rock cycle. Earlier you learned that the water of the hydrosphere is constantly cycled through the atmosphere, biosphere, and solid Earth. Time and time again, the same water

evaporates from the oceans into the atmosphere, precipitates upon the land, and flows in rivers and underground back to the sea. However, when precipitation falls at high elevations or high latitudes, the water may not immediately make its way toward the sea. Instead, it may become part of a glacier. Although the ice will eventually melt, allowing the water to continue its path to the sea, water can be stored as glacial ice for many tens, hundreds, or even thousands of years. During the time that the water is locked up in a glacier, it can be a powerful erosional force. Erosional processes are an important part of the rock cycle. Like rivers and other erosional processes, the moving ice modifies the landscape as it transports and deposits sediment.

Types of Glaciers

 External Processes
↳ Glaciers

A **glacier** is a thick ice mass that originates on land from the accumulation, compaction, and recrystallization of snow. Because glaciers are agents of erosion, they must *flow*. Although glaciers are found in many parts of the world today, most are located in high altitudes and high latitudes.

Valley (Alpine) Glaciers

Literally thousands of relatively small glaciers exist in lofty mountain areas, where they usually follow valleys that were originally occupied by streams. Unlike the rivers that previously flowed in these valleys, the glaciers advance slowly, perhaps only a few centimetres per day. Because of their setting, these moving ice masses are termed **valley glaciers** or **alpine glaciers** (Figure 12.2). Each glacier actually is a stream of ice, bounded by precipitous rock walls, that flows down-valley from an accumulation centre near its head. Like rivers, valley glaciers can be long or short, wide or narrow, single or with branching tributaries. Generally, the widths of alpine glaciers are small compared to their lengths. Some extend for just a fraction of a kilometre, whereas others go on for many tens of kilometres. The west branch of the Hubbard Glacier, for example, runs through 112 kilometres of mountainous terrain in Alaska and the Yukon Territory.

Ice Sheets

In contrast to valley glaciers, **ice sheets** exist on a much larger scale. The low total annual solar radiation reaching the poles makes these regions prone to great ice accumulations. Although many ice sheets have existed in the past, just two achieve this status at present (Figure 12.3). In the area near the North Pole, Greenland is covered by an imposing ice sheet that occupies 1.7 million square kilometres, or about 80 percent of this large island. Averaging nearly 1500 metres thick, the ice extends 3000 metres above the island's bedrock floor in some places.

In the South Polar realm, the huge Antarctic Ice Sheet attains a maximum thickness of nearly 4300 metres and covers an area of more than 13.9 million square kilometres. Because of the proportions of these huge features, they often are called *continental ice sheets*. Indeed, the combined areas of present-day continental ice sheets represent almost 10 percent of Earth's land area.

These enormous ice masses flow out in all directions from one or more snow-accumulation centres and completely obscure all but the highest areas of underlying terrain. Even sharp variations in the topography

◆ **Figure 12.2**   Aerial view of Biafo Glacier, a valley glacier in the mountains of Pakistan. (Photo by Galen Rowell/Mountain Light Photography, Inc.)

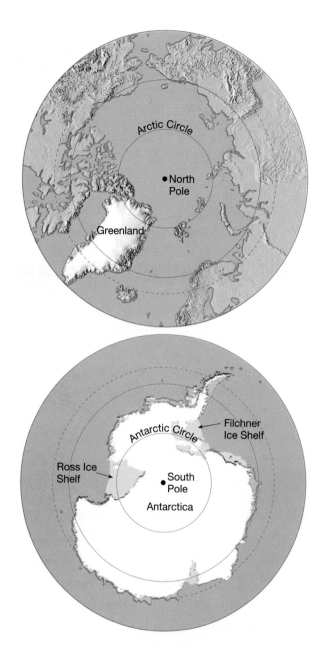

These are large, relatively flat, masses of floating ice that extend seaward from the coast but remain attached to the land along one or more sides. The shelves are thickest on their landward sides, and they become thinner seaward. They are sustained by ice from the adjacent ice sheet as well as being nourished by snowfall and the freezing of seawater to their bases. Antarctica's ice shelves extend over approximately 1.4 million square kilometres. The Ross and Filchner ice shelves are the largest, with the Ross Ice Shelf alone covering an area approximately the size of Manitoba (Figure 12.3).

Other Types of Glaciers

In addition to valley glaciers and ice sheets, other types of glaciers are also identified. Covering some uplands and plateaus are masses of glacial ice called **ice caps**. Like ice sheets, ice caps completely bury the underlying landscape but are much smaller than the continental-scale features. Ice caps occur in many places, including Iceland and several of the large islands in Canada's Arctic Archipelago (Figure 12.4).

◆ **Figure 12.3**   The only present-day continental ice sheets are those covering Greenland and Antarctica. Their combined areas represent almost 10 percent of Earth's land area. Greenland's ice sheet occupies 1.7 million square kilometres, or about 80 percent of the island. The area of the Antarctic Ice Sheet is almost 14 million square kilometres. Ice shelves occupy an additional 1.4 million square kilometres adjacent to the Antarctic Ice Sheet.

beneath the glacier usually appear as relatively subdued undulations on the surface of the ice. Such topographic differences, however, do affect the behaviour of the ice sheets, especially near their margins, by guiding flow in certain directions and creating zones of faster and slower movement.

Along portions of the Antarctic coast, glacial ice flows into bays, creating features called **ice shelves**.

◆ **Figure 12.4**   The ice cap in this satellite image is the Vatnajokull in southeastern Iceland (*jükull* means "ice cap" in Danish). In 1996 the Grimsvötn Volcano erupted beneath the ice cap, producing large quantities of melted glacial water that created floods.
(*Landsat* image from NASA)

Often ice caps and ice sheets feed **outlet glaciers**. These tongues of ice flow down valleys extending outward from the margins of these larger ice masses. The tongues are essentially valley glaciers that are avenues for ice movement from an ice cap or ice sheet through mountainous terrain to the sea. Where they encounter the ocean, some outlet glaciers spread out as floating ice shelves and produce large numbers of icebergs.

Piedmont glaciers occupy broad lowlands at the bases of steep mountains and form when one or more alpine glaciers emerge from the confining walls of mountain valleys. Here the advancing ice spreads out to form a broad sheet. The size of individual piedmont glaciers varies greatly. Among the largest in Canada is the broad piedmont lobe west of Lake Hazen on Ellesmere Island.

What If the Ice Melted?

How much water is stored as glacial ice? It is estimated by glacialogists that only slightly more than 2 percent of the world's water is tied up in glaciers. But even 2 percent of a vast quantity is still large. The total volume of just valley glaciers is about 210,000 cubic kilometres, comparable to the combined volume of the world's largest saline and freshwater lakes.

As for ice sheets, the Antarctic ice sheet includes 80 percent of the world's ice and nearly two-thirds of Earth's fresh water and covers an area similar to the total land area of Canada. If this ice melted, sea level would rise an estimated 60 to 70 metres and the ocean would inundate many densely populated coastal areas (Figure 12.5).

The hydrologic importance of the Antarctic ice can be illustrated in another way. If the ice sheet were melted at a uniform rate, it could feed (1) the Mississippi River for more than 50,000 years, (2) all the rivers in the United States for about 17,000 years, (3) the Amazon River for approximately 5000 years, or (4) all the rivers of the world for about 750 years.

Formation of Glacial Ice

Snow is the raw material from which glacial ice originates; therefore, glaciers form in areas where more snow falls in winter than melts during the summer. Before a glacier is created, snow must be converted into glacial ice. This transformation is shown in Figure 12.6.

When temperatures remain below freezing following a snowfall, the fluffy accumulation of delicate hexagonal crystals soon changes. As air infiltrates the spaces between the crystals, the extremities of the crystals evaporate and the water vapour condenses near the centres of the crystals. In this manner snowflakes become smaller, thicker, and more spherical, and the large pore spaces disappear. By this process air is forced

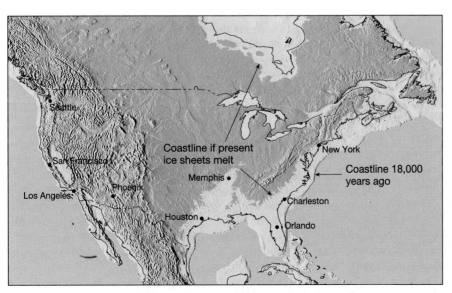

◆ **Figure 12.5** This map of a portion of North America shows the present-day coastline compared to the coastline that existed during the last ice age (18,000 years ago) and the coastline that would exist if present ice sheets on Greenland and Antarctica melted. (After R. H. Dott, Jr., and R. L. Battan, *Evolution of the Earth*, New York: McGraw Hill, 1971. Reprinted by permission of the publisher.)

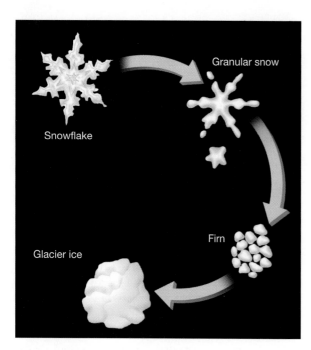

◆ **Figure 12.6**   The conversion of freshly fallen snow into dense, crystalline glacial ice.

out and what was once light, fluffy snow is recrystallized into a much denser mass of small grains having the consistency of coarse sand. This granular recrystallized snow is called **firn** and is commonly found making up old snowbanks near the end of winter. As more snow is added, the pressure on the lower layers increases, compacting the ice grains at depth. Once the thickness of ice and snow exceeds 50 metres, the weight is sufficient to fuse firn into a solid mass of interlocking ice crystals. Glacial ice has now been formed.

Movement of a Glacier

 External Processes
     ↳ Glaciers

The movement of glacial ice is generally referred to as *flow*. The fact that glacial movement is described in this way seems paradoxical—how can a solid flow? The way in which ice flows is complex and is of three basic types. The first of these, **plastic flow**, involves movement *within* the ice. Ice behaves as a brittle solid until the pressure upon it is equivalent to the weight of about 50 metres of ice. Once that load is surpassed, ice behaves as a plastic material and flow begins. Such flow occurs because of the molecular structure of ice. Glacial ice consists of layers of molecules stacked one upon the other. The bonds between layers are weaker than those within each layer. Therefore, when a stress exceeds the strength of the bonds between the layers, the layers remain intact and slide over one another.

A second and often equally important mechanism of glacial movement consists of the entire ice mass slipping along the ground. With the exception of some glaciers located in polar regions where the ice is probably frozen to the solid bedrock floor, most glaciers are thought to move by this sliding process, called **basal slip**. In this process, meltwater probably acts as a hydraulic jack and perhaps as a lubricant helping the ice over the rock. The source of the liquid water is related in part to the fact that the melting point of ice decreases as pressure increases. Therefore, deep within a glacier the ice may be at the melting point even though its temperature is below 0°C.

The third mechanism is by the underlying material deforming by friction of the overriding glacier. Water-saturated sediment and weak bedrock such as shale are easily deformed and offer little resistance to the power of the glacier. Figure 12.7 illustrates the effects of these three basic types of glacial motion. This vertical profile through a glacier also shows that not all the ice flows forward at the same rate. Frictional drag with the bedrock floor causes the lower portions of the glacier to move more slowly.

In contrast to the lower portion of the glacier, the upper 50 metres or so is not under sufficient pressure to exhibit plastic flow. Rather, the ice in this uppermost zone is brittle and is appropriately referred to as the **zone of fracture**. The ice in the zone of fracture is carried along "piggyback" style by the ice below. When the glacier moves over irregular terrain, the zone of fracture is subjected to tension, resulting in cracks called **crevasses** (Figure 12.8). These gaping cracks can make travel across glaciers dangerous and may extend to depths of 50 metres. Below this depth, plastic flow seals them off.

Students Sometimes Ask...

I've heard that icebergs might be used as a source of water in deserts. Is that possible?

It is true that people in arid lands have seriously studied the possibility of towing icebergs from Antarctica to serve as a source of fresh water. There are certainly ample supplies. Each year, in the waters surrounding Antarctica, about 1000 cubic kilometres of glacial ice breaks off and creates icebergs. However, there are significant technological problems that are not likely to be overcome soon. For instance, vessels capable of towing huge icebergs (1 to 2 kilometres across) have not yet been developed. In addition, there would be a substantial loss of ice to melting and evaporation that would take place as an iceberg is slowly rafted (for up to one year) through warm ocean waters.

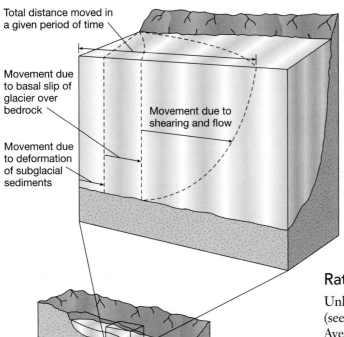

Total distance moved in a given period of time

Movement due to basal slip of glacier over bedrock

Movement due to deformation of subglacial sediments

Movement due to shearing and flow

◆ **Figure 12.7**   Cross-section through a glacier to illustrate ice movement, divided into three components. 1) Ice in the zone of fracture is carried "piggyback" on the underlying mass of ice. 2) Ice below approximately 50 metres behaves plastically and flows—it may even slip over the bed if wet. 3) Soft material under the ice may deform by frictional drag of the ice. In fact, the rate of movement is slowest at the base of the glacier where frictional drag is greatest.

Rates of Glacial Movement

Unlike streamflow, glacial movement is not obvious (see Box 12.1). How rapidly does glacial ice move? Average velocities vary considerably from one glacier to another. Some move so slowly that trees and other vegetation may become well established in the debris that has accumulated on the glacier's surface, whereas others may move at rates of up to several metres per day. For example, Byrd Glacier, an outlet glacier in Antarctica that was the subject of a 10-year study using satellite images, moved at an average rate of 750 to 800 metres per year (about 2 metres per day). Other glaciers in the study advanced at one-quarter that rate.

The advance of some glaciers is characterized by periods of extremely rapid movements called **surges**. Glaciers that exhibit such movement may flow along in an apparently normal manner, then speed up for a relatively short time before returning to the normal rate again. The flow rates during surges are as much as 100 times the normal rate. Evidence indicates that many glaciers may be of the surging type (Box 12.2).

Budget of a Glacier

Snow is the raw material from which glacial ice originates; therefore, glaciers form in areas where more snow falls in winter than melts during the summer. Glaciers are constantly gaining and losing ice. Snow accumulation and ice formation occur in the **zone of accumulation**. Its outer limits are defined by the **equilibrium line**. The elevation of the snowline varies greatly. In polar regions it may be sea level, whereas in tropical areas the equilibrium line exists only high in mountain areas, often at altitudes exceeding 4500 metres. Above the equilibrium line, in the zone of accumulation, the addition of snow thickens the glacier and promotes movement. The equilibrium line separates the zone of accumulation from the **zone of**

◆ **Figure 12.8**   Crevasses form in the brittle ice of the zone of fracture. They do not continue down into the zone of flow. (Photo by Steve Hicock)

◆ **Figure 12.23** Well-developed
lateral moraines deposited by the
shrinking Athabaska Glacier in
Jasper National Park.
(Photo by David Barnes/
The Stock Market)

provide fresh supplies of sediment to the terminus.
In this manner a large quantity of till is deposited as
the ice melts away, creating a rock-strewn, undulating plain. This gently rolling layer of till deposited as
the ice front advanced then receded is termed **ground
moraine**. Ground moraine has a levelling effect, filling in low spots and clogging old stream channels,
often leading to a derangement of the existing drainage
system. In areas where this layer of till is still relatively fresh, such as the northern Great Lakes region,
poorly drained, swampy lands are quite common.

Periodically, a glacier will retreat to a point where
ablation and nourishment once again balance. When
this happens, the ice front stabilizes and a new end
moraine forms.

The pattern of end moraine formation and
ground moraine deposition may be repeated many
times before the glacier has completely vanished. Such
a pattern is illustrated by Figure 12.25. It should be
pointed out that the outermost end moraine marks
the limit of the glacial advance. Because of its special
status, this end moraine is also called the **terminal
moraine**. On the other hand, the end moraines that
were created as the ice front occasionally stabilized
during retreat are termed **recessional moraines**. Note

◆ **Figure 12.24** Medial moraines form when the lateral
moraines of merging valley glaciers join. Kluane National
Park, Yukon.
(Photo by Martin Miller)

that both terminal and recessional moraines are identical in origin; the only difference between them is their
relative positions.

End moraines deposited during the last glaciation are prominent features in many parts of Ontario. A well-known example north of Lake Ontario is the Oak Ridges Moraine. This hilly strip of glacial sediment extends westward from Trenton to Orangeville and is an important groundwater resource for the greater Toronto area (Figure 12.26). The moraine formed between the Simcoe and Ontario glacial lobes (Figure 12.25) and contains thick channel fills of sand and gravel that comprise a large aquifer within and between till aquitards. It also stores and discharges clean cool water that is vital to aquatic life in the area.

Figure 12.27 represents a hypothetical area during glaciation and following the retreat of ice sheets. It shows the moraines that were described in this section, as well as the depositional features that are discussed in the sections that follow. This figure depicts landscape features similar to what might be encountered if you were travelling in southern Canada during the last Ice Age or present-day Nunavut or Northwest Territories. As you read upcoming sections dealing with other glacial deposits, you will be referred to this figure several times.

Drumlins

Moraines are not the only landforms deposited by glaciers. In some areas that were once covered by continental ice sheets, a special variety of glacial landscape exists—one characterized by smooth, teardrop-shaped, parallel hills called **drumlins** (Figure 12.27). Good examples occur on islands off the Saanich Peninsula of southern Vancouver Island, in the Guelph and Peterborough areas of Ontario, Gaetz Head near Halifax, and on the Morley Flats near Calgary (Figure 12.27).

These streamlined asymmetrical hills are composed largely of till. They range in height from about 15 to 50 metres and may be up to 1 kilometre long. The steep side of the hill faces the direction from which the ice advanced, whereas the gentler, longer slope points in the direction the ice moved. Drumlins are not found as isolated landforms but rather occur in clusters called *drumlin fields* (Figure 12.28). Two such clusters occur in Peterborough and Guelph, Ontario. The Peterborough drumlin field alone is estimated to contain about 3000 drumlins. Although drumlin formation is not fully understood, their streamlined shape indicates that they were moulded under the zone

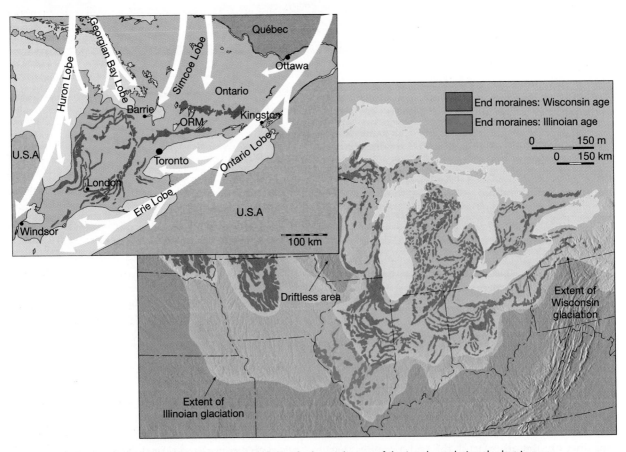

◆ **Figure 12.25**   End moraines of North America recording farthest advance of the ice sheet during the last ice age. Moraines deposited during the most recent (Wisconsinan) stage are most prominent. Inset: Close-up view of eastern Great Lakes region showing flow directions of major ice lobes. ORM is the Oak Ridges Moraine in Figure 12.26.

◆ **Figure 12.26**   Oak Ridges Moraine (pink) depicted in a digital elevation model image. The moraine is an important aquifer and aggregate reserve for the greater Toronto area. (Image courtesy of D.R. Sharpe, Geological Survey of Canada)

Students Sometimes Ask...

Are any glacial deposits valuable?

Yes. In glaciated regions, landforms made of stratified drift, such as eskers, are often excellent sources of sand and gravel. Although the value per ton is low, huge quantities of these materials are used in the construction industry. In addition, glacial sands and gravels are valuable because they make excellent aquifers and thus are significant sources of groundwater in some areas. Clays from former glacial lakes have been used in the manufacture of bricks.

of plastic flow of an active glacier. It is believed that many drumlins originate when glaciers advance over previously deposited drift and reshape the material.

Landforms Made of Stratified Drift

As the name implies, stratified drift is sorted according to the size and weight of the particles. Because ice is not capable of such sorting activity, these materials are not deposited directly by the glacier as till is, but instead reflect the sorting action of glacial meltwater. Accumulations of stratified drift commonly comprise sand and gravel—that is, bed-load material—because the finer rock flour remains suspended and therefore is commonly carried far from the glacier by the meltwater streams.

Outwash Plains and Valley Trains

At the same time that an end moraine is forming, water from the melting glacier cascades over the till, sweeping some of it out in front of the growing ridge of poorly sorted debris. Meltwater generally emerges from the ice in rapidly moving streams that are commonly choked with suspended material and carry a substantial bed load as well. As the water leaves the glacier, it moves onto the relatively flat surface beyond and rapidly loses velocity. As a consequence, much of its bed load is dropped and the meltwater begins weaving a complex pattern of braided channels (Figure 12.27). In this way, a broad, ramplike surface composed of stratified drift is built adjacent to the downstream edge of most end moraines. When the feature is formed in association with an ice sheet, it is termed an **outwash plain**, and when largely confined to a mountain valley, it is usually referred to as a **valley train**.

Outwash plains and valley trains commonly are pockmarked with basins or depressions known as **kettles** (Figure 12.27). Kettles also occur in deposits of till. Kettles are formed when blocks of stagnant ice become wholly or partly buried in drift and eventually melt, leaving pits in the glacial sediment. Although most kettles do not exceed 2 kilometres in diameter, some with diameters exceeding 10 kilometres are known. In many cases water eventually fills the depression and forms a pond or lake. A good example is Sifton Bog in London, Ontario, the most southerly large acidic (sphagnum) bog in Canada.

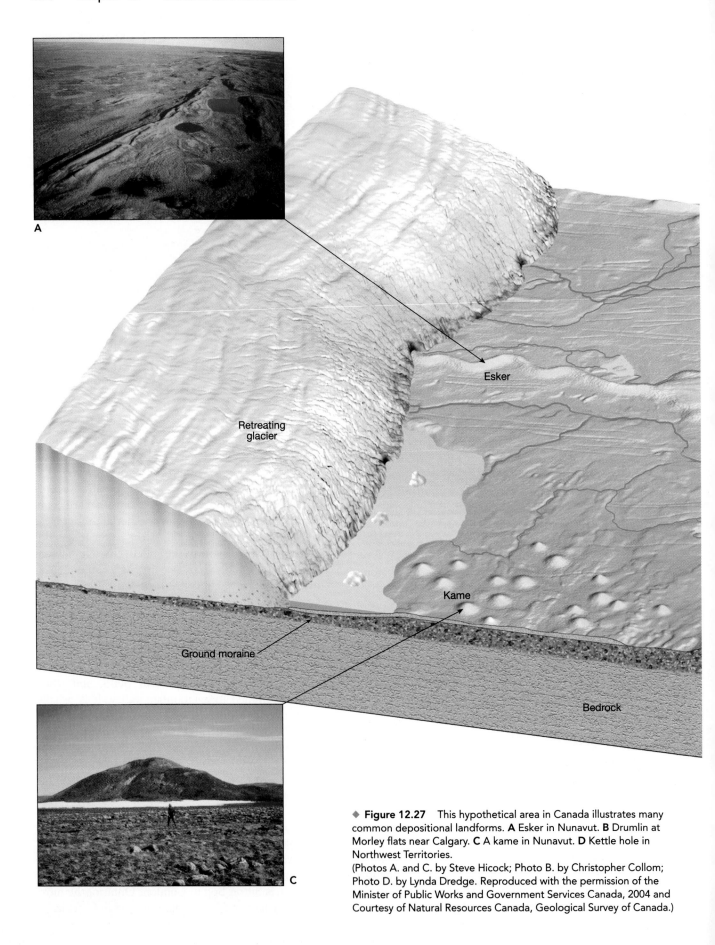

◆ **Figure 12.27** This hypothetical area in Canada illustrates many common depositional landforms. **A** Esker in Nunavut. **B** Drumlin at Morley flats near Calgary. **C** A kame in Nunavut. **D** Kettle hole in Northwest Territories.
(Photos A. and C. by Steve Hicock; Photo B. by Christopher Collom; Photo D. by Lynda Dredge. Reproduced with the permission of the Minister of Public Works and Government Services Canada, 2004 and Courtesy of Natural Resources Canada, Geological Survey of Canada.)

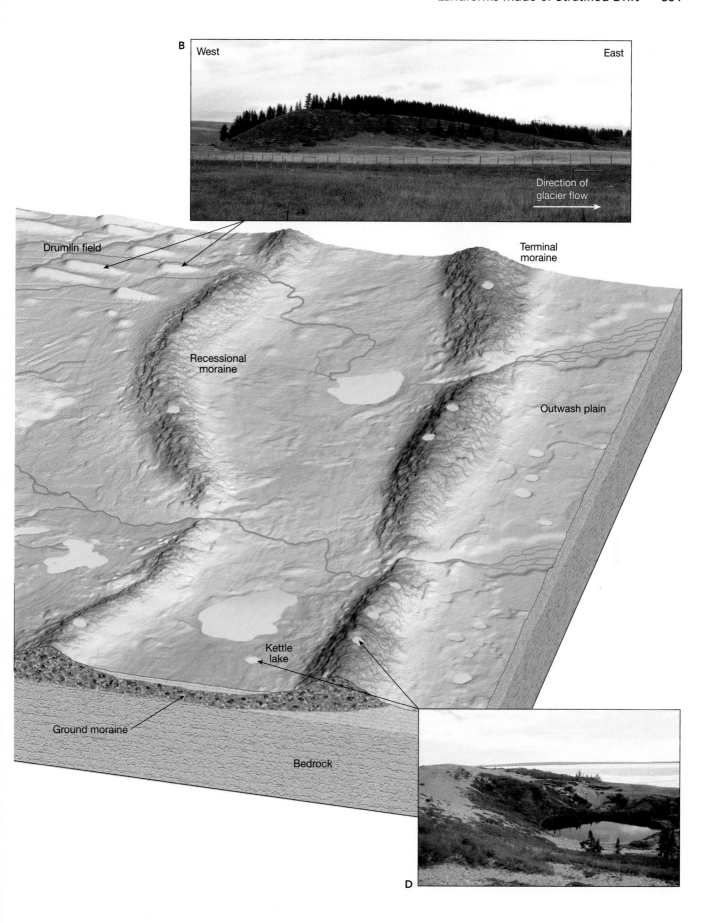

B   West   East

Direction of
glacier flow

Drumlin field   Terminal
                                                 moraine

Recessional
moraine

                                                 Outwash plain

Kettle
lake

Ground moraine

Bedrock

D

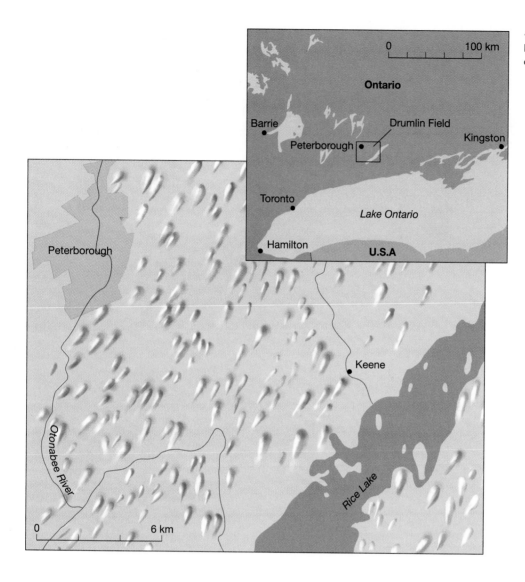

Ice-Contact Deposits

When the melting terminus of a glacier shrinks to a critical point, flow virtually stops and the ice becomes stagnant. Meltwater that flows over, within, and at the base of the motionless ice lays down deposits of stratified drift. Then, as the supporting ice melts away, the stratified sediment is left behind in the form of hills, terraces, and ridges. Such accumulations are collectively termed **ice-contact deposits** and are classified according to their shapes.

When the ice-contact stratified drift is in the form of a mound or steep-sided hill, it is called a **kame** (Figure 12.27). Some kames represent bodies of sediment deposited by meltwater in openings within, or depressions on top of, the ice. Others originate as deltas or fans built outward from the ice by meltwater streams. Later, when the stagnant ice melts away, these various accumulations of sediment collapse to form isolated, irregular mounds.

When glacial ice occupies a valley, **kame terraces** may be built along the sides of the valley. These features commonly are narrow masses of stratified drift laid down between the glacier and the side of the valley by streams that drop debris along the margins of the shrinking ice mass.

A third type of ice-contact deposit is a long, narrow, sinuous ridge composed largely of sand and gravel. Some are more than 100 metres high with lengths in excess of 100 kilometres. The dimensions of many others are far less spectacular. Known as **eskers**, these ridges are deposited by meltwater rivers flowing within, on top of, and beneath a mass of motionless, stagnant glacial ice (Figure 12.27). Many sediment sizes are carried by the torrents of meltwater in the ice-banked channels, but only the coarser material can settle out of the turbulent stream.

The Glacial Theory and the Ice Age

In the preceding pages we mentioned the Ice Age, a time when ice sheets and alpine glaciers were far more extensive than they are today. As noted, there was a time when the most popular explanation for what we now know to be glacial deposits was that the materials had been drifted in by means of icebergs or perhaps simply swept across the landscape by a catastrophic flood. What convinced geologists that an extensive ice age was responsible for these deposits and many other glacial features?

In 1821 a Swiss engineer, Ignaz Venetz, presented a paper suggesting that glacial landscape features occurred at considerable distances from the existing glaciers in the Alps. This implied that the glaciers had once been larger, and occupied positions farther downvalley. Another Swiss scientist, Louis Agassiz, doubted the proposal of widespread glacial activity put forth by Venetz. He set out to prove that the idea was not valid. Ironically, his 1836 fieldwork in the Alps convinced him of the merits of his colleague's hypothesis. In fact, a year later Agassiz hypothesized a great ice age that had extensive and far-reaching effects—an idea that was to give Agassiz widespread fame.

The proof of the glacial theory proposed by Agassiz and others constitutes a classic example of applying the principle of uniformitarianism. Realizing that certain features are produced by no other known process but glacial action, they were able to begin reconstructing the extent of now-vanished ice sheets based on the presence of features and deposits found far beyond the margins of present-day glaciers. In this manner, the development and verification of the glacial theory continued during the nineteenth century, and through the efforts of many scientists, a knowledge of the nature and extent of former ice sheets became clear.

By the beginning of the twentieth century, geologists had largely determined the areal extent of the Ice Age glaciation. Further, during the course of their investigations, they had discovered that many glaciated regions had not one layer of drift but several. Moreover, close examination of these older deposits showed well-developed zones of chemical weathering and soil formations as well as the remains of plants that require warm temperatures. The evidence was clear; there had been not just one glacial advance but many, each separated by extended periods when climates were as warm or warmer than the present. The Ice Age had not simply been a time when the ice advanced over the land, lingered for a while, and then receded. Rather, the period was a very complex event, characterized by a number of advances and withdrawals of glacial ice.

By the early twentieth century a fourfold division of the Ice Age had been established for both North America and Europe. The divisions were based largely on studies of glacial deposits. In North America each of the four major stages was named for the midwestern state of the U.S. where deposits of that stage were well exposed and/or were first studied. These are, in order of occurrence, the Nebraskan, Kansan, Illinoian, and Wisconsinan. These traditional divisions remained in place until relatively recently, when it was learned that sediment cores from the ocean floor contain a much more complete record of climate change during the Ice Age.*

Unlike the glacial record on land, which is punctuated by many unconformities, seafloor sediments provide an uninterrupted record of climatic cycles for this period. Studies of these seafloor sediments showed that glacial/interglacial cycles had occurred about every 100,000 years. About 20 such cycles of cooling and warming were identified for the span we call the Ice Age.

During the glacial age, ice left its imprint on almost 30 percent of Earth's land area, including about 10 million square kilometres of North America, 5 million square kilometres of Europe, and 4 million square kilometres of Siberia (Figure 12.29). The amount of glacial ice in the Northern Hemisphere was roughly twice that of the Southern Hemisphere. The primary reason is that the southern polar ice could not spread far beyond the margins of Antarctica. By contrast, North America and Eurasia provided great expanses of land for the spread of ice sheets.

Today we know that the Ice Age began between 2 million and 3 million years ago. This means that most of the major glacial stages occurred during a division of the geologic time scale called the **Pleistocene Epoch**. Although the Pleistocene is commonly used as a synonym for the Ice Age, note that this epoch does not encompass all of the last glacial period. The Antarctic ice sheet, for example, probably formed at least 20 million years ago.

Some Indirect Effects of Ice-Age Glaciers

In addition to the massive erosional and depositional work carried on by Pleistocene glaciers, the ice sheets had other effects, sometimes profound, on the landscape. For example, as the ice advanced and retreated, animals and plants were forced to migrate. This led to stresses that some organisms could not tolerate. Hence, a number of plants and animals became extinct.

---

*For more on this topic, see Box 6.2, "Using Seafloor Sediments to Unravel Past Climates."

◆ **Figure 12.29**   Maximum extent of ice sheets in the Northern Hemisphere during the Ice Age.

Furthermore, many present-day stream courses bear little resemblance to their preglacial routes. Other rivers that today carry only a trickle of water but nevertheless occupy broad channels are testimony to the fact that they once carried torrents of glacial meltwater.

In areas that were centres of ice accumulation, such as Scandinavia and the Canadian Shield, the land has been slowly rising over the past several thousand years. Uplifting of almost 300 metres has occurred in the Hudson Bay region. This, too, is the result of the continental ice sheets. But how can glacial ice cause such vertical crustal movement? We now understand that the land is rising because the added weight of the 3-kilometre-thick mass of ice caused downwarping of Earth's crust. Following the removal of this immense load, the crust has been adjusting by gradually rebounding upward ever since (Figure 12.30).*

---

*For a more complete discussion of this concept, termed *isostatic adjustment*, see the section on isostasy in Chapter 20.

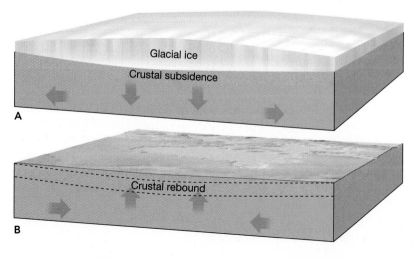

Glacial ice

Crustal subsidence

A

Crustal rebound

B

◆ **Figure 12.30**   Simplified illustration showing crustal subsidence and rebound resulting from the addition and removal of continental ice sheets. **A.** In northern Canada and Scandinavia, where the greatest accumulation of glacial ice occurred, the added weight caused downwarping of the crust. **B.** Ever since the ice melted, there has been gradual uplift or rebound of the crust.

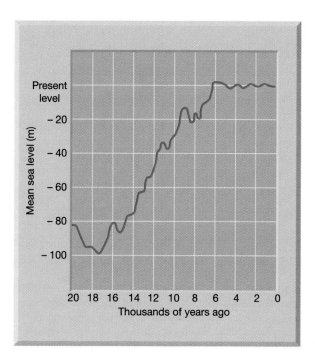

◆ **Figure 12.31** Changing sea level during the past 20,000 years. The lowest level shown on the graph represents the time about 18,000 years ago when the most recent ice advance was at a maximum.

Certainly, one of the most interesting and perhaps dramatic effects of the Ice Age was the fall and rise of sea level that accompanied the advance and retreat of the glaciers. Earlier in the chapter it was pointed out that sea level would rise by an estimated 60 or 70 metres if the water locked up in the Antarctic Ice Sheet were to melt completely. Such an occurrence would flood many densely populated coastal areas.

Although the total volume of glacial ice today is great, exceeding 25 million cubic kilometres, during the Ice Age the volume of glacial ice amounted to about 70 million cubic kilometres, or 45 million cubic kilometres more than at present. Because we know that the snow from which glaciers are made ultimately comes from the evaporation of ocean water, the growth of ice sheets must have caused a worldwide drop in sea level (Figure 12.31). Indeed, estimates suggest that sea level was as much as 100 metres lower than today. Thus, land that is presently flooded by the oceans was dry. The Atlantic Coast of Canada lay more than 100 kilometres to the east of Nova Scotia; France and Britain were joined where the famous English Channel is today; Alaska and Siberia were connected across the Bering Strait; and Southeast Asia was tied by dry land to the islands of Indonesia.

While formation and growth of ice sheets was an obvious response to significant changes in the climate, the existence of the glaciers triggered important climatic changes in the regions beyond their margins. In central Canada, glacial Lake Agassiz was the largest lake in North America (Figure 12.32) and its eventual discharge down the St. Lawrence River into the North Atlantic contributed to the Younger Dryas cooling event about 11,000 years ago. It was dammed by the retreating margin of the Laurentide Ice Sheet and drained down the Mississippi into the Gulf of Mexico for most of its life. Lake Winnipeg is the largest remnant of Lake Agassiz.

Causes of Glaciation

A great deal is known about glaciers and glaciation. Much has been learned about glacier formation and movement, the extent of glaciers past and present, and the features created by glaciers, both erosional and depositional. However, a widely accepted theory for the causes of glacial ages has not yet been established. Although more than 160 years have elapsed since Louis Agassiz proposed his theory of a great Ice Age, no complete agreement exists as to the causes of such events.

Although widespread glaciation has been rare in Earth's history, the Ice Age that encompassed the Pleistocene Epoch is not the only glacial period for which a record exists. Earlier glaciations are indicated by deposits called **tillite**, a sedimentary rock formed when glacial till becomes lithified. Such deposits, found in strata of several different ages, usually contain striated rock fragments, and some overlie grooved and polished bedrock surfaces or are associated with sandstones and conglomerates that show features of outwash deposits. Two Precambrian glacial episodes have been identified in the geologic record, the first approximately 2 billion years ago and the second about 600 million years ago (Box 12.3). Further, a well-documented record of an earlier glacial age is found in late Paleozoic rocks that are about 250 million years old, which exist on several landmasses.

Any theory that attempts to explain the causes of glacial ages must successfully answer two basic questions. (1) *What causes the onset of glacial conditions?* For continental ice sheets to have formed, average temperature must have been somewhat lower than at present and perhaps substantially lower than throughout much of geologic time. Thus, a successful theory would have to account for the cooling that finally leads to glacial conditions. (2) *What caused the alternation of glacial and interglacial stages that have been documented for the Pleistocene Epoch?* The first question deals with long-term trends in temperature on a scale of millions of years, but this second question relates to much shorter-term changes.

BOX 12.3

Canadian Profile
Snowball Earth: Canadian Cryospheric Controversy

The recognition that glaciers covered huge areas of the northern hemisphere in recent Earth's history was an intellectual triumph. Once criteria for the recognition of "recent" glacial activity were developed it was quickly realized that these features could also be sought in ancient rocks. The Precambrian represents almost 90 percent of the 4.5 billion years of Earth history but extensive glacial deposits have been recognized only at the beginning and end of the Proterozoic Eon (from 2.5 billion years ago to about 540 million years ago). The oldest extensive glacial deposits occur in Early Proterozoic rocks of North America, Finland, South Africa, and Western Australia, which are about 2.3 billion years old (Fig. 12.C). Late Proterozoic glaciogenic deposits (deposits produced under glacial influence) are present on every continent.

The widespread nature of Late Proterozoic glaciogenic deposits suggested to Canadian scientist Paul Hoffman (currently at Harvard University) that the entire surface of Earth could have been frozen (the snowball Earth hypothesis or SEH). Hoffman interpreted that the Earth, as its surface was progressively frozen, lost heat by reflection of solar radiation from ice-covered areas, causing rapid cooling (runaway albedo), eventually leading to the planet being encapsulated in ice. Escape from the "deep freeze" is explained by release of greenhouse gases into the atmosphere from volcanoes, reversing the "deep freeze" into a "deep fry" interval. The appeal of this idea lies in its simplicity.

There are, however, many problems with the SEH. For example, the SEH calls for global-scale glaciation at about the same time but among the few glaciogenic deposits that have been accurately dated, there are significant differences in age. In addition, the thickness (many kilometres in some cases), and subaqueous origin of many glaciogenic Precambrian successions are difficult to reconcile with a totally frozen planet because moving ice requires an active hydrologic cycle. Yet another problem concerns the limestone and dolostone "cap carbonates," that occur above many Late Proterozoic glacial successions. According to the SEH, it is interpreted that high levels of atmospheric CO_2 caused extreme global warming and increased weathering of silicate minerals, leading to saturation of the world's oceans with bicarbonate ions and widespread deposition of carbonate sediments. Mudstones deposited immediately after the glaciations ended should show signs of intense weathering. Available evidence, however, indicates the contrary—the glacial muds that immediately overlie the cap carbonates show less evidence of weathering than in higher levels of the successions.

Remnant magnetism in some rocks can be used to determine the latitude at which they formed. Several ancient glacial deposits appear to have formed at sea level in equatorial latitudes. Does this indicate that the entire planet was glaciated? The presence of varved deposits (mudstones with different summer and winter layers) and fossil ice-wedge structures is puzzling because both require significant seasonal temperature variations, which are not found in equatorial regions today. One possible explanation is the "big tilt hypothesis" (BTH), which invokes a highly tilted Earth (axis at greater than 54° to the orbital plane) in the Precambrian. If the Earth's axis were tilted at an extreme angle, equatorial regions would have experienced greater seasonality, possibly initiating glaciation on landmasses at low latitudes. Unfortunately, a mechanism that adequately explains *how* such a radical change in tilt could have been accomplished and independent evidence substantiating this hypothesis remains to be seen.

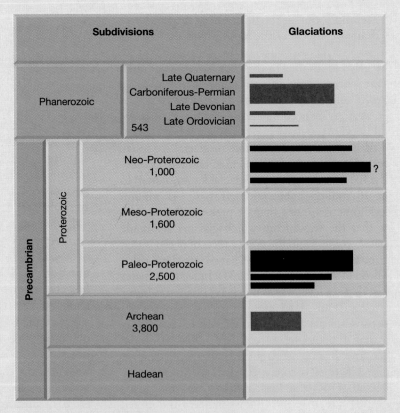

◆ **Figure 12.C** Major glacial episodes in Earth history as indicated by horizontal bars.

The snowball Earth concept has stimulated lively discussions on other aspects of Precambrian history. As indicated by studies of Canadian paleontologist Guy Narbonne and others, the evolution of the complex life forms that eventually colonized the Earth may be closely tied to these climatic and environmental upheavals. In addition, Paul Hoffman and colleagues have suggested that Late Proterozoic banded iron-formations (BIF) may be explained by the SEH. As discussed in Chapter 6, in the presence of abundant oxygen (as today) iron is almost insoluble but under reducing conditions it will easily dissolve in water. Isolation of the Late Proterozoic atmosphere from the oceans by ice would have created a reduced ocean saturated with iron delivered to the ocean by volcanic vents. According to the SEH, when the ice melted, atmospheric O_2 entered the ocean and precipitated BIF in the world's oceans on a global scale. Contrary to this prediction, BIF deposits are only sporadically developed in Precambrian successions. Furthermore, some of the BIF occurs underneath, rather than above the glaciogenic deposits, in contradiction to the suggestion that they formed after the ice disappeared according to the SEH.

Thus the great glaciations of the Proterozoic are poorly understood. We do not understand why they happened.

We do not know whether they affected the entire planet (SEH) or occurred at different times in different regions. We do not understand the enigmatic associations of warm and cold climate indicators (cap carbonates and glaciogenic sedimentary rocks respectively) and no definitive explanation (apart from the speculative BTH) has been offered for the enigmatic evidence of strong seasonality in equatorial regions. We have come a long way in terms of recognizing ancient glaciations but many of the deposits still have to be accurately dated and most are far from being adequately explained.

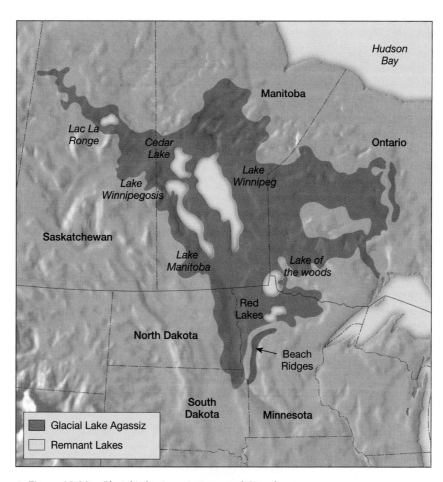

◆ **Figure 12.32** Glacial Lake Agassiz, in central Canada.
(Image courtesy of the Minnesota Department of Natural Resources)

BOX 12.4

Understanding Earth
Climate Change Recorded in Glacial Ice

Vertical cores taken from the Greenland and Antarctic ice sheets are important sources of data about climate change during and following the most recent cycle of glaciation. Scientists collect samples with a drilling rig, like a small version of an oil drill. A hollow shaft follows the drill head into the ice and an ice core is extracted. In this way, cores that sometimes exceed 2000 metres in length and may represent more than 200,000 years of climate history are acquired for study.

The ice provides a detailed record of changing air temperatures and snowfall. Air bubbles trapped in the ice record variations in atmospheric composition. Changes in carbon dioxide and methane are linked to fluctuating temperatures. The cores also include atmospheric fallout such as windblown dust, volcanic ash, pollen, and modern-day pollution.

Past temperatures are determined by *oxygen isotope analysis*. This technique is based on precise measurement of the ratio between two isotopes of oxygen: O^{16}, which is the most common, and the heavier O^{18}. More O^{18} is evaporated from the oceans when temperatures are high, and less is evaporated when temperatures are low. Therefore, the heavier isotope is more abundant in the precipitation of warm eras and less abundant during colder periods. Using this principle, scientists are able to produce a record of past temperature changes.

In 2001–2002, Canadian expeditions to Mount Logan, Canada's highest peak, in the southwest corner of the Yukon, drilled over 200 metres through a glacier below the summit to get a record of climate in recent millennia (Figure 12.D). Scientists are now analyzing the ice core that will yield a wealth of information, including changes in air temperature, snow accumulation rates, and atmospheric pollution. It will provide new insights into the interaction of atmosphere and ocean in a region that plays a key role in regulating North America's climate, past and present (Figure 12.E). It will help to place present climate, under human-induced influences, in the context of naturally varying climate. This detailed record of the last several thousand years will allow more accurate predictions of future climate changes in the region and how they will affect people—especially those living in coastal areas affected by sea level fluctuations.

◆ **Figure 12.D** Canada's ICE 2001 expedition to drill through ice on the glacial side of Mount Logan. Geologists are studying the climate record in ice cores several thousand years old.
(Photo by H Samuelsson. Reproduced with the permission of the Minister of Public Works and Government Services Canada, 2004 and Courtesy of Natural Resources Canada, Geological Survey of Canada).

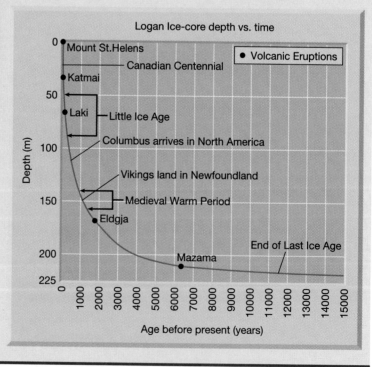

◆ **Figure 12.E** Graph puting the ice core into context of geologic and historic events. (David Lemieux. Used with permission, courtesy of Natural Resources Canada).

Although the literature of science contains a vast array of hypotheses relating to the possible causes of glacial periods, we will discuss only a few major ideas to summarize current thought.

Plate Tectonics

Probably the most attractive proposal for explaining the fact that extensive glaciations have occurred only a few times in the geologic past comes from the theory of plate tectonics.*

Because glaciers can form only on land, we know that landmasses must exist somewhere in the higher latitudes before an ice age can commence. Many scientists suggest that ice ages have occurred only when Earth's shifting crustal plates have carried the continents from tropical latitudes to more poleward positions.

Glacial features in present-day Africa, Australia, South America, and India indicate that these regions, which are now tropical or subtropical, experienced

_____
*A brief overview of the theory appears in Chapter 1, and a more extensive discussion is presented in Chapter 19.

an Ice Age near the end of the Paleozoic Era, about 250 million years ago. However, there is no evidence that ice sheets existed during this same period in what are today the higher latitudes of North America and Eurasia. For many years this puzzled scientists. Was the climate in these relatively tropical latitudes once like it is today in Greenland and Antarctica? Why did glaciers not form in North America and Eurasia? Until the plate tectonics theory was formulated, there had been no reasonable explanation.

Today scientists understand that the areas containing these ancient glacial features were joined together as a single supercontinent located at latitudes far to the south of their present positions. Later this landmass broke apart and its pieces, each moving on a different plate, migrated toward their present locations (Figure 12.33). Now we know that during the geologic past, plate movements accounted for many dramatic climatic changes as landmasses shifted in relation to one another and moved to different latitudinal positions. Changes in oceanic circulation also must have occurred, altering the transport of heat and moisture and consequently the climate as well. Because the rate of plate movement is very slow—

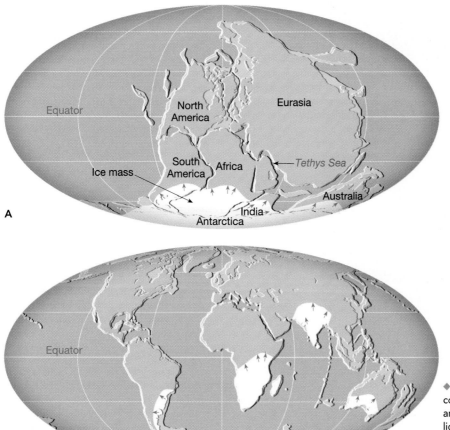

◆ **Figure 12.33** **A**. The supercontinent Pangaea, showing the area covered by glacial ice 300 million years ago. **B**. The continents as they are today. The white areas indicate where evidence of the old ice sheets exists.

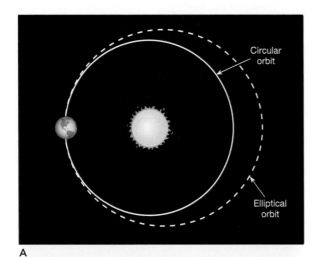

A

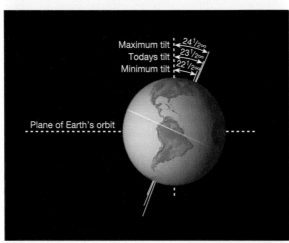

B

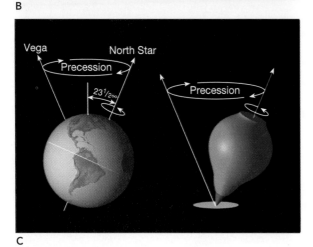

C

◆ **Figure 12.34** Orbital variations. **A.** The shape of Earth's orbit changes during a cycle that spans about 100,000 years. It gradually changes from nearly circular to one that is more elliptical and then back again. This diagram greatly exaggerates the amount of change. **B.** Today the axis of rotation is tilted about 23.5° to the plane of Earth's orbit. During a cycle of 41,000 years, this angle varies from 21.5° to 24.5°. **C.** Precession. Earth's axis wobbles like that of a spinning top. Consequently, the axis points to different spots in the sky during a cycle of about 26,000 years.

a few centimetres annually—appreciable changes in the positions of the continents occur only over great spans of geologic time. Thus, climatic changes triggered by shifting plates are extremely gradual and happen on a scale of millions of years.

Variations in Earth's Orbit

Because climatic changes brought about by moving plates are extremely gradual, the plate tectonics theory cannot be used to explain the alternation between glacial and interglacial climates that occurred during the Pleistocene Epoch. Therefore, we must look to some other triggering mechanism that might cause climatic change on a scale of thousands rather than millions of years (see Box 12.4). Many scientists today believe, or strongly suspect, that the climatic oscillations that characterized the Pleistocene may be linked to variations in Earth's orbit. This hypothesis was first developed by James Croll in the 19th century and later strongly advocated by the Yugoslavian scientist Milutin Milankovitch and is based on the premise that variations in incoming solar radiation are a principal factor in controlling Earth's climate.

Milankovitch formulated a comprehensive mathematical model based on the following elements (Figure 12.34):

1. Variations in the shape (*eccentricity*) of Earth's orbit about the Sun;
2. Changes in *obliquity*; that is, changes in the angle that the axis makes with the plane of Earth's orbit;
3. The wobbling of Earth's axis, called *precession*.

Using these factors, Milankovitch calculated variations in the receipt of solar energy and the corresponding surface temperature of Earth back into time in an attempt to correlate these changes with the climatic fluctuations of the Pleistocene. In explaining climatic changes that result from these three variables, note that they cause little or no variation in the total solar energy reaching the ground. Instead, their impact is felt because they change the degree of contrast between the seasons. Somewhat milder winters in the middle to high latitudes means greater snowfall totals, whereas cooler summers would bring a reduction in snowmelt.

Among the studies that have added credibility to the astronomical theory of Milankovitch is one in which deep-sea sediments containing certain climatically sensitive microorganisms were analyzed to establish a chronology of temperature changes going back nearly one-half million years.*

---

*J. D. Hays, John Imbrie, and N. J. Shackelton, "Variations in the Earth's Orbit: Pacemaker of the Ice Ages," *Science* 194 (1976): 1121–32.

This time scale of climatic change was then compared to astronomical calculations of eccentricity, obliquity, and precession to determine if a correlation did indeed exist. Although the study was very involved and mathematically complex, the conclusions were straightforward. The researchers found that major variations in climate over the past several hundred thousand years were closely associated with changes in the geometry of Earth's orbit; that is, cycles of climatic change were shown to correspond closely with the periods of obliquity, precession, and orbital eccentricity. More specifically, the authors stated: "It is concluded that changes in the earth's orbital geometry are the fundamental cause of the succession of Quaternary ice ages."*

---

*J. D. Hays et al., p. 1131. The term *quaternary* refers to the period on the geologic time scale that encompasses the last 1.6 million years.

Let us briefly summarize the ideas that were just described. The theory of plate tectonics provides us with an explanation for the widely spaced and nonperiodic onset of glacial conditions at various times in the geologic past, whereas the theory proposed by Milankovitch and supported by the work of J. D. Hays and his colleagues furnishes an explanation for the alternating glacial and interglacial episodes *within* the Pleistocene.

In conclusion, we emphasize that the ideas just discussed do not represent the only possible explanations for glacial ages. Although interesting and attractive, these proposals are certainly not without critics, nor are they the only possibilities currently under study. Other factors may be, and probably are, involved.

Chapter Summary

- A *glacier* is a thick mass of ice originating on land as a result of the compaction and recrystallization of snow, and it shows evidence of past or present flow. Today *valley* or *alpine glaciers* are found in mountain areas where they usually follow valleys that were originally occupied by streams. *Ice sheets* exist on a much larger scale, covering most of Greenland and Antarctica.

- Near the surface of a glacier, in the *zone of fracture*, ice is brittle. However, below about 50 metres, pressure is great, causing ice to *flow* like *plastic material*. A second important mechanism of glacial movement consists of the entire ice mass *slipping* along the ground.

- The average velocity of glacial movement is generally quite slow but varies considerably from one glacier to another. The advance of some glaciers is characterized by periods of extremely rapid movements called *surges*.

- Glaciers form in areas where more snow falls in winter than melts during summer. Snow accumulation and ice formation occur in the *zone of accumulation*. Its lower limit is defined by the *equilibrium line*. Beyond the equilibrium line is the *zone of wastage*, where there is a net loss to the glacier. The *glacial budget* is the balance, or lack of balance, between accumulation at the upper end of the glacier, and loss, called *ablation*, at the lower end.

- Glaciers erode land by *plucking* (lifting pieces of bedrock out of place) and *abrasion* (grinding and scraping of a rock surface). Erosional features produced by valley glaciers include *glacial troughs*, *hanging valleys*, *fjords*, *cirques*, *arêtes*, *horns*, and *roches moutonnées*.

- Any sediment of glacial origin is called *drift*. The two distinct types of glacial drift are (1) *till*, which is poorly sorted sediment deposited directly by the ice; and (2) *stratified drift*, which is relatively well-sorted sediment laid down by glacial meltwater.

- The most widespread features created by glacial deposition are layers or ridges of till, called *moraines*. Associated with valley glaciers are *lateral moraines*, formed along the sides of the valley, and *medial moraines*, formed between two valley glaciers that have joined. *End moraines*, which mark the former position of the front of a glacier, and *ground moraines*, undulating layers of till deposited as the ice front advances then retreats, are common to both valley glaciers and ice sheets. An *outwash plain* is often associated with the end moraine of an ice sheet. A *valley train* may form when the glacier is confined to a valley. Other depositional features include *drumlins* (streamlined asymmetrical hills composed mainly of till), *eskers* (sinuous ridges composed largely of sand and gravel deposited by streams flowing in tunnels beneath the ice, near the terminus of a glacier), and *kames* (steep-sided hills consisting of sand and gravel).

- The *Ice Age*, which began about 2 million years ago, was a very complex period characterized by a number of advances and withdrawals of glacial ice. Most of the major glacial episodes occurred during a division of the geologic time scale called the *Pleistocene Epoch*. Perhaps the most convincing evidence for the occurrence of several glacial advances during the Ice Age is the widespread existence of *multiple layers of drift* and an uninterrupted record of climate cycles preserved in *seafloor sediments*.

- In addition to massive erosional and depositional work, other effects of Ice Age glaciers included the *forced migration of organisms, changes in stream courses, adjustment of the crust* by rebounding after the removal of the immense load of ice, and *climate changes* caused by the existence of the glaciers themselves. In the sea, the most far-reaching effect of the Ice Age was the *worldwide change* in *sea level* that accompanied each advance and retreat of the ice sheets.

- Any theory that attempts to explain the causes of glacial ages must answer two basic questions: (1) What causes the onset of glacial conditions? and (2) What caused the alternating glacial and interglacial stages that have been documented for the Pleistocene Epoch? Two of the many hypotheses for the cause of glacial ages involve (1) plate tectonics and (2) variations in Earth's orbit.

Review Questions

1. Where are glaciers found today? What percentage of Earth's land area do they cover? How does this compare to the area covered by glaciers during the Pleistocene?

2. Describe how glaciers fit into the hydrologic cycle. What role do they play in the rock cycle?

3. Each statement below refers to a particular type of glacier. Name the type of glacier.
   (a) The term *continental* is often used to describe this type of glacier.
   (b) This type of glacier is also called an *alpine glacier*.
   (c) This is a stream of ice leading from the margin of an ice sheet through the mountains to the sea.
   (d) This is a glacier formed when one or more valley glaciers spreads out at the base of a steep mountain front.
   (e) Greenland is the only example in the Northern Hemisphere.

4. Describe the three components of glacial flow. At what rates do glaciers move? In a valley glacier, does all of the ice move at the same rate? Explain.

5. Why do crevasses form in the upper portion of a glacier but not below 50 metres?

6. Under what circumstances will the front of a glacier advance? Retreat? Remain stationary?

7. Describe the processes of glacial erosion.

8. How does a glaciated mountain valley differ in appearance from a mountain valley that was not glaciated?

9. List and describe the erosional features you might expect to see in an area where valley glaciers exist or have recently existed.

10. What is glacial drift? What is the difference between till and stratified drift? What general effect do glacial deposits have on the landscape?

11. List the four basic moraine types. What do all moraines have in common? What is the significance of terminal and recessional moraines?

12. Why are medial moraines proof that valley glaciers must move?

13. How do kettles form?

14. What direction was the ice sheet moving that affected the area shown in Figure 12.28? Explain how you were able to determine this.

15. What are ice-contact deposits? Distinguish between kames and eskers.

16. The development of the glacial theory is a good example of applying the principle of uniformitarianism. Explain briefly.

17. During the Pleistocene Epoch the amount of glacial ice in the Northern Hemisphere was about twice as great as in the Southern Hemisphere. Briefly explain why this was the case.

18. List three indirect effects of Ice Age glaciers.

19. How might plate tectonics help explain the cause of ice ages? Can plate tectonics explain the alternation between glacial and interglacial climates during the Pleistocene?

Key Terms

ablation (p. 336)
abrasion (p. 339)
alpine glacier (p. 331)
arête (p. 343)
basal slip (p. 334)
calving (p. 336)
cirque (p. 341)
col (p. 342)
crevasse (p. 334)
drumlin (p. 348)
end moraine (p. 346)
esker (p. 352)
equilibrium line (p. 335)
fjord (p. 342)
firn (p. 334)

glacial budget (p. 336)
glacial drift (p. 345)
glacial erratic (p. 345)
glacial striations (p. 339)
glacial trough (p. 341)
glacier (p. 330)
ground moraine (p. 347)
hanging valley (p. 341)
horn (p. 343)
ice cap (p. 332)
ice-contact deposit (p. 352)
ice sheet (p. 331)
ice shelf (p. 332)
kame (p. 352)
kame terrace (p. 352)

kettle (p. 349)
lateral moraine (p. 346)
medial moraine (p. 346)
outlet glacier (p. 333)
outwash plain (p. 349)
piedmont glacier (p. 333)
plastic flow (p. 334)
Pleistocene Epoch
  (p. 353)
plucking (p. 339)
recessional moraine
  (p. 347)
roche moutonnée (p. 343)
rock flour (p. 339)
stratified drift (p. 345)

surge (p. 335)
tarn (p. 341)
terminal moraine (p. 347)
till (p. 345)
tillite (p. 355)
truncated spur (p. 341)
valley glacier (p. 331)
valley train (p. 349)
zone of accumulation
  (p. 335)
zone of fracture (p. 334)
zone of wastage (p. 336)

Web Resources

 The *Earth* Web site uses the resources and flexibility of the Internet to aid in your study of the topics in this chapter. Written and developed by geology instructors, this site will help improve your understanding of geology. Visit **http://www.pearson.ca/tarbuck** and click on the cover of the text to find:

■ Online review quizzes.
■ Web-based critical thinking and writing exercises.
■ Links to chapter-specific Web resources.
■ Internet-wide key-term searches.

http://www.pearson.ca/tarbuck

Chapter 13
Deserts and Winds

Snow covers the rocky ground in the Sonoran Desert near Tucson, Arizona. As this scene demonstrates, deserts are not necessarily always hot, lifeless, dune-covered expanses. (Photo by T. Wiewandt/DRK Photo)

Climate has a strong influence on the nature and intensity of Earth's external processes. This was clearly demonstrated in the preceding chapter on glaciers. Another excellent example of the strong link between climate and geology is seen when we examine the development of arid landscapes. The word desert literally means deserted or unoccupied. For many dry regions this is a very appropriate description, although where water is available in deserts, plants and animals thrive. Nevertheless, the world's dry regions are probably the least familiar land areas on Earth outside of the polar realm.

Desert landscapes frequently appear stark. Their profiles are not softened by a carpet of soil and abundant plant life. Instead, barren rocky outcrops with steep, angular slopes are common. At some places the rocks are tinted orange and red. At others they are grey and brown and streaked with black. For many visitors, desert scenery exhibits a striking beauty; to others, the terrain seems bleak. No matter which feeling is elicited, it is clear that deserts are very different from the more humid places where most people live.

As you shall see, arid regions are not dominated by a single geologic process. Rather, the effects of tectonic forces, running water, and wind are all apparent. Because these processes combine in different ways from place to place, the appearance of desert landscapes varies a great deal as well (Figure 13.1).

Distribution and Causes of Dry Lands

External Processes
↳ Deserts

The dry regions of the world encompass about 42 million square kilometres, a surprising 30 percent of Earth's land surface. No other climatic group covers so large a land area. Within these water-deficient regions, two climatic types are commonly recognized: **desert**, or arid, and **steppe**, or semi-arid. The two share many features; their differences are primarily a matter of degree (see Box 13.1). The steppe is a marginal and more humid variant of the desert and is a transition

◆ **Figure 13.1**   A scene in southern Utah near the San Juan River. The appearance of desert landscapes varies a great deal from place to place.

BOX 13.1

Understanding Earth
What Is Meant by "Dry"?

Albuquerque, New Mexico, in the southwestern United States, receives an average of 20.7 centimetres of rainfall annually. As you might expect, because Albuquerque's precipitation total is modest, the station is classified as a desert when the commonly used Köppen climate classification is applied. The Russian city of Verkhoyansk is a remote station located near the Arctic Circle in Siberia. The yearly precipitation total there averages 15.5 centimetres, about 5 centimetres less than Albuquerque's. Although Verkhoyansk receives less precipitation than Albuquerque, its classification is that of a humid climate. How can this occur?

We all recognize that deserts are dry places, but just what is meant by the term *dry*? That is, how much rain defines the boundary between humid and dry regions? Sometimes it is arbitrarily defined by a single rainfall figure; for example, 25 centimetres per year of precipitation. However, the concept of dryness is a relative one that refers to any situation in which a water deficiency exists. Hence, climatologists define *dry climate* as one in which yearly precipitation is not as great as the potential loss of water by evaporation. Dryness then not only is related to annual rainfall totals but is also a function of evaporation, which in turn is closely dependent upon temperature.

As temperatures climb, potential evaporation also increases. Fifteen to

TABLE 13.A   Average Annual Precipitation Defining the Boundary Between Dry and Humid Climates

Average Annual Temp. (C°)	Winter Rainfall Maximum (centimetres)	Even Distribution (centimetres)	Summer Rainfall Maximum (centimetres)
5	10	24	38
10	20	34	48
15	30	44	58
20	40	54	68
25	50	64	78
30	60	74	88

25 centimetres of precipitation may be sufficient to support coniferous forests in northern Scandinavia or Siberia, where evaporation into the cool, humid air is slight and a surplus of water remains in the soil. However, the same amount of rain falling on New Mexico or Iran supports only a sparse vegetative cover because evaporation into the hot, dry air is great. So, clearly, no specific amount of precipitation can serve as a universal boundary for dry climates.

To establish the boundary between dry and humid climates, the widely used Köppen classification system uses formulas that involve three variables: average annual precipitation, average annual temperature, and seasonal distribution of precipitation. The use of average annual temperature reflects its importance as an index of evaporation. The amount of rainfall defining the humid–dry boundary will be larger where mean annual temperatures are high, and smaller where temperatures are low. The use of seasonal precipitation distribution as a variable is also related to this idea. If rain is concentrated in the warmest months, loss to evaporation is greater than if the precipitation is concentrated in the cooler months.

Table 13.A summarizes the precipitation amounts that divide dry and humid climates. Notice that a station with an annual mean of 20°C and a summer rainfall maximum of 68 centimetres is classified as dry. If the rain falls primarily in winter, however, the station need receive only 40 centimetres or more to be considered humid. If the precipitation is more evenly distributed, the figure defining the humid–dry boundary is in between the other two.

---

zone that surrounds the desert and separates it from bordering humid climates. The world map showing the distribution of desert and steppe regions reveals that dry lands are concentrated in the subtropics and in the middle latitudes (Figure 13.2).

Low-Latitude Deserts

The heart of the low-latitude dry climates lies in the vicinities of the tropics of Cancer and Capricorn. Figure 13.2 shows a virtually unbroken desert environment stretching for more than 9300 kilometres from the Atlantic coast of North Africa to the dry lands of northwestern India. In addition to this single great

expanse, the Northern Hemisphere contains another, much smaller, area of tropical desert and steppe in northern Mexico and the southwestern United States.

In the Southern Hemisphere, dry climates dominate Australia. Almost 40 percent of the continent is desert, and much of the remainder is steppe. In addition, arid and semi-arid areas occur in southern Africa and make a limited appearance in coastal Chile and Peru.

What causes these bands of low-latitude desert? The answer is the global distribution of air pressure and winds. Figure 13.3, an idealized diagram of Earth's general circulation, helps the visualization of the relationship. Heated air in the pressure belt known as the *equatorial low* rises to great heights (usually between

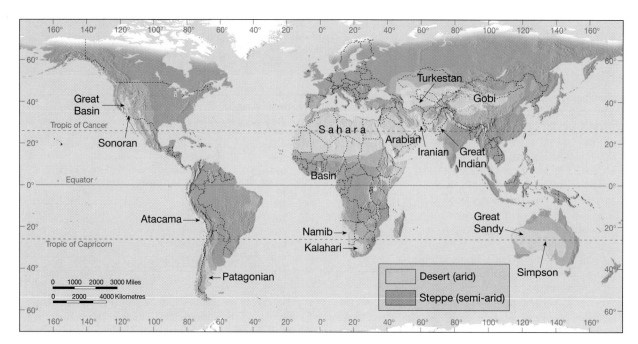

◆ **Figure 13.2**   Arid and semi-arid climates cover about 30 percent of Earth's land surface. No other climate group covers so large an area.

15 and 20 kilometres) and then spreads out. As the upper-level flow reaches 20° to 30° latitude, north or south, it sinks toward the surface. Air that rises through the atmosphere expands and cools, a process that leads to the development of clouds and precipitation. For this reason, the areas under the influence of the equatorial low are among the rainiest on Earth. Just the opposite is true for the regions in the vicinity of 30° north and south latitude, where high pressure predominates. Here, in the zones known as the *subtropical highs*, air is subsiding. When air sinks, it is compressed and warmed. Such conditions are just the opposite

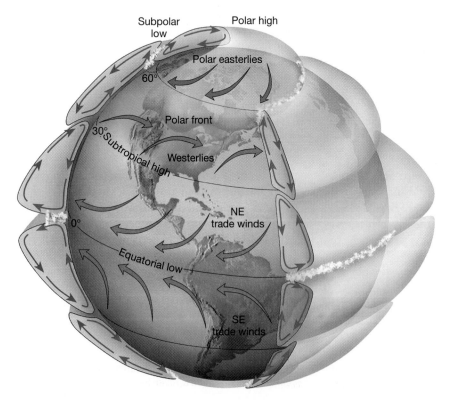

◆ **Figure 13.3**   Idealized diagram of Earth's general circulation. The deserts and steppes that are centred in the latitude belt between 20° and 30° north and south coincide with the subtropical high-pressure belts. Here dry, subsiding air inhibits cloud formation and precipitation. By contrast, the pressure belt known as the equatorial low is associated with areas that are among the rainiest on Earth.

dunes resembles the shape of barchans except that their tips point into the wind rather than downwind (Figure 13.15E). Parabolic dunes often form along coasts where there are strong onshore winds and abundant sand. If the sand's sparse vegetative cover is disturbed at some spot, deflation creates a blowout. Sand is then transported out of the depression and deposited as a curved rim, which grows higher as deflation enlarges the blowout.

Star Dunes   Confined largely to parts of the Sahara and Arabian deserts, **star dunes** are isolated hills of sand that exhibit a complex form (Figure 13.15F). Their name is derived from the fact that the bases of these dunes resemble multipointed stars. Usually three or four sharp-crested ridges diverge from a central high point that in some cases may approach a height of 90 metres. As their form suggests, star dunes develop where wind directions are variable.

Loess (Silt) Deposits

In some parts of the world the surface topography is mantled with deposits of windblown silt, called **loess**. Over periods of perhaps thousands of years, dust storms deposited this material. When loess is breached by streams or road cuts, it tends to maintain vertical cliffs and lacks any visible layers, as you can see in Figure 13.16.

The distribution of loess worldwide indicates that there are two primary sources for this sediment: deserts and glacial outwash deposits. The thickest and most extensive deposits of loess on Earth occur in western and northern China. They were blown there from the extensive desert basins of central Asia. Accumulations of 30 metres are common, and thicknesses of more than 100 metres have been measured. It is this fine buff-coloured sediment that gives the Yellow River (Huang Ho) its name.

In North America, deposits of loess are significant in many areas, including South Dakota, Nebraska, Iowa, Missouri, and Illinois, as well as portions of the Columbia Plateau in the Pacific Northwest. The correlation between the distribution of loess and important farming regions in the Midwest and eastern Washington State is not just a coincidence, because soils derived from this wind-deposited sediment are among the most fertile in the world.

Unlike the deposits in China, which originated in deserts, the loess in North America (and Europe) is an indirect product of glaciation. Its source is deposits of stratified drift and till. During the retreat of the ice sheets, many river valleys were choked with sediment deposited by meltwater, and ground moraine was exposed to the air. Strong westerly winds sweeping across the barren floodplains and till plains picked up the finer sediment and dropped it as a blanket on the eastern sides of the valleys. Such an origin is confirmed by the fact that loess deposits are thickest and coarsest on the lee sides of such major glacial drainage outlets as the Mississippi and Illinois rivers and rapidly thin with increasing distance from the valleys. Furthermore, the angular mechanically weathered particles composing the loess are essentially the same as the rock flour produced by the grinding action of glaciers.

◆ **Figure 13.16** This vertical loess bluff near the Mississippi River in southern Illinois is about 3 metres high. (Photo by James E. Patterson)

Chapter Summary

- The *concept of dryness is relative*; it refers to any situation in which a water deficiency exists. Dry regions encompass about 30 percent of Earth's land surface. Two climatic types are commonly recognized: *desert*, which is arid, and *steppe* (a marginal and more humid variant of desert), which is semi-arid. *Low-latitude deserts* coincide with the zones of subtropical highs in lower latitudes. On the other hand, *middle-latitude deserts* exist principally because of their positions in the deep interiors of large landmasses far removed from the ocean.

- The same geologic processes that operate in humid regions also operate in deserts, but under contrasting climatic conditions. In dry lands *rock weathering of any type is greatly reduced* because of the lack of moisture and the scarcity of organic acids from decaying plants. Much of the weathered debris in deserts is the result of *mechanical weathering*. Practically all desert streams are dry most of the time and are said to be *ephemeral*. Stream courses in deserts are seldom well integrated and lack an extensive system of tributaries. Nevertheless, *running water is responsible for most of the erosional work in a desert*. Although wind erosion is more significant in dry areas than elsewhere, the main role of wind in a desert is in the transportation and deposition of sediment.

- The transport of sediment by wind differs from that by running water in two ways. First, wind has a low density compared to water; thus, it is not capable of picking up and transporting coarse materials. Second, because wind is not confined to channels, it can spread sediment over large areas.

- The *bed load* of wind consists of sand grains skipping and bouncing along the surface in a process termed *saltation*. Fine dust particles can be carried great distances by the wind as *suspended load*.

- Compared to running water and glaciers, wind is a relatively insignificant erosional agent. *Deflation*, the lifting and removal of loose material, often produces shallow depressions called *blowouts*. In portions of many deserts the surface, called *desert pavement*, is a layer of coarse pebbles and gravels too large to be moved by the wind. Wind also erodes by *abrasion*, often creating interestingly shaped stones called *ventifacts*. *Yardangs* are narrow streamlined wind-sculpted ridges that can be up to 90 metres high and 100 kilometres long.

- Wind deposits are of two distinct types: (1) *mounds and ridges of sand*, called *dunes*, which are formed from sediment that is carried as part of the wind's bed load; and (2) extensive *blankets of silt*, called *loess*, that once were carried by wind in *suspension*. The profile of a dune shows an asymmetrical shape with the leeward (sheltered) slope being steep and the windward slope more gently inclined. The *types of sand dunes* include (1) *barchan dunes*; (2) *transverse dunes*; (3) *barchanoid dunes*; (4) *longitudinal dunes*; (5) *parabolic dunes*; and (6) *star dunes*. The thickest and most extensive deposits of loess occur in western and northern China. Unlike the deposits in China, which originated in deserts, the loess in North America and Europe is an indirect product of glaciation.

Review Questions

1. How extensive are the desert and steppe regions of Earth?

2. What is the primary cause of subtropical deserts? Of middle-latitude deserts?

3. In which hemisphere (Northern or Southern) are middle-latitude deserts most common?

4. Why is the amount of precipitation that is used to determine whether a place has a dry climate or a humid climate a variable figure? (See Box 13.1, p. 367)

5. *Deserts are hot, lifeless, sand-covered landscapes shaped largely by the force of wind.* The preceding statement summarizes the image of arid regions that many people hold, especially those living in more humid places. Is it an accurate view?

6. Why is rock weathering reduced in deserts?

7. As a permanent stream such as the Nile River crosses a desert, does discharge increase or decrease? How does this compare to a river in a humid region?

8. What is the most important erosional agent in deserts?

9. Why is sea level (ultimate base level) not a significant factor influencing erosion in desert regions?

10. Describe two places where true desert conditions occur in Canada.

11. Describe the way in which wind transports sand. During very strong winds, how high above the surface can sand be carried?

12. Why is wind erosion relatively more important in arid regions than in humid areas?

13. What factor limits the depths of blowouts?

14. How do sand dunes migrate?

15. List three factors that influence the form and size of a sand dune.

16. Six major dune types are recognized. Indicate which type of dune is associated with each of the following statements.

(a) dunes whose tips point into the wind

(b) long sand ridges oriented at right angles to the wind

(c) dunes that often form along coasts where strong winds create a blowout

(d) solitary dunes whose tips point downwind

(e) long sand ridges that are oriented more or less parallel to the prevailing wind

(f) an isolated dune consisting of three or four sharp-crested ridges diverging from a central high point

(g) scalloped rows of sand oriented at right angles to the wind

17. Although sand dunes are the best-known wind deposits, accumulations of loess are very significant in some parts of the world. What is loess? Where are such deposits found? What are the origins of this sediment?

Key Terms

abrasion (p. 376)
barchan dune (p. 379)
barchanoid dune (p. 380)
bed load (p. 374)
blowout (p. 376)
cross beds (p. 378)

deflation (p. 376)
desert (p. 366)
desert pavement (p. 376)
dune (p. 378)
ephemeral stream (p. 373)
loess (p. 381)

longitudinal dune (p. 380)
parabolic dune (p. 380)
rainshadow desert (p. 369)
saltation (p. 374)
slip face (p. 378)
star dune (p. 381)

steppe (p. 366)
suspended load (p. 375)
transverse dune (p. 379)
ventifact (p. 376)
yardang (p. 377)

Web Resources

 The *Earth* Web site uses the resources and flexibility of the Internet to aid in your study of the topics in this chapter. Written and developed by geology instructors, this site will help improve your understanding of geology. Visit **http://www.pearson.ca/tarbuck** and click on the cover of the text to find:

■ Online review quizzes.
■ Web-based critical thinking and writing exercises.
■ Links to chapter-specific Web resources.
■ Internet-wide key-term searches.

http://www.pearson.ca/tarbuck

Chapter 14
Shorelines

The rocky shoreline at Big Sur, California. (Photo by David Muench/David Muench Photography, Inc.)

A

◆ **Figure 14.1   A.** The satellite image includes the outline of Cape Cod. Boston is to the upper left. The two large islands off the south shore of Cape Cod are Martha's Vineyard (left) and Nantucket (right). Although the work of waves constantly modifies this coastal landscape, shoreline processes are not responsible for creating it. Rather, the present size and shape of Cape Cod results from the positioning of moraines and other glacial materials (see inset photo) deposited during the Pleistocene Epoch. **B.** High-altitude image of the Point Reyes area north of San Francisco, California. The 5.5-kilometre-long south-facing cliffs at Point Reyes (lower left corner) are exposed to the full force of the waves from the Pacific Ocean. Nevertheless, this promontory retreats slowly because the bedrock (see inset photo) from which it formed is very resistant. (Photo A. Satellite image courtesy of Earth Satellite Corporation/Science Photo Library/Photo Researchers, Inc.; inset photo A. by Stephen J. Krasemann/DRK Photo; Photo B. High-altitude image courtesy of USDA-ASCS; inset photo B. by David Muench/David Muench Photography, Inc.)

The restless waters of the ocean are constantly in motion. Winds generate surface currents, the gravity of the Moon and Sun produces tides, and density differences create deep-ocean circulation. Further, waves carry the energy from storms to distant shores, where their impact erodes the land.

Shorelines are dynamic environments. Their topography, geologic makeup, and climate vary greatly from place to place. Continental and oceanic processes converge along coasts to create landscapes that fre-quently undergo rapid change. When it comes to the deposition of sediment, they are transition zones between marine and continental environments.

The Shoreline: A Dynamic Interface

Nowhere is the restless nature of the ocean's water more noticeable than along the shore—the dynamic interface (common boundary) among air, land, and

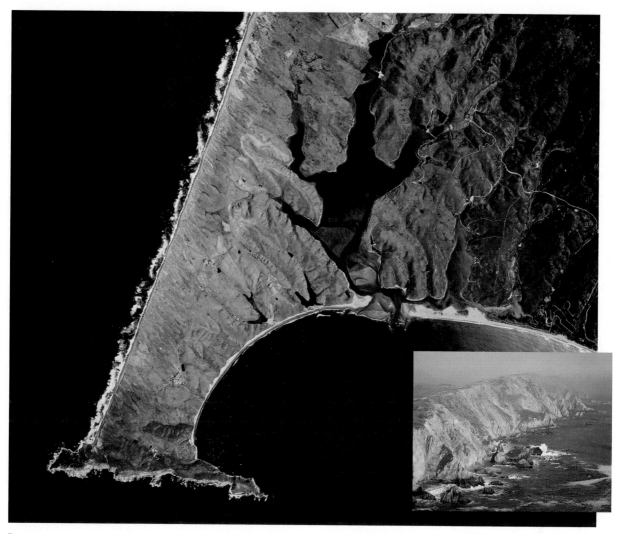

B

sea. Here we can see the rhythmic rise and fall of tides and observe waves constantly rolling in and breaking. Sometimes the waves are low and gentle. At other times they pound the shore with awesome fury. Although it may not be obvious, the shoreline is constantly being modified (Figure 14.1).

The nature of present-day shorelines is not just the result of the relentless attack on the land by the sea. Indeed, the shore has a complex character that results from multiple geologic processes. For example, practically all coastal areas were affected by the worldwide rise in sea level that accompanied the melting of glaciers at the close of the Pleistocene Epoch. As the sea encroached landward, the shoreline retreated, modifying existing landscapes.

Today the coastal zone is greatly affected by human activity. We will examine the implications of this human activity in detail on page 403 under the heading "Shorelines and Human Activities."

Waves

External Processes
↳ Coastal Processes

Wind-generated waves provide most of the energy that shapes and modifies shorelines. Where the land and sea meet, waves that may have travelled unimpeded for hundreds or thousands of kilometres suddenly encounter a barrier that will not allow them to advance farther and must absorb their energy. The conflict that results is never-ending and sometimes dramatic.

Characteristics of Waves

The undulations of the water surface (that we call waves) derive their energy and motion from the wind. If a breeze of less than 3 kilometres per hour starts to blow across still water, small wavelets appear almost

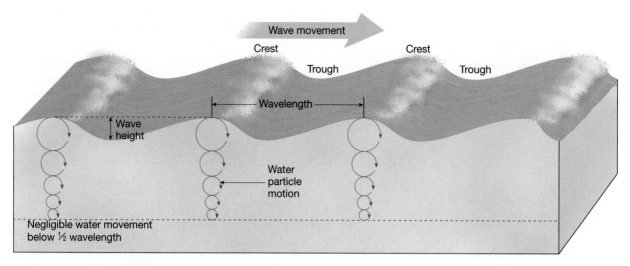

◆ **Figure 14.2**   This diagram illustrates the basic parts of a wave as well as the movement of water particles with the passage of the wave. Negligible water movement occurs below a depth equal to one half the wavelength (the level of the dashed line).

instantly. When the breeze dies, the ripples disappear as suddenly as they formed. However, if the wind exceeds 3 kilometres per hour, more stable waves gradually form and advance with the wind.

All waves have the characteristics illustrated in Figure 14.2. The tops of the waves are *crests*, which are separated by *troughs*. The vertical distance between trough and crest is the **wave height**. The horizontal distance separating successive crests is the **wavelength**. The **wave period** is the time interval between the passage of two successive crests at a stationary point.

The height, length, and period that are eventually achieved by a wave depend on three factors: (1) the wind speed, (2) the length of time the wind has blown, and (3) the **fetch**, or distance that the wind has travelled across open water. As the quantity of energy transferred from the wind to the water increases, the height and steepness of the waves increases as well. Eventually a critical point is reached where waves grow so tall that they topple over, forming breakers called *whitecaps*.

For a particular wind speed, there is a maximum fetch and duration of wind beyond which waves no longer increase in size. By this stage, waves lose as much energy through the breaking of whitecaps as they are receiving from the wind.

When wind stops or changes direction, or if waves leave the stormy area where they were created, they continue on without relation to local winds. The waves also undergo a gradual change to *swells*, which are lower and longer and may carry a storm's energy to distant shores. Because many independent wave systems exist at the same time, the sea surface acquires a complex, irregular pattern. Hence, the sea waves we watch from the shore are often a mixture of swells from faraway storms and waves created by local winds.

Types of Waves

When observing waves, always remember that you are watching *energy* travel through a medium (water). If you make waves by tossing a pebble into a pond, by splashing in a pool, or by blowing across the surface of an aquarium, you are transferring *energy* to the water, and the waves you see are just the visible evidence of the energy passing through.

In open waters, it is the wave energy that moves forward, not the water itself. Each water particle moves in a circular path during the passage of a wave (Figure 14.2). As a wave passes, a water particle returns almost to its original position. The circular orbits followed by the water particles at the surface have a diameter equal to the wave height. When water is part of the wave crest, it moves in the same direction as the advancing waveform. In the trough, the water moves in the opposite direction. This is demonstrated by observing the behaviour of a floating object as a wave passes. The toy boat in Figure 14.3 merely seems to bob up and down and sway slightly to and fro without advancing appreciably from its original position. (The wind does drag the water forward slightly, causing the surface circulation of the oceans.) Because of this, waves in the open sea are called **waves of oscillation**.

The energy contributed by the wind to the water is transmitted not only along the surface of the sea but also downward. However, beneath the surface, the circular motion rapidly diminishes until, at a depth equal to about half the wavelength, the movement of water particles becomes negligible. This is shown by the rapidly diminishing diameters of water-particle orbits in Figure 14.2.

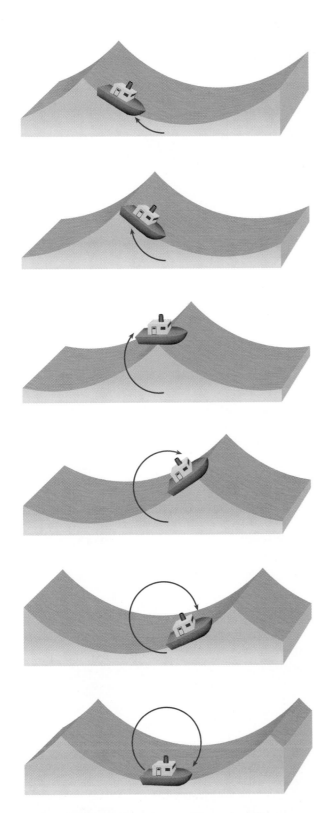

◆ Figure 14.3 The movements of the toy boat illustrate that the wave energy advances but the water itself advances only slightly from its original position. In this sequence, the wave moves from left to right as the boat (and the water in which it is floating) rotates in a circular motion. The boat moves slightly to the left up the front of the approaching wave, then, after reaching the crest, slides to the right down the back of the wave.

Students Sometimes Ask...

What are tidal waves?
Tidal waves, more accurately known as *tsunami* (*tsu* = harbour, *nami* = wave), have nothing to do with the tides! They are long-wavelength, fast-moving, often large, and sometimes destructive waves that originate from sudden changes in the topography of the seafloor. They are caused by underwater fault slippage, underwater avalanches, or underwater volcanic eruptions. Since the mechanisms that trigger tsunami are frequently seismic events, tsunami are appropriately termed *seismic sea waves*. For more information about characteristics of tsunami and their destructive effects, see Chapter 16, "Earthquakes."

As long as a wave is in deep water, it is unaffected by water depth (Figure 14.4, left). However, when a wave approaches the shore, the water becomes shallower and influences wave behaviour. The wave begins to "feel bottom" at a water depth equal to about half its wavelength. Such depths interfere with water movement at the base of the wave and slow its advance (Figure 14.4, centre). As a wave advances toward the shore, the slightly faster waves farther out to sea catch up, decreasing the wavelength. As the speed and length of the wave diminish, the wave steadily grows higher. Finally a critical point is reached when the steep wave front collapses, or *breaks* (Figure 14.4, right). What had been a wave of oscillation now becomes a **wave of translation,** in which the water advances up the shore.

The turbulent water created by breaking waves is called **surf**. On the landward margin of the surf zone the turbulent sheet of water from collapsing breakers, called *swash*, moves up the slope of the beach. When the energy of the swash has been expended, the water flows back down the beach toward the surf zone as *backwash*.

Wave Erosion

 External Processes
↳ Coastal Processes

During calm weather, wave action is minimal. However, just as streams do most of their work during floods, so too do waves accomplish most of their work during storms. The impact of high storm-induced waves against the shore can be awesome in its violence (Figure 14.5). Each breaking wave may hurl thousands of tonnes of water against the land, sometimes causing the ground to literally tremble. The pressures exerted by Atlantic waves in wintertime, for example, average nearly 10,000

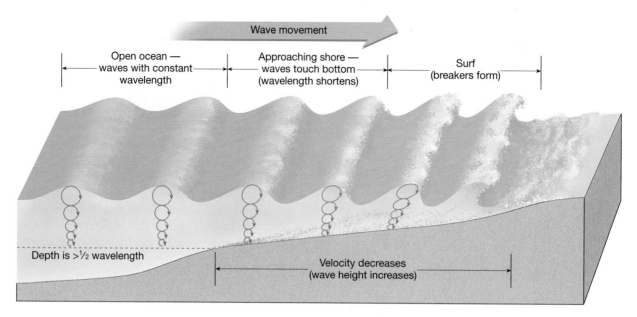

Wave movement

Open ocean —
waves with constant
wavelength

Approaching shore —
waves touch bottom
(wavelength shortens)

Surf
(breakers form)

Depth is >½ wavelength

Velocity decreases
(wave height increases)

◆ **Figure 14.4**   Changes that occur when a wave moves onto shore.

kilograms per square metre. The force during storms is even greater. During one such storm a 1350-tonne portion of a steel-and-concrete breakwater was ripped from the rest of the structure and moved to a useless position toward the shore at Wick Bay, Scotland. Five years later the 2600-tonne unit that replaced it met a similar fate.

There are many such stories that demonstrate the great force of breaking waves. It is no wonder that cracks and crevices are quickly opened in cliffs, sea-walls, breakwaters, and anything else that is subjected to these enormous shocks. Water is forced into every opening, causing air in the cracks to become highly compressed by the thrust of crashing waves. When the wave subsides, the air expands rapidly, dislodging rock fragments and enlarging and extending fractures.

In addition to the erosion caused by wave impact and pressure, **abrasion**—the sawing and grinding action of the water armed with rock fragments—is also important. In fact, abrasion is probably more intense in the surf zone than in any other environment. Smooth, rounded stones and pebbles along the shore are obvious reminders of the relentless grinding action of rock against rock in the surf zone. Further, such fragments are used as "tools" by the waves as they cut horizontally into the land (Figure 14.6).

Along shorelines composed of unconsolidated material rather than hard rock, the rate of erosion by breaking waves can be extraordinary. In parts of Britain, where waves have the easy task of eroding glacial deposits of sand, gravel, and clay, the coast has been worn back 3 to 5 kilometres since Roman times (2000 years ago), sweeping away many villages and ancient landmarks.

◆ **Figure 14.5**   When waves break against the shore, the force of the water can be powerful and the erosional work that is accomplished can be great. Marin headlands, Golden Gate National Recreation Area, California. (Photo by Galen Rowell/Mountain Light Photography, Inc.)

◆ **Figure 14.6** Cliff undercut by wave erosion along the Oregon coast.
(Photo by E. J. Tarbuck)

◆ **Figure 14.7** Wave bending around the end of a beach at Stinson Beach, California.
(Photo by James E. Patterson)

forward at its full speed. The net result is a wave front that may approach nearly parallel to the shore regardless of the original direction of the wave.

Because of refraction, wave impact is concentrated against the sides and ends of headlands that project into the water, whereas wave attack is weakened in bays. This differential wave attack along irregular coastlines is illustrated in Figure 14.8. As the waves reach the shallow water in front of the headland sooner than they do in adjacent bays, they are bent more nearly parallel to the protruding land and strike it from all three sides. By contrast, refraction in the bays causes waves to diverge and expend less energy. In these zones of weakened wave activity, sediments can accumulate and form sandy beaches. Over a long period, erosion of the headlands and deposition in the bays will straighten an irregular shoreline.

Longshore Transport

Beaches are sometimes called "rivers of sand." The reason is that the energy from breaking waves often causes large quantities of sand to move along the beach and in the surf zone.

Wave Refraction

The bending of waves, called **wave refraction**, plays an important part in shoreline processes (Figure 14.7). It affects the distribution of energy along the shore and thus strongly influences where and to what degree erosion, sediment transport, and deposition will take place.

Waves seldom approach the shore straight on and most move toward the shore at an angle. When they reach the shallow water of a smoothly sloping bottom, they are bent and tend to become parallel to the shore. Such bending occurs because the part of the wave nearest the shore reaches shallow water and slows first, whereas the end that is still in deep water continues

Beach Drift and Longshore Currents

Although waves are refracted, most still reach the shore at some angle, however slight. Consequently, the uprush of water from each breaking wave (the swash) is oblique. However, the backwash is straight down the slope of the beach. The effect of this pattern of water movement is to transport sediment in a zigzag pattern along the beach (Figure 14.9). This movement is called **beach drift**, and it can transport sand and pebbles hundreds or even thousands of metres each day.

Oblique waves also produce currents within the surf zone that flow parallel to the shore and move

As with coastal zones dominated by land-derived sediments, reefs contain sedimentary environments that reflect variations in water agitation. It must be emphasized, however, that the sediments produced in a reef setting are produced by living things and are therefore also influenced by factors controlling the life processes of the organisms involved.

The spectrum of sedimentary environments within reef settings is best illustrated in barrier reef systems (Figure 14.21). The hub of carbonate production lies in the reef crest, which is the highest part of the reef and stands nearest to the surface of the sea. The reef crest receives most of the wind and wave energy in the reef system and typically contains the most robust frame builders of the reef. Seaward of the reef crest is the reef front, typified by less robust reef builders and lots of carbonate debris shed from the reef crest. Still farther seaward is the fore reef, where low light levels discourage the establishment of reef-building organisms, and carbonate debris is abundant.

Landward of the reef crest is the reef flat, a plateau-like zone that is scoured by waves and where coarse-grained and well-sorted carbonate tends to accumulate. Still more landward is the lagoon, which is protected from wave action by the reef crest and reef flat, and is therefore an area where fine-grained carbonate sediment can settle out. Much of the carbonate mud within the lagoon is derived from types of algae that contain tiny grains of calcite and aragonite in their tissues. If the climate is arid, evaporite deposits can accumulate on **tidal flats** of the mainland adjacent to the lagoon.

Shorelines and Human Activities

As mentioned before, today the coastal zone teems with human activity. Unfortunately, people often treat the shoreline as if it were a stable platform on which structures can be built safely. This attitude jeopardizes both people and the shoreline. Many coastal landforms are relatively fragile, short-lived features that are easily damaged by development. And as anyone who has endured a tropical storm knows, the shoreline is not always a safe place to live.

Compared with natural hazards such as earthquakes, volcanic eruptions, and landslides, shoreline erosion is often perceived to be a more continuous and predictable process that appears to cause relatively modest damage to limited areas. In reality, the shoreline is a dynamic place that can change rapidly in response to natural forces. Exceptional storms are capable of eroding beaches and cliffs at rates that are far in excess of the long-term average. Such bursts of accelerated erosion can have a significant impact on the natural evolution of a coast; it can also have a profound impact on people who reside in the coastal zone (Figure 14.22).

Erosion along our coasts causes significant property damage. Large sums are spent annually not only to repair damage but also to prevent or control erosion. Already a problem at many sites, shoreline erosion is certain to become an increasingly serious problem as extensive coastal development continues.

Although the same processes cause change along every coast, not all coasts respond in the same way. Interactions among different processes and the relative importance of these processes depends on local factors. The factors include (1) the proximity of a coast to sediment-laden rivers, (2) the degree of tectonic activity, (3) the topography and composition of the land, (4) prevailing winds and weather patterns, and (5) the configuration of the coastline and nearshore areas.

Over the past 100 years, growing affluence and increasing demands for recreation have brought unprecedented development to many coastal areas. As both the number and the value of buildings have

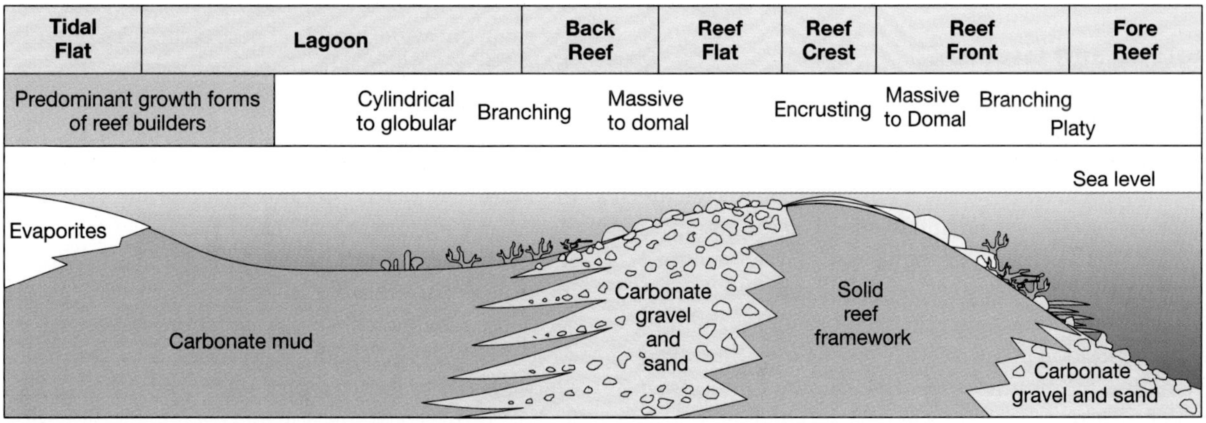

◆ **Figure 14.21** Principal zones of a barrier reef system.

◆ **Figure 14.22**   This beach house on Oak Island, North Carolina, was destroyed by Hurricane Floyd in November, 1999. (Photo by Robert Miller/*The News & Observer*/AP/Wide World Photos)

increased, so too have efforts to protect property from storm waves. Also, controlling the natural migration of sand is an ongoing struggle in many coastal areas. Such interference can result in unwanted changes that are difficult and expensive to correct.

Jetties and Groins

From relatively early in North America's history a principal goal in coastal areas was the development and maintenance of harbours. In many cases, this involved the construction of jetty systems. **Jetties** are usually built in pairs and extend into the ocean at the entrances to rivers and harbours. With the flow of water confined to a narrow zone, the ebb and flow caused by the rise and fall of the tides keeps the sand in motion and prevents deposition in the channel. However, as illustrated in Figure 14.23, the jetty may act as a dam against which the longshore current and beach drift deposit sand. At the same time, wave activity removes sand on

the other side. Because the other side is not receiving any new sand, there is soon no beach at all.

To maintain or widen beaches that are losing sand, groins are sometimes constructed. A **groin** is a barrier built at a right angle to the beach to trap sand that is moving parallel to the shore. The result is an irregular but wider beach. These structures often do their job so effectively that the longshore current beyond the groin becomes sand-starved. As a result, the current erodes sand from the beach on the leeward side of the groin. Such a situation is illustrated in Figure 14.24. To offset this effect, property owners downcurrent from the structure may erect groins on their property. In this manner, the number of groins multiplies. An example of such proliferation is the shoreline of New Jersey, where hundreds of these structures have been built. Many times, however, groins do not provide a satisfactory solution, so they are no longer the preferred method of keeping beach erosion in check.

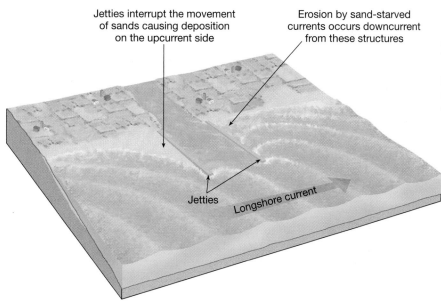

Jetties interrupt the movement of sands causing deposition on the upcurrent side

Erosion by sand-starved currents occurs downcurrent from these structures

Jetties

Longshore current

◆ **Figure 14.23**   Jetties are built at the entrances to rivers and harbours and are intended to prevent deposition in the navigation channel. Jetties interrupt the movement of sand by beach drift and longshore currents. Beach erosion often results downcurrent from the site of the structure.

Breakwaters and Seawalls

In some coastal areas a **breakwater** may be constructed parallel to the shoreline to protect boats from the force of large breaking waves. But while a breakwater creates

◆ **Figure 14.24**   Groins along the New Jersey shore at Cape May.
(Photo by John S. Shelton)

a quiet water zone near the shore, the reduced wave activity along the shore behind the structure may promote sand accumulation. A marina behind a breakwater can eventually fill with sand while the downstream beach erodes and retreats. At Santa Monica, California, where the building of a breakwater created such a problem, the city had to install a dredge to remove sand from the protected quiet water zone and deposit it down the beach where longshore currents and beach drift could recirculate the sand (Figure 14.25).

As development moves ever closer to a beach, seawalls are sometimes built to further defend property from breaking waves. Like breakwaters, **seawalls** are massive barriers intended to prevent waves from reaching the areas behind the wall. Waves expend much of their energy as they move across an open beach. Seawalls cut this process short by reflecting the force of unspent waves seaward. As a consequence, the beach to the seaward side of the seawall experiences significant erosion and may in some instances be eliminated entirely (Figure 14.26). Once the width of the beach is reduced, the seawall is subjected to even greater pounding by the waves, causing the wall to fail. A larger, more expensive wall must therefore be built to take its place.

The wisdom of building temporary protective structures along shorelines is increasingly questioned. The feelings of many coastal scientists and engineers are expressed in the following excerpt from a position paper that grew out of a conference on America's eroding shoreline:

It is now clear that halting the receding shoreline with protective structures benefits only a few and seriously degrades or destroys the natural beach and the value it holds for the majority. Protective structures divert the ocean's energy temporarily from private properties, but usually refocus that energy on the adjacent natural beaches. Many interrupt the natural sand flow in coastal currents, robbing many beaches of vital sand replacement.*

---

*"Strategy for Beach Preservation Proposed," *Geotimes* 30 (No. 12, December 1985): 15.

◆ **Figure 14.25**   Aerial view of a breakwater at Santa Monica, California. The breakwater appears as a faint line in the water behind which many boats are anchored. The construction of the breakwater disrupted longshore transport and caused the seaward growth of the beach. (Photo by John S. Shelton)

Beach Nourishment

Another approach to stabilizing shoreline sands is **beach nourishment**. As the term implies, this practice simply involves the artificial replenishment addition of sand to the beach system. Building the beaches seaward improves beach quality and storm protection, but only temporarily. The same processes that removed the original sand in the first place will eventually remove the replacement sand as well. Beach nourishment is also very expensive. For example, the latest beach nourishment project in Miami Beach, Florida cost $64 million for a mere 24 kilometres of coastline, and such restoration must be redone every 10 to 12 years (Figure 14.27). Not every project has been successful either. For example, when Ocean City, New Jersey, restored a stretch of beach, the sand lasted less than three months before a series of storms eroded away the $5 million investment.

In some instances, beach nourishment can lead to unwanted environmental effects. For example, beach replenishment at Waikiki Beach, Hawaii, involved replacing coarse calcareous sand with softer, muddier calcareous sand. Destruction of the soft beach sand by breaking waves increased the water's turbidity and killed offshore coral reefs. Beach nourishment appears to be an economically viable long-range solution to the beach preservation problem only in areas where there exists dense development, large supplies of sand, relatively low wave energy, and reconcilable environmental issues. Unfortunately, few areas possess all these attributes.

◆ **Figure 14.26**   Seabright in northern New Jersey once had a broad, sandy beach. A seawall 5 to 6 metres high and 8 kilometres long was built to protect the town and the railroad that brought tourists to the beach. As you can see, after the wall was built, the beach narrowed dramatically. (Photo by Raphael Macia/Photo Researchers, Inc.)

A

B

◆ **Figure 14.27**   Miami Beach, Florida. **A.** Before beach nourishment and **B.** After beach nourishment. (Photos courtesy of the U.S. Army Corps of Engineers, Vicksburg District)

Abandonment and Relocation

So far two basic responses to shoreline erosion problems have been considered: (1) the building of structures such as seawalls to hold the shoreline in place and (2) the addition of sand to replenish eroding beaches. However, a third option is also available. Many coastal scientists and planners are calling for a policy shift from defending and rebuilding beaches and coastal property in high hazard areas to removing storm-damaged buildings in those places and letting nature reclaim the beach.

Such proposals, of course, are controversial. People with significant nearshore investments shudder at the thought of not rebuilding and defending coastal developments from the erosional wrath of the sea. Others, however, argue that with sea level rising, the impact of coastal storms will only get worse in the decades to come (Box 14.2). This group advocates that oft-damaged structures be abandoned or relocated to improve personal safety and to reduce costs. Such ideas will no doubt be the focus of much study and debate as governments and communities evaluate and revise coastal land-use policies.

Other Consequences of Human Activities

Humans have not only precipitated their own losses by attempting to directly manipulate the sediment dynamics of coastlines, but have also inadvertently augmented these problems indirectly by failing to recognize coastal areas as components of larger-scale geologic systems. This holds true for both the Pacific and Atlantic coasts, although specific issues pertaining to each one differ according to geologic circumstance.

The Pacific Coast

The rugged Pacific Coast, located along a tectonically active continental margin, is characterized by relatively narrow beaches that are backed by steep cliffs and mountain ranges (see Figure 14.1B). Much of the beach sand is weathered material that is transported to the coast by streams that originate in the mountains. Humans have interrupted the supply of sand to the coast by constructing dams for irrigating and flood control, thus preventing the natural replenishment of sand on the beaches. If more sand is lost by erosion than is gained from river sources, the beach becomes progressively narrower, and this has obvious economic implications for developed areas. In addition, the loss of beaches that would otherwise buffer the coastline from storm waves can result in the acceleration of sea cliff erosion. Furthermore, if sea level rises at an increasing rate in the years to come, increased shoreline erosion and sea-cliff retreat should be expected along many parts of the Pacific Coast.

The Atlantic and Gulf Coasts

The Atlantic and Gulf Coasts are located on a passive margin with wide continental shelf areas bordered by land areas that tend to be less rugged than on the

BOX 14.2

People and the Environment
Coastal Vulnerability to Sea-Level Rise

Human activities, especially the combustion of fossil fuels, have been adding vast amounts of carbon dioxide and other trace gases to the atmosphere for 200 years or more. The prospect is that emissions of these gases will continue to increase during the twenty-first century.* One consequence of this change in the composition of the atmosphere is an enhancement of Earth's greenhouse effect with a resulting increase in global temperatures. During the twentieth century, average global temperatures increased by about 0.6°C. During the twenty-first century, the increase is projected to be considerably greater.

One probable impact of a human-induced global warming is a rise in sea level. How is a warmer atmosphere related to a global rise in sea level? The most obvious connection—the melting of glaciers—is important but *not* the most significant factor. More significant is that a warmer atmosphere causes an increase in ocean volume due to thermal expansion. Higher air temperatures warm the adjacent upper layers of the ocean, which in turn causes the water to expand and sea level to rise.

Research indicates that sea level has risen between 20 and 25 centimetres over the past century and that the trend will continue. Some models indicate that the rise in the next 100 years may approach or even exceed 50 centimetres (Figure 14.B). Such a change may seem modest, but scientists realize that any

*For more on this, see the section on "Carbon Dioxide and Global Warming" in Chapter 21. Also related is Box 4.3, p. 123.

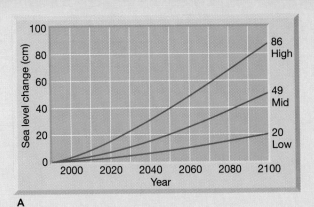

A

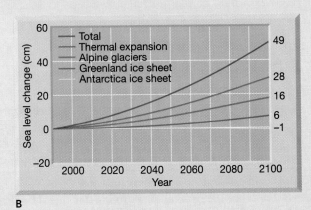

B

◆ **Figure 14.B** **A.** High, middle, and low projections of global sea-level rise, 1990–2100. **B.** Projected individual contributions to global sea-level change, 1990–2100, for the middle projection in part A. Thermal expansion is responsible for 28 centimetres of the sea-level rise. Melting alpine glaciers and ice caps contribute another 16 centimetres. The impact shown for Antarctica indicates that this massive ice sheet is projected to grow larger during this period.

rise in sea level along a *gently* sloping shoreline, such as the Atlantic and Gulf coasts of the United States, will lead to significant erosion and severe permanent inland flooding (Figure 14.C). If this happens, many beaches and wetlands will be eliminated, and coastal civilization would be severely disrupted.

Because rising sea level is a gradual phenomenon, it may be overlooked by

Pacific Coast (see Figure 14.1A). Low coastal gradients, high sediment supply, and wide shallow-water areas along the eastern coastline of the United States have promoted the formation of broad beaches and barrier islands in that region. The scenic beauty of barrier islands has encouraged significant coastal development, but not without cost.

Because barrier islands face the open ocean, they receive the full force of major storms that affect the coast and absorb wave energy through sand transport.

Waves can move sand from the beach to offshore areas or into the backshore dunes or lagoon. Alternatively, waves may transport sand from the dunes to the beach or farther seaward. The common factor is movement. Just as a flexible reed may survive a wind that destroys an oak tree, so the barriers survive hurricanes and nor'easters not through unyielding strength but by giving before the storm.

When a barrier is developed for homes or a resort, storm waves that previously rushed harmlessly through

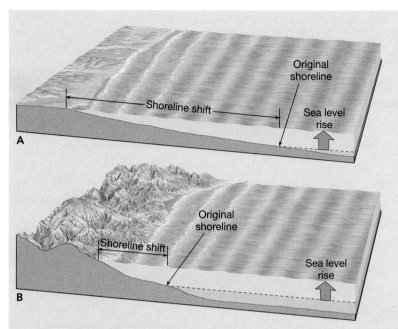

◆ **Figure 14.C** The slope of a shoreline is critical to determining the degree to which sea-level changes will affect it. **A.** When the slope is gentle, small changes in the sea level cause a substantial shift. **B.** The same sea-level rise along a steep coast results in a small shoreline shift.

coastal residents as an important contributor to shoreline erosion problems. Rather, the blame may be assigned to other forces, especially storm activity. Although a given storm may be the immediate cause, the magnitude of its destruction may result from the relatively small sea-level rise that allowed the storm's power to cross a much greater land area.

One of the most challenging problems for coastal scientists today is determining the physical response of the coastline to sea-level rise. Predicting shoreline retreat and land-loss rates is critical to formulating coastal management strategies. To date, long-term planning for our shorelines has been done piecemeal, if at all. Consequently, development continues without adequate consideration of the potential costs of erosion, flooding, and storm damage.

gaps between the dunes now encounter buildings and roadways. Since the dynamic nature of the barriers is most apparent during storms, homeowners tend to attribute damage to a "freak" storm, rather than to the natural mobility of coastal barriers. Local landowners are more likely to try to keep the sand in place and the waves at bay than they are to acknowledge that development was improperly placed from the beginning.*

Emergent and Submergent Coasts

Recent interest in the response of shorelines to changes in relative sea level has encouraged geologists to classify shorelines into emergent and submergent types. **Emergent coasts** develop either because an area experiences uplift or as a result of a drop in sea level. Conversely, **submergent coasts** are created when sea level rises or the land adjacent to the sea subsides.

In some areas the coast is clearly emergent because rising land or a falling water level exposes wave-cut cliffs and platforms above sea level (see Figure 14.14).

For example, seven different terrace levels exist in the Palos Verdes Hills, south of Los Angeles, indicating seven episodes of uplift in the recent geological past. In Canada, emergent coasts occur in regions that were once weighed down by great ice sheets and rebounded when the ice melted. The Hudson Bay region is such an area, portions of which are still rising at a rate of more than a centimetre per year.

Coastal regions that have undergone submergence in the relatively recent past are often highly irregular due to the partial drowning of river valleys, the headlands of the coast representing intervening ridges between the valleys. Estuaries, such as the Bay of Fundy, that occupy drowned river mouths along coastlines of the Canadian Maritimes are the products of post-glacial sea level rise (see Figure 14.10).

Keep in mind that most coasts have complicated geologic histories. With respect to sea level, many have at various times emerged and then submerged. Each time they may retain some of the features created during the previous situation.

---

*Frank Lowenstein, "Beaches or Bedrooms—The Choice as Sea Level Rises," *Oceanus* 28 (No. 3, Fall 1985): 22.

Chapter Summary

- The three factors that influence the *height*, *wavelength*, and *period* of a wave are (1) *wind speed*, (2) *length of time the wind has blown*, and (3) *fetch*, the distance that the wind has travelled across the open water.

- The two types of wind-generated waves are (1) *waves of oscillation*, which are waves in the open sea where the wave form advances as the water particles move in circular orbits, and (2) *waves of translation*, the turbulent advance of water formed near the shore as waves of oscillation collapse, or *break*, and form *surf*.

- Wave erosion is caused by *wave impact pressure* and *abrasion* (the sawing and grinding action of water armed with rock fragments). The bending of waves is called *wave refraction*. Owing to refraction, wave impact is concentrated against the sides and ends of headlands.

- Most waves reach the shore at an angle. The uprush (swash) and backwash of water from each breaking wave moves the sediment in a zigzag pattern along the beach. This movement, called *beach drift*, can transport sand hundreds or even thousands of metres each day. Oblique waves also produce *longshore currents* within the surf zone that flow parallel to the shore.

- Features produced by *shoreline erosion* include *wave-cut cliffs* (which originate from the cutting action of the surf against the base of coastal land), *wave-cut platforms* (relatively flat, benchlike surfaces left behind by receding cliffs), *sea arches* (formed when a headland is eroded and two caves from opposite sides unite), and *sea stacks* (formed when the roof of a sea arch collapses).

- *Tides*, the daily rise and fall in the elevation of the ocean surface, are caused by the *gravitational attraction* of the Moon and, to a lesser extent, the Sun. Near the times of new and full moons, the Sun and Moon are aligned, and their gravitational forces are added together to produce especially high and low tides. These are called the *spring tides*. Conversely, at about the times of the first and third quarters of the Moon, when the gravitational forces of the Moon and Sun are at right angles, the daily tidal range is less. These are called *neap tides*.

- *Tidal currents* are horizontal movements of water that accompany the rise and fall of tides. *Tidal flats* are the areas that are affected by the advancing and retreating tidal currents. When tidal currents slow after emerging from narrow inlets, they deposit sediment that may eventually create *tidal deltas*.

- Some of the depositional features that form when sediment is moved by beach drift and longshore currents are *spits* (elongated ridges of sand that project from the land into the mouth of an adjacent bay), *baymouth bars* (sandbars that completely cross a bay), and *tombolos* (ridges of sand that connect an island to the mainland or to another island). Along the Atlantic and Gulf Coastal Plains of the U.S., the shore zone is characterized by *barrier islands*, low ridges of sand that parallel the coast at distances of 3 to 30 kilometres offshore.

- Local factors that influence shoreline erosion are (1) the proximity of a coast to sediment-laden rivers, (2) the degree of tectonic activity, (3) the topography and composition of the land, (4) prevailing winds and weather patterns, and (5) the configuration of the coastline and nearshore areas.

- Three basic responses to shoreline erosion problems are (1) building *structures* such as *groins* (short walls built at a right angle to the shore to trap moving sand), *breakwaters* (structures built parallel to the shoreline to protect it from the force of large breaking waves), and *seawalls* (barriers constructed to prevent waves from reaching the area behind the wall) to hold the shoreline in place; (2) *beach nourishment*, which involves the addition of sand to replenish eroding beaches; and (3) *relocating* buildings away from the beach.

- Because of basic geological differences, the *nature of shoreline erosion problems along North America's Pacific and Atlantic coasts is very different*. Much of the development along the Atlantic and Gulf coasts has occurred on barrier islands, which receive the full force of major storms. Much of the Pacific Coast is characterized by narrow beaches backed by steep cliffs and mountain ranges. A major problem facing the Pacific shoreline is a narrowing of beaches caused by the natural flow of materials to the coast being interrupted by dams built for irrigation and flood control.

- One frequently used classification of coasts is based upon changes that have occurred with respect to sea level. *Emergent coasts*, often with wave-cut cliffs and wave-cut platforms above sea level, develop either because an area experiences uplift or as a result of a drop in sea level. Conversely, *submergent coasts*, with their drowned river mouths, called *estuaries*, are created when sea level rises or the land adjacent to the sea subsides.

Review Questions

1. List three factors that determine the height, length, and period of a wave.

2. Describe the motion of a water particle as a wave passes.

3. Explain what happens when a wave breaks.

4. Describe two ways in which waves cause erosion.

5. What is wave refraction? What is the effect of this process along irregular coastlines?

6. Discuss the origin of ocean tides.

7. How does the Sun influence tides?

8. Distinguish between flood current and ebb current.

9. How has the construction of artificial levees and dams on the Mississippi River and its tributaries contributed to a shrinking of the Mississippi's delta and its extensive wetlands (see Box 14.1)?

10. Describe the formation of the following features: wave-cut cliff, wave-cut platform, sea stack, spit, baymouth bar, and tombolo.

11. Explain how barrier islands are formed.

12. How can shoreline zones be recognized from sedimentary features of ancient deposits?

13. How might a seawall lead to increased beach erosion?

14. What are the drawbacks of beach nourishment?

15. How is a warmer atmosphere related to a global rise in sea level?

16. Relate the damming of rivers to the shrinking of beaches at many locations along the West Coast of the United States. Why do narrower beaches lead to accelerated sea-cliff retreat?

17. What observable features would lead you to classify a coastal area as emergent?

18. Are estuaries associated with submergent or emergent coasts? Why?

Key Terms

abrasion (p. 390)
barrier island (p. 398)
baymouth bar (p. 398)
beach drift (p. 391)
beach nourishment (p. 406)
breakwater (p. 405)
ebb current (p. 396)
emergent coast (p. 409)

estuaries (p. 396)
fetch (p. 388)
flood current (p. 396)
groin (p. 404)
jetty (p. 404)
longshore current (p. 392)
neap tide (p. 395)
sea arch (p. 396)
sea stack (p. 396)

seawall (p. 405)
spit (p. 398)
spring tide (p. 395)
submergent coast (p. 409)
surf (p. 389)
tidal current (p. 396)
tidal delta (p. 399)
tidal flats (p. 403)
tide (p. 392)

tombolo (p. 398)
wave-cut cliff (p. 396)
wave-cut platform (p. 396)
wave height (p. 388)
wavelength (p. 388)
wave of oscillation (p. 388)
wave of translation (p. 389)
wave period (p. 388)
wave refraction (p. 391)

Web Resources

 The *Earth* Web site uses the resources and flexibility of the Internet to aid in your study of the topics in this chapter. Written and developed by geology instructors, this site will help improve your understanding of geology. Visit **http://www.pearson.ca/tarbuck** and click on the cover of the text to find:

■ Online review quizzes.

■ Web-based critical thinking and writing exercises.

■ Links to chapter-specific Web resources.

■ Internet-wide key-term searches.

http://www.pearson.ca/tarbuck

Chapter 15
Crustal Deformation

24

Uplift along high angle faults, Mount Sneffels, Colorado Rockies.
(Photo by Art Wolfe, Inc.)

Earth is a dynamic planet. In the preceding chapters, you learned that weathering, mass wasting, and erosion by water, wind, and ice continually sculpt the landscape. In addition, tectonic forces deform rocks in the crust. Evidence demonstrating the operation of enormous forces within Earth includes thousands of kilometres of rock layers that are bent, contorted, overturned, and sometimes riddled with fractures (Figure 15.1). In the Canadian Rockies, for example, some rock units have been thrust for hundreds of kilometres over other layers. On a smaller scale, crustal movements of a few metres occur along faults during major earthquakes. In addition, rifting (spreading) and extension of the crust produce elongated depressions and over long spans of geologic time may even create ocean basins.

Structural Geology: A Study of Earth's Architecture

The results of tectonic activity are strikingly apparent in Earth's major mountain belts (see chapter opening photo). Here, rocks containing fossils of marine organisms may be found thousands of metres above sea level, and massive rock units show evidence of having been intensely fractured and folded, as though they were made of putty. Even in the stable interiors of the continents, rocks reveal a history of deformation that shows they have been uplifted from much deeper levels in the crust.

Structural geologists study the architecture of Earth's crust and "how it got this way" insofar as it resulted from deformation. By studying the orientations of faults and folds, as well as small-scale features of deformed rocks, structural geologists can often reconstruct the original geologic setting and the nature of the forces that generated these rock structures. In this way, the complex events of Earth's geologic history are unravelled.

An understanding of rock structures is not only important in deciphering Earth history, it is also basic to our economic well-being. For example, most occurrences of oil and natural gas are associated with geologic structures that act to trap these fluids in valuable "reservoirs" (see Chapter 21). Furthermore, rock fractures are sites of hydrothermal mineralization, which means they can be sites of metallic ore deposits. Moreover, the orientation of fracture surfaces, which represent zones of weakness in rocks, must be considered when selecting sites for major construction projects such as bridges, hydroelectric dams, and nuclear power plants. In short, a working knowledge of rock structures is essential to our modern way of life.

◆ **Figure 15.1** Uplifted and folded sedimentary strata at Stair Hole, near Lulworth, Dorset, England. These layers of Jurassic-age rock, originally deposited in horizontal beds, have been folded as a result of the collision between the African and European crustal plates. (Photo by Tom & Susan Bean, Inc.)

In this chapter we will examine the forces that deform rock, as well as the structures that result. The basic geologic structures associated with deformation are folds, faults, joints, and foliation (including rock cleavage). Because rock cleavage and foliation were examined in Chapter 7, this chapter will be devoted to the remaining rock structures and the tectonic forces that produce them.

Deformation

Internal Processes
↳ Crustal Deformation

Every body of rock, no matter how strong, has a point at which it will fracture or flow. **Deformation** (*de* = out, *forma* = form) is a general term that refers to all changes in the original form and/or size of a rock body. Deformation may also produce changes in the location and orientation of rock. Most crustal deformation occurs along plate margins. Recall from Chapter 1 that the lithosphere consists of large segments (plates) that move relative to one another. Plate motions and the interactions along plate boundaries generate tectonic forces that cause rock units to deform. The three types of plate boundaries are *divergent boundaries*, where plates move apart; *convergent boundaries*, where plates move together; and *transform fault boundaries*, where plates grind past each other.

Force, Stress, and Strain

Force is that which tends to put stationary objects in motion or change the motions of moving bodies. From everyday experience you know that if a door is stuck (stationary), you apply force to open it (get it in motion).

To describe the forces that deform rocks, structural geologists use the term **stress**—the amount of force applied to a given area. The magnitude of stress is not simply a function of the amount of force applied but also relates to the area on which the force acts. For example, if you are walking barefoot on a hard, flat surface, the force (weight) of your body is distributed across your entire foot, so the stress acting on any one point of your foot is low. However, if you step on a small pointed rock (ouch!), the stress concentration at a single point on your foot will be high. Thus, you can think of stress as a measure of how concentrated the force is. As you saw in Chapter 7, stress may be applied uniformly from all directions (*uniform stress*) or non-uniformly (*differential stress*).

Types of Stress

When stress is applied unequally from different directions, it is termed differential stress. **Differential stress**

that shortens a rock body is known as **compressional** (*com* = together, *premere* = to press) **stress**. Compressional stresses associated with plate collisions tend to shorten and thicken Earth's crust by folding, flowing, and faulting (Figure 15.2B). Recall from our discussion of metamorphic rocks that compressional stress is more concentrated at points where mineral grains are in contact, causing mineral matter to migrate from areas of high stress to areas of low stress (see Figure 7.5). As a result, the mineral grains (and the rock unit) tend to shorten in the direction parallel to the plane of maximum stress and elongate perpendicular to the direction of greatest stress.

When stress tends to elongate or pull apart a rock unit, it is known as **tensional** (*tendere* = to stretch) **stress** (Figure 15.2C). Where plates are being rifted apart (divergent plate boundaries), tensional stresses tend to lengthen those rock bodies located in the upper crust by displacement along faults. At depth, on the other hand, displacement is accomplished by a type of puttylike flow.

Differential stress can also cause rock to **shear** (Figure 15.2D). Shearing is similar to the slippage that occurs between individual playing cards when the top of the deck is moved relative to the bottom (Figure 15.3). In near-surface environments, shearing often occurs on closely spaced parallel surfaces of weakness, such as bedding planes, foliation, and microfaults. Further, at transform fault boundaries, shearing stresses produce large-scale offsets along major fault zones. By contrast, at great depths where temperatures and confining pressures are high, shearing is accomplished by solid-state flow.

Strain   Perhaps the easiest type of deformation to visualize occurs along small fault surfaces where differential stress causes rocks to move relative to each other in such a way that their original size and shape are preserved. Stress can also cause an irreversible change in the shape and size of a rock body, referred to as **strain**. Like the circle shown in Figure 15.3B, *strained bodies do not retain their original configuration during deformation*. Figure 15.1 illustrates the strain (deformation) exhibited by rock units near Dorset, England. When studying strained rock units like those shown in Figure 15.1, geologists ask, "What do these deformed structures indicate about the original arrangement of these rocks, and how have they been deformed?"

How Rocks Deform

When rocks are subjected to stresses greater than their own strength, they begin to deform, usually by folding, flowing, or fracturing (Figure 15.4). It is easy to visualize how rocks break, because we normally think

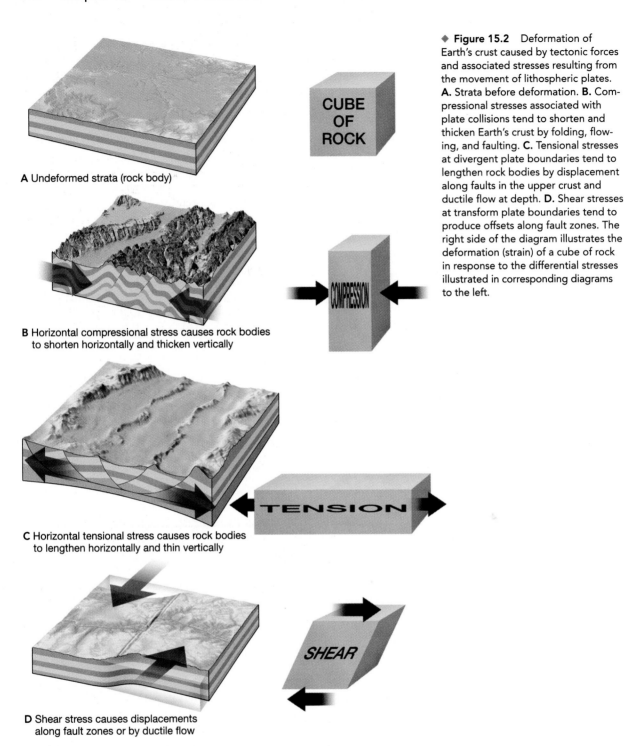

A Undeformed strata (rock body)

B Horizontal compressional stress causes rock bodies to shorten horizontally and thicken vertically

C Horizontal tensional stress causes rock bodies to lengthen horizontally and thin vertically

D Shear stress causes displacements along fault zones or by ductile flow

CUBE OF ROCK

COMPRESSION

TENSION

SHEAR

◆ **Figure 15.2** Deformation of Earth's crust caused by tectonic forces and associated stresses resulting from the movement of lithospheric plates. **A.** Strata before deformation. **B.** Compressional stresses associated with plate collisions tend to shorten and thicken Earth's crust by folding, flowing, and faulting. **C.** Tensional stresses at divergent plate boundaries tend to lengthen rock bodies by displacement along faults in the upper crust and ductile flow at depth. **D.** Shear stresses at transform plate boundaries tend to produce offsets along fault zones. The right side of the diagram illustrates the deformation (strain) of a cube of rock in response to the differential stresses illustrated in corresponding diagrams to the left.

of them as being brittle. But how can rock units be *bent* into intricate folds without being broken during the process? To answer this question, structural geologists performed laboratory experiments in which rocks were subjected to differential stress under conditions that simulated those existing at various depths within the crust (Figure 15.5).

Although each rock type deforms somewhat differently, the general characteristics of rock deformation were determined from these experiments. Geologists discovered that when stress is gradually applied, rocks first respond by deforming elastically. Changes that result from *elastic deformation* are recoverable; that is, like a rubber band, the rock will return to nearly its original size and shape when the stress is removed. (As you will see in the next chapter, the energy for most earthquakes comes from stored elastic energy that is released as rock snaps back to its original shape.)

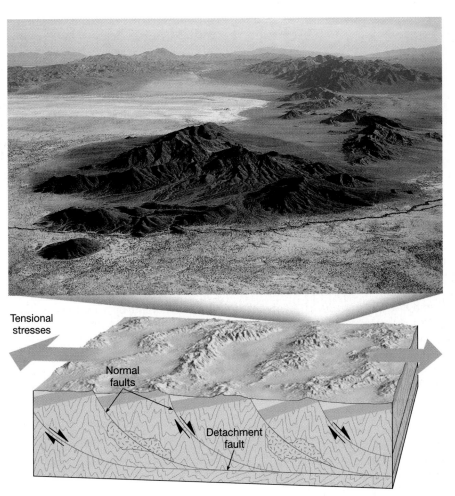

◆ **Figure 15.22** Normal faulting in the Basin and Range Province. Here, tensional stresses have elongated and fractured the crust into numerous blocks. Movement along these fractures has tilted the blocks, producing parallel mountain ranges called fault-block mountains. (Photo by Michael Collier)

by other types of faulting. Thrust faults, on the other hand, exist at all scales. Small thrust faults exhibit displacement on the order of millimetres to a few metres. Some large thrust faults have displacements on the order of tens to hundreds of kilometres.

Whereas normal faults occur in tensional environments, thrust faults result from strong compressional stresses. In these settings, crustal blocks are displaced *toward* one another, with the hanging wall being displaced upward relative to the footwall. Thrust faulting is most pronounced in subduction zones and

other convergent boundaries where plates are colliding. Compressional forces generally produce folds as well as faults and result in a thickening and shortening of the material involved.

In mountainous regions such as the Alps, Northern Rockies, Himalayas, and Appalachians, thrust faults have displaced strata as far as 50 kilometres over adjacent rock units. The result of this large-scale movement is that older strata end up overlying younger rocks. A classic site of thrust faulting occurs in the Crowsnest Pass of Alberta (Figure 15.25). Here, mountain peaks

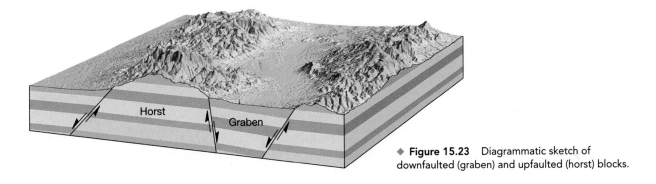

◆ **Figure 15.23** Diagrammatic sketch of downfaulted (graben) and upfaulted (horst) blocks.

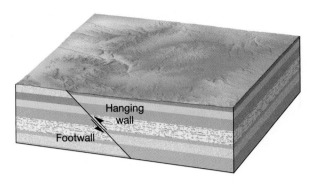

◆ **Figure 15.24** Block diagram showing the relative movement along a reverse fault.

Students Sometimes Ask...

Has anyone ever seen a fault scarp forming?
Amazingly, yes. There have been several instances where people have been at the fortuitously appropriate place and time to observe the creation of a fault scarp—and have lived to tell about it. In Idaho a large earthquake in 1983 created a 3-metre fault scarp that was witnessed by several people, many of whom were knocked off their feet. More often, though, fault scarps are noticed *after* they form. For example, a 1999 earthquake in Taiwan created a fault scarp that formed a new waterfall and destroyed a nearby bridge.

that provide the area's majestic scenery have been carved from Paleozoic rocks that were displaced over much younger Cretaceous strata. At the eastern edge of the Crowsnest Pass, there is an outlying peak called Crowsnest Mountain. This structure is an isolated remnant of the thrust sheet that was severed by the erosional forces of glacial ice and running water. An isolated block, such as Crowsnest Mountain, is called a **klippe** (*klippe* = cliff) (Figure 15.26).

Strike-Slip Faults

Faults in which the dominant displacement is horizontal and parallel to the strike of the fault surface are called **strike-slip faults**. Because of their large size and linear nature, many strike-slip faults produce a trace that is visible over a great distance (Figure 15.27). Rather than a single fracture along which movement

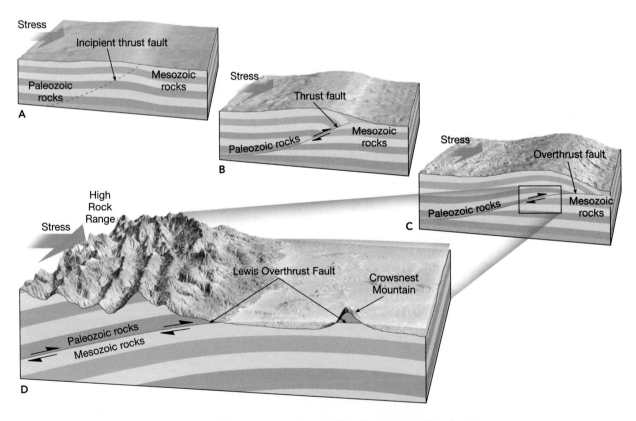

◆ **Figure 15.25** Idealized development of the Lewis Overthrust fault in Glacier National Park. **A.** Geologic setting prior to deformation. **B., C.** Large-scale movement along a thrust fault displaced Paleozoic rock over Mesozoic strata in the region of the Crowsnest Pass, Alberta. **D.** Erosion by glacial ice and running water sculptured the thrust sheet into a majestic landscape and isolated remnants of the thrust sheet such as Crowsnest Mountain.

◆ **Figure 15.26**   Crowsnest Mountain in the Crowsnest Pass area of southwestern Alberta, is a klippe.
(Photo by Thelma Pirot)

◆ **Figure 15.27**   Aerial view of strike-slip (right-lateral) fault in Southern Nevada.
(Photo by Martin G. Miller)

takes place, large strike-slip faults consist of a zone of roughly parallel fractures. The zone may be up to several kilometres in width. The most recent movement, however, is often along a strand only a few metres wide, which may offset features such as stream channels (Figure 15.28). Furthermore, crushed and broken rocks produced during faulting are more easily eroded, often producing linear valleys or troughs that mark the locations of strike-slip faults.

The earliest scientific records of strike-slip faulting were made following surface ruptures that produced

Students Sometimes Ask...

Do faults exhibit only strike-slip or dip-slip motion?
No. Strike-slip faults and dip-slip faults are on the opposite ends of a spectrum of fault structures. Faults that exhibit a combination of dip-slip and strike-slip movements are called *oblique-slip faults*. Although most faults could technically be classified as oblique-slip, they predominately exhibit either strike-slip or dip-slip motion.

large earthquakes. One of the most noteworthy of these was the great San Francisco earthquake of 1906. During this strong earthquake, structures such as fences that

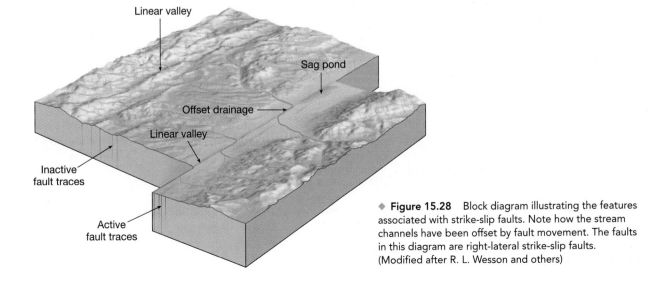

◆ **Figure 15.28**   Block diagram illustrating the features associated with strike-slip faults. Note how the stream channels have been offset by fault movement. The faults in this diagram are right-lateral strike-slip faults.
(Modified after R. L. Wesson and others)

were built across the San Andreas Fault were displaced by as much as 5 metres. Because the movement along the San Andreas causes the crustal block on the opposite side of the fault to move to the right as you face the fault, it is called a *right-lateral* strike-slip fault. The Great Glen fault in Scotland is a well-known example of a *left-lateral* strike-slip fault, which exhibits the opposite sense of displacement. The total displacement along the Great Glen fault is estimated to exceed 100 kilometres. Also associated with this fault trace are numerous lakes, including Loch Ness, the home of the legendary monster.

Many major strike-slip faults cut through the lithosphere and accommodate motion between two large crustal plates. This special kind of strike-slip fault is called a **transform** (*trans* = across *forma* = form) **fault**. Numerous transform faults cut the oceanic lithosphere and link spreading oceanic ridges. Others accommodate displacement between continental plates that move horizontally with respect to each other. One of the best-known transform faults is California's San Andreas Fault (Box 15.2). This plate-bounding fault can be traced for about 950 kilometres from the Gulf of California to a point along the Pacific Coast north of San Francisco, where it heads out to sea. Since its formation, about 29 million years ago, displacement along the San Andreas Fault has exceeded

560 kilometres. This movement has accommodated the northward displacement of southwestern California and the Baja Peninsula of Mexico in relation to the remainder of North America. The nature of these important structures will be discussed in more detail in Chapter 19.

Joints

 Internal Processes
↳ Crustal Deformation

Among the most common rock structures are fractures called **joints**. Unlike faults, joints are fractures along which no appreciable displacement has occurred. Although some joints have a random orientation, most occur in roughly parallel groups (see Figure 5.6).

We have already considered two types of joints. Earlier we learned that *columnar joints* form when igneous rocks cool and develop shrinkage fractures that produce elongated, pillarlike columns (Figure 15.29). Also recall that sheeting produces a pattern of gently curved joints that develop more or less parallel to the surface of large exposed igneous bodies such as batholiths. Here the jointing results from the gradual expansion that occurs when erosion removes the overlying load.

◆ **Figure 15.29** This felsic igneous intrusion exposed on the Burlington Peninsula of Newfoundland exhibits columnar joints.
(Photo courtesy of RW Hodder)

ROAD
CLOSED

People and the Environment
The San Andreas Fault System

The San Andreas, the best-known and largest fault system in North America, first attracted wide attention after the great 1906 San Francisco earthquake and fire. Following this devastating event, geologic studies demonstrated that a displacement of as much as 5 metres along the fault had been responsible for the earthquake. It is now known that this dramatic event is just one of many thousands of earthquakes that have resulted from repeated movements along the San Andreas throughout its 29-million-year history.

Where is the San Andreas fault system located? As shown in Figure 15.B, it trends in a northwesterly direction for nearly 1300 kilometres through much of western California. At its southern end, the San Andreas connects with a spreading centre located in the Gulf of California. In the north, the fault enters the Pacific Ocean at Point Arena, where it is thought to continue its

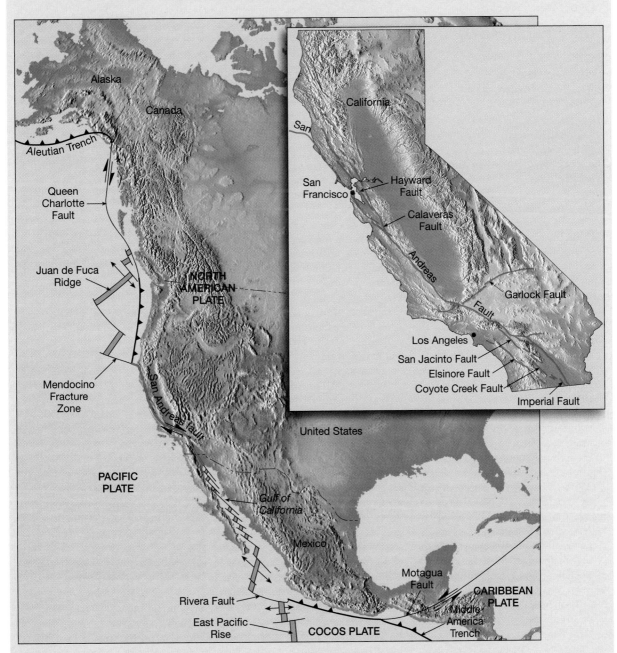

◆ **Figure 15.B**   Map showing the extent of the San Andreas fault system. Inset shows only a few of the many splinter faults that are part of this great fault system.

TABLE 15.A   Major Earthquakes on the San Andreas Fault System

Date	Location	Magnitude	Remarks
1812	Wrightwood, CA	7	Church of San Juan Capistrano collapsed, killing 40 worshippers.
1812	Santa Barbara channel	7	Churches and other buildings wrecked in and around Santa Barbara.
1838	San Francisco peninsula	7	At one time thought to have been comparable to the great earthquake of 1906.
1857	Fort Tejon, CA	8.25	One of the greatest U.S. earthquakes. Occurred near Los Angeles, then a city of 4000.
1868	Hayward, CA	7	Rupture of the Hayward fault caused extensive damage in San Francisco Bay area.
1906	San Francisco, CA	8.25	The great San Francisco earthquake. As much as 80 percent of the damage caused by fire.
1940	Imperial Valley	7.1	Displacement on the newly discovered Imperial fault.
1952	Kern County	7.7	Rupture of the White Wolf fault. Largest earthquake in California since 1906. Sixty million dollars in damage and 12 people killed.
1971	San Fernando Valley	6.5	One-half billion dollars in damage and 58 lives lost.
1989	Santa Cruz Mountains	7.1	Loma Prieta earthquake. Six billion dollars in damage, 62 lives lost, and 3757 people injured.
1994	Northridge (Los Angeles area)	6.9	Over 15 billion dollars in damage, 51 lives lost, and over 5000 injured.

northwesterly trend, eventually joining the Mendocino fracture zone. In the central section, the San Andreas is relatively simple and straight. However, at its two extremities, several branches spread from the main trace, so that in some areas the fault zone exceeds 100 kilometres in width.

Over much of its extent, a linear trough reveals the presence of the San Andreas Fault. When the system is viewed from the air, linear scars, offset stream channels, and elongated ponds mark the trace in a striking manner. On the ground, however, surface expressions of the faults are much more difficult to detect. Some of the most distinctive landforms include long, straight escarpments, narrow ridges, and sag ponds formed by settling of blocks within the fault zone. Further, many stream channels characteristically bend sharply to the right where they cross the fault (Figure 15.C).

With the advent of the theory of plate tectonics, geologists began to realize the significance of this great fault system. The San Andreas Fault is a transform boundary separating two crustal plates that move very slowly. The Pacific plate, located to the west, moves northwestward relative to the North American plate, causing earthquakes along the fault (Table 15.A).

The San Andreas is undoubtedly the most studied of any fault system in the world. Although many questions remain unanswered, geologists have learned that each fault segment exhibits somewhat different behaviour. Some portions of the San Andreas exhibit a slow creep with little noticeable seismic activity. Other segments regularly slip, producing small earthquakes, while still other segments seem to store elastic energy for hundreds of years and rupture in great earthquakes. This knowledge is useful when assigning earthquake hazard potential to a given segment of the fault zone.

Because of the great length and complexity of the San Andreas Fault, it is more appropriately referred to as a "fault system." This major fault system consists primarily of the San Andreas Fault and several major branches, including the Hayward and Calaveras faults of central California and the San Jacinto and Elsinore faults of southern California (see Figure 15.B). These major segments, plus a vast number of smaller faults that include the Imperial Fault, San Fernando Fault, and the Santa Monica Fault, collectively accommodate the relative motion between the North American and Pacific plates.

Blocks on opposite sides of the San Andreas Fault move horizontally in opposite directions, such that if a person stood on one side of the fault, the block on the opposite side would appear to move to the right when slippage occurred. This type of displacement is known as *right-lateral strike-slip* by geologists (see Figure 15.C).

Ever since the great San Francisco earthquake of 1906, when as much as 5 metres of displacement occurred, geologists have attempted to establish the cumulative displacement along this fault over its 29-million-year history. By matching rock units across the fault, geologists have determined that the total accumulated displacement from earthquakes and creep exceeds 560 kilometres.

◆ **Figure 15.C** Aerial view showing offset stream channel across the San Andreas Fault on the Carrizo Plain west of Taft, California. (Photo by Michael Collier/DRK Photo)

In contrast to the situations just described, most joints are produced when rocks in the outermost crust are deformed. Here tensional and shearing stresses associated with crustal movements cause the rock to fail by brittle fracture. For example, when folding occurs, rocks situated at the axes of the folds are elongated and pulled apart to produce tensional joints. Extensive joint patterns can also develop in response to relatively subtle and often barely perceptible regional upwarping and downwarping of the crust. In many cases, the cause for jointing at a particular locale is not readily apparent.

Many rocks are broken by two or even three sets of intersecting joints that slice the rock into numerous regularly shaped blocks. These joint sets often exert a strong influence on other geologic processes. For example, chemical weathering tends to be concentrated along joints, and in many areas groundwater movement and the resulting dissolution in soluble rocks is controlled by the joint pattern (Figure 15.30). Moreover, a system of joints can influence the direction that stream courses follow. The rectangular drainage pattern described in Chapter 10 is such a case.

Joints may also be significant from an economic standpoint. Some of the world's largest and most important mineral deposits were emplaced along joint systems. Hydrothermal solutions, which are basically mineralized fluids, can migrate into fractured host rocks and precipitate economically important amounts of copper, silver, gold, zinc, lead, and uranium.

Further, highly jointed rocks present a risk to the construction of engineering projects, including highways and dams. On June 5, 1976, 14 lives were lost and nearly 1 billion dollars in property damage occurred when the Teton Dam in Idaho failed. This earthen dam was constructed of very erodible clays and silts and was situated on highly fractured volcanic rocks. Although attempts were made to fill the voids in the jointed rock, water gradually penetrated the subsurface fractures and undermined the dam's foundation. Eventually the moving water cut a tunnel into the easily erodible clays and silts. Within minutes the dam failed, sending a 20-metre-high wall of water down the Teton and Snake rivers.

◆ **Figure 15.30** These Silurian dolostones exposed in a roadcut south of Meaford, Ontario, have broken along joints to produce sheer vertical surfaces.
(Photo by C. Tsujita)

Chapter Summary

- *Deformation* refers to changes in the shape and/or volume of a rock body and is most pronounced along plate margins. To describe the forces that deform rocks, geologists use the term *stress*, which is the amount of force applied to a given area. Stress that is uniform in all directions is called *confining pressure*, whereas *differential stresses* are applied unequally in different directions. Differential stresses that shorten a rock body are *compressional stresses*; those that elongate a rock unit are *tensional stresses*. *Strain* is the change in size and shape of a rock unit caused by stress.

- Rocks deform differently depending on the environment (temperature and confining pressure), the composition and texture of the rock, and the length of time stress is maintained. Rocks first respond by deforming *elastically* and will return to their original shape when the stress is removed. Once their elastic limit (strength) is surpassed, rocks either deform by ductile flow or they fracture. *Ductile deformation* is a solid-state flow that results in a change in size and shape of an object without fracturing. Ductile flow may be accomplished by gradual slippage and recrystallization along planes of weakness within the crystal lattice of mineral grains. Ductile deformation occurs in a high-temperature/high-pressure environment. In a near-surface environment, most rocks deform by *brittle failure*.

- The orientation of rock units or fault surfaces is established with measurements called strike and dip. *Strike* is the compass direction of a line produced by the intersection of an inclined rock layer or fault with a horizontal plane. *Dip* is the angle of inclination of the surface of a rock unit or fault measured from a horizontal plane.

- The most basic geologic structures associated with rock deformation are *folds* (flat-lying sedimentary and volcanic rocks bent into a series of wavelike undulations) and *faults*. The two most common types of folds are *anticlines*, formed by the upfold-ing, or arching, of rock layers, and *synclines*, which are downfolds. Most folds are the result of horizontal *compressional stresses*. Folds can be *symmetrical*, *asymmetrical*, or, if one limb has been tilted beyond the vertical, *overturned*. *Domes* (upwarped structures) and *basins* (downwarped structures) are circular or somewhat elongated folds formed by vertical displacements of strata.

- Faults are fractures in the crust along which appreciable displacement has occurred. Faults in which the movement is primarily vertical are called *dip-slip faults*. Dip-slip faults include both *normal* and *reverse faults*. Low-angle reverse faults are called *thrust faults*. Normal faults indicate *tensional stresses* that pull the crust apart. Along the spreading centres of plates, divergence can cause a central block called a *graben*, bounded by normal faults, to drop as the plates separate.

- Reverse and thrust faulting indicate that *compressional forces* are at work. Large *thrust faults* are found along subduction zones and other convergent boundaries where plates are colliding. In mountainous regions such as the Alps, Northern Rockies, Himalayas, and Appalachians, thrust faults have displaced strata as far as 50 kilometres over adjacent rock units.

- *Strike-slip faults* exhibit mainly horizontal displacement parallel to the strike of the fault surface. Large strike-slip faults, called *transform faults*, accommodate displacement between plate boundaries. Most transform faults cut the oceanic lithosphere and link spreading centres. The San Andreas Fault cuts the continental lithosphere and accommodates the northward displacement of southwestern California.

- *Joints* are fractures along which no appreciable displacement has occurred. Joints generally occur in groups with roughly parallel orientations and are the result of brittle failure of rock units located in the outermost crust.

Review Questions

1. What is rock deformation? How does a rock body change during deformation?

2. List five (5) geologic structures associated with deformation.

3. How is *stress* related to *force*?

4. Contrast compressional and tensional stresses.

5. Describe how shearing can deform a rock in a near-surface environment.

6. Compare strain and stress.

7. How is brittle deformation different from ductile deformation?

8. List three factors that determine how rocks will behave when exposed to stresses that exceed their strength. Briefly explain the role of each.

9. What is an outcrop?

10. What two measurements are used to establish the orientation of deformed strata? Distinguish between them.

11. Distinguish between anticlines and synclines, domes and basins, anticlines and domes.

12. How is a monocline different from an anticline?

13. Crowsnest Mountain in southwestern Alberta is an example of what type of geologic feature?

14. Contrast the movements that occur along normal and reverse faults. What type of stress is indicated by each fault?

15. Is the fault shown in Figure 15.18 a normal or a reverse fault?

16. Describe a horst and a graben. Explain how a graben valley forms and name one.

17. What type of faults are associated with fault-block mountains?

18. How are reverse faults different from thrust faults? In what way are they the same?

19. The San Andreas Fault is an excellent example of a _____ fault.

20. With which of the three types of plate boundaries does normal faulting predominate? Reverse faulting? Strike-slip faulting?

21. How are joints different from faults?

Key Terms

anticline (p. 422)
basin (p. 425)
brittle failure (p. 417)
brittle deformation (p. 417)
compressional stress (p. 415)
deformation (p. 415)
detachment fault (p. 428)
differential stress (p. 415)
dip (p. 419)

dip-slip fault (p. 427)
dome (p. 424)
ductile deformation (p. 418)
fault (p. 425)
fault-block mountain (p. 427)
fault scarp (p. 427)
fold (p. 421)
force (p. 415)

graben (p. 428)
hogback (p. 425)
horst (p. 428)
joint (p. 432)
klippe (p. 430)
monocline (p. 424)
normal fault (p. 427)
reverse fault (p. 427)
rock structure (p. 419)
shear (p. 415)

strain (p. 415)
stress (p. 415)
strike (p. 419)
strike-slip fault (p. 430)
syncline (p. 422)
tensional stress (p. 415)
thrust fault (p. 428)
transform fault (p. 432)

Web Resources

The *Earth* Web site uses the resources and flexibility of the Internet to aid in your study of the topics in this chapter. Written and developed by geology instructors, this site will help improve your understanding of geology. Visit **http://www.pearson.ca/tarbuck** and click on the cover of the text to find:

■ Online review quizzes.

■ Web-based critical thinking and writing exercises.

■ Links to chapter-specific Web resources.

■ Internet-wide key-term searches.

http://www.pearson.ca/tarbuck

Chapter 16
Earthquakes

Damage caused by a major earthquake
near Ismit, Turkey in August 1999.
(Photo by Hurriyet/AP/Wide World Photos)

On October 17, 1989, at 5:04 p.m. Pacific Daylight Time, millions of television viewers around the world were settling in to watch the third game of the World Series. Instead, they saw their television sets go black as tremors hit San Francisco's Candlestick Park. Although the earthquake was centred in a remote section of the Santa Cruz Mountains, 100 kilometres to the south, major damage occurred in the Marina District of San Francisco.

The most tragic result of the violent shaking was the collapse of some double-decked sections of Interstate 880, also known as the Nimitz Freeway. The ground motions caused the upper deck to sway, shattering the concrete support columns along a mile-long section of the freeway. The upper deck then collapsed onto the lower roadway, flattening cars as if they were aluminum cans. This earthquake, named the Loma Prieta quake for its point of origin, claimed 67 lives.

In mid-January 1994, less than five years after the Loma Prieta earthquake devastated portions of the San Francisco Bay area, a major earthquake struck the Northridge area of Los Angeles. Although not the fabled "Big One," this moderate 6.7-magnitude earthquake left 51 dead, over 5000 injured, and tens of thousands of households without water and electricity.

The damage exceeded $40 billion and was attributed to an unknown fault that ruptured 18 kilometres beneath Northridge.

The Northridge earthquake began at 4:31 a.m. and lasted roughly 40 seconds. During this brief period the quake terrorized the entire Los Angeles area. In the three-story Northridge Meadows apartment complex, 16 people died when sections of the upper floors collapsed onto the first-floor units. Nearly 300 schools were seriously damaged and a dozen major roadways buckled. Among these were two of California's major arteries—the Golden State Freeway (Interstate 5), where an overpass collapsed completely and blocked the roadway, and the Santa Monica Freeway. Fortunately, these roadways had practically no traffic at this early morning hour.

In nearby Granada Hills, broken gas lines were set ablaze while the streets flooded from broken water mains. Seventy homes burned in the Sylmar area. A 64-car freight train derailed, including some cars carrying hazardous cargo. But it is remarkable that the destruction was not greater. Unquestionably, the upgrading of structures to meet the requirements of building codes developed for this earthquake-prone area helped minimize what could have been a much greater human tragedy.

◆ **Figure 16.1**   Damage to La Pulperie, (Chicoutimi, Québec) caused by the November 25, 1988 magnitude Mw 5.9 Saguenay earthquake.
(Reproduced with the permission of the Minister of Public Works and Government Services Canada, 2004 and courtesy of Natural Resources Canada, Geological Survey of Canada. Photo by D.E. Allen.)

◆ **Figure 16.2**   San Francisco in flames after the 1906 earthquake.
(Reproduced from the collection of the Library of Congress)

Over 30,000 earthquakes that are strong enough to be felt occur worldwide annually (Figure 16.1). Fortunately, most are minor tremors and do very little damage. Generally, only about 75 significant earthquakes take place each year, and many of these occur in remote regions. However, occasionally a large earthquake occurs near a large population centre. Under these conditions, an earthquake is among the most destructive natural forces on Earth.

The shaking of the ground, coupled with the liquefaction of some soils, wreaks havoc on buildings and other structures. In addition, when a quake occurs in a populated area, power and gas lines are often ruptured, causing numerous fires. In the famous 1906 San Francisco earthquake, much of the damage was caused by fires (Figure 16.2). They quickly became uncontrollable when broken water mains left firefighters with only trickles of water.

What Is an Earthquake?

Internal Processes
↳ Earthquakes

An **earthquake** is the vibration of Earth produced by the rapid release of energy. Most often, earthquakes are caused by slippage along a fault in Earth's crust. The energy released radiates in all directions from its source, the **focus** (*focus* = a point), in the form of waves. These waves are analogous to those produced when a stone is dropped into a calm pond (Figure 16.3). Just as the impact of the stone sets water waves in motion, an earthquake generates seismic waves that radiate throughout Earth. Even though the energy dissipates rapidly with increasing distance from the focus, sensitive instruments located around the world record the event.

Earthquakes and Faults

The tremendous energy released by atomic explosions or by volcanic eruptions can produce an earthquake, but these events are relatively weak and infrequent. What mechanism produces a destructive earthquake? Ample evidence exists that Earth is not a static planet. We know that Earth's crust has been uplifted at times, because we have found numerous ancient wave-cut benches many metres above the level of the highest tides. Other regions exhibit evidence of extensive subsidence. In addition to these vertical displacements, offsets in fence lines, roads, and other structures indicate that horizontal movement is common (Figure 16.4).

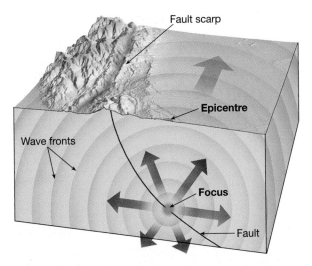

◆ **Figure 16.3** Earthquake focus and epicentre. The focus is the zone within Earth where the initial displacement occurs. The epicentre is the surface location directly above the focus.

These movements are usually associated with large fractures in Earth's crust called **faults**.

Most of the motion along faults can be satisfactorily explained by plate tectonic theory. This theory states that large slabs of Earth's crust are in continual slow motion. These mobile plates interact with neighbouring plates, straining and deforming the rocks at their edges. In fact, it is along faults associated with plate boundaries that most earthquakes occur. Furthermore, earthquakes are repetitive: As soon as one is over, the continuous motion of the plates resumes, adding strain to the rocks until they fail again.

Elastic Rebound

The actual mechanism of earthquake generation eluded geologists until H. F. Reid of Johns Hopkins University conducted a study following the great 1906 San Francisco earthquake. The earthquake was accompanied by horizontal surface displacements of several metres along the northern portion of the San Andreas Fault. This 1300-kilometre fracture runs north-south through southern California. It is a large fault zone that separates two great sections of Earth's crust, the North American plate and the Pacific plate. Field investigations determined that during this single earthquake, the Pacific plate lurched as much as 5 metres northward past the adjacent North American plate.

The mechanism for earthquake formation that Reid deduced from this information is illustrated in Figure 16.5. In part A of the figure, you see an existing fault, or break in the rock. In part B, tectonic forces ever so slowly deform the crustal rocks on both sides of the fault, as demonstrated by the bent features. Under these conditions, rocks are bending and storing elastic energy, much like a wooden stick does if bent. Eventually, the frictional resistance holding the rocks together is overcome. As slippage occurs at the weakest point (the focus), displacement will exert stress farther along the fault, where additional slippage will occur until most of the built-up strain is released (Figure 16.5C). This slippage allows the deformed rock to "snap back." The vibrations we know as an earthquake occur as the rock elastically returns to its original shape. The "springing back" of the rock was termed **elastic rebound** by Reid, because the rock

◆ **Figure 16.4** This fence was offset 2.5 metres during the 1906 San Francisco earthquake. (Photo by G. K. Gilbert, U.S. Geological Survey)

Deformation of rocks

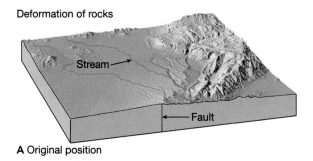

A Original position

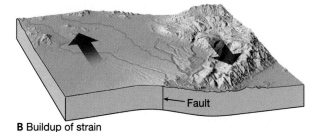

B Buildup of strain

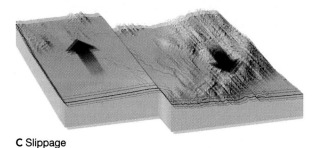

C Slippage

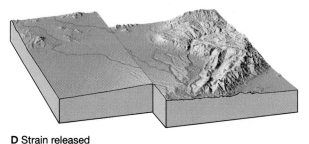

D Strain released

◆ **Figure 16.5**   Elastic rebound. As rock is deformed, it bends, storing elastic energy. Once strained beyond its breaking point, the rock cracks, releasing the stored-up energy in the form of earthquake waves.

behaves elastically, much like a stretched rubber band does when it is released.

In summary, most earthquakes are produced by the rapid release of elastic energy stored in rock that has been subjected to great stress. Once the strength of the rock is exceeded, it suddenly ruptures, causing the vibrations of an earthquake. Earthquakes most often occur along existing faults whenever the frictional forces on the fault surfaces are overcome.

Foreshocks and Aftershocks

The intense vibrations of the 1906 San Francisco earthquake lasted about 40 seconds. Although most of the displacement along the fault occurred in this rather short period, additional movements along this and other nearby faults lasted for several days following the main quake. The adjustments that follow a major earthquake often generate smaller earthquakes called **aftershocks**. Although these aftershocks are usually much weaker than the main earthquake, they can sometimes destroy already badly weakened structures. This occurred, for example, during a 1988 earthquake in Armenia. A large aftershock of magnitude 5.8 collapsed many structures that had been weakened by the main tremor.

In addition, small earthquakes called **foreshocks** often precede a major earthquake by days or, in some cases, by as much as several years. Monitoring of these foreshocks has been used as a means of predicting forthcoming major earthquakes, with mixed success. We will consider the topic of earthquake prediction in a later section of this chapter.

San Andreas Fault: An Active Earthquake Zone

The San Andreas is undoubtedly the most studied fault system in the world. Over the years, investigations have shown that displacement occurs along discrete segments that are 100 to 200 kilometres long. Further, each fault segment behaves somewhat differently from the others. Some portions of the San Andreas exhibit a slow, gradual displacement known as **fault creep**, which occurs relatively smoothly and therefore with little noticeable seismic activity. Other segments regularly slip, producing small earthquakes.

Still other segments remain locked and store elastic energy for hundreds of years before rupturing in great earthquakes. The latter process is described as *stick-slip* motion, because the fault exhibits alternating periods of locked behaviour followed by sudden slippage. It is estimated that great earthquakes should occur about every 50 to 200 years along those sections of the San Andreas Fault that exhibit stick-slip motion. This knowledge is useful when assigning a potential earthquake risk to a given segment of the fault zone.

The tectonic forces along the San Andreas fault zone that were responsible for the 1906 San Francisco earthquake are still active. Currently, laser beams are used to measure the relative motion between the opposite sides of this fault. These measurements reveal a displacement of 2 to 5 centimetres per year. Although this seems slow, it produces substantial movement

◆ **Figure 16.19**   Effects of liquefaction. This tilted building rests on unconsolidated sediment that behaved like quicksand during the 1985 Mexican earthquake. (Photo by James L. Beck)

felt for about 40 seconds, and the strong vibrations of the 1989 Loma Prieta earthquake lasted less than 15 seconds. But the Alaska quake reverberated for 3 to 4 minutes.

Amplification of Seismic Waves   Although the region within 20 to 50 kilometres of the epicentre will experience about the same intensity of ground shaking, the destruction varies considerably within this area. This difference is mainly attributable to the nature of the ground on which the structures are built. Soft sediments, for example, generally amplify the vibrations more than solid bedrock. Thus, the buildings located in Anchorage, which were situated on unconsolidated sediments, experienced heavy structural damage. By contrast, most of the town of Whittier, although much nearer the epicentre, rests on a firm foundation of granite and hence suffered much less damage. However, Whittier was damaged by a seismic sea wave (described in the next section).

The 1985 Mexican earthquake gave seismologists and engineers a vivid reminder of what had been learned from the 1964 Alaskan earthquake. The Mexican coast, where the earthquake was centred, experienced unusually mild tremors despite the strength of the quake. As expected, the seismic waves became progressively weaker with increasing distance from the epicentre. However, in the central section of Mexico City, nearly 400 kilometres from the source, the vibrations intensified to five times that experienced in outlying districts. Much of this amplified ground motion can be attributed to soft sediments, remnants of an ancient lakebed, that underlie portions of the city (see Box 16.2).

Liquefaction   In areas where unconsolidated materials are saturated with water, earthquake vibrations can generate a phenomenon known as **liquefaction** (*liqueo* = to be fluid, *facio* = to make). Under these conditions, what had been a stable soil turns into a mobile fluid that is not capable of supporting buildings or other structures (Figure 16.19). As a result, underground objects such as storage tanks and sewer lines may literally float toward the surface of their newly liquefied environment. Buildings and other structures may settle and collapse. During the 1989 Loma Prieta earthquake, in San Francisco's Marina District, foundations failed and geysers of sand and water shot from the ground, indicating that liquefaction had occurred (Figure 16.20).

◆ **Figure 16.20**   These "mud volcanoes" were produced by the Loma Prieta earthquake of 1989. They formed when geysers of sand and water shot from the ground, an indication that liquefaction had occurred. (Photo by Richard Hilton, courtesy of Dennis Fox)

BOX 16.2

Understanding Earth
Wave Amplification and Seismic Risks

Much of the damage and loss of life in the 1985 Mexico City earthquake occurred because downtown buildings were constructed on lake sediment that greatly amplified the ground motion. To understand why this happens, recall that as seismic waves pass through Earth, they cause the intervening material to vibrate much as a tuning fork does when it is struck. Although most objects can be "forced" to vibrate over a wide range of frequencies, each has a natural period of vibration that is preferred. Different Earth materials, like different-length tuning forks, also have different natural periods of vibration.*

Ground-motion amplification results when the supporting material has a natural period of vibration (frequency) that matches that of the seismic waves. A common example of this phenomenon occurs when a parent pushes a child on a swing. When the parent periodically pushes the child in rhythm with the frequency of the swing, the child moves back and forth in a greater and greater arc (amplitude). By chance, the column of sediment beneath Mexico City had a natural period of vibration of about 2 seconds, matching that of the strongest seismic waves. Thus, when the seismic waves began shaking the soft sediments, a *resonance* developed, which greatly increased the amplitude of the vibrations. This amplification resulted in vibrations that exhibited 40 centimetres of back-and-forth ground motion every 2 seconds for nearly 2 minutes. Such movement was too intense for many poorly designed buildings in the city. In addition, intermediate-height structures (5 to 15 stories) sway back and forth with a period of about 2 seconds. Thus, resonance also developed between these buildings and the ground, with the result that most of the

building failures occurred in structures in this height range (Figure 16.B).

Sediment-induced wave amplification is also thought to have contributed significantly to the failure of the two-tiered Cypress section of Interstate 880 during the 1989 Loma Prieta earth-

quake (Figure 16.C). Studies conducted on the 1.4-kilometre section that did collapse showed that it was built on San Francisco Bay mud. Another section of this interstate that was damaged but did not collapse was constructed on firmer alluvial materials.

◆ **Figure 16.B**   During the 1985 Mexican earthquake, multi-storey buildings swayed back and forth as much as 1 metre. Many, including the hotel shown here, collapsed or were seriously damaged.
(Photo by James L. Beck)

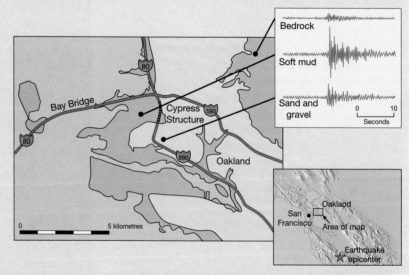

◆ **Figure 16.C**   The portion of the Cypress Freeway structure in Oakland, California, that stood on soft mud (dashed red line) collapsed during the 1989 Loma Prieta earthquake. Adjacent parts of the structure (solid red) that were built on firmer ground remained standing. Seismograms from an aftershock (upper right) show that shaking is greatly amplified in the soft mud as compared to the firmer materials.

---

*To demonstrate the natural period of vibration of an object, hold a ruler over the edge of a desk so that most of it is not supported by the desk. Start it vibrating, and notice the noise it makes. Changing the length of the unsupported portion of the ruler will change the natural period of vibration.

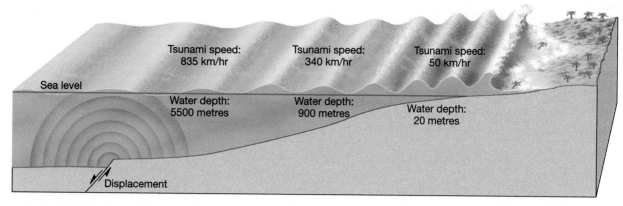

◆ **Figure 16.21**   Schematic drawing of a tsunami generated by displacement of the ocean floor. The speed of a wave correlates with ocean depth. As shown, waves moving in deep water advance at speeds in excess of 800 kilometres per hour. Speed gradually slows to 50 kilometres per hour at depths of 20 metres. Decreasing depth slows the movement of the wave. As waves slow in shallow water, they grow in height until they topple and rush onto shore with tremendous force. The size and spacing of these swells are not to scale.

Seiches   The effects of great earthquakes may be felt thousands of kilometres from their source. Ground motion may generate *seiches*, the rhythmic sloshing of water in lakes, reservoirs, and enclosed basins such as the Gulf of Mexico. The 1964 Alaskan earthquake, for example, generated 2-metre waves off the coast of Texas, which damaged small craft, while much smaller waves were noticed in swimming pools in both Texas and Louisiana.

Seiches can be particularly dangerous when they occur in reservoirs retained by earthen dams. These waves have been known to slosh over reservoir walls and weaken the structure, thereby endangering the lives of those downstream.

Tsunami

Most deaths associated with the 1964 Alaskan quake were caused by **seismic sea waves**, or **tsunami.*** (*tsu* = harbour, *nami* = waves). These destructive waves often are called "tidal waves" by the media. However, this name is inappropriate, for these waves are generated by earthquakes, not the tidal effect of the Moon or Sun.

Most tsunami result from vertical displacement along a fault located on the ocean floor, or from a large underwater landslide triggered by an earthquake (Figure 16.21). Once created, a tsunami resembles the ripples formed when a pebble is dropped into a pond. In contrast to ripples, tsunamis advance across the ocean at amazing speeds between 500 and 950 kilometres per hour. Despite this striking char-

*Seismic sea waves were given the name *tsunami* by the Japanese, who have suffered a great deal from them. The term *tsunami* is now used worldwide.

acteristic, a tsunami in the open ocean can pass undetected because its height is usually less than 1 metre and the distance between wave crests is great, ranging from 100 to 700 kilometres. However, upon entering shallower coastal waters, these destructive waves are slowed down and the water begins to pile up to heights that occasionally exceed 30 metres (Figure 16.21). As the crest of a tsunami approaches the shore, it appears as a rapid rise in sea level with a turbulent and chaotic surface. Tsunami can be very destructive (Figure 16.22).

Usually the first warning of an approaching tsunami is a relatively rapid withdrawal of water from beaches. Coastal residents have learned to heed this warning and move to higher ground; for about 5 to 30 minutes later, the retreat of water is followed by a surge capable of extending hundreds of metres inland. In a successive fashion, each surge is followed by rapid oceanward retreat of the water.

The tsunamis generated in the 1964 Alaskan earthquake inflicted heavy damage to the communities in the vicinity of the Gulf of Alaska, completely destroying the town of Chenega. Kodiak was also heavily damaged and most of its fishing fleet destroyed when a seismic sea wave carried many vessels into the business district. The deaths of 107 persons have been attributed to this tsunami. By contrast, only nine people died in Anchorage as a direct result of the vibrations.

Tsunami damage following the Alaskan earthquake extended along much of the west coast of North America, and despite a one-hour warning, 12 people perished in Crescent City, California, where all of the deaths and most of the destruction were caused by the fifth wave. The first wave crested about 4 metres above low tide and was followed by three progressively smaller waves. Believing that the tsunami had ceased, people returned to the shore, only to be met by the

◆ **Figure 16.22**   A man stands before a wall of water about to engulf him at Hilo, Hawaii, on April 1, 1946. This tsunami, which originated in the Aleutian Islands near Alaska, was still powerful enough when it hit Hawaii to rise 9 to 16 metres. The *S.S. Brigham Victory*, from which this photograph was taken, managed to survive the onslaught, but 159 people in Hawaii, including the man seen here, were killed. (Photo courtesy of Water Resources Center Archives, University of California, Berkeley)

fifth and most devastating wave, which, superimposed upon high tide, crested about 6 metres higher than the level of low tide.

On July 17, 1998, four villages on New Guinea's north coast were almost entirely swept away. Here an otherwise ordinary earthquake with a magnitude of 7.1 is thought to have triggered a massive underwater landslide. (Tremors of at least this size strike the globe every three weeks.) Within 5 to 10 minutes a plateau of water averaging 10 metres in height and perhaps 1 or 2 kilometres in width swept over the shore for more than a minute. Two similar waves followed at intervals of several minutes. Officially, the tsunami, the worst in more than two decades, killed 2134 people, but there were many who were unaccounted for (Box 16.3).

Landslides and Ground Subsidence

In the 1964 Alaskan earthquake, the greatest damage to structures was from landslides and ground subsidence triggered by the vibrations. At Valdez and Seward, the violent shaking caused deltaic materials

to experience liquefaction; the subsequent slumping carried both waterfronts away. Because of the threat of recurrence, the entire town of Valdez was relocated about 7 kilometres away on more stable ground. In Valdez, 31 people on a dock died when it slid into the sea.

Most of the damage in the city of Anchorage was also attributed to landslides. Many homes were destroyed in Turnagain Heights when a layer of clay lost its strength and over 80 hectares of land slid toward the ocean (Figure 16.23). A portion of this spectacular landslide was left in its natural condition as a reminder of this destructive event. The site was appropriately named "Earthquake Park." Downtown Anchorage was also disrupted as sections of the main business district dropped by as much as 3 metres.

Fire

The 1906 earthquake in San Francisco reminds us of the formidable threat of fire. The central city contained mostly large, older wooden structures and brick

BOX 16.3

People and the Environment
Tsunami Warning System

Tsunami traverse large stretches of ocean before their energy is fully dissipated. The tsunami generated by a 1960 Chilean earthquake, in addition to destroying villages along an 800-kilometre stretch of coastal South America, travelled 17,000 kilometres across the Pacific to Japan. Here, about 22 hours after the quake, considerable damage occurred in southern coastal villages. For several days afterward, tidal gauges in Hilo, Hawaii, detected these diminishing waves as they reverberated like echoes about the Pacific.

In 1946 a large tsunami struck the Hawaiian Islands without warning. A wave more than 15 metres high left several coastal villages in shambles. This destruction motivated the National Oceanic and Atmospheric Administration to establish a tsunami warning system for coastal areas of the Pacific. From seismic observatories throughout the region, large earthquakes are reported to the Pacific Tsunami Warning Center at Ewa Beach (near Honolulu), Hawaii. Scientists at the Center use tidal gauges to determine whether a tsunami has been formed. Within an hour a warning is issued. Although tsunamis travel very rapidly, there is

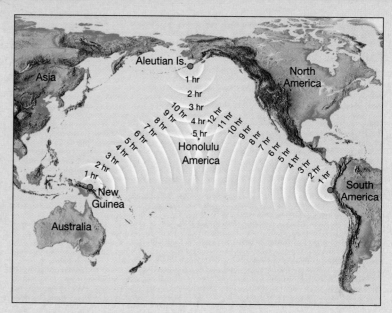

◆ **Figure 16.D**   Tsunami travel times to Honolulu, Hawaii, from selected locations throughout the Pacific.
(Data from NOAA)

sufficient time to evacuate all but the region nearest the epicentre (Figure 16.D). For example, a tsunami generated near the Aleutian Islands would take 5 hours to reach Hawaii, and one generated near the coast of Chile would travel 15 hours before reaching Hawaii.

Fortunately, most earthquakes do not generate tsunamis. On the average, only about 1.5 destructive tsunamis are generated worldwide annually. Of these, only about one every 10 years is catastrophic.

buildings. Although many of the unreinforced brick buildings were extensively damaged by vibrations, the greatest destruction was caused by fires, which started when gas and electrical lines were severed. The fires raged out of control for three days and devastated over 500 blocks of the city (see Figure 16.2). The problem was compounded by the initial ground shaking, which broke the city's water lines into hundreds of unconnected pieces.

The fire was finally contained when buildings were dynamited along a wide boulevard to provide a firebreak, the same strategy used in fighting a forest fire. Although only a few deaths were attributed to the San Francisco fire, such is not always the case. A 1923 earthquake in Japan triggered an estimated 250 fires, which devastated the city of Yokohama and destroyed more than half the homes in Tokyo. Over 100,000 deaths were attributed to the fires, which were driven by unusually high winds.

Can Earthquakes Be Predicted?

The vibrations that shook Northridge, California, in 1994 inflicted 57 deaths and about $40 billion in damage. This was from a brief earthquake (about 40 seconds) of moderate rating (M_W 6.7). Seismologists warn that earthquakes of comparable or greater strength will occur along the San Andreas Fault, which cuts a 1300-kilometre path through the state. The obvious question is: Can earthquakes be predicted?

Short-Range Predictions

The goal of short-range earthquake prediction is to provide a warning of the location and magnitude of a large earthquake within a narrow time frame (Figure 16.24). Substantial efforts to achieve this objective are being put forth in Japan, the United States, China, and Russia—countries where earthquake

Students Sometimes Ask...

I've heard that the safest place to be in a house during an earthquake is in a doorframe. Is that really the best place to be while an earthquake is occurring?

It depends. If you're on the road, stay away from tunnels, underpasses, and overpasses. Stop in a safe area and stay in your vehicle until the shaking stops. If you happen to be outside during an earthquake, stand away from buildings, trees, and telephone and electric lines. If you're inside, remember to *duck, cover, and hold.* When you feel an earthquake, *duck* under a desk or sturdy table. Stay away from windows, bookcases, file cabinets, heavy mirrors, hanging plants, and other heavy objects that could fall. Stay under *cover* until the shaking stops. And, *hold* on to the desk or table: If it moves, move with it.

An enduring earthquake image of California is a collapsed adobe home with the doorframe as the only standing part. From this came the belief that a doorway is the safest place to be during an earthquake. This is true only if you live in an old, unreinforced adobe house. In modern homes, doorways are no stronger than any other part of the house and usually have doors that will swing and can injure you. You'd be safer under a table.

risks are high (Table 16.4). This research has concentrated on monitoring possible *precursors*—phenomena that precede and thus provide a warning of a forthcoming earthquake. In California, for example, seismologists are measuring uplift, subsidence, and strain in the rocks near active faults. Some Japanese scientists are studying anomalous animal behaviour that may precede a quake. Other researchers are monitoring changes in groundwater levels, while still others are trying to predict earthquakes based on changes in the electrical conductivity of rocks.

Among the most ambitious earthquake experiments is one being conducted along a segment of the San Andreas Fault near the town of Parkfield in central California. Here, earthquakes of moderate intensity have occurred on a regular basis about once every 22 years since 1857. The most recent rupture was a 5.6-magnitude quake that occurred in 1966. With the next event already significantly "overdue," the U.S. Geological Survey has established an elaborate monitoring network. Included are creepmeters, tiltmeters, and bore-hole strain meters that are used to measure the accumulation and release of strain. Moreover, 70 seismographs of various designs have been installed to record foreshocks as well as the

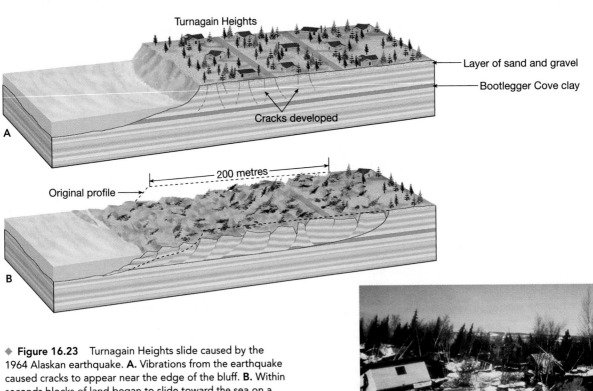

◆ **Figure 16.23** Turnagain Heights slide caused by the 1964 Alaskan earthquake. **A.** Vibrations from the earthquake caused cracks to appear near the edge of the bluff. **B.** Within seconds blocks of land began to slide toward the sea on a weak layer of clay. In less than 5 minutes, as much as 200 metres of the Turnagain Heights bluff area had been destroyed. **C.** Photo of a small portion of the Turnagain Heights slide. (Photo courtesy of U.S. Geological Survey)

TABLE 16.4 Some Notable Earthquakes

Year	Location	Deaths (est.)	Magnitude†	Comments
1556	Shensi, China	830,000		Possibly the greatest natural disaster.
1755	Lisbon, Portugal	70,000		Tsunami damage extensive.
1811–1812	New Madrid, Missouri	Few	7.9	Three major earthquakes.
1886	Charleston, South Carolina	60		Greatest historical earthquake in the eastern United States.
1906	San Francisco, California	1500	7.8	Fires caused extensive damage.
1908	Messina, Italy	120,000		
1923	Tokyo, Japan	143,000	7.9	Fire caused extensive destruction.
1960	Southern Chile	5700	9.6	Possibly the largest-magnitude earthquake ever recorded.
1964	Alaska	131	9.2	Greatest North American earthquake.
1970	Peru	66,000	7.8	Great rockslide.
1971	San Fernando, California	65	6.5	Damage exceeded $1 billion.
1975	Liaoning Province, China	1328	7.5	First major earthquake to be predicted.
1976	Tangshan, China	240,000	7.6	Not predicted.
1985	Mexico City	9500	8.1	Major damage occurred 400 km from epicentre.
1988	Armenia	25,000	6.9	Poor construction practices.
1989	Loma Prieta	62	6.9	Damage exceeded $6 billion.
1990	Iran	50,000	7.3	Landslides and poor construction practices caused great damage.
1993	Latur, India	10,000	6.4	Located in stable continental interior.
1994	Northridge, California	57	6.7	Damage in excess of $40 billion.
1995	Kobe, Japan	5472	6.9	Damage estimated to exceed $100 billion.
1999	Izmit, Turkey	17,127	7.4	Nearly 44,000 injured and more than 250,000 displaced
1999	Chi-Chi, Taiwan	2300	7.6	Severe destruction; 8700 injuries.
2001	El Salvador	1000	7.6	Triggered many landslides.
2001	Bhuj, India	20,000+	7.9	1 million or more homeless.

†Widely differing magnitudes have been estimated for some of these earthquakes. When available, moment magnitudes are used.
Source: U.S. Geological Survey

main event. Finally, a network of distance-measuring devices that employ lasers measures movement across the fault (Figure 16.25). The object is to identify ground movements that may precede a sizable rupture.

One claim of a successful short-range prediction was made by Chinese seismologists after the February 4, 1975, earthquake in Liaoning Province. According to reports, very few people were killed, although more than 1 million lived near the epicentre, because the earthquake was predicted and the population was evacuated. Recently, some Western seismologists have questioned this claim and suggest instead that an intense swarm of foreshocks, which began 24 hours before the main earthquake, may have caused many people to evacuate spontaneously. Further, an official Chinese government report issued 10 years later stated that 1328 people died and 16,980 injuries resulted from this earthquake.

One year after the Liaoning earthquake, at least 240,000 people died in the Tangshan, China, earthquake, which was not predicted. The Chinese have also issued false alarms. In a province near Hong Kong,

people reportedly left their dwellings for over a month, but no earthquake followed. Clearly, whatever method the Chinese employ for short-range predictions, it is *not* reliable.

In order for a prediction scheme to warrant general acceptance, it must be both accurate and reliable. Thus, *it must have a small range of uncertainty as regards to location and timing, and it must produce few failures, or false alarms.* Can you imagine the debate that would precede an order to evacuate a large city in North America, such as Vancouver or Los Angeles? The cost of evacuating millions of people, arranging for living accommodations, and providing for their lost work time and wages, would be staggering.

Currently, *no reliable method exists* for making short-range earthquake predictions. In fact, except for a brief period of optimism during the 1970s, the leading seismologists of the past 100 years have generally concluded that short-range earthquake prediction is *not* feasible. To quote Charles Richter, developer of the well-known magnitude scale, "Prediction provides a happy hunting ground for amateurs, cranks, and

◆ **Figure 16.24**   Damage to Interstate 5 caused by the January 17, 1994, Northridge earthquake. (Photo by Bill Nation/Sygma)

◆ **Figure 16.25**   Lasers used to measure movement along the San Andreas Fault. (Photo by John K. Nakata/U.S. Geological Survey)

outright publicity-seeking fakers." This statement was validated in 1990 when Iben Browning, a self-proclaimed expert, predicted that a major earthquake on the New Madrid fault would devastate an area around southeast Missouri on December 2 or 3. Many people in Missouri, Tennessee, and Illinois rushed out to buy earthquake insurance. Some schools and factories closed, while people as far away as northern Illinois stayed home rather than risk travelling to work. The designated date passed without even the slightest tremor.

Long-Range Forecasts

In contrast to short-range predictions, which aim to predict earthquakes within a time frame of hours or at most days, long-range forecasts give the probability of a certain magnitude earthquake occurring on a time scale of 30 to 100 years or more. Stated another way, these forecasts give statistical estimates of the expected intensity of ground motion for a given area over a specified time frame. Although long-range forecasts may not be as informative as we might like, this data is important for updating the Uniform Building Code, which contains nationwide standards for designing earthquake-resistant structures.

Long-range forecasts are based on the premise that earthquakes are repetitive or cyclical, like the weather. In other words, as soon as one earthquake is over, the continuing motions of Earth's plates begin to build strain in the rocks again until they fail once more. This has led seismologists to study historical records of earthquakes to see if there are any discernible patterns so that the probability of recurrence might be established.

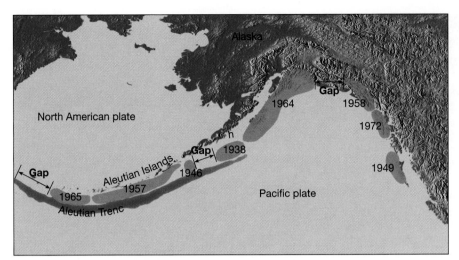

◆ **Figure 16.26** The distribution of rupture areas of large, shallow earthquakes from 1930 to 1979 along the southwestern coast of Alaska and the Aleutian Islands. The three seismic gaps denote the most likely locations for the next large earthquakes along this plate boundary. (After J. C. Savage et al., U.S. Geological Survey)

With this concept in mind, a group of seismologists plotted the distribution of rupture zones associated with great earthquakes that have occurred in the seismically active regions of the Pacific Basin. The maps revealed that individual rupture zones tended to occur adjacent to one another without appreciable overlap, thereby tracing out a plate boundary. Recall that most earthquakes are generated along plate boundaries by the relative motion of large crustal blocks. Because plates are in constant motion, the researchers predicted that over a span of one or two centuries, major earthquakes would occur along each segment of the Pacific plate boundary.

When the researchers studied historical records, they discovered that some zones had not produced a great earthquake in more than a century. These quiet zones, called **seismic gaps**, were identified as probable sites for major earthquakes in the next few decades (Figure 16.26). In the 25 years since the original studies were conducted, some of these gaps have ruptured (Box 16.4). Included in this group is the zone that produced the earthquake that devastated portions of Mexico City in September 1985.

Another method of long-term forecasting, known as *paleoseismology* (*palaios* = ancient, *seismos* = shake, *ology* = the study of), has been implemented. One technique involves the study of layered deposits that were offset by prehistoric seismic disturbances. To date, the most complete investigation that employed this method focused on a segment of the San Andreas Fault about 50 kilometres northeast of Los Angeles. Here the drainage of Pallet Creek has been repeatedly disturbed by successive ruptures along the fault zone. Ditches excavated across the creek bed have exposed sediments that had apparently been displaced by nine large earthquakes over a period of 1400 years. From these data it was determined that a great earthquake

occurs here an average of once every 140 to 150 years. The last major event occurred along this segment of the San Andreas Fault in 1857. Thus, roughly 140 years have elapsed. If earthquakes are truly cyclic, a major event in southern California seems imminent. Such information led the U.S. Geological Survey to predict that there is a 50 percent probability that an earthquake of magnitude 8.3 will occur along the southern San Andreas Fault within the next 30 years.

Using other paleoseismology techniques, researchers recently discovered strong evidence that very powerful earthquakes (magnitude of 8 or larger) have repeatedly struck the Pacific Northwest over the past several thousand years. The most recent event occurred about 300 years ago. As a result of these findings, public officials have taken steps to strengthen some of the region's existing dams, bridges, and water systems. Even the private sector responded. The U.S. Bancorp building in Portland, Oregon, was strengthened at a cost of $8 million and now exceeds the standards of the Uniform Building Code.

Another U.S. Geological Survey study gives the probability of a rupture occurring along various segments of the San Andreas Fault for the 30 years between 1988 and 2018 (Figure 16.27). From this investigation, the Santa Cruz Mountains region was given a 30 percent probability of producing a 6.5-magnitude earthquake during this time period. In fact, the region experienced the Loma Prieta quake in 1989, of 6.9 magnitude.

The region along the San Andreas Fault given the highest probability (90 percent) of generating a quake is the Parkfield section. This area has been called the "Old Faithful" of earthquake zones because activity here has been very regular since record-keeping began in 1857. (Although this section has had an earthquake on the average of once every 22 years, the last

BOX 16.4

Understanding Earth
A Major Earthquake in Turkey

On August 17, 1999, at 3:02 a.m. local time, northwestern Turkey was shaken by a moment magnitude (M_W) 7.4 earthquake, catching most people in their sleep. The epicentre was 10 kilometres southeast of Ismit in a region that is the industrial centre and most densely populated part of the country. Istanbul and its 13 million people is just 70 kilometres to the west.

According to official government estimates, the earthquake killed more than 17,000 people and injured nearly 44,000. More than 250,000 people were forced out of their damaged homes and were sheltered in 120 makeshift "tent cities." Estimates of property losses by the World Bank approached $7 billion. Liquefaction and ground shaking were the dominant causes of damage, but surface faulting and landslides were also responsible for substantial death and destruction. It was the most devastating earthquake to strike Turkey in 60 years.

Turkey is a geologically active region that frequently experiences large earthquakes. Most of the country is part of a small block of continental lithosphere known as the Turkish microplate. This small plate is caught between the northward-moving Arabian and African plates and the relatively stable Eurasian plate (Figure 16.E). The August 1999 earthquake occurred along the western end of the 1500-kilometre-long North Anatolian fault system. This fault has much in common with California's San Andreas Fault. Both are right-lateral strike-slip faults having similar lengths and similar long-term rates of movement.* Also like its North American

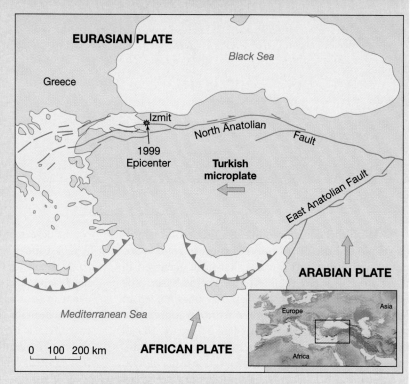

◆ **Figure 16.E** Earthquakes in Turkey are caused by the northward movement of the Arabian and African plates against the Eurasian plate, squeezing the small Turkish microplate westward. Movement is accommodated along two major strike-slip faults—the North Anatolian Fault and the East Anatolian Fault.

counterpart, the North Anatolian Fault is a transform plate boundary.

The fact that a large earthquake occurred along this portion of the North Anatolian Fault did not come as a complete surprise. Based on historical

---

*Recall that if a person is looking across a right-lateral strike-slip fault during an earthquake, that person would see the opposite side move to the right.

records, the region of the epicentre had been identified as a *seismic gap*, a "quiet zone" along the fault where strain had been building for perhaps 300 years. Moreover, during the preceding 60 years an interesting pattern of seismic activity had developed. Beginning in 1939 with a (M_W) 7.9 quake that produced about 350 kilometres of ground rupture, seven earthquakes had broken the fault progressively from east to west as shown in Figure 16.F.

---

one occurred in 1966—it is overdue by more than 16 years and counting!) Another section between Parkfield and the Santa Cruz Mountains is given a very low probability of generating an earthquake. This area has experienced very little seismic activity in historical times; rather, it exhibits a slow, continual movement known as *fault creep*. Such movement is beneficial because it prevents strain from building to high levels in the rocks.

In summary, it appears that the best prospects for making useful earthquake predictions involve forecasting magnitudes and locations on time scales of years or perhaps even decades. These forecasts are important because they provide information used to develop the Uniform Building Code and to assist in land-use planning.

Researchers now understand that as each earthquake occurred, it loaded the zone to the west with additional stress. That is, as a quake released stress on the section of the fault it broke, it transferred stress to adjacent segments. The next segment in line to break is west of Ismit, near Istanbul. It could

happen relatively soon. In the sequence since 1939, no earthquake waited longer than 22 years, and some came within a year of the one before.

The 1999 earthquake near Ismit, Turkey, demonstrated the awesome power of a large earthquake and the immense human suffering that can occur

when an earthquake strikes an urban area. Although no one knows for sure where or when the next major quake will occur in the region, it appears that the 1999 earthquake near Ismit increased the risk for those living in and near Istanbul.

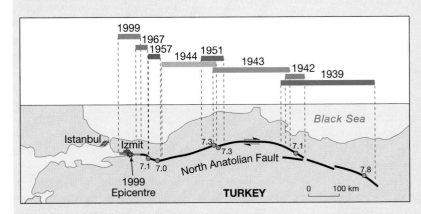

◆ **Figure 16.F**   This map depicts the sequential westward progression of large earthquakes along the North Anatolian Fault between 1939 and 1999. The epicentre and magnitude of each is noted. The length of each coloured segment shows the extent of surface rupture along the fault for each event.

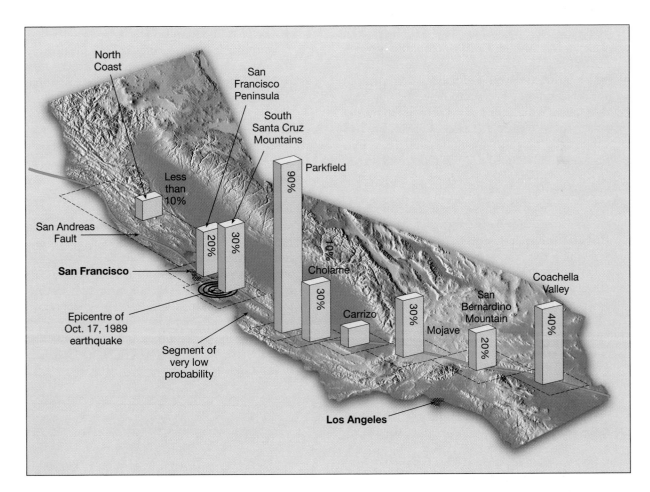

◆ **Figure 16.27**   Probabilities of a major earthquake between 1988 and 2018 along the San Andreas Fault.

Chapter Summary

- *Earthquakes* are vibrations of Earth produced by the rapid release of energy from rocks that rupture because they have been subjected to stresses beyond their limit. This energy, which takes the form of waves, radiates in all directions from the earthquake's source, called the *focus*. The movements that produce most earthquakes occur along large fractures, called *faults*, that are usually associated with plate boundaries.

- Along a fault, rocks store energy as they are bent. As slippage occurs at the weakest point (the focus), displacement will exert stress farther along a fault, where additional slippage will occur until most of the built-up strain is released. An earthquake occurs as the rock elastically returns to its original shape. The "springing back" of the rock is termed *elastic rebound*. Small earthquakes, called *foreshocks*, often precede a major earthquake. The adjustments that follow a major earthquake often generate smaller earthquakes called *aftershocks*.

- Two main types of *seismic waves* are generated during an earthquake: (1) *surface waves*, which travel along the outer layer of Earth, and (2) *body waves*, which travel through Earth's interior. Body waves are further divided into *primary*, or *P*, *waves*, which push (compress) and pull (expand) rocks in the direction the wave is travelling, and *secondary*, or *S*, *waves*, which "shake" the particles in rock at right angles to their direction of travel. P waves can travel through solids, liquids, and gases. Fluids (gases and liquids) will not transmit S waves. In any solid material, P waves travel about 1.7 times faster than do S waves.

- The location on Earth's surface directly above the focus of an earthquake is the *epicentre*. An epicentre is determined using the difference in velocities of P and S waves. Using the difference in arrival times between P and S waves, the distance separating a recording station from the earthquake can be determined. When the distances are known from three or more seismic stations, the epicentre can be located using a method called *triangulation*.

- *A close correlation exists between earthquake epicentres and plate boundaries.* The principal earthquake epicentre zones are along the outer margin of the Pacific Ocean, known as the *circum-Pacific belt*, and through the world's oceans along the *oceanic ridge system*.

- Seismologists use two fundamentally different measures to describe the size of an earthquake—intensity and magnitude. *Intensity* is a measure of the degree of ground shaking at a given locale based on the amount of damage. The *Modified Mercalli Intensity Scale* uses damage to buildings in California to estimate the intensity of ground shaking for a local earthquake. *Magnitude* is calculated from seismic records and estimates the amount of energy released at the source of an earthquake. Using the *Richter scale*, the magnitude of an earthquake is estimated by measuring the *amplitude* (maximum displacement) of the largest seismic wave recorded. A logarithmic scale is used to express magnitude, in which a tenfold increase in ground shaking corresponds to an increase of 1 on the magnitude scale. *Moment magnitude* is currently used to estimate the size of moderate and large earthquakes. It is calculated using the average displacement of the fault, the area of the fault surface, and the sheer strength of the faulted rock.

- The most obvious factors determining the amount of destruction accompanying an earthquake are the magnitude of the earthquake and the proximity of the quake to a populated area. Structural damage attributable to earthquake vibrations depends on several factors, including (1) wave amplitudes, (2) the duration of the vibrations, (3) the nature of the material upon which the structure rests, and (4) the design of the structure. Secondary effects of earthquakes include *tsunami*, landslides, ground subsidence, and fire.

- Substantial research to predict earthquakes is under way in Japan, the United States, China, and Russia—countries where earthquake risk is high. No reliable method of short-range prediction has yet been devised. Long-range forecasts are based on the premise that earthquakes are repetitive or cyclical. Seismologists study the history of earthquakes for patterns so their occurrences might be predicted. Long-range forecasts are important because they provide information used to develop the Uniform Building Code and to assist in land-use planning.

Review Questions

1. What is an earthquake? Under what circumstances do earthquakes *occur?*

2. How are faults, foci, and epicentres related?

3. Who was first to explain the actual mechanism by which earthquakes are generated?

4. Explain what is meant by *elastic rebound.*

5. Faults that are experiencing no active creep may be considered "safe." Rebut or defend this statement.

6. Describe the principle of a seismograph.

7. List the major differences between P and S waves.

8. P waves move through solids, liquids, and gases, whereas S waves move only through solids. Explain.

9. Which type of seismic wave causes the greatest destruction to buildings?

10. Using Figure 16.11, determine the distance between an earthquake and a seismic station if the first S wave arrives 3 minutes after the first P wave.

11. Most strong earthquakes occur in a zone on the globe known as the _____.

12. Deep-focus earthquakes occur several hundred kilometres below what prominent features on the deep-ocean floor?

13. Distinguish between the Mercalli scale and the Richter scale.

14. For each increase of 1 on the Richter scale, wave amplitude increases _____ times.

15. An earthquake measuring 7 on the Richter scale releases about _____ times as much energy as an earthquake with a magnitude of 6.

16. List three reasons why the moment magnitude scale has gained popularity among seismologists.

17. List four factors that affect the amount of destruction caused by seismic vibrations.

18. What factor contributed most to the extensive damage that occurred in the central portion of Mexico City during the 1985 earthquake?

19. The 1988 Armenian earthquake had a Richter magnitude of 6.9, less than the 1994 Northridge California earthquake. Nevertheless, the loss of life was far greater in the Armenian event. Why?

20. In addition to the destruction created directly by seismic vibrations, list three other types of destruction associated with earthquakes.

21. What is a tsunami? How is one generated?

22. Cite some reasons why an earthquake with a moderate magnitude might cause more extensive damage than a quake with a high magnitude.

23. Can earthquakes be predicted?

24. What is the value of long-range earthquake forecasts?

Key Terms

aftershock (p. 443)
body wave (p. 445)
earthquake (p. 441)
elastic rebound (p. 442)
epicentre (p. 447)
fault (p. 442)
fault creep (p. 443)
focus (p. 441)

foreshock (p. 443)
inertia (p. 444)
intensity (p. 451)
liquefaction (p. 457)
long (L) waves (p. 447)
magnitude (p. 451)
Modified Mercalli
   Intensity Scale (p. 452)

moment magnitude
  (p. 455)
primary (P) waves
  (p. 445)
Richter scale (p. 453)
secondary (S) waves
  (p. 445)
seismic gaps (p. 465)

seismic sea wave (p. 459)
seismogram (p. 444)
seismograph (p. 444)
seismology (p. 444)
surface wave (p. 445)
tsunami (p. 459)
Wadati–Benioff zones
  (p. 451)

Web Resources

The *Earth* Web site uses the resources and flexibility of the Internet to aid in your study of the topics in this chapter. Written and developed by geology instructors, this site will help improve your understanding of geology. Visit **http://www.pearson.ca/tarbuck** and click on the cover of the text to find:

■ Online review quizzes.

■ Web-based critical thinking and writing exercises.

■ Links to chapter-specific Web resources.

■ Internet-wide key-term searches.

http://www.pearson.ca/tarbuck

Chapter 17
Earth's Interior

An oil rig probes Earth's crust off the coast of southern California.
(Photo by Mark Lewis/Stone)

arth's interior lies just below us. However, direct access to it remains very limited. Wells drilled into the crust in search of oil, gas, and other natural resources have generally been confined to the upper 7 kilometres—only a small fraction of Earth's 6370-kilometre radius. Even the Kola Well, a super-deep research well located in a remote northern outpost of Russia, has penetrated to a depth of only 12.3 kilometres. Although volcanic activity is considered a window into Earth's interior because materials are brought up from below, it allows a glimpse of only the outer 200 kilometres of our planet.

Fortunately, geologists have learned a great deal about Earth's composition and structure through computer modelling, high-pressure laboratory experiments, and from samples of the solar system (meteorites) that collide with Earth. In addition, many clues to the physical conditions inside our planet have been acquired through the study of seismic waves generated by earthquakes, nuclear explosions, and weaker, human-produced vibrations. As seismic waves pass through Earth, they carry information to the surface about the materials through which they were transmitted. Hence, when carefully analyzed, seismic records provide an image of materials below the surface. The Canadian LITHOPROBE project, for example, has yielded exciting insights on the deep lithosphere from the processing of seismic data (Box 17.1).

Probing Earth's Interior

Much of our knowledge of Earth's interior comes from the study of earthquake waves that penetrate Earth and emerge at some distant point. Simply stated, the technique involves accurately measuring the time required for P (*compressional*) and S (*shear*) waves to travel from an earthquake or nuclear explosion to a seismographic station. Because the time required for P and S waves to travel through Earth depends on the properties of the materials encountered, seismologists search for variations in travel times that cannot be accounted for simply by differences in the distances travelled. These variations correspond to changes in the properties of the materials encountered.

One major problem is that to obtain accurate travel times, seismologists must establish the exact location and time of an earthquake. This is often a difficult task because most earthquakes occur in remote areas. By contrast, the time and location of a nuclear test explosion is always known precisely. Despite the limitations of studying seismic waves generated by earthquakes, seismologists in the first half of the twentieth century were able to use them to detect the major layers of Earth. It was not until the early 1960s, when nuclear

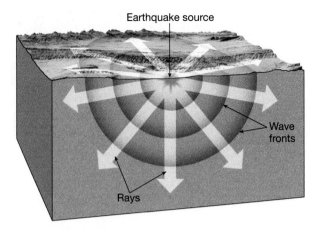

◆ **Figure 17.1**   Seismic energy travels in all directions from an earthquake source (focus). The energy can be portrayed as expanding wave fronts or as rays drawn perpendicular to the wave fronts.

testing was in its heyday and networks consisting of hundreds of sensitive seismographs were deployed, that the finer structures of Earth's interior were established with certainty.

The Nature of Seismic Waves

To examine Earth's composition and structure, we must first study some of the basic properties of wave transmission, or *propagation*. As stated in Chapter 16, seismic energy travels out from its source in all directions as waves. (For purposes of description, the common practice is to consider the paths taken by these waves as *rays*, or lines drawn perpendicular to the wave front, as shown in Figure 17.1.) Significant characteristics of seismic waves include the following:

1. The velocity of seismic waves depends on the density and elasticity of the intervening material. Seismic waves travel most rapidly in rigid materials that elastically spring back to their original shapes when the stress caused by a seismic wave is removed. For instance, crystalline rock transmits seismic waves more rapidly than does a layer of unconsolidated mud.

2. Within a given layer the speed of seismic waves generally increases with depth because pressure increases and squeezes the rock into a more compact elastic material.

3. Compressional waves (P waves), which vibrate back and forth in the same plane as their direction of travel, are able to propagate through liquids as well as solids because, when compressed, these materials behave elastically; that is, they resist a change in volume and, like a rubber band, return to their original shape as a wave passes (Figure 17.2A).

BOX 17.1

Canadian Profile
LITHOPROBE: Probing the Depths of Canada

David Eaton*

Unlike the oceanic crust, continents preserve a record of Earth's early history, including how ancient mountains formed and how the continents grew and developed over time. Geoscientists are faced with a daunting challenge, however: to reconstruct this tectonic record from fragmentary evidence provided by rocks exposed at the surface. Canada's LITHOPROBE project is a bold 20-year endeavour, which commenced in 1984, to unravel the tectonic evolution of North America with a focus on the critical third dimension of geology—depth. With collaboration by more than 900 scientists working in academia, government agencies, and the petroleum and mining industries, LITHOPROBE has involved a significant segment of Canada's Earth Science community in a coordinated research effort. The research program was divided into a series of transects that collectively span the evolution of the North American continent in time (almost 4 billion years) and space.

In order to achieve adequate resolution of geological features down to the base of the crust (approximately 40 kilometres) and the upper mantle, LITHOPROBE research has been spearheaded by multichannel seismic-reflection (MSR) profiling. Adapted from the petroleum industry, where it is used extensively for exploration of the top 10 kilometres of sedimentary basins, this technique makes use of P waves, usually generated mechanically using "vibroseis trucks" to record faint echoes produced at buried geologic contacts, faults, and shear zones. Much like medical imaging techniques, the seismic data are processed by computer to enhance the seismic reflections, producing cross-sectional images that can be compared to surficial geological data to reveal the underlying anatomy of the lithosphere. The great success of LITHOPROBE can be attributed to the use of a coordinated, multidisciplinary approach. Specialists in all areas of geology, geophysics, and geochemistry have collaborated on major geotectonic problems using techniques that include basic mapping, precise geochronology, electromagnetic studies, seismic refraction, and paleomagnetism. Through such a collective effort, the scientific study of these problems is enhanced far beyond the level of individual contributions.

LITHOPROBE's pilot study on Vancouver Island in 1984 demonstrated that MSR profiling methods could be used to map layers near the top of the descending Juan de Fuca slab (Figure 17.A). Particularly significant are underplated slabs of lithosphere (underplates) that have been welded to continental lithosphere in western British Columbia. Results from this modern subduction system (Cascadia) have provided a template for understanding ancient subduction systems, leading to remarkable new insights. For example, subduction "scars" have been detected in the mantle beneath the Archean (2.7 billion years) Opatica plutonic subprovince of the Superior craton in Quebec, and the Proterozoic (1.8 billion years) Great Bear Arc in the Northwest Territories. These results are important for our understanding of Earth history, since they establish that the subduction occurred in some areas in the late Archean and support the concept that subducted slabs can be

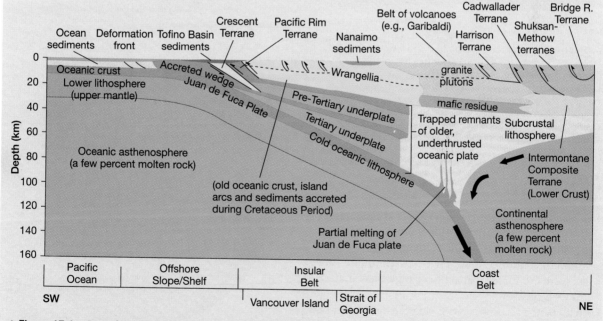

◆ Figure 17.A Part of the Canadian Lithoprobe project involved the study of the lithospheric structure of western Canada. Among the exciting discoveries was the detection of lithospheric fragments (underplates) that remain lodged under the west coast region and have contributed to the growth of mountain roots.
(modified from image courtesy of R. M. Clowes)

accreted to the base of continental plates as part of the process of formation of deep lithospheric roots.

A large fraction of LITHOPROBE's research was focused on ancient shield and platform regions, where North America is known to have grown via a series of mountain building events called orogenies. A wealth of crustal reflectivity revealed by LITHOPROBE profiles has allowed geoscientists to trace structural contacts down to the base of the crust (Moho). This work has led to the adaptation of seismic methods from mineral exploration, and the discovery of previously unrecog- nized orogenic belts that lie buried beneath undeformed sedimentary rocks.

For further information visit **http:// www.geop.ubc.ca/Lithoprobe/**

---

*David Eaton is a professor in the Department of Earth Sciences at the University of Western Ontario.

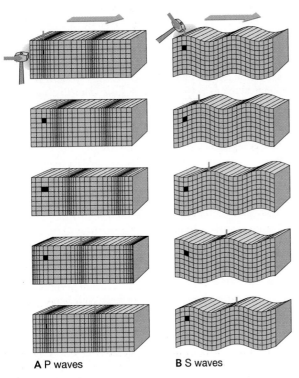

A P waves   **B** S waves

◆ **Figure 17.2**   The transmission of P and S waves through a solid. **A.** The passage of P waves causes the intervening material to experience alternate compressions and expansions. **B.** The passage of S waves causes a change in shape without changing the volume of the material. Because liquids behave elastically when compressed (they spring back when the stress is removed), they will transmit P waves. However, since liquids do not resist changes in shape, S waves cannot be transmitted through liquids.
(After O. M. Phillips, *The Heart of the Earth*, San Francisco: Freeman, Cooper and Co., 1968)

4. Shear waves (S waves), which vibrate at right angles to their direction of travel, cannot propagate through liquids because, unlike solids, liquids have no shear strength (Figure 17.2B). That is, when liquids are subjected to forces that act to change their shapes, they simply flow.

5. In all materials, P waves travel faster than S waves.

6. When seismic waves pass from one material to another, the path of the wave is refracted (bent).* In addition, some of the energy is reflected from the **discontinuity** (the boundary between the two dissimilar materials). This is similar to what happens to light when it passes from air into water.

Thus, depending on the nature of the layers through which they pass, seismic waves speed up or slow down and may be refracted (bent) or reflected. These measurable changes in seismic wave motions enable seismologists to probe Earth's interior.

Seismic Waves and Earth's Structure

Introduction
↳ Earth's Layered Structure

If Earth were a perfectly homogeneous body, seismic waves would spread through it in all directions, as shown in Figure 17.3. Such seismic waves would travel in a straight line at a constant speed. However, this is not the case for Earth. It so happens that the

---

*Refraction occurs provided that the ray is not travelling perpendicular to the boundary.

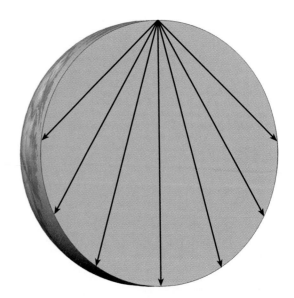

◆ **Figure 17.3**   Seismic waves would travel through a hypo- thetical planet with uniform properties along straight-line paths and at constant velocities. Contrast with Figure 17.4.

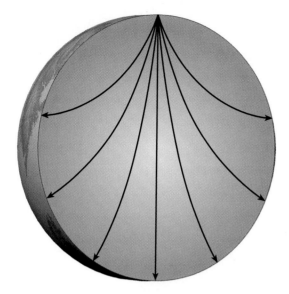

◆ **Figure 17.4** Wave paths through a planet where velocity increases with depth.

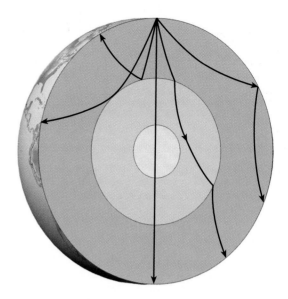

◆ **Figure 17.5** A few of the many possible paths that seismic rays take through Earth.

seismic waves reaching seismographs located farther from an earthquake travel at faster average speeds than those recorded at locations closer to the event. This general increase in speed with depth is a consequence of increased pressure, which enhances the elastic properties of deeply buried rock. As a result, the paths of seismic rays through Earth are refracted in the manner shown in Figure 17.4.

As more sensitive seismographs were developed, it became apparent that in addition to gradual changes in seismic-wave velocities, rather abrupt velocity changes also occur at particular depths. Because these discontinuities were detected worldwide, seismologists concluded that Earth must be composed of distinct shells having varying compositions and/or mechanical properties (Figure 17.5).

Layers Defined by Composition

Compositional layering in Earth's interior likely resulted from density sorting that took place during an early period of partial melting. During this period the heavier elements, principally iron and nickel, sank as the lighter rocky components floated upward. This segregation of material is still occurring, but at a much reduced rate. Because of this chemical differentiation, Earth's interior is not homogeneous. Rather, it consists of three major regions that have markedly different chemical compositions (Figure 17.6).

The principal compositional layers of Earth include:

• the **crust**, Earth's comparatively thin outer skin that ranges in thickness from 3 kilometres at the oceanic

ridges to over 70 kilometres in some mountain belts, such as the Andes and Himalayas;
• the **mantle**, a solid rocky (silica-rich) shell that extends to a depth of about 2900 kilometres;
• the **core**, an iron-rich sphere having a radius of 3486 kilometres.

We will consider the composition and structure of these major divisions of Earth's interior later in the chapter.

Layers Defined by Physical Properties

Earth's interior is characterized by a gradual increase in temperature, pressure, and density with depth. Estimates put the temperature at a depth of 100 kilometres at between 1200°C and 1400°C, whereas the temperature at Earth's centre may exceed 6700°C. Clearly, Earth's interior has retained much of the energy acquired during its formative years, despite the fact that heat is continuously flowing toward the surface, where it is lost to space. The increase in pressure with depth causes a corresponding increase in rock density.

The gradual increase in temperature and pressure with depth affects the physical properties and hence the mechanical behaviour of Earth materials. When a substance is heated, its chemical bonds weaken and its mechanical strength (resistance to deformation) is reduced. If the temperature exceeds the melting point of an Earth material, the material's chemical bonds break and melting ensues. If temperature were the only factor that determined whether a substance melted, our planet would be a molten ball covered with a thin, solid outer shell. However, pressure also

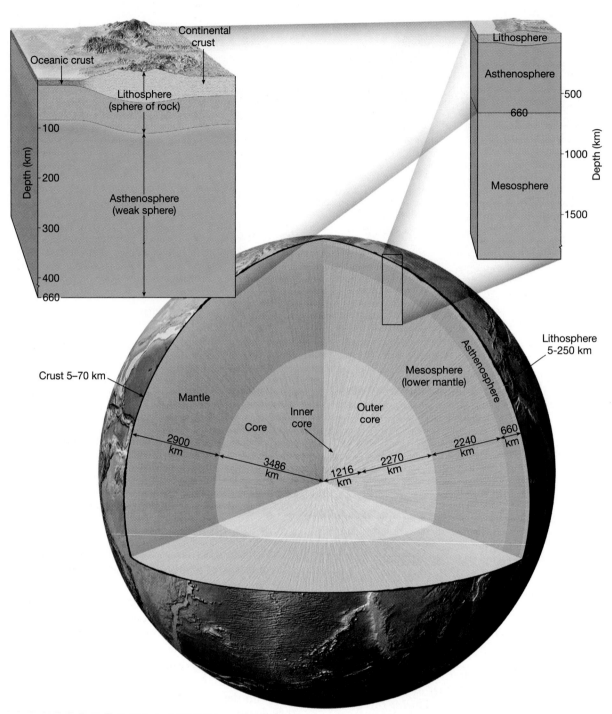

◆ **Figure 17.6**   Views of Earth's layered structure. The left side of the large cross-section shows that Earth's interior is divided into three different layers based on compositional differences—the crust, mantle, and core. The right side of the large cross-section depicts the five main layers of Earth's interior based on physical properties and hence mechanical strength—the lithosphere, asthenosphere, mesosphere, outer core, and inner core. The block diagrams above the large cross-section show an enlarged view of the upper portion of Earth's interior.

increases with depth and tends to increase rock strength. Furthermore, because melting is accompanied by an increase in volume, it occurs at higher temperatures at depth because of greater confining pressure. Thus, depending on the physical environment (temperature and pressure), a particular Earth material may behave like a brittle solid, deform in a puttylike manner, or even melt and become liquid.

Earth can be divided into five main layers based on their physical properties and hence mechanical strength—the *lithosphere, asthenosphere, mesosphere (lower mantle), outer core,* and *inner core.*

Lithosphere and Asthenosphere   Based on physical properties, Earth's outermost layer consists of the crust and uppermost mantle and forms a relatively cool, rigid shell. Although this layer is composed of materials with markedly different chemical compositions, it tends to act as a unit that exhibits rigid behaviour—mainly because it is cool and thus strong. This layer, called the **lithosphere** (*sphere of rock*), averages about 100 kilometres in thickness but may be 250 kilometres thick or more below the older portions of the continents (Figure 17.6). Within the ocean basins, the lithosphere is only a few kilometres thick along the oceanic ridges but increases to perhaps 100 kilometres in regions of older and cooler oceanic crust.

Beneath the lithosphere in the upper mantle (to a depth of about 660 kilometres) lies a soft, comparatively weak layer known as the **asthenosphere** (*weak sphere*). The top portion of the asthenosphere has a temperature/pressure regime that results in a small amount of melting. Within this very weak zone, the lithosphere is mechanically detached from the layer below. The result is that the lithosphere is able to move independently of the asthenosphere, a topic we will consider in the next chapter.

It is important to emphasize that the strength of various Earth materials is a function of both their composition and of the temperature and pressure of their environment. You should not get the idea that the entire lithosphere behaves like a brittle solid similar to rocks found on the surface. Rather, the rocks of the lithosphere get progressively hotter and weaker (more easily deformed) with increasing depth. At the depth of the uppermost asthenosphere, the rocks are close enough to their melting temperatures (some melting may actually occur) that they are very easily deformed. Thus, the uppermost asthenosphere is weak because it is near its melting point, just as hot wax is weaker than cold wax.

Mesosphere or Lower Mantle   Below the zone of weakness in the uppermost asthenosphere, increased pressure counteracts the effects of higher temperature and the rocks gradually strengthen with depth. Between the depths of 660 kilometres and 2900 kilometres a more rigid layer called the **mesosphere** (*middle sphere*) or **lower mantle** is found (Figure 17.6). Despite their strength, the rocks of the mesosphere are still very hot and capable of very gradual flow.

Inner and Outer Core   The core, which is composed mostly of an iron-nickel alloy, is divided into two regions that exhibit very different mechanical strengths (Figure 17.6). The **outer core** is a *liquid layer* 2270 kilometres thick. It is the convective flow of metallic iron within this zone that generates Earth's magnetic field. The **inner core** is a sphere having a radius of 3486 kilome-

tres. Despite its higher temperature, the material in the inner core is stronger (because of immense pressure) than the outer core and behaves like a *solid*.

Discovering Earth's Major Boundaries

Over the past century, seismological data gathered from many seismographic stations have been compiled and analyzed. From this information, seismologists have developed a detailed image of Earth's interior (Figure 17.6). This model is continually being fine-tuned as more data become available and as new seismic techniques are employed. Furthermore, laboratory studies that experimentally determine the properties of various Earth materials under the extreme environments found deep in Earth add to this body of knowledge.

The Moho

In 1909 a pioneering Yugoslavian seismologist, Andrija Mohorovicic, presented the first convincing evidence for layering within Earth. The boundary he discovered separates crustal materials from rocks of different composition in the underlying mantle and was named the **Mohorovicic discontinuity** in his honour. The name for this boundary was quickly shortened to **Moho**.

By carefully examining the seismograms of shallow earthquakes, Mohorovicic found that seismographic stations located more than 200 kilometres from an earthquake obtained appreciably faster average travel velocities for P waves than did stations located nearer the quake (Figure 17.7). In particular, P waves that reached the closest stations first had velocities that averaged about 6 kilometres per second. By contrast, the seismic energy recorded at more distant stations travelled at speeds that approached 8 kilometres per second. This abrupt jump in velocity did not fit the general pattern that had been previously observed. From these data, Mohorovicic concluded that below 50 kilometres there exists a layer with properties markedly different from those of Earth's outer shell.

Figure 17.7 illustrates how Mohorovicic reached this important conclusion. Notice that the first wave to reach the seismograph located 100 kilometres from the epicentre traveled the shortest route directly through the crust. However, at the seismograph located 300 kilometres from the epicentre, the first P wave to arrive traveled through the mantle, a zone of higher velocity. Thus, although this wave traveled a greater distance, it reached the recording instrument sooner than did the rays taking the more direct route. This is because a large portion of its journey was through a region having a composition where seismic waves travel more

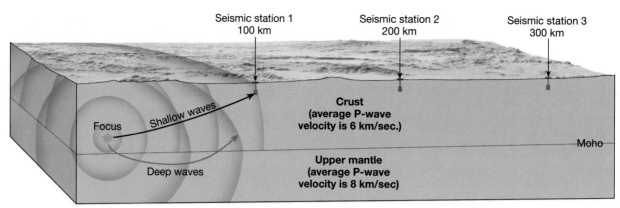

A Time 1 – Slower shallow waves arrive at seismic station 1 first.

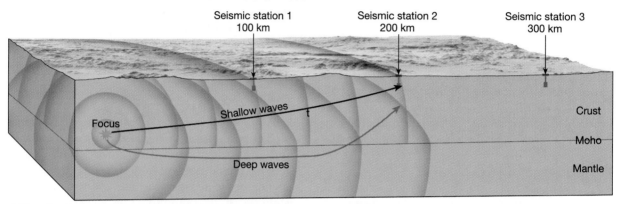

B Time 2 – Slower shallow waves arrive at seismic station 2 first.

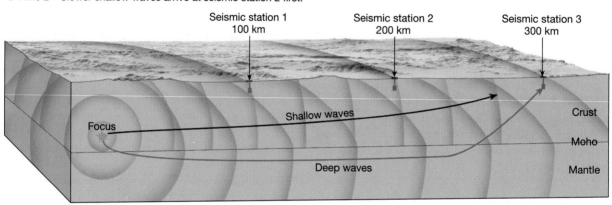

C Time 3 – Faster deeper waves arrive at seismic station 3 first.

◆ **Figure 17.7**   Idealized paths of seismic waves travelling from an earthquake focus to three seismographic stations. In parts **A** and **B**, you can see that the two nearest recording stations receive the slower waves first because the waves travelled a shorter distance. However, as shown in part **C**, beyond 200 kilometres, the first waves received passed through the mantle, which is a zone of higher velocity.

rapidly. This principle is analogous to a driver taking a bypass route around a large city during rush hour. Although this alternate route is longer, it may be faster.

The Core–Mantle Boundary

A few years later, in 1914, the location of another major boundary was established by the German seismologist Beno Gutenberg.* This discovery was based

primarily on the observation that P waves diminish and eventually die out completely about 105° from an earthquake (Figure 17.8). Then, about 140° away, the P waves reappear, but about two minutes later

---

*The core–mantle boundary had been predicted by R. D. Oldham in 1906, but his arguments for a central core did not receive wide acceptance.

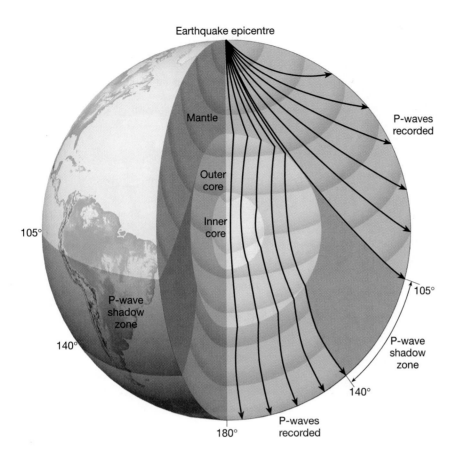

Earthquake epicentre

Mantle

Outer core

Inner core

P-waves recorded

105°

140°

P-wave shadow zone

105°

140°

P-wave shadow zone

180°

P-waves recorded

◆ **Figure 17.8** The abrupt change in physical properties at the core–mantle boundary causes the wave paths to bend sharply, resulting in a shadow zone for P waves between about 105° and 140°.

than would be expected based on the distance travelled. This belt, where direct seismic waves are absent, is about 35° wide and has been named the **P wave shadow zone*** (Figure 17.8).

Gutenberg, and others before him, realized that the P wave shadow zone could be explained if Earth contained a core that was composed of material unlike the overlying mantle. The core, which Gutenberg calculated to be located at a depth of 2900 kilometres, must somehow hinder the transmission of P waves similar to the way in which light rays are blocked by an object that casts a shadow. However, rather than actually stopping the P waves, the shadow zone is produced by bending (refracting) of the P waves, which enter the core as shown in Figure 17.8.

It was further determined that S waves do not travel through the core. This fact led geologists to conclude that at least a portion of this region is liquid (Figure 17.9). This conclusion was further supported by the observation that P-wave velocities suddenly decrease by about 40 percent as they enter the core. Because melting reduces the elasticity of rock, this evidence pointed to the existence of a liquid layer below the rocky mantle.

Discovery of the Inner Core

In 1936 the last major subdivision of Earth's interior was predicted by Inge Lehmann, a Danish seismologist. Lehmann discovered a new region of seismic reflection and refraction within the core. Hence, a core within a core was discovered. The size of the inner core was not precisely established until the early 1960s when underground nuclear tests were conducted in Nevada. Because the exact location and time of the explosions were known, echoes from seismic waves that bounced off the inner core provided a precise means of determining its size (Figure 17.10).

From these data, the inner core was found to have a radius of about 1216 kilometres. Furthermore, P waves passing through the inner core have appreciably faster average velocities than do those penetrating only the outer core. The apparent increase in the elasticity of the inner core is evidence that this innermost region is solid.

Over the past few decades, advances in seismology and rock mechanics have allowed for much refinement of the gross view of Earth's interior that has been presented to this point. Some of these refinements, as well as other properties of the major divisions, including their densities and compositions, will be considered next.

---

*As more sensitive instruments were developed, weak and delayed P waves that enter this zone via reflection were detected.

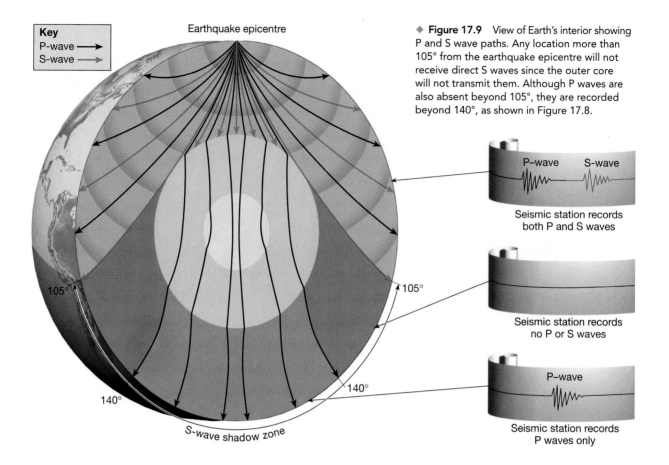

Key
P-wave ⟶
S-wave ⟶

Earthquake epicentre

105°

140°

S-wave shadow zone

105°

140°

P-wave   S-wave

Seismic station records
both P and S waves

Seismic station records
no P or S waves

P-wave

Seismic station records
P waves only

◆ **Figure 17.9**   View of Earth's interior showing P and S wave paths. Any location more than 105° from the earthquake epicentre will not receive direct S waves since the outer core will not transmit them. Although P waves are also absent beyond 105°, they are recorded beyond 140°, as shown in Figure 17.8.

The Crust

Earth's crust averages less than 20 kilometres thick, making it the thinnest of Earth's divisions (see Figure 17.6). Along this eggshell-thin layer, great variations in thickness exist. Crustal rocks of the stable continental interiors average about 35–40 kilometres thick. However, in a few exceptionally prominent mountainous regions, the crust reaches its greatest thickness, exceeding 70 kilometres. The oceanic crust is much thinner, ranging from 3 to 15 kilometres thick and averaging about 7 kilometres thick. Further, crustal rocks of the deep-ocean basins are compositionally different from those of the continents.

Continental rocks have an average density of about 2.7 g/cm³ and some have been discovered that exceed 4 billion years in age. From both seismic studies and direct observations, the average composition of upper continental rocks is estimated to be comparable to the felsic igneous rock *granodiorite*. Like granodiorite, the continental crust is enriched in the elements potassium, sodium, and silicon. Although numerous felsic intrusions and chemically equivalent metamorphic rocks are very abundant, large quantities of mafic and andesitic rocks are also commonly found on the continents. Further, the lowermost crust is thought to have a composition similar to basalt.

The rocks of the oceanic crust are younger (180 million years or less) and denser (about 3.0 g/cm³) than continental rocks. The deep-ocean basins lie beneath 4 kilometres of seawater as well as hundreds of metres of sediment. Thus, until recently, geologists had to rely on indirect evidence (such as slivers of what was thought to be oceanic crust that were thrust onto land) to estimate the composition of this inaccessible region. With the development of deep-sea drilling ships, the recovery of core samples from the ocean floor became possible. As predicted, the samples obtained were predominantly *basalt*. Recall that volcanic eruptions of mafic lavas are known to have generated many islands, such as the Hawaiian chain, located within the deep-ocean basins.

The Mantle

Over 82 percent of Earth's volume is contained within the mantle, a nearly 2900-kilometre-thick shell of silicate rock extending from the base of the crust (Moho) to the liquid outer core. Our knowledge of the mantle's composition comes from experimental data and from the examination of material carried to the surface by

*Liquid water has a density of 1 g/cm³; therefore, crustal rocks have a density nearly three times that of water.

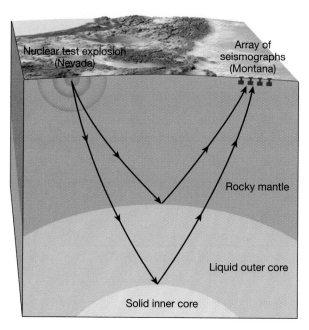

Nuclear test explosion
(Nevada)

Array of
seismographs
(Montana)

Rocky mantle

Liquid outer core

Solid inner core

◆ **Figure 17.10**   Travel times of seismic waves generated from nuclear test explosions were used to measure the depth of the inner core accurately. An array of seismographs located in Montana detected the "echoes" that bounced back from the boundary of the inner core.

volcanic activity. In particular, the rocks composing kimberlite pipes, in which diamonds are sometimes found, are often thought to have originated at depths approaching 200 kilometres, well within the mantle. Kimberlite deposits are composed of *peridotite*, a rock that contains iron and magnesium-rich silicate minerals, mainly olivine and pyroxene, plus lesser amounts of garnet. Further, because S waves readily travel through the mantle, we know that it behaves like an elastic solid. Thus, the mantle is described as a solid rocky layer, the upper portion of which has the composition of the ultramafic rock peridotite.

The mantle is divided into the *mesosphere*, or *lower mantle*, which extends from the core–mantle boundary up to a depth of 660 kilometres, and the asthenosphere or *upper mantle*, which continues up to the base of the crust. In addition, other subdivisions have been identified. At a depth of about 410 kilometres a relatively abrupt increase in seismic velocity occurs (Figure 17.11). Whereas the crust–mantle boundary represents a compositional change, the zone of seismic-velocity increase at the 410-kilometre level is the result of a *phase change*. (A phase change occurs when the crystalline structure of a mineral is altered in response to changes in temperature and/or pressure.) Laboratory studies show that magnesium-rich *olivine* ($MgSiO_4$), which is one of the main constituents in the rock peridotite, will collapse to the more compact, high-pressure mineral *spinel*, at the pressures experienced at this depth (Figure 17.12). This

change to a denser crystal form explains the increased seismic velocities observed.

Another boundary within the mantle has been detected from variations in seismic velocity at a depth of 660 kilometres (Figure 17.11). At this depth the mineral spinel is believed to undergo a transformation to the mineral perovskite $(Mg,Fe)SiO_3$. Because perovskite is thought to be pervasive in the lower mantle, it is perhaps Earth's most abundant mineral.

In the lowermost roughly 200 kilometres of the mantle, there exists an important region known as the **D″ layer**. Recently, researchers reported that seismic waves travelling through some parts of the D″ layer experience a sharp decrease in P-wave velocities. So far, the best explanation for this phenomenon is that the lowermost layer of the mantle is partially molten, at least in some places.

If these zones of partially melted rock exist, they are very important because they would be capable of transporting heat from the core to the lower mantle much more efficiently than solid rock. A high rate of heat flow would in turn cause the solid mantle located above these partly molten zones to be warmed sufficiently to become buoyant and slowly rise toward the surface. Such rising plumes of superheated rock may be the source of volcanic activity associated with hotspots, such as that experienced at Hawaii and Iceland. If these observations are accurate, some of the volcanic activity that is seen at the surface is a manifestation of processes occurring 2900 kilometres below.

Students Sometimes Ask...

As compared to continental crust, the ocean crust is quite thin. Has there ever been an attempt to drill through it to obtain a sample of the mantle?

Yes. "Project Mohole" was initiated in 1958 to retrieve a sample of material from Earth's mantle by drilling a hole through Earth's crust to the *Mohorovicic discontinuity*, or *Moho*. The plan was to drill to the Moho to gain valuable information on Earth's age, makeup, and internal processes. Despite a successful test phase, drilling was halted as control of the project was shifted from one organization to another until Congress, objecting to increasing costs, discontinued the project toward the end of 1966, before Phase II could be implemented. Although Project Mohole failed in its intended purpose, it did show that deep-ocean drilling was a viable means of obtaining geological samples. Since Mohole's demise, a number of related programs have been undertaken, the most recent one being the Ocean Drilling Program, which provides valuable information about Earth's history.

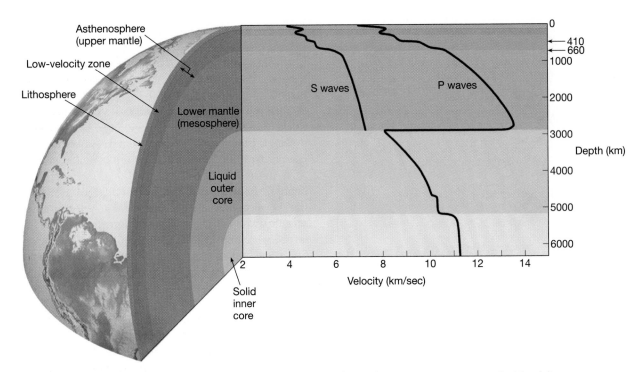

◆ **Figure 17.11**   Variations in P and S wave velocities with depth. Abrupt changes in average wave velocities delineate the major features of Earth's interior. At a depth of about 100 kilometres, the sharp decrease in wave velocity corresponds to the top of the low-velocity zone. Two other bends in the velocity curves occur in the upper mantle at depths of about 410 and 660 kilometres. These variations are thought to be caused by minerals that have undergone phase changes, rather than resulting from compositional differences. The abrupt decrease in P-wave velocity and the absence of S waves at 2900 kilometres marks the core–mantle boundary. The liquid outer core will not transmit S waves, and the propagation of P waves is slowed within this layer. As the P waves enter the solid inner core, their velocity once again increases. (Data from Bruce A. Bolt)

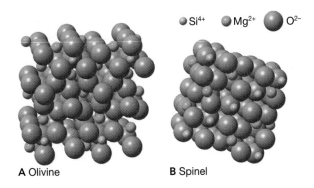

◆ **Figure 17.12**   Comparison of the crystalline structures of olivine and spinel. At a depth of about 410 kilometres, magnesium-rich olivine is thought to collapse into spinel— a mineral that exhibits a more compact structure and hence greater density.

The Core

Larger than the planet Mars, the core is Earth's dense central sphere, with a radius of 3486 kilometres. Extending from the inner edge of the mantle to the centre of Earth, the core constitutes about one-sixth of Earth's volume and nearly one-third of its total mass. Pressure at the centre is millions of times greater than the air pressure at Earth's surface, and the temperatures can exceed 6700°C. As more precise seismic data became available, the core was found to consist of a liquid outer layer about 2270 kilometres thick and a solid inner sphere with a radius of 1216 kilometres.

Density and Composition

One of the more interesting characteristics of the core is its great density. The average density of the core is nearly 11 g/cm³, and at Earth's centre it approaches 14 times the density of water. Even under the extreme pressures at these depths, the common silicate minerals found in the crust (with surface densities of 2.6 to 3.5g/cm³) would not be compacted enough to account for the density calculated for the core. Consequently, attempts were undertaken to determine the material that could account for this property.

Surprisingly enough, meteorites provided an important clue to Earth's internal composition. Because meteorites are part of the solar system, they are assumed to be representative samples of the material from which Earth originally accreted. Their composition ranges from metallic types, made primarily of iron and lesser amounts of nickel, to stony meteorites composed of rocky substances that closely resemble the rock peridotite. Because Earth's crust and mantle contain a much smaller percentage of iron than is found in the debris of the solar system, geologists concluded that the interior of Earth must be enriched in this heavy metal. Further, iron is by far the most abundant substance found in the solar system that possesses the seismic properties and density resembling that measured for the core. Current estimates suggest that the core is mostly iron, with 5 to 10 percent nickel and lesser amounts of lighter elements, including perhaps sulphur and oxygen.

Origin

Although the existence of a metallic central core is well established, explanations of the core's origin are more speculative. The most widely accepted explanation suggests that the core formed early in Earth's history from what was originally a relatively homogeneous body. During the period of accretion, the entire Earth was heated by energy released by the collisions of infalling material. Sometime late in this period of growth, Earth's internal temperature was sufficiently high to melt and mobilize the accumulated material. Blobs of heavy iron-rich materials collected and sank toward the centre. Simultaneously, lighter substances may have floated upward to generate the crust. In a short time, geologically speaking, Earth took on a layered configuration, not significantly different from what we find today.

In its formative stage, the entire core was probably liquid. Further, this liquid iron alloy was in a state of vigorous mixing. However, as Earth began to cool, iron in the core began to crystallize and the inner core began to form. As the core continues to cool, the inner core should grow at the expense of the outer core.

Earth's Magnetic Field

Our picture of the core with its solid inner sphere surrounded by a mobile liquid shell is further supported by the existence of Earth's magnetic field. This field behaves as though a large bar magnet were situated deep within Earth. However, we know that the source of the magnetic field cannot be permanently magnetized material, because Earth's interior is too hot for any material to retain its magnetism. The most widely accepted explanation of Earth's magnetic field requires that the core be made of a material that conducts electricity, such as iron, and one that is mobile (Box 17.2). Both of these conditions are met by the model of Earth's core that was established on the basis of seismological data.

One recently discovered consequence of Earth's magnetic field is that it affects the rotation of the

solid inner core. Current estimates indicate that the inner core rotates in a west-to-east direction about 1° a year *faster* than the Earth's surface. Thus, the core makes one extra rotation about every 400 years. Further, the axis of rotation for the inner core is offset about 10° from Earth's rotational poles.

Earth's Internal Heat Engine

As we discussed in Chapter 3, temperature gradually increases with an increase in depth at a rate known as the **geothermal gradient** (Figure 17.13). The geothermal gradient varies considerably from place to place. In the crust, temperatures increase rapidly, averaging between 20°C and 30°C per kilometre. However, the rate of increase is much less in the mantle and core. At a depth of 100 kilometres, the temperature is estimated to exceed 1200°C, whereas the temperature at the core–mantle boundary is estimated to be 3500°–4500°C, and it may exceed 6700°C at Earth's centre (hotter than the surface of the Sun!).

Three major processes have contributed to Earth's internal heat: (1) heat emitted by radioactive decay of isotopes of uranium (U), thorium (Th), and potassium (K); (2) heat released as iron crystallized to form the solid inner core; and (3) heat released by colliding particles during the formation of our planet. Although the first two processes are still operating, their rate of heat generation is much less than in the geologic past. Today our planet is radiating more of its internal heat out to space than is being replaced by these mechanisms. Therefore, Earth is slowly, but continuously, cooling.

Heat Flow in the Crust

Heat flow in the crust occurs by the familiar process called **conduction**. Anyone who has attempted to pick

BOX 17.2

Understanding Earth
Why Does Earth Have a Magnetic Field?

Anyone who has used a compass to find direction knows that Earth's magnetic field has a north pole and a south pole. In many respects our planet's magnetic field resembles that produced by a simple bar magnet. Invisible lines of force pass through Earth and out into space while extending from one pole to the other (Figure 17.B). A compass needle, which is a small magnet free to move about, becomes aligned with these lines of force and points toward the magnetic poles. It should be noted that Earth's magnetic poles do not coincide exactly with the geographic poles. The north magnetic pole is located in northeastern Canada, near Hudson Bay, while the south magnetic pole is located near Antarctica in the Indian Ocean south of Australia.

In the early 1960s, geophysicists learned that Earth's magnetic field periodically (every million years or so) reverses polarity; that is, the north magnetic pole becomes the south magnetic pole, and vice versa. The cause of these reversals is apparently linked to the fact that Earth's magnetic field experiences long-term fluctuations in intensity. Recent calculations indicate that the magnetic field has weakened by about 5 percent over the past century. If this trend continues for another 1500 years, Earth's magnetic field will become very weak or even nonexistent.

It has been suggested that the decline in magnetic intensity is related to changes in the convective flow in the core. In a similar manner, magnetic reversals may be triggered when something disturbs the main convection pattern of the fluid core. After a reversal takes place, the flow is re-established and builds a magnetic field with opposite polarity.

Magnetic reversals are not unique to Earth. The Sun's magnetic field regularly reverses polarity, having an aver-

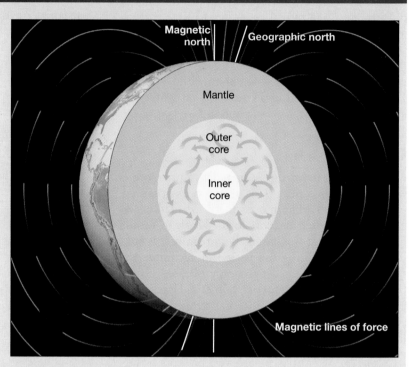

◆ **Figure 17.B** Earth's magnetic field is thought to be generated by vigorous convection of molten iron alloy in the liquid outer core.

age period of about 22 years. These solar reversals are closely tied to the well-known 11-year sunspot cycle.

When Earth's magnetic field was first described in 1600, it was thought to originate from permanently magnetized materials located deep within Earth. We have since learned that except for the upper crust, the planet is much too hot for magnetic materials to retain their magnetism. Furthermore, permanently magnetized materials are not known to vary their intensity in a manner that would account for the observed long-term waxing and waning of Earth's magnetic field.

The details of how Earth's magnetic field is produced are still poorly understood. Nevertheless, most investigators agree that the gradual flow of molten iron in the outer core is an

important part of the process. The most widely accepted view proposes that the core behaves like a self-sustaining *dynamo*, a device that converts mechanical energy into magnetic energy. The driving forces of this system are Earth's rotation and the unequal distribution of heat in the interior, which propels the highly conductive molten iron in the outer core. As the iron churns in the outer core, it interacts with Earth's magnetic field. This interaction generates an electric current, just as moving a wire past a magnet creates a current in the wire. Once established, the electric current produces a magnetic field that reinforces Earth's magnetic field. As long as the flow within the molten iron outer core continues, electric currents will be produced and Earth's magnetic field will be sustained.

---

up a metal spoon left in a hot pan is quick to realize that heat was conducted through the spoon. *Conduction*, which is the transfer of heat through matter by molecular activity, occurs at a relatively slow rate in crustal

rocks. Thus, the crust tends to act as an insulator (cool on top and hot on the bottom), which helps account for the steep temperature gradient exhibited by the crust.

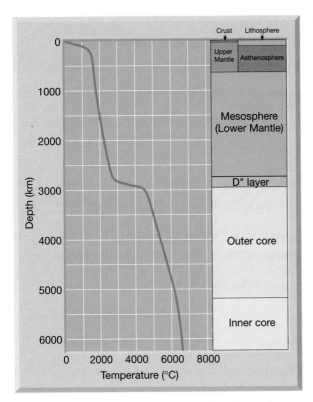

Figure 17.13 Estimated geothermal gradient for Earth. Temperatures in the mantle and core are based on several assumptions and may vary ±500°C. (Data from Kent C. Condie)

Certain regions of Earth's crust have much higher rates of heat flow than do others. In particular, along the axes of mid-ocean ridges where the crust is only a few kilometres thick, heat flow rates are relatively high. By contrast, a relatively low heat flow is observed in ancient shields (such as the Canadian and Baltic Shields). This may occur because these zones have a thick lithospheric root that effectively insulates the crust from the asthenospheric heat below. Other crustal regions exhibit a high heat flow because of shallow igneous intrusions or because of higher-than-average concentrations of radioactive materials.

Mantle Convection

Any working model of the mantle must explain the temperature distribution calculated for this layer. Whereas a large increase in temperature with depth occurs within the crust, this same trend does not continue downward through the mantle. Rather, the temperature increase with depth in the mantle is much more gradual. This means that the mantle must have an effective method of transmitting heat from the core outward. Because rocks are relatively poor conductors of heat, most researchers conclude that some form of mass transport (convection) of rock must exist within the mantle. **Convection** (*con* = with, *vect* = carried) is the transfer of heat by the mass movement or circulation in a substance. Consequently, the rock of the mantle must be capable of flow.

Convective flow in the mantle—in which warm, less dense rock rises, and cooler, denser material sinks— is the most important process operating in Earth's interior. This thermally driven flow is the force that propels the rigid lithospheric plates across the globe, and so ultimately generates Earth's mountain belts and worldwide earthquake and volcanic activity. Recall that plumes of superheated rock are thought to form near the core–mantle boundary and slowly rise toward the surface (Figure 17.14). These buoyant plumes would be the warm, upward flowing arms in the convective mechanism at work in the mantle (see Box 17.3). Downwelling is thought to occur at convergent plate boundaries where cool, dense slabs of lithosphere are being subducted (Figure 17.14). Some studies predict that this dense, cool material may eventually descend all the way to the core–mantle boundary.

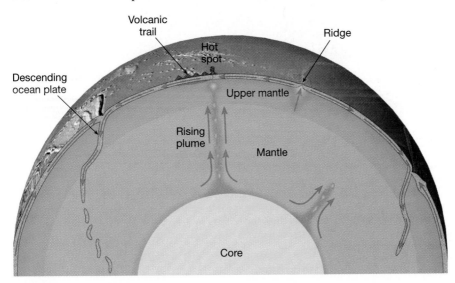

◆ **Figure 17.14** A proposed model for convective flow in the mantle. The upward arms of convective flow are concentrated mainly in mantle plumes that rise from the core–mantle boundary. The downward flow of cool material is accomplished by the descent of oceanic lithosphere.

BOX 17.3

Canadian Profile
Imaging and Understanding the Heat Engine in the Mantle

Alessandro Forte*

Efforts to understand the process of thermal convection in the mantle have been impeded by the obvious difficulty in directly "seeing" the structure of our planet deep below the surface. Rapid progress occurred in the early 1980s when seismologists developed a variety of techniques, called "seismic tomography," which provide direct images of the three-dimensional (3-D) structures deep inside Earth's mantle. Seismic tomography is based on the detailed study of waves radiated by earthquakes that are recorded by a worldwide network of seismic stations. These seismic waves travel in criss-crossing paths across the mantle, analogous with X-ray CAT scans used in medical imaging. In effect, the earthquakes behave like multiple light sources and "illuminate" deep Earth structures with the seismic waves that pass through or are reflected

by them. Laboratory experiments on rocks show that the waves generated by earthquakes will travel faster in rocks that are colder and will travel slower in rocks that are hotter. A detailed analysis of the arrival times of seismic waves which travel across multiple paths in the mantle can therefore be used to deduce the lateral variations in temperature in Earth's mantle.

A seismic tomographic reconstruction of the very large-scale structure in the bottom half of the mantle is shown in Figure 17.C. The blue-coloured regions show accumulated slabs of cold, dense oceanic plates that were subducted into the mantle below the deep oceanic trenches around the periphery of the Pacific basin. Especially large slab accumulations appear below the west Pacific margin, in a broad arc extending from northern Japan to the Indonesian archipelago, and below the east Pacific

margin extending from central North America to South America. This figure also shows the presence of mushroom-shaped blobs of presumably hotter-than-average rock, with horizontal dimensions of several thousand kilometres. These remarkable, immense structures appear to sit at the bottom of the mantle and tower upwards to heights exceeding 1000 kilometres. There are two such "mega-plumes" of hot rock, one below Africa and one on the other side of the Earth below the south-central Pacific Ocean. Recent, more detailed tomographic images clearly show the African mega-plume extending right to the top of the mantle. Very recent, computer-based numerical models of the mantle convective flow, which is predicted on the basis of the structures shown in Figure 17.C, have shown that the two hot mega-plumes rise at with speeds

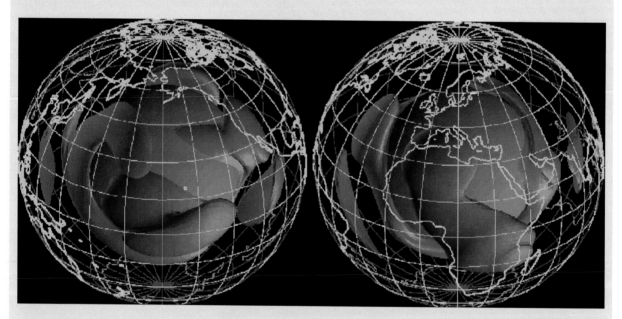

◆ **Figure 17.C**   The dominant large-scale structure in Earth's lower mantle revealed by seismic tomography (from data provided by A.M. Forte, R.L. Woodward and A.M. Dziewonski). The Earth's surface is shown transparently (except for the lines of latitude and longitude—represented by the white lines—and the continental boundaries—represented by the yellow lines). All structures in the top half of the mantle have been removed in order to visualize the deeper structures shown here. This image has been mathematically smoothed and only shows structures with horizontal length scales greater than about 4500 kilometres. The blue colours in this figure show the regions of the mantle in which earthquake waves are sped up, because of the colder temperatures in these regions. The red colours show the regions of the mantle where earthquake waves are slowed down, suggesting that these huge mushroom-shaped features or "mega-plumes" have hotter temperatures. The large grey sphere represents the top of Earth's (liquid) metallic core.
(Image courtesy of Alessandro Forte, GEOTOP – UQAM)

of about 2 to 3 centimetres per year at mid-mantle depths. This is similar to the predicted speed with which the cold slabs are sinking at these depths, and it is comparable to the average speed of horizontal drift of the continents at the surface, about 4 centimetres per year.

These numerical models of the mantle convective flow demonstrate that the internal heat engine in the deep mantle functions with four immense "pistons": two pistons are associated with the cold descending slabs (below the western and eastern Pacific margins) and the other two pistons are associated with the hot rising mega-plumes below Africa and the south-central Pacific. Earth's mantle is therefore being cooled from above, with the descent of cold slabs, and also from within, with the active upward transport of heat in the buoyant mega-plumes.

————————

*Alessandro Forte is a professor of Earth and Atmosphere Sciences at l'Université du Québec à Montréal.

If this convective mechanism exists, how does the rocky mantle transmit S waves, which can travel only through solids, and at the same time flow like a fluid? This apparent contradiction can be resolved if the mantle behaves like a solid under certain conditions and like a fluid under other conditions. Geologists generally describe material of this type as exhibiting *plastic* behaviour. When a material that exhibits plastic behaviour encounters short-lived stresses, such as those produced by seismic waves, the material behaves like an elastic solid. However, in response to stresses applied over very long time periods, this same material will flow.

This behaviour explains why S waves can penetrate the mantle, yet this rocky layer is able to flow. Plastic behaviour is not restricted to mantle rocks. Manufactured substances such as Silly Putty and some taffy candies also exhibit this behaviour. When struck with a hammer, these materials shatter like a brittle solid. However, when slowly pulled apart, they deform by flowing. From this analogy, do not get the idea that the mantle is composed of soft, puttylike material. Rather, it is composed of hot, solid rock, which, under extreme confining pressures unknown on the surface of Earth, is able to flow.

Chapter Summary

- Much of our knowledge of Earth's interior comes from the study of earthquake waves that penetrate Earth and emerge at some distant point. In general, seismic waves travel faster in solid elastic materials and slower in weaker layers. Further, seismic energy is reflected and refracted (bent) at boundaries between compositionally or mechanically different materials. By carefully measuring the travel times of seismic waves, seismologists have been able to determine the major divisions of Earth's interior.

- The principal compositional layers of Earth include (1) the *crust*, Earth's comparatively thin outer skin, which ranges in thickness from 3 kilometres at the oceanic ridges to over 70 kilometres in some mountainous belts such as the Andes and Himalayas; (2) the *mantle*, a solid rocky shell that extends to a depth of about 2900 kilometres; and (3) the *core*, an iron-rich sphere having a radius of 3486 kilometres.

- Earth's outer mechanical layer, including the uppermost mantle and crust, forms a relatively cool, rigid shell known as the *lithosphere* (sphere of rock). Averaging about 100 kilometres in thickness, the lithosphere may be 250 kilometres or more in thickness below older portions (shields) of the continents. Within the ocean basins the lithosphere ranges from a few kilometres thick along the oceanic ridges to perhaps 100 kilometres thick in regions of older and cooler crustal rocks.

- Beneath the lithosphere (to a depth of about 660 kilometres) lies a soft, relatively weak layer located in the upper mantle known as the *asthenosphere* ("weak sphere"). The upper 150 kilometres or so of the asthenosphere has a temperature/pressure regime in which a small amount of melting takes place (perhaps 1 to 5 percent). Within this very weak zone, the lithosphere is effectively detached from the asthenosphere located below.

- The crust, the rigid outermost layer of Earth, is divided into oceanic and continental crust. Oceanic crust ranges from 3 to 15 kilometres in thickness and is composed of mafic igneous rocks. By contrast, the continental crust consists of a large variety of rock types having an average composition of felsic rock called granodiorite. The rocks of the oceanic crust are younger (180 million years or less) and denser (about 3.0 g/cm^3) than continental rocks. Continental rocks have an average density of about 2.7 g/cm^3, and some have been discovered that exceed 4 billion years in age.

- Over 82 percent of Earth's volume is contained in the mantle, a rocky shell about 2900 kilometres thick. The boundary between the crust and mantle represents a change in composition. Although the mantle behaves like a solid when transmitting earthquake waves, mantle rocks are able to flow at an infinitesimally slow rate. Some of the rocks in the lowermost mantle (D″ layer) are thought to be partially molten.

- The core is composed mostly of iron, with lesser amounts of nickel and other elements. At the extreme pressure found in the core, this iron-rich material has an average density of nearly 11 g/cm^3 and approaches 14 times the density of water at Earth's centre. The inner and outer cores are compositionally similar; however, the outer core is liquid and capable of flow. It is the circulation within the core of our rotating planet that generates Earth's magnetic field.

- Temperature gradually increases with depth in our planet's interior. Three processes contribute to Earth's internal heat: (1) heat emitted by radioactivity; (2) heat released as iron solidifies in the core; and (3) heat released by colliding particles during the formative years of our planet.

- Convective flow in the mantle is thought to consist of buoyant plumes of hot rock and downward flow of cool, dense slabs of lithosphere. This thermally generated convective flow is the driving force that propels lithospheric plates across the globe.

Review Questions

1. List six major characteristics of seismic waves.
2. What are the three compositional layers of Earth?
3. List the five main layers of Earth's interior defined by differences in physical properties. How is the inner core different from the outer core?
4. Describe the lithosphere. In what important way is it different from the asthenosphere?
5. How does the boundary between the crust and mantle (Moho) differ from the boundary that occurs between the lithosphere and asthenosphere?
6. Briefly describe how the Moho was discovered.
7. What evidence did Beno Gutenberg use for the existence of Earth's central core?
8. Suppose the shadow zone for P waves was located between 120° and 160°, rather than between 105° and 140°. What would this indicate about the size of the core?
9. Describe the method first used to accurately measure the size of the inner core.
10. Which of Earth's three compositional layers is the most voluminous?
11. What is thought to cause the increase in seismic velocity that occurs at depths of 410 and 660 kilometres?
12. Where is the D″ layer located, and what role is it thought to play in transporting heat within Earth?
13. What evidence is provided by seismology to indicate that the outer core is liquid?
14. Why are meteorites considered important clues to the composition of Earth's interior?
15. Describe the chemical (mineral) makeup of the four principal compositional layers of Earth—crust (both continental and oceanic), mantle, and core.
16. List three processes that have contributed to Earth's internal heat.
17. Describe the process of conduction.
18. Briefly explain how heat is transported through the mantle.

Key Terms

asthenosphere (p. 477)
conduction (p. 483)
convection (p. 485)
core (p. 475)
crust (p. 475)
D″ layer (p. 481)

discontinuity (p. 474)
geothermal gradient
   (p. 483)
inner core (p. 477)
lithosphere (p. 476)
lower mantle (p. 477)

mantle (p. 475)
mesosphere (p. 477)
Mohorovicic discontinuity,
   or Moho (p. 477)

outer core (p. 477)
P wave shadow zone
   (p. 478)

Web Resources

 The *Earth* Web site uses the resources and flexibility of the Internet to aid in your study of the topics in this chapter. Written and developed by geology instructors, this site will help improve your understanding of geology. Visit **http://www.pearson.ca/tarbuck** and click on the cover of the text to find:

■ Online review quizzes.
■ Web-based critical thinking and writing exercises.
■ Links to chapter-specific Web resources.
■ Internet-wide key-term searches.

http://www.pearson.ca/tarbuck

Chapter 18

The Ocean Floor

20

A coral reef in the tropical Pacific Ocean.
(Photo by Steve Wolper/DRK Photos)

If all water were drained from the ocean basins, a great diversity of features would be seen, including linear chains of volcanoes, deep canyons, rift valleys, and large submarine plateaus. The scenery would be nearly as varied as that on the continents.

An understanding of the ocean floor's varied topography did not unfold until the historic three-year voyage of the H.M.S. *Challenger* (Figure 18.1). From December 1872 to May 1876, the *Challenger* expedition made the first, and still perhaps most comprehensive, study of the global ocean ever attempted by one agency. The 110,000-kilometre trip took the ship and its crew of scientists to every ocean except the Arctic. Throughout the voyage they sampled the depth of the water by laboriously lowering a weighted line overboard. Not many years later, the knowledge gained by the *Challenger* of the ocean's great depths and varied topography was further expanded with the laying of transatlantic cables. However, as long as ocean depth had to be measured with weighted line, our knowledge of the seafloor remained slight.

Mapping the Ocean Floor

In the 1920s a technological breakthrough occurred with the invention of electronic depth-sounding equipment (**echo sounder**). (The echo sounder is also referred to as *sonar*, an acronym for *so*und *n*avigation *a*nd *r*anging.) The echo sounder works by transmitting sound waves toward the ocean bottom (Figure 18.2A). A delicate receiver intercepts the echo reflected from the bottom, and a clock precisely measures the time interval to fractions of a second. By knowing the velocity of sound waves in water (about 1500 metres per second) and the time required for the energy pulse to reach the ocean floor and return, depth can be established. The depths determined from continuous monitoring of these echoes are normally plotted so that a profile of the ocean floor is obtained. By laboriously combining profiles from several adjacent traverses, a chart of the seafloor is produced. Although much more complete and detailed than anything available before, these charts only show the largest topographic features of the ocean floor (see Figure 18.7).

In the last few decades, researchers have designed even more sophisticated echo sounders to map the ocean floor. (The major impetus for these developments comes from U.S. concerns for national security.) In contrast to simple echo sounders, *multibeam sonar* employs an array of sound sources and listening devices. Thus, rather than obtaining the depth of a single point every few seconds, this technique obtains a profile of a narrow strip of seafloor (Figure 18.2B). These profiles are recorded every few seconds as the research vessel advances, building a continuous swath of relatively detailed coverage. (A ship will map a section of seafloor traveling in a back-and-forth pattern like that used when mowing a lawn.)

◆ **Figure 18.1**   The H.M.S. *Challenger*.
(From C.W. Thompson and Sir John Murray, *Report on the Scientific Results of the Voyage of the H.M.S. Challenger*, Vol. 1. Great Britain: Challenger Office, 1895, Plate 1, Library of Congress)

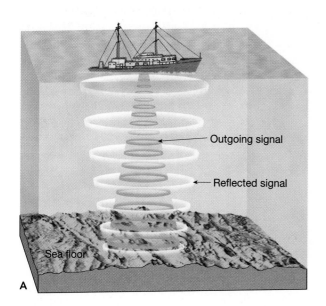
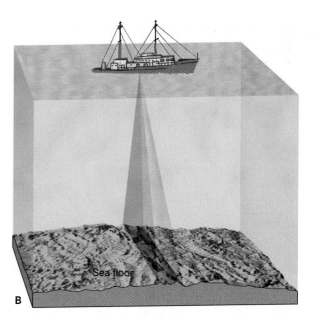

Outgoing signal

Reflected signal

Sea floor

A

Sea floor

B

◆ **Figure 18.2** Echo sounders. **A.** An echo sounder determines the water depth by measuring the time interval required for an acoustic wave to travel from a ship to the seafloor and back. The speed of sound in water is 1500 m/sec. Therefore, depth = 1/2 (1500 m/sec × echo travel time). **B.** Modern multibeam sonar obtains a profile of a narrow swath of seafloor every few seconds.

Students Sometimes Ask...

Are satellites used to map the ocean floor?

Yes. Remarkably, the technology exists to map the ocean floor from Earth-orbiting satellites in spite of their great distance above Earth and an average depth of water of nearly 4 kilometres. How does a satellite, which can only view the ocean's *surface*, obtain a picture of the ocean *floor?* The answer lies in the fact that ocean-floor features directly influence Earth's gravitational field. Deep areas such as trenches correspond to a lower gravitational attraction, and large undersea objects such as tall volcanoes exert an extra gravitational pull. These differences affect sea level, which cause ocean surfaces to bulge outward and sink inward, mimicking the relief of the ocean floor. A 2000-metre-high volcano, for example, exerts a small but measurable gravitational pull on the water around it, creating a bulge 2 metres high on the ocean surface. These irregularities are easily detectable by satellites, which use microwave beams to measure sea level to an accuracy within 4 centimetres. After corrections are made for waves, tides, currents, and atmospheric effects, the resulting pattern of lumps and bulges at the ocean surface can be used to indirectly reveal ocean-floor topography (see Figure 18.7).

Despite their greater efficiency and enhanced detail, research vessels equipped with multibeam sonar travel at a mere 10 to 20 kilometres per hour. It would take at least 100 vessels outfitted with this equipment hundreds of years to map the entire seafloor. By contrast, in 1991 and 1992 the *Magellan* spacecraft, travelling at 19,500 kilometres per hour, used radar to map more than 90 percent of the surface of Venus. Someday we hope to be able to view the ocean floor with the same detail we presently have for the Moon and some of the planets.

Oceanographers studying the topography of the ocean floor have delineated three major units: *continental margins, deep-ocean basins,* and *mid-ocean ridges.* The map in Figure 18.3 outlines these provinces for the North Atlantic, and the profile at the bottom illustrates the varied topography. Such profiles usually have their vertical dimension exaggerated many times—40 times in this case—to make topographic features more conspicuous. Because of this, the slopes shown in the seafloor profile appear to be *much* steeper than they actually are.

Continental Margins

Two main types of **continental margins** have been identified—passive and active. Passive margins are found along most of the coastal areas that surround the Atlantic Ocean, including the east coasts of North and South America, as well as the coastal areas of western Europe and Africa. Passive margins are *not* associated with plate boundaries and therefore experience

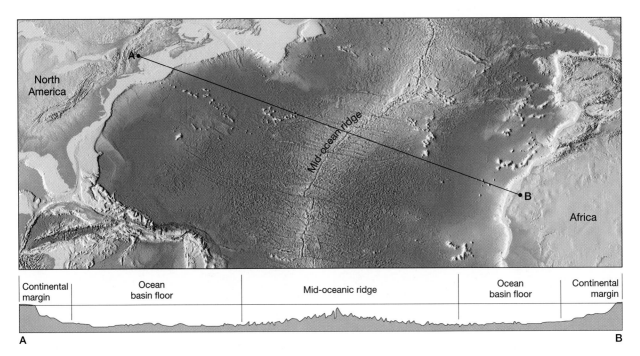

◆ **Figure 18.3**   Major topographic divisions of the North Atlantic and a profile from New England to the coast of North Africa.

very little volcanism and few earthquakes. Here, weathered materials that eroded from the adjacent landmass accumulate to form a thick, broad wedge of relatively undisturbed sediments.

By contrast, active continental margins occur where oceanic lithosphere is being subducted beneath the edge of a continent. The result is a relatively narrow margin, consisting of highly deformed sediments that were scraped from the descending lithospheric slab. Active continental margins are common around the Pacific Rim, where they parallel deep oceanic trenches.

Passive Continental Margins

The features comprising a **passive continental margin** include the continental shelf, the continental slope, and the continental rise (Figure 18.4).

Continental Shelf   The **continental shelf** is a gently sloping submerged surface extending from the shoreline toward the deep-ocean basin. Because it is underlain by continental crust, it is clearly a flooded extension of the continents. The continental shelf varies greatly in width. Almost nonexistent along some continents,

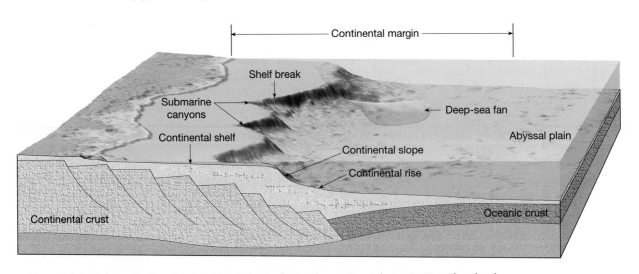

◆ **Figure 18.4**   Schematic view showing the provinces of a passive continental margin. Note that the slopes shown for the continental shelf and continental slope are greatly exaggerated. The continental shelf has an average slope of one-tenth of 1 degree, while the continental slope has an average slope of about 5 degrees.

the shelf may extend seaward as far as 1500 kilometres along others. On average, the continental shelf is about 80 kilometres wide and 130 metres deep at the seaward edge. The average inclination of the continental shelf is only about one-tenth of 1 degree, a drop of about 2 metres per kilometre. The slope is so slight that it would appear to an observer to be a horizontal surface.

Although continental shelves represent only 7.5 percent of the total ocean area, they have economic and political significance because they contain important mineral deposits, including large reservoirs of petroleum and natural gas, as well as huge sand and gravel deposits. The waters of the continental shelf also contain many important fishing grounds that are significant sources of food.

Although the continental shelf is relatively featureless, some areas are mantled by extensive glacial deposits and are thus quite rugged. The most profound features are long valleys running from the coastline into deeper waters. Many of these valleys are the seaward extensions of river valleys on the adjacent landmass. Such valleys appear to have been excavated during the Pleistocene Epoch (the last ice age).

During this time great quantities of water were tied up in vast ice sheets on the continents. This caused sea level to drop by 100 metres or more, exposing large areas of the continental shelves (see Figure 12.5 on p. 333). Because of this drop in sea level, rivers extended their courses, and land-dwelling plants and animals inhabited the newly exposed portions of the continents. Dredging off the east coast of North America has produced the remains of numerous land dwellers, including mammoths, mastodons, and horses, adding to the evidence that portions of the continental shelves were once above sea level.

Most passive continental shelves, such as those along the East Coast of Canada, consist of thick accumulations of shallow-water sediments. These sediments are frequently several kilometres thick and are interbedded with limestones that formed during earlier periods of coral reef building, a process that occurs only in shallow water. Such evidence led researchers to conclude that these thick accumulations of sediment are produced along a gradually subsiding continental margin.

Continental Slope   Marking the seaward edge of the continental shelf is the **continental slope**, a relatively steep structure (as compared with the shelf) that marks the boundary between continental crust and oceanic crust (Figure 18.4). Although the inclination of the continental slope varies greatly from place to place, it averages about 5 degrees and in places may exceed 25 degrees. Further, the continental slope is a relatively narrow feature, averaging only about 20 kilometres in width.

Continental Rise   In regions where trenches do not exist, the continental slope merges into a more gradual incline known as the **continental rise**. Here the slope drops to about one-third of a degree, or about 6 metres per kilometre. Whereas the width of the continental slope averages about 20 kilometres, the continental rise may extend for hundreds of kilometres into the deep-ocean basin.

The continental rise consists of a thick accumulation of sediment that moved downslope from the continental shelf to the deep-ocean floor. The sediments are delivered to the base of the continental slope by *turbidity currents* that follow submarine canyons (see Figure 18.6). (We will discuss these shortly.) When these muddy currents emerge from the mouth of a canyon onto the relatively flat ocean floor, they deposit sediment that forms a **deep-sea fan**. Deep-sea fans have the same basic shape as alluvial fans, which form at the foot of steep mountain slopes. As fans from adjacent submarine canyons grow, they coalesce to produce the continuous apron of sediment at the base of the continental slope that we call the continental rise.

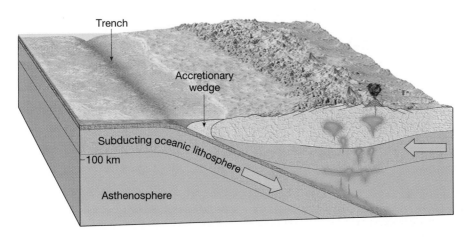

◆ **Figure 18.5**   Active continental margin. Here sediments from the ocean floor are scraped from the descending plate and added to the continental crust as an accretionary wedge.

Active Continental Margins

Along some coasts the continental slope descends abruptly into a deep-ocean trench located between the continent and ocean basin. Thus, the landward wall of the trench and the continental slope are essentially the same feature. In such locations, the continental shelf is very narrow, if it exists at all.

Active continental margins are located primarily around the Pacific Ocean in areas where the leading edge of a continent is overrunning oceanic lithosphere (Figure 18.5). Here sediments from the ocean floor and pieces of oceanic crust are scraped from the descending oceanic plate and plastered against the edge of the overriding continent. This chaotic accumulation of deformed sediment and scraps of oceanic crust is called an **accretionary** (*ad* = toward, *crescere* = grow) **wedge**. Prolonged subduction, along with the accretion of sediments on the landward side of the trench, can produce a large accumulation of sediments along a continental margin. For example, a large accretionary wedge is found along the northern coast of Japan's Honshu island.

Some subduction zones have little or no accumulation of sediments, indicating that ocean sediments are being carried into the mantle with the subducting plate. Here the continental margin is very narrow as the trench may lie a mere 50 kilometres offshore.

Submarine Canyons and Turbidity Currents

Deep, steep-sided valleys known as **submarine canyons** are cut into the continental slope and may extend across the entire continental rise to the deep-ocean basin (Figure 18.6). Although some of the canyons appear to be the seaward extensions of river valleys such as the Amazon River, others are not directly associated with existing river systems. Furthermore, because these canyons extend to depths far below the lowest level of the sea during the Ice Age, we cannot attribute their formation to stream erosion. These features must be created by some process that operates far below the ocean surface.

Most available information favours the view that submarine canyons have been eroded, at least in part, by turbidity currents (Figure 18.6). **Turbidity currents** are downslope movements of dense, sediment-laden water. They are created when sand and mud on the continental shelf and slope are dislodged, perhaps by an earthquake, and are thrown into suspension. Because such mud-choked water is denser than normal seawater, it flows downslope, eroding and accumulating more sediment. The erosional work repeatedly carried on by these muddy torrents is thought to be the major force in the growth of most submarine canyons.

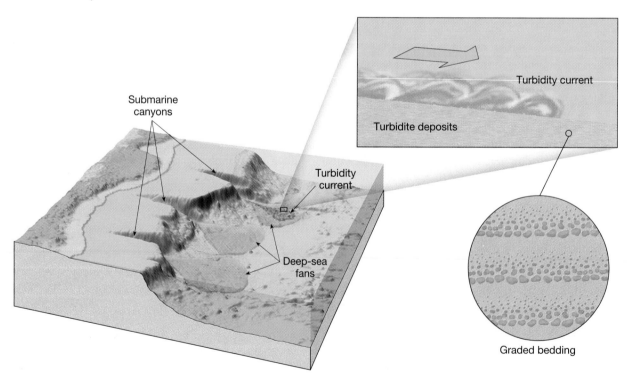

◆ **Figure 18.6**   Turbidity currents move downslope, eroding the continental margin to enlarge submarine canyons. These sediment-laden high-density currents eventually lose momentum and deposit their loads of sediment as deep-sea fans. Beds deposited by these currents are called turbidites. Each event produces a single bed characterized by a decrease in sediment size from bottom to top, a feature known as a graded bed.

Canadian Profile
Grand Banks Earthquake and Turbidity Current

BOX 18.1

On November 18, 1929, an earthquake measuring 7.2 on the Richter Scale occurred at about 5:00 p.m. on the Laurentian Slope near the eastern edge of the Grand Banks. The quake was felt throughout the Maritimes but produced only minor damage. At 7:30 p.m., however, a tsunami hit the Maritimes, causing great devastation, particularly in the Burin Peninsula of Newfoundland, where 27 people drowned and boats and buildings were destroyed.

The earthquake occurred at a time when long-distance communication was limited to telegraph, a system that required extensive cable systems. Based on the timing of telegraph communication breaks, it was realized that a number of cables were severed at the precise time of the earthquake, near the epicentre. More difficult to explain was the sequence of cable breaks east of the epicentre (Figure 18.A); the last, and most easterly cable, some 600 kilometres away, was cut 13 hours after the earthquake. It was not until the 1950s, when a lobe-shaped deposit (see Figure 18.A) of graded gravel and sand up to 1 metre thick was discovered on the deep ocean floor, that the reason for the successive cable breaks was realized. It was eventually deduced that the instantaneous cable breaks occurred due to slumping near the epicentre, and that the later successive cable breaks resulted from the downslope movement of a turbidity current that deposited the sand.

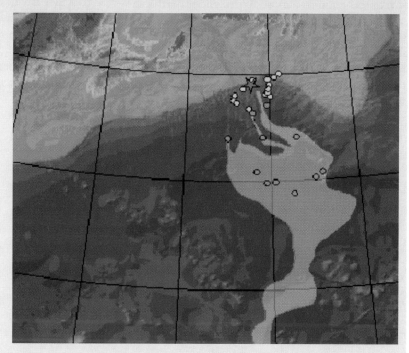

◆ **Figure 18.A**   The distribution of the turbidite (light purple) deposited after the Grand Banks Earthquake of 1929. The orange star indicates the location of the earthquake epicentre. Yellow dots mark locations of telegraph cable breaks that occurred during the earthquake, whereas orange dots indicate locations of telegraph cable breaks recorded after the earthquake.
(Image from *The Last Billion Years*, Atlantic Geoscience Society.)

Timing of the cable breaks in the deep sea provided the first indication of how fast turbidity currents can flow. It is estimated that the Grand Banks turbidity current reached speeds of up to 20 metres per second on the steepest part of the continental slope, slowing to 11 kilometres per second on the abyssal plain. The total area of deposition from this single event is estimated to be up to 250,000 square kilometres.

Narrow continental margins, such as the one located along the California coast, are dissected by numerous submarine canyons. Here, headward erosion has extended many of these canyons landward into shallow water where longshore currents are active. As a result, sediments carried to the coasts by rivers are transported along the shore until they reach a submarine canyon. This steady supply of sediment collects until it becomes unstable and moves as a massive landslide (turbidity current) to the deep-ocean floor.

Turbidity currents eventually lose momentum and come to rest along the floor of the ocean basin (Figure 18.6). As these currents slow, the suspended sediments begin to settle out. First, the coarser, heavier sand is dropped, followed by successively finer deposits of silt and then clay. Consequently, these deposits, called **turbidites**, are characterized by a decrease in sediment grain size from bottom to top, producing **graded bedding** (Figure 18.6).

Although there is still more to be learned about the complex workings of turbidity currents, it has been well established that they are an important mechanism of sediment transport in the ocean (Box 18.1). By the action of turbidity currents, submarine canyons are created and sediments are carried to the deep-ocean floor.

Features of the Deep-Ocean Basin

Between the continental margin and the oceanic ridge system lies the **deep-ocean basin** (Box 18.2). The size of this region—almost 30 percent of Earth's surface—is roughly comparable to the percentage of the surface that projects above sea level as land. Here we find remarkably flat regions known as abyssal plains, broad volcanic peaks called seamounts, and deep-ocean trenches, which are extremely deep linear depressions in the ocean floor.

Deep-Ocean Trenches

Deep-ocean **trenches** are long, relatively narrow features that form the deepest parts of the ocean. Most trenches are located in the Pacific Ocean where some approach or exceed 10,000 metres in depth, and at least a portion of one, the Challenger Deep in the Mariana trench, is more than 11,000 metres below sea level (Figure 18.7). Table 18.1 presents dimensional characteristics of some of the larger trenches.

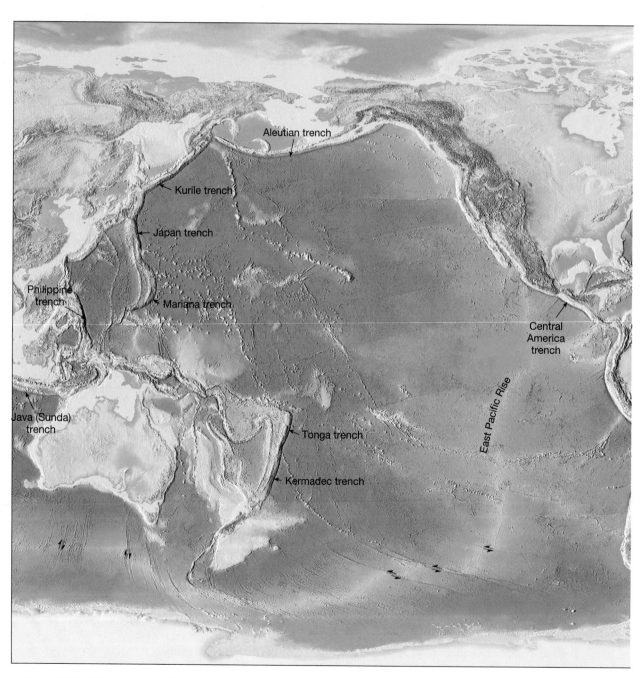

◆ **Figure 18.7**   The topography of Earth's solid surface is shown on these two pages.

Although deep-ocean trenches represent only a very small portion of the area of the ocean floor, they are nevertheless significant geologic features. Trenches are the sites where moving lithospheric plates plunge back into the mantle. In addition to the earthquakes created as one plate descends beneath another, volcanic activity is also associated with these regions. Thus, trenches are often paralleled by volcanic island arcs. Furthermore, volcanic mountains, such as those making up portions of the Andes, are located parallel to trenches that lie adjacent to continental margins. The release of volatiles (water) from a descending plate triggers melting in the wedge of asthenosphere above it. This buoyant material slowly migrates upward and gives rise to volcanic activity at the surface.

Abyssal Plains

Abyssal (*a* = without, *byssus* = bottom) **plains** are incredibly flat features; in fact, these regions are

BOX 18.2

Understanding Earth
Sampling the Ocean Floor

The *JOIDES Resolution* is the drilling ship of the Ocean Drilling Program. The "*JOIDES*" in the ship's name stands for Joint Oceanographic Institutions for Deep Earth Sampling and reflects the international commitment from the program's 19 member countries. The "*Resolution*" honours the H.M.S. *Resolution*, commanded more than 200 years ago by the well-known explorer Captain James Cook.

During cruises, holes are drilled deep into the seafloor. The cores of sediment and rock that are recovered represent millions of years of Earth history and are used by scientists to study many aspects of Earth science, including changes in global climate.

The ship is 143 metres long and 21 metres wide with a derrick that towers 62 metres above the waterline. A computer-controlled positioning system maintains the ship over a specific location. The vessel can drill in water depths up to 8200 metres and can deploy as much as 9100 metres of drill pipe (Figure 18.B).

The *JOIDES Resolution* has living quarters for 50 scientists and technicians, plus a crew of 65. The 12 onboard laboratories are equipped with the largest and most varied array of sea-going research equipment in the world.

Since 1985, the ship has drilled in the Atlantic, Pacific, Indian, and Arctic oceans as well as several seas. The result has been the recovery of more than 168,000 metres of core samples.

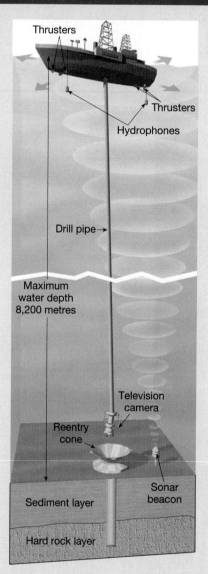

◆ **Figure 18.B** A *JOIDES Resolution* drill can re-enter holes in the seafloor years after initial drilling. The ship's dynamic positioning system consists of powerful thrusters (small propellers) that allow it to remain in place. Previous drill sites are located by bouncing sound waves between the ship's hydrophones and sonar beacons. A remote television camera aids in positioning the drill pipe into the re-entry cone.

likely the most level places on Earth. The abyssal plain found off the coast of Argentina, for example, has less than 3 metres of relief over a distance exceeding 1300 kilometres (Figure 18.7, dark area off southeast coast of South America). The monotonous topography of abyssal plains will occasionally be interrupted by the protruding summit of a partially buried volcanic structure.

Using seismic profilers, instruments whose signals penetrate far below the ocean floor, researchers have determined that abyssal plains owe their relatively featureless topography to thick accumulations of sediment that have buried an otherwise rugged ocean floor. The nature of the sediment indicates that these plains consist primarily of sediments transported far out to sea by turbidity currents.

Abyssal plains are found in all the oceans. However, because the Atlantic Ocean has fewer trenches to act as traps for the sediments carried down the continental slope, it has more extensive abyssal plains than does the Pacific.

TABLE 18.1 Dimensions of Some Deep-Ocean Trenches

Trench	Depth (kilometres)	Average Width (kilometres)	Length (kilometres)
Aleutian	7.7	50	3700
Japan	8.4	100	800
Java	7.5	80	4500
Kurile–Kamchatka	10.5	120	2200
Mariana	11.0	70	2550
Central America	6.7	40	2800
Peru–Chile	8.1	100	5900
Philippine	10.5	60	1400
Puerto Rico	8.4	120	1550
South Sandwich	8.4	90	1450
Tonga	10.8	55	1400

Students Sometimes Ask...

Have humans ever visited the deepest ocean trenches? Could anything live there?

Humans have indeed visited the deepest part of the oceans—where there is crushing high pressure, lack of sunlight, and near-freezing water temperatures—over 40 years ago! In January 1960, U.S. Navy Lt. Don Walsh and explorer Jacques Piccard descended to the bottom of the Challenger Deep region of the Mariana Trench—the world's deepest trench—in the deep diving bathyscaphe (*bathos* = depth, *scaphe* = a small ship) *Trieste*. At 9906 metres, the men heard a loud cracking sound that shook the cabin. They were unable to see that the 7.6-centimetre Plexiglas viewing port on the entranceway had cracked (miraculously, it held for the rest of the dive). Over five hours after leaving the surface, they reached the bottom at 10,912 metres—a record depth of human descent that has not been broken since. Not surprisingly, they did see some life forms there: a small flatfish from the sole family, a shrimp, and some jellyfish, all of which are adapted to life in the deep.

Seamounts

Dotting the ocean floors are isolated volcanic peaks called **seamounts**. These features may rise hundreds of metres above the surrounding topography. Although these broad conical peaks have been discovered in all the oceans, the greatest number have been identified in the Pacific.

Many of these undersea volcanoes form near oceanic ridges, regions of seafloor spreading. If a volcano grows rapidly, it may emerge as an island. Examples of volcanic islands in the Atlantic include the Azores, Ascension, Tristan da Cunha, and St. Helena.

While they exist as islands, some of these volcanoes are eroded to near sea level by running water and wave action. Over a span of millions of years the islands gradually sink as the moving plate slowly carries them from the oceanic ridge area. These submerged flat-topped seamounts are called **tablemounts** or **guyots**. In other instances, guyots may be remnants of eroded volcanic islands that were formed away from the ridge crest, possibly by hot-spot activity. Here subsidence occurs after the volcanic activity ceases and the seafloor cools and contracts.

Coral Reefs and Atolls

Coral reefs are among the most picturesque features found in the ocean (see chapter-opening photo). They are constructed primarily from the calcareous (calcite-rich) skeletal remains and secretions of corals and certain algae. The term *coral reef* is somewhat misleading in that it makes no mention of the skeletons of many small animals and plants found inside the branching framework built by the corals, nor does it reveal that secretions of algae help bind the entire structure together.

Coral reefs are confined largely to the warm, clear waters of the Pacific and Indian oceans, although a few occur elsewhere. Reef-building corals grow best in waters with an average annual temperature of about 24°C. They can survive neither sudden temperature changes nor prolonged exposure to temperatures below 18°C. In addition, these reef-builders require clear sunlit water. Consequently, the limiting depth of active reef growth is only about 45 metres. Clear blue waters such as those in the Bahamas support active reef building.

---

*The term guyot is named after Princeton University's first geology professor. It is pronounced "GEE-oh" with a hard *g* as in "give."

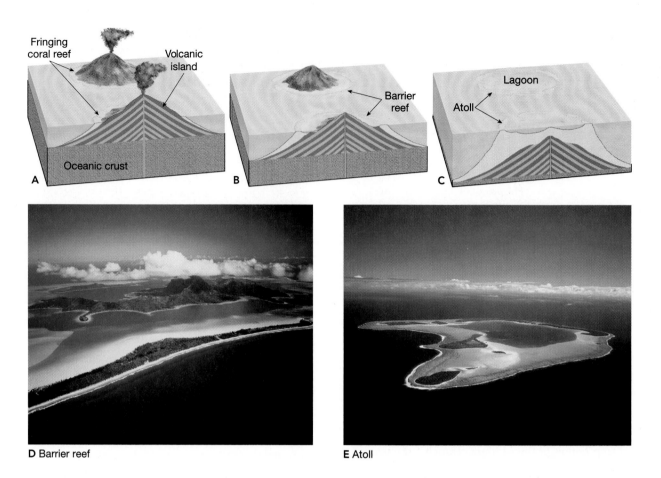

D Barrier reef   **E** Atoll

◆ **Figure 18.8**   Formation of a coral atoll due to the gradual sinking of oceanic crust and upward growth of the coral reef. **A.** Fringing reef forms around volcanic island. **B.** As the volcanic island sinks, the fringing reef gradually becomes a barrier reef. **C.** Eventually, the volcano is completely submerged and an atoll remains. **D.** Aerial view of Bora Bora with surrounding barrier reef. **E.** An aerial view of Tetiaroa Atoll in the Pacific. The light blue waters of the relatively shallow lagoon contrast with the dark blue colour of the deep ocean surrounding the atoll. (Photo D by Tom Till/DRK Photo; Photo E by Douglas Peebles Photography)

In 1831 the naturalist Charles Darwin set out aboard the British ship H.M.S. *Beagle* on its famous five-year expedition that circumnavigated the globe. One outcome of Darwin's studies was the development of an hypothesis on the formation of coral islands, called **atolls**. As Figure 18.8E illustrates, atolls consist of a nearly continuous ring of coral reef surrounding a central lagoon. Darwin's hypothesis explained what seemed to be a paradox; that is, how can corals, which require warm, shallow, sunlit water no deeper than a few dozen metres to live, create structures that reach thousands of metres to the floor of the ocean? Commenting on this in his book *The Voyage of the Beagle*, Darwin observed:

> ... from the fact of the reef-building corals not living at great depths, it is absolutely certain that throughout these vast areas, wherever there is now an atoll, a foundation must have originally existed within a depth of from 20 to 30 fathoms from the surface.*

The essence of Darwin's hypothesis was that coral reefs form on the flanks of sinking volcanic islands. As the island slowly sinks, the corals continue to build the reef complex upward (Figure 18.8).

As Darwin further noted:

> For as mountain after mountain, and island after island slowly sank beneath the water, fresh bases would be successively afforded for the growth of the corals.

Thus, atolls, like guyots, are thought to owe their existence to the gradual sinking of oceanic crust. In succeeding years there were numerous challenges to

---

*One fathom (*fathme* = out stretched arms) equals 1.8 metres. This term is derived from how depth-sounding lines were brought back on board a vessel—by hauling in the line and counting the number of arm-lengths collected.

Darwin's proposal. These arguments were not completely put to rest until after the Second World War when the United States made extensive studies of two atolls (Eniwetok and Bikini), which became sites for testing atomic bombs. Drilling operations at these atolls revealed that volcanic rock did indeed underlie the thick coral reef structure. This finding was a striking confirmation of Darwin's explanation.

Seafloor Sediments

Except for a few areas, such as near the crests of mid-ocean ridges, the ocean floor is mantled with sediment. Part of this material has been deposited by turbidity currents, and the rest has slowly settled to the bottom from above. The thickness of this carpet of debris varies greatly. In some trenches, which act as traps for sediments originating on the continental margin, accumulations may exceed 9 kilometres. In general, however, sediment accumulations are considerably less. In the Pacific Ocean, uncompacted sediment measures about 600 metres or less, while on the floor of the Atlantic, the thickness varies from 500 to 1000 metres.

Although deposits of sand-size particles are found on the deep-ocean floor, mud is the most common sediment covering this region. Muds also predominate on the continental shelf and slopes; however, the sediments in these areas are coarser overall because of greater quantities of sand.

Sampling has revealed that sand deposits are most prevalent as beach deposits along the shore. In some cases, though, coarse sediment, which we normally expect to be deposited near the shore, occurs in irregular patches at greater depths near the seaward limits of the continental shelves. Some sand may have been deposited by strong localized currents capable of moving coarse sediment far from shore, but much of it appears to be the result of sand deposition on ancient beaches. Such beaches formed during the Ice Age, when sea level was much lower than today.

Types of Seafloor Sediments

Seafloor sediments can be classified according to their origin into three broad categories: (1) **terrigenous** (*terra* = the earth, *generare* = to produce); (2) **biogenous** (*bio* = life, *generare* = to produce); and (3) **hydrogenous** (*hydro* = water, *generare* = to produce). Although each category is discussed separately, remember that all seafloor sediments are mixtures. No body of sediment comes from a single source.

Terrigenous Sediments   Terrigenous sediment consists primarily of mineral grains that were weathered from continental rocks and transported to the ocean.

The sand-size particles settle near shore. However, because the very smallest particles take years to settle to the ocean floor, they may be carried for thousands of kilometres by ocean currents. As a consequence, virtually every part of the ocean receives some terrigenous sediment. The rate at which this sediment accumulates on the deep-ocean floor, though, is very slow. From 5000 to 50,000 years are necessary for a 1-centimetre layer to form. Conversely, on the continental margins near the mouths of large rivers, terrigenous sediment accumulates rapidly. In the Gulf of Mexico, for example, the sediment carried by the Mississippi River has reached a depth of many kilometres.

Because fine particles remain suspended in the water for a very long time, ample opportunity exists for chemical reactions to occur. Because of this, the colours of the deep-sea sediments are often red or brown. This results when iron in the particle or in the water reacts with dissolved oxygen in the water and produces a coating of iron oxide (rust).

Biogenous Sediments   Biogenous sediment consists of shells and skeletons of marine animals and plants (Figure 18.9). This debris is produced mostly by microscopic organisms living in the sunlit waters near the ocean surface. Once these organisms die, their remains constantly "rain" down on the seafloor.

The most common biogenous sediments are known as *calcareous* ($CaCO_3$) *oozes*, and as their name implies, they have the consistency of thick mud. These sediments are produced by microscopic organisms that inhabit warm surface waters. When calcareous hard parts slowly sink through a cool layer of water, they begin to dissolve. This results because cold seawater is rich in carbon dioxide and is thus more acidic than warm water. In seawater deeper than about 4500 metres, calcareous shells will completely dissolve before they reach the bottom. Consequently, calcareous ooze does not accumulate in the deep-ocean basins.

Other biogenous sediments include *siliceous* (SiO_2) *oozes* and phosphate-rich materials. The former are composed primarily of opaline skeletons of diatoms (single-celled algae-like plants) and radiolarians (single-celled animals) (Figure 18.9), whereas the latter are derived from the bones, teeth, and scales of fish and other marine organisms.

Hydrogenous Sediments   Hydrogenous sediment consists of minerals that crystallize directly from seawater through various chemical reactions. For example, some limestones are formed when calcium carbonate precipitates directly from the water; however, most limestone is composed of biogenous sediment.

Evaporation is often the mechanism triggering deposition of hydrogenous sediments. Minerals commonly precipitated in this fashion include halite (sodium

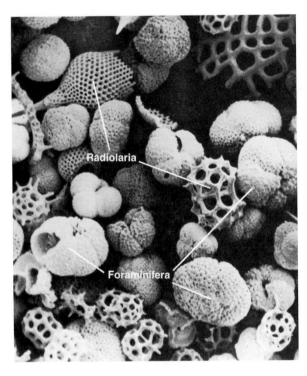

◆ **Figure 18.9** Microscopic hard parts of radiolarians and foraminifera are examples of biogenous sediments. This photomicrograph has been enlarged hundreds of times. (Photo courtesy of Deep Sea Drilling Project, Scripps Institution of Oceanography, University of California, San Diego)

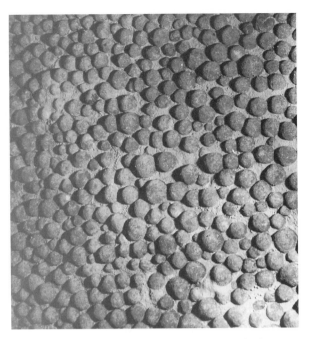

◆ **Figure 18.10**   Manganese nodules photographed at a depth of 5323 metres beneath the *Robert Conrad* south of Tahiti.
(Photo courtesy of Lawrence Sullivan, Lamont-Doherty Earth Observatory/Columbia University)

chloride), the chief component of rock salt; gypsum (hydrous calcium sulphate); and the mineral sylvite, which is used as a source of potassium for fertilizer.

Among the more interesting sediments on the ocean floor in terms of economic potential are **manganese nodules** (Figure 18.10). These rounded blackish lumps are composed of a complex mixture of minerals that form very slowly on the floors of the ocean basins. *Sulphide deposits* are another potential source of deep-ocean mineral resources. The richest submarine metallic sulfide deposits yet known have been discovered in a series of deep basins along the axis of the Red Sea. The largest deposit ranks with the largest deposits on land. It may contain 100 million metric tonnes of potential ore with 29 percent iron, over 3 percent zinc, 1 percent copper, and substantial gold and silver.

Mid-Ocean Ridges

Our knowledge of **mid-ocean ridges** comes from soundings taken of the ocean floor, core samples obtained from deep-sea drilling, visual inspection using deep-diving submersibles, and even firsthand inspection of slices of ocean floor that have been shoved up onto dry land (Figure 18.11). The ocean-ridge system is characterized by an elevated position, extensive faulting, and numerous volcanic structures that have developed on the newly formed crust (Figure 18.11).

The interconnected ocean-ridge system is the longest topographic feature on Earth's surface, exceeding 70,000 kilometres in length. Representing 20 percent of Earth's surface, the ocean-ridge system winds through all major oceans in a manner similar to the seam on a baseball (see Figure 18.7). The term *ridge* may be misleading, as these features are not narrow but have widths of from 1000 to 4000 kilometres and, in places, may occupy as much as half the total area of the ocean floor. An examination of Figure 18.7 shows that the ridge system is broken into segments that are offset by large transform faults. Further, along the axis of some segments are deep down-faulted structures called **rift valleys**.

Although ocean ridges stand 2 to 3 kilometres above the adjacent deep-ocean basins, they are much different from mountains found on the continents. Rather than consisting of thick sequences of folded and faulted sedimentary rocks, ocean ridges consist of layer upon layer of basaltic rocks that have been faulted and uplifted.

The flanks of most ridges are topographically relatively featureless and rise very gradually (slope less than 1 degree) toward the ridge crest. Approaching the ridge crest, the topography begins to exhibit greater

◆ **Figure 18.11** The deep-diving submersible *Alvin* is 7.6 metres long, weighs 16 tonnes, has a cruising speed of 1 knot, and can reach depths as great as 4000 metres. A pilot and two scientific observers are along during a normal 6- to 10-hour dive. (Photo courtesy of Rod Catanachy/ Woods Hole Oceanographic Institute)

Students Sometimes Ask...

Are diatoms an ingredient in diatomaceous earth, which is used in swimming-pool filters?

Not only are diatoms used as swimming-pool filters, they are also used in a variety of everyday products, including toothpaste (yes, you're brushing your teeth with the remains of dead microscopic organisms!). Diatoms secrete walls of silica in a great variety of forms that accumulate as sediments in enormous quantities. Because it is lightweight, chemically stable, has high surface area, and is highly absorbent, diatomaceous earth has many practical uses. The main uses of diatoms include: filters (for refining sugar, straining yeast from beer, and filtering swimming-pool water); mild abrasives (in household cleaning and polishing compounds and facial scrubs); and absorbents (for chemical spills and as pest control). Other products from diatoms include optical-quality glass (because of the pure silica content of diatoms), space-shuttle tiles (because they are lightweight and provide good insulation), an additive in concrete, and a filler and anti-caking agent (they're in automobile tires and paints).

local relief as volcanic structures and faulted valleys become more prominent. The most rugged topography is found on those ridges that exhibit large rift valleys.

Partly because of its accessibility to both American and European scientists, the Mid-Atlantic Ridge has been studied more thoroughly than other ridge systems (see Figure 18.7). The Mid-Atlantic Ridge is a broad, submerged structure standing 2500 to 3000 metres above the adjacent ocean basin floor. In a few places, such as Iceland, the ridge has actually grown above sea

level. Throughout most of its length, however, this divergent plate boundary lies 2500 metres below sea level. Another prominent feature of the Mid-Atlantic Ridge is its deep linear rift valley extending along the ridge axis. In places this rift valley is deeper than the Grand Canyon of the Colorado River and two or three times as wide. The name *rift valley* has been applied to this feature because it is so strikingly similar to continental rift valleys such as the East African Rift Valley.

Seafloor Spreading

The concept of seafloor spreading was formulated in the early 1960s by Harry Hess of Princeton University. Later geologists were able to verify Hess's contention that seafloor spreading occurs along relatively narrow zones, called **rift zones**, located at the crests of ocean ridges (see Box 18.3). As plates move apart, magma wells up into the newly created fractures and generates new slivers of oceanic lithosphere (Figure 18.12). This continually operating process generates new lithosphere that moves away from the ridge crest in a conveyor-belt fashion.

Active rift zones are characterized by frequent but generally weak earthquakes and a rate of heat flow that is greater than most other crustal segments. Here vertical displacement of large slabs of oceanic crust caused by faulting and the growth of volcanic piles contribute to the characteristically rugged topography of the oceanic ridge system. Further, the rocks along the ridge axis appear very fresh and are nearly devoid of sediment. Away from the ridge axis the topography becomes more subdued, and the thickness of the sediments and the depth of the water increase. Gradually the ridge system grades into the flat, sediment-laden abyssal plains of the deep-ocean basin.

BOX 18.3

Canadian Profile
Deep-Sea Hydrothermal Vents in Canada's Backyard

Richard Léveillé*

Sitting in a computer-filled, darkened room, a group of geologists, biologists, and chemists peer intently at video monitors showing astonishing imagery of giant smoke-billowing, chimney-like rock formations and an abundance of bizarre animals. Is this a scene from Hollywood's latest sci-fi blockbuster? No, it's a typical scene from an actual research vessel located 250 kilometres southwest of Vancouver Island. The images are being relayed to the ship by the Canadian remotely-controlled vehicle *ROPOS* (*Remotely Operated Platform for Ocean Science*), which is busy working more than 2 kilometres below the surface along the Juan de Fuca Ridge. This seafloor mountain chain is actively being created by the rifting and pulling apart of the Pacific and Juan de Fuca tectonic plates and the production of new oceanic crust by upwelling magma.

Along ridges like the Juan de Fuca, cold seawater is entrained several hundreds of metres into the highly fractured basaltic crust, where it is then heated at depth by magmatic sources. Along the way, the heated water strips metals and elements like sulphur from the surrounding rocks. This heated fluid tends to rise upwards due to convection, following conduits and fractures that are concentrated along the ridges. When it reaches the surface of the crust, the fluid can be over 400°C, but it does not boil because of the extremely high pressures caused by the water column above the vents. When this hydrothermal fluid comes into contact with the much colder seawater, minerals rapidly precipitate and form a shimmering smoke-like cloud, which gives "*black smokers*" their name (Figure 18.C). Some minerals immediately solidify and contribute to the formation of spectacular chimney-like structures, which can be as tall as a 15-storey building, and are appropriately given names like *Godzilla* and *Inferno*. In some cases, these chimneys and related deposits contain concentrated amounts of iron, copper, zinc, lead, silver, and even gold.

◆ **Figure 18.C** A black smoker spewing hot, mineral-rich water along the East Pacific Rise. As heated solutions meet cold seawater, sulphides of copper, iron, and zinc precipitate immediately, forming mounds of minerals around these vents. (Photo by Dudley Foster, Woods Hole Oceanographic Institute)

The Juan de Fuca vents are also remarkable for the biology that they support. In these environments completely devoid of sunlight, microorganisms take advantage of the mineral-rich hydrothermal fluid to perform what is known as chemosynthesis. The microbial communities, in turn, support larger, more complex animals such as fish, crabs, worms, mussels, and clams. Some species at these vents are not found anywhere else on Earth. The most famous of these, and perhaps the most unique, is the tubeworm (Figure 18.D). With their white chitinous tubes and bright red plumes, these conspicuous creatures rely entirely on bacteria growing in their *trophosome*, an internal organ designed for harvesting bacteria. These *symbiotic* bacteria rely on the tubeworm to provide them with a suitable habitat and, in return, they provide carbon-based building blocks to the tubeworms.

Concern over the possibility of damaging these unique ecosystems by sampling activities of scientists, increasing

◆ **Figure 18.D** Tubeworms are among the organisms found in the extreme environment of mid-ocean ridges where sunlight is nonexistent. These organisms derive nourishment and energy from processes associated with hydrothermal vents. (Photo by Al Giddings/Al Giddings Images, Inc.)

ecotourism, or the potential exploitation of biological and mineral resources, has recently led the Canadian government to label part of the Juan de Fuca hydrothermal vents as Canada's first Marine Protected Area (MPA). This designation puts in place enforceable regulations to preserve and protect the area and its marine organisms, while encouraging continued scientific study of this unique and remarkable ecosystem located in Canada's backyard.

For further information visit http:// www.ropos.com; http://www.pac.

dfo-mpo.gc.ca/oceans/mpa/ Endeavour_e.htm; and http://www. er.uqam.ca/nobel/oasis/index2.html

*Richard Léveillé is a Research Scientist at l'Université du Québec à Montréal.

◆ **Figure 18.12** A photograph taken from the *Alvin* during project FAMOUS shows lava extrusions in the rift valley of the Mid-Atlantic Ridge. Large toothpastelike extrusions such as this were common features. A mechanical arm is sampling an adjacent blisterlike extrusion. (Photo courtesy of Woods Hole Oceanographic Institution)

The primary reason for the elevated position of a ridge system is the fact that newly created oceanic crust is hot, and therefore it occupies more volume than cooler rocks of the deep-ocean basin. As the young lithosphere travels away from the spreading centre, it gradually cools and contracts. This thermal contraction accounts in part for the greater ocean depths that exist away from the ridge. It takes almost 100 million years before cooling and contraction cease completely. By this time, rock that was once part of an elevated ocean-ridge system is located in the deep-ocean basin, where it is mantled by thick accumulations of sediment.

During seafloor spreading, new material is equally added to the two diverging plates; hence, we would expect new ocean floor to grow symmetrically on each side of a centrally located ridge crest. Indeed, the ridge systems of the Atlantic and Indian oceans are located near the middle of these water bodies and as a conse-

quence are named mid-ocean ridges. However, the East Pacific Rise is situated far from the centre of the Pacific Ocean. Despite uniform spreading along the East Pacific Rise, much of the Pacific Basin that once lay east of this spreading centre has been overridden by the westward migration of the American plate.

Structure of the Oceanic Crust

Although most oceanic crust forms out of view, far below sea level, geologists have been able to examine the structure of the ocean floor firsthand. In such locations as Newfoundland, Cyprus, and California, slivers of oceanic crust have been thrust high above sea level. From these outcrops, researchers conclude that the ocean floor consists of three distinct layers (Figure 18.13A). The upper layer is composed mainly of basaltic **pillow lavas**. The middle layer is made up of

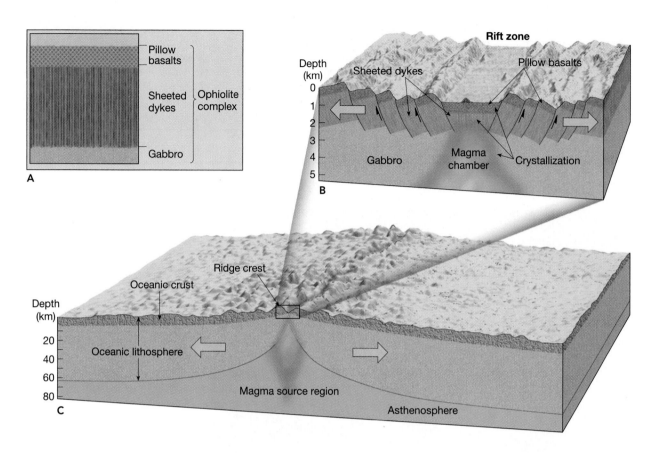

◆ **Figure 18.13   A.** The structure of oceanic crust is thought to be equivalent to the ophiolite complexes that have been discovered elevated above sea level in such places as California and Newfoundland. **B.** The formation of the three units of an ophiolite complex in the rift zone of an oceanic ridge. **C.** Diagram illustrating the site where new ocean crust is generated.

numerous interconnected dykes called **sheeted dykes**. Finally, the lower layer is made up of gabbro, the coarse-grained equivalent of basalt, that crystallized at depth. This sequence of rocks is called an **ophiolite complex** (Figure 18.13, 18.14). From studies of various ophiolite complexes and related data, geologists have pieced together a scenario for the formation of the ocean floor.

The magma that migrates upward to create new ocean floor originates from partially melted peridotite in the asthenosphere. In the region of the rift zone, this magma source may lie no more than 35 kilometres below the seafloor. Being molten and less dense than the surrounding solid rock, the magma gradually moves upward to where it enters large reservoirs located only a few kilometres below the ridge crest (Figure 18.13A). As the ocean floor moves apart, numerous fractures develop in the crust, permitting this molten rock to migrate to the surface. (Eventually magma in these vertical fractures will crystallize, generating the zone of sheeted dykes.)

During each eruptive phase, the initial lava flows are quite fluid and spread over the rift zone in broad, thin sheets. As new flows are added to the ocean floor, each is cut by fractures that allow additional lava to migrate upward and form overlying layers. Later in each eruptive cycle, as the magma in the shallow reservoir cools and thickens, shorter flows with a more characteristic pillow form occur. Recall that pillow lava has the appearance of large, elongate sand bags stacked one atop another (Figure 18.14A). Depending on the rate of flow, the thick pillow lavas may build into volcano-size mounds. These mounds will eventually be cut off from their supply of magma and be carried away from the ridge crest by seafloor spreading.

The magma that remains in the subterranean chamber will crystallize at depth to generate thick units of coarse-grained gabbro. This lowest rock unit forms as crystallization takes place along the walls and floor of the magma chamber. In this manner the processes at work along a ridge system generate the sequence of rocks found in an ophiolite complex (Figure 18.14B).

A

B

◆ **Figure 18.14** **A.** Ancient pillow lava at Trinity Bay, Newfoundland. **B.** A rare glimpse of the Moho provided by the deepest levels of an ophiolite complex exposed in Table Mountain, Gros Morne National Park, Newfoundland. The uppermost mantle and lower crust are represented by peridotite and gabbro respectively. (Photo A courtesy of the Geological Survey of Canada, photo no. 152581; Photo B by W.R. Church)

Chapter Summary

- Ocean depths are determined using an *echo sounder*, a device carried by a ship that bounces sound off the ocean floor. The time it takes for the sound waves to make the round-trip to the bottom and back to the ship is directly related to the depth. Continuous data from the echoes are plotted to produce a profile of the ocean floor. Although much of the ocean floor has been mapped using echo sounders, only the large topographic features are shown.

- Oceanographers studying the topography of the ocean basins have delineated three major units: *continental margins*, *deep-ocean basins*, and *mid-ocean ridges*.

- The zones that collectively make up a *passive continental margin* include the *continental shelf* (a gently sloping, submerged surface extending from the shoreline toward the deep-ocean basin); *continental slope* (the true edge of the continent, which has a steep slope that leads from the continental shelf into deep water); and in regions where trenches do not exist, the steep continental slope merges into a more gradual incline known as the *continental rise*. The continental rise consists of sediments that have moved downslope from the continental shelf to the deep-ocean floor.

- *Active continental margins* are located primarily around the Pacific Ocean in areas where the leading edge of a continent is overrunning oceanic lithosphere. Here sediment scraped from the descending oceanic plate is plastered against the continent to form a collection of sediments called an *accretionary wedge*. An active continental margin generally has a narrow continental shelf, which grades into a deep-ocean trench.

- *Submarine canyons* are deep steep-sided valleys that originate on the continental slope and may extend to depths of 3 kilometres. Some of these canyons appear to be the seaward extensions of river valleys. However, most information seems to favour the view that many submarine canyons are excavated by *turbidity currents* (downslope movements of dense, sediment-laden water). *Turbidites*, sediments deposited by turbidity currents, are characterized by a decrease in sediment grain size from bottom to top, a phenomenon known as *graded bedding*.

- The deep-ocean basin lies between the continental margin and the mid-ocean ridge system. Its features include *deep-ocean trenches* (long, narrow depressions that are the deepest parts of the ocean and where moving crustal plates descend back into the mantle); *abyssal plains* (among the most level places on Earth, consisting of thick accumulations of sediments that were deposited atop the low, rough portions of the ocean floor by turbidity currents); and *seamounts* (isolated, steep-sided volcanic peaks on the ocean floor that originate near oceanic ridges or in association with volcanic hot spots).

- *Coral reefs*, which are confined largely to the warm, sunlit waters of the Pacific and Indian oceans, are constructed over thousands of years primarily from the accumulation of skeletal remains and secretions of corals and certain algae. A coral island, called an *atoll*, consists of a continuous or broken ring of coral reef surrounding a central lagoon. Atolls form from corals that grow on the flanks of sinking volcanic islands, where the corals continue to build the reef complex upward as the island slowly sinks.

- There are three broad categories of seafloor sediments. *Terrigenous sediment* consists primarily of mineral grains that were weathered from continental rocks and transported to the ocean. *Biogenous sediment* consists of shells and skeletons of marine animals and plants. *Hydrogenous sediment* includes minerals that crystallize directly from seawater through various chemical reactions.

- *Mid-ocean ridges*, the sites of seafloor spreading, are found in all major oceans and represent more than 20 percent of Earth's surface. These broad features are certainly the most prominent features in the oceans, for they form an almost continuous mountain range. Ridges are characterized by an *elevated position, extensive faulting*, and *volcanic structures* that have developed on newly formed oceanic crust. Most of the geologic activity associated with ridges occurs along a narrow region on the ridge crest, called the *rift zone*, where magma from the asthenosphere moves upward to create new slivers of oceanic crust.

- New oceanic crust is formed in a continuous manner by the process of seafloor spreading. The upper crust is composed of *pillow lavas* of basaltic composition. Underlying this layer are numerous interconnected dykes (*sheeted dykes*) that are connected to a layer of gabbro. This entire sequence of rock is called an *ophiolite complex*.

Review Questions

1. Assuming that the average speed of sound waves in water is 1500 metres per second, determine the water depth if the signal sent out by an echo sounder requires 6 seconds to strike bottom and return to the recorder (see Figure 18.2).

2. List the three major features that comprise the continental margin. Which of these features is considered a flooded extension of the continent? Which has the steepest slope?

3. How does an active continental margin differ from a passive continental margin? Give examples of each type.

4. Defend or rebut the following statements: "Most submarine canyons found on the continental slope and rise were formed during the Ice Age when rivers extended their valleys seaward."

5. What are turbidites? What evidence has been used to calculate the velocity of a turbidity current?

6. Why are abyssal plains more extensive on the floor of the Atlantic than on the floor of the Pacific?

7. What is an atoll? Describe Darwin's proposal on the origin of atolls. Was it ever confirmed?

8. Distinguish among the three basic types of seafloor sediment.

9. If you were to examine recently deposited biogenous sediment taken from a depth in excess of 4500 metres, would it more likely be rich in calcareous materials or siliceous materials? Explain.

10. How are mid-ocean ridges related to seafloor spreading?

11. What is the primary reason for the elevated position of the oceanic ridge system?

Key Terms

abyssal plain (p. 499)
accretionary wedge (p. 496)
active continental margin (p. 496)
atoll (p. 502)
biogenous sediment (p. 503)
continental margin (p. 493)
continental rise (p. 495)

continental shelf (p. 494)
continental slope (p. 495)
coral reef (p. 501)
deep-ocean basin (p. 498)
deep-sea fan (p. 495)
echo sounder (p. 492)
graded bedding (p. 497)
guyot (p. 501)
hydrogenous sediment (p. 503)

manganese nodule (p. 504)
mid-ocean ridge (p. 504)
ophiolite complex (p. 508)
passive continental margin (p. 494)
pillow lavas (p. 508)
rift valley (p. 504)
rift zone (p. 505)
seamount (p. 501)

sheeted dyke (p. 508)
submarine canyon (p. 496)
tablemount (p. 501)
terrigenous sediment (p. 503)
trench (p. 498)
turbidite (p. 497)
turbidity current (p. 496)

Web Resources

The *Earth* Web site uses the resources and flexibility of the Internet to aid in your study of the topics in this chapter. Written and developed by geology instructors, this site will help improve your understanding of geology. Visit **http://www.pearson.ca/tarbuck** and click on the cover of the text to find:

■ Online review quizzes.

■ Web-based critical thinking and writing exercises.

■ Links to chapter-specific Web resources.

■ Internet-wide key-term searches.

http://www.pearson.ca/tarbuck

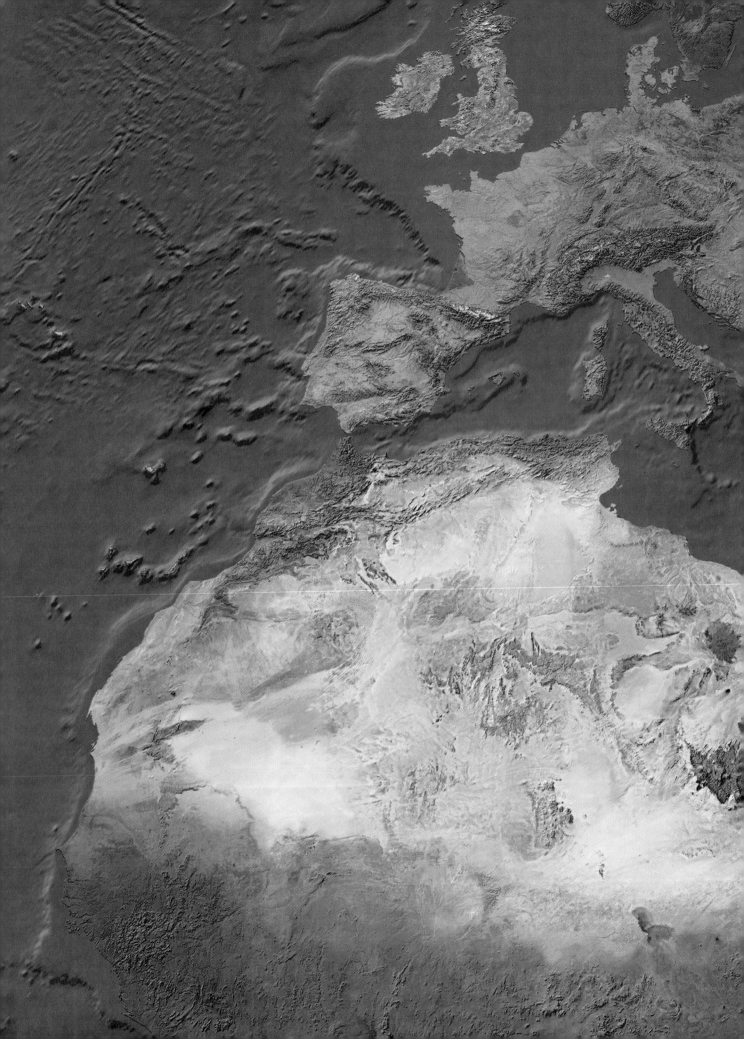

Chapter 19
Plate Tectonics

Composite satellite image of Europe, North Africa, and the Arabian Peninsula. (Image by Worldsat International, Inc.)

Early in the twentieth century, geologic thought about the age of the ocean basins was dominated by a belief in their antiquity. Moreover, most geologists accepted the geographic permanency of the oceans and continents. Mountains were thought to result from contractions caused by gradual cooling from a once-molten state. As the interior cooled and contracted, Earth's solid outer skin was deformed by folding to fit the shrinking planet. Mountains were therefore regarded as analogous to the wrinkles on a dried-out piece of fruit. This model of Earth's tectonic processes, however inadequate, was firmly entrenched in geologic thought.*

Since the 1960s, vast amounts of new data have dramatically changed our understanding of the nature and workings of our planet. Earth scientists now realize that the continents gradually migrate across the globe. Where landmasses split apart, new ocean basins are created between the diverging blocks. Meanwhile, older portions of the seafloor are being carried back down into the mantle in regions where trenches occur in the deep-ocean floor. Because of this movement, blocks of continental material eventually collide and form Earth's great mountain ranges (Figure 19.1). A revolutionary new model of Earth's tectonic processes has emerged.

This profound reversal of scientific opinion has been appropriately described as a scientific revolution. Like other scientific revolutions, considerable time elapsed between the idea's inception and its general acceptance. The revolution began early in the twentieth century as a relatively straightforward proposal that the continents drift about the face of Earth. After many years of heated debate, the idea of drifting continents was rejected by the vast majority of Earth scientists.

The concept of a mobile Earth was particularly distasteful to North American geologists, perhaps because much of the supporting evidence for movement had been gathered from the southern continents, with which most North American geologists were essentially unfamiliar. However, during the 1950s and 1960s, new evidence began to rekindle interest in this nearly abandoned proposal. By 1968, these new developments led to the unfolding of a far more encompassing theory that incorporated aspects of continental drift and seafloor spreading—a theory known as *plate tectonics*.

Tectonics refers to the deformation of Earth's crust and results in the formation of structural features such as mountains.

◆ **Figure 19.1**   Converging lithospheric plates were responsible for creating mountain ranges of the Canadian Cordillera. Shown are mountains of the Babine Range, Central British Columbia. (Photo courtesy of R.W. Hodder)

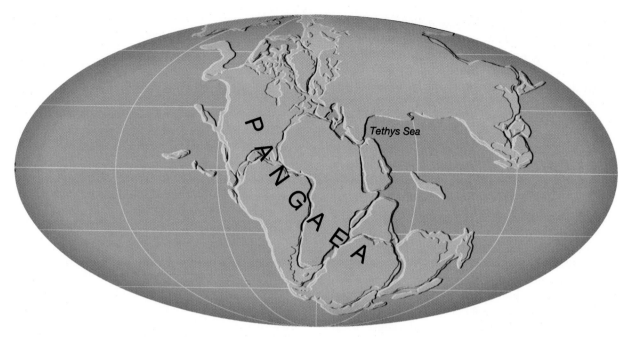

◆ **Figure 19.2**   Reconstruction of Pangaea as it is thought to have appeared 200 million years ago.

In this chapter we examine the events that led to this dramatic reversal of scientific opinion in an attempt to provide some insight into how science works. We will briefly trace the developments that took place from the inception of the concept of continental drift through to the general acceptance of the theory of plate tectonics. Evidence gathered to support the concept of a mobile Earth will also be provided.

Continental Drift: An Idea Before Its Time

The idea that continents, particularly South America and Africa, fit together like pieces of a jigsaw puzzle originated with the development of reasonably accurate world maps. However, little significance was given this idea until 1915, when Alfred Wegener, a German meteorologist and geophysicist, published *The Origin of Continents and Oceans.** In this book, Wegener set forth the basic outline of his radical hypothesis of **continental drift**.

Wegener suggested that a single *supercontinent* called **Pangaea** (*pan* = all, *gaea* = Earth) once existed (Figure 19.2). He further hypothesized that about 200 million years ago this supercontinent began breaking into smaller continents, which then "drifted" to their present positions.

---

*Wegener's ideas were actually preceded by those of an American geologist, F. B. Taylor, who in 1910 published a paper on continental drift. Taylor's paper provided little corroborating evidence for continental drift, which may have been the reason that it had a relatively small impact on the scientific community.

Students Sometimes Ask...

If all the continents were joined during the time of Pangaea, what did the rest of Earth look like?
When all the continents were together, there must also have been one huge ocean surrounding them. This ocean is called *Panthalassa* (*pan* = all, *thalassa* = sea). Panthalassa had several smaller seas, one of which was the centrally located and shallow *Tethys Sea* (see Figure 19.2). About 180 million years ago the supercontinent of Pangaea began to split apart, and the various continental masses we know today started to drift toward their present geographic positions. Today all that remains of Panthalassa is the Pacific Ocean, which has been decreasing in size since the breakup of Pangaea.

Wegener and others who advocated this hypothesis collected substantial evidence to support these claims. The fit of South America and Africa and the geographic distribution of fossils, rock structures, and ancient climates all seemed to support the idea that these now separate landmasses were once joined. Let us examine their evidence.

Fit of the Continents

Like a few others before him, Wegener first suspected that the continents might once have been joined when he noticed the remarkable similarity between the coastlines on opposite sides of the South Atlantic

(Figure 19.3). However, his use of present-day shorelines to make a fit of the continents was challenged immediately by other Earth scientists. These opponents correctly argued that shorelines are continually modified by erosional and depositional processes. Even if continental displacement had taken place, a good fit today would be unlikely. Furthermore, abundant fossil evidence indicates that most of the world's land areas have experienced periods of either uplift or subsidence in the recent geologic past. This would have markedly altered the position of the global coastlines. Wegener appeared to be aware of these problems, and in fact his original jigsaw fit of the continents was only very crude.

A much better approximation of the true outer boundary of the continents is the continental shelf. Today the seaward edge of the continental shelf lies submerged, a few hundred metres below sea level. In the early 1960s, Sir Edward Bullard and two associates produced a map that attempted to fit the edges of the continental shelves of South America and Africa at a depth of 900 metres. The remarkable fit that was obtained is shown in Figure 19.3. Although the continents overlap in a few places, these are regions where streams have deposited large quantities of sediment,

thus enlarging the continental shelves. The overall fit was even better than the supporters of continental drift suspected it would be.

Fossil Evidence

Although Wegener was intrigued by the remarkable similarities of the shorelines on opposite sides of the Atlantic, he at first thought the idea of a mobile Earth improbable. Not until he came across an article citing fossil evidence for the existence of a land bridge connecting South America and Africa did he begin to take his own idea seriously. Through a search of the literature, Wegener learned that most paleobotanists (scientists who study the fossilized remains of plants) were in agreement that some type of land connection was needed to explain the existence of identical fossils on widely separated landmasses. This requirement was particularly true for late Paleozoic and early Mesozoic life forms.

Mesosaurus   To add credibility to his argument for the existence of the supercontinent Pangaea, Wegener cited documented cases of several fossil organisms that had been found on different landmasses but which could not have crossed the vast oceans presently separating the continents. The classic example is *Mesosaurus*, a presumably aquatic, snaggle-toothed reptile whose fossil remains are limited to eastern South America and southern Africa (Figure 19.4). If *Mesosaurus* had been able to swim well enough to cross the vast South Atlantic Ocean, its remains should be more widely distributed. As this was not the case, Wegener argued that South America and Africa must have been joined somehow.

How did scientists during Wegener's era explain the discovery of identical fossil organisms in places separated by thousands of kilometres of open ocean? The idea of land bridges was the most widely accepted solution to the problem of migration (Figure 19.5). We know, for example, that during the recent Ice Age, the lowering of sea level allowed animals to cross the narrow Bering Strait between Asia and North America. Was it possible that land bridges once connected Africa and South America? We are now quite certain that land bridges of this magnitude did not exist. If they had, their remnants should still lie below sea level, but they are nowhere to be found.

Glossopteris   Wegener also cited the distribution of the fossil fern *Glossopteris* as evidence for the existence of Pangaea. This plant, identified by its large seeds that could not be blown very far, was known to be widely dispersed among Africa, Australia, India, and South America during the late Paleozoic Era. Later, fossil remains of *Glossopteris* were discovered in

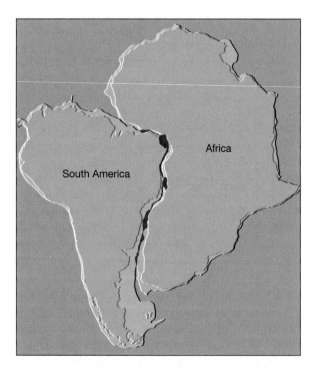

◆ **Figure 19.3**   This shows the best fit of South America and Africa along the continental slope at a depth of about 900 metres. The areas where continental blocks overlap appear in brown.
(After A. G. Smith, "Continental Drift." In *Understanding the Earth*, edited by I. G. Gass)

various stages between 200 million and 165 million years ago. This time span can be used as the birth date for this section of the North Atlantic.

Rifting that formed the South Atlantic began about 130 million years ago near the tips of what are now Africa and South America. As this rift propagated northward, it gradually opened the South Atlantic (compare Figure 19.A, parts B and C). Continued rifting of the southern landmass of Gondwanaland sent India on a northward journey. By the early Cenozoic, about 50 million years ago, Australia began to separate from Antarctica, and the South Atlantic had emerged as a full-fledged ocean (Figure 19.A, part D).

A modern map (Figure 19.A, part E) shows that India eventually collided with Asia, an event that began about 45 million years ago and created the Himalayas as well as the Tibetan Highlands. About the same time, the separation of Greenland from Eurasia completed the breakup of the northern landmass (Laurasia). Also, notice the recent formation of the Baja Peninsula along with the Gulf of California. This event began less than 10 million years ago.

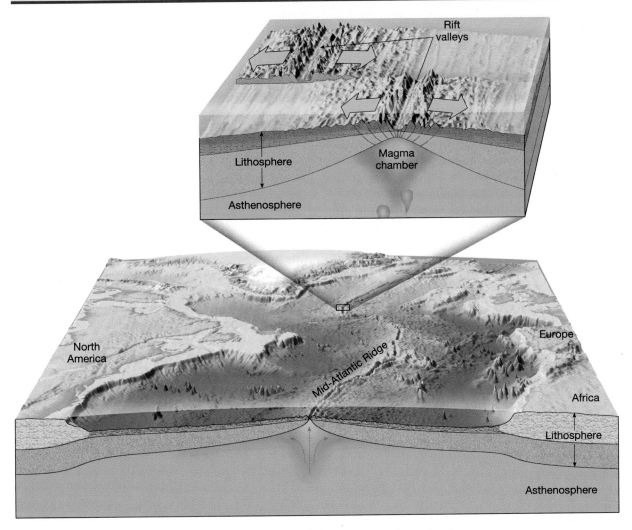

◆ **Figure 19.18**   Most divergent plate boundaries are situated along the crests of oceanic ridges.

The mechanism that operates along the oceanic ridge system to create new seafloor is appropriately called **seafloor spreading**. Typical rates of spreading average around 5 centimetres per year. Comparatively slow spreading rates of 2 centimetres per year are found in the North Atlantic, whereas spreading rates exceeding 15 centimetres have been measured along sections of the East Pacific Rise. Although these rates of lithospheric production are slow on a human time scale,

they are nevertheless rapid enough so that all of Earth's ocean basins could have been generated within the last 200 million years. In fact, none of the ocean floor that has been dated exceeds 180 million years in age.

During seafloor spreading, the magma that is injected into newly developed fractures forms dykes that tend to cool from their outer borders inward toward their centres. Because the warm interiors of these newly formed dykes are weak, continued spreading produces new fractures that split these young rocks roughly in half. As a result, new material is added about equally to the two diverging plates. Consequently, new ocean floor grows symmetrically on each side of a centrally located ridge crest. Indeed, the ridge systems of the Atlantic and Indian oceans are located near the middle of these water bodies and as a consequence are called mid-ocean ridges. However, the East Pacific Rise is situated far from the centre of the Pacific Ocean. Despite uniform spreading along the East Pacific Rise, much of the Pacific Basin that once lay east of this spreading centre has been overridden by the westward migration of the American plates.

Spreading Rates and Ridge Topography

When various segments of the oceanic ridge system were studied in detail, some topographic differences came to light. Most of these differences appear to be controlled by spreading rates. At comparatively slow spreading rates of 1 to 5 centimetres per year, such as occur at the Mid-Atlantic and Mid-Indian ridges, a prominent rift valley develops along the ridge crest. This structure is usually 30 to 50 kilometres across and from 1500 to 3000 metres deep. Here, the displacement of large slabs of oceanic crust along nearly vertical faults and outpourings of pillow lavas contribute to the characteristically rugged topography of these rift valleys.

Along the Galapagos ridge and the northernmost section of the East Pacific Rise, an intermediate spreading rate of 5 to 9 centimetres per year is the norm. In these settings the rift valleys that develop are shallow, often less than 200 metres deep, and their topography is relatively smooth.

At faster spreading rates (greater than 9 centimetres per year), such as those which occur along much of the East Pacific Rise, no median rift valleys develop. Here oceanic ridges are usually narrow (roughly 10 kilometres wide) topographic highs. These elevated structures are extensively faulted and exhibit topography consisting of numerous horsts and grabens (see Chapter 15). Faulting is the primary cause of topographic relief along fast-spreading centres; the buildup of volcanic structures is significant at slow spreading centres.

Continental Rifts

Spreading centres can also develop within a continent, in which case the landmass may split into two or more smaller segments, as Alfred Wegener had proposed for the breakup of the Pangaea. Examples of active continental rifts include the East African rift valleys, the Baikal Rift in south-central Siberia, the Rhine Valley (Northwest Europe), the Rio Grande Rift, and the Basin and Range province in western North America. Whether any of these rifts will continue to develop and eventually split a continent is a matter of much speculation.

The most widely accepted model for continental splitting suggests that extensional forces must be acting on the lithospheric plate. These forces are thought to arise from the "pull" of cold lithospheric plates as they subduct along the margins of a continent. It appears that by themselves these extensional forces are not great enough to actually tear the lithosphere apart. Rather, the rupture of the lithosphere is initiated only in those settings where plumes of hot rock rise from the mantle. The effect of this hot-spot activity is weakening of the lithosphere and doming of the crust directly above the hot rising plume. Uplifting stretches and thins the crust, as shown in Figure 19.19A. Extension is accompanied by alternating episodes of faulting and volcanism that result in the development of a rift valley that resembles those found along the axes of some ridges (Figure 19.19B).

The East African rift valleys may represent the initial stage in the breakup of a continent as described above (Figure 19.20). Large volcanic mountains such as Kilimanjaro and Mount Kenya exemplify the extensive volcanic activity that accompanies continental rifting.

In those settings where the extensional forces are maintained, the rift valley will lengthen and deepen, eventually extending out to the margin of the plate, and split it in two (Figure 19.19C). At this point the rift becomes a narrow linear sea with an outlet to the ocean, similar to the Red Sea (Figure 19.20). The Red Sea formed when the Arabian Peninsula rifted from Africa, an event that began about 20 million years ago. Consequently, the Red Sea provides oceanographers with a view of how the Atlantic Ocean may have looked in its infancy.

Not all rift valleys develop into full-fledged spreading centres. Running through central North America is a failed rift extending from Lake Superior to Oklahoma. This once active rift valley is filled with volcanic rock that was extruded onto the crust more than a billion years ago.

Why one rift valley develops into an active spreading centre while others are abandoned may be related to the degree in which rift systems are connected. As found in the Afar region of East Africa (see Figure 19.20), lithospheric doming tends to produce three rift arms that diverge from a central area called a *triple junction*, in this case located in the Afar lowlands. Where two of the three rift arms in a particular triple junction system meet rift arms of other triple junction systems, they can form a composite rift zone, along which spreading can continue to produce a linear sea, and an ocean basin. This idea appears to hold for the opening of the Atlantic Ocean; many inactive rift zones

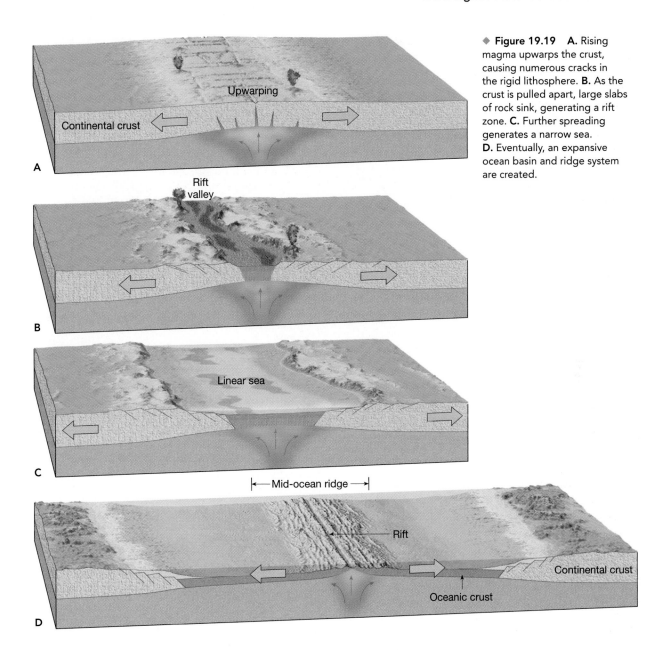

◆ **Figure 19.19** **A.** Rising magma upwarps the crust, causing numerous cracks in the rigid lithosphere. **B.** As the crust is pulled apart, large slabs of rock sink, generating a rift zone. **C.** Further spreading generates a narrow sea. **D.** Eventually, an expansive ocean basin and ridge system are created.

occur on the continental margins bordering the ocean, and probably represent the orphaned arms of triple junction systems. The arms that do not connect with others eventually stop spreading, and therefore become failed rifts that simply become filled with sediment. In eastern Africa, it is probable that spreading of the Red Sea and Gulf of Aden will form a new ocean basin, and that the East African rift will become a failed rift.

Convergent Plate Boundaries

Internal Processes
↳ Plate Tectonics

Although new lithosphere is constantly being added at the oceanic ridges, our planet is not growing larger—

its total surface area remains constant. To accommodate the newly created lithosphere, older portions of oceanic plates return to the mantle along **convergent** (*con* = together, *vergere* = to move) **plate boundaries**. (Because lithosphere is "destroyed" at convergent boundaries, they are also called *destructive plate margins*.) As two plates slowly converge, the leading edge of one is bent downward, allowing it to slide beneath the other. The surface expression produced by the descending plate is an ocean **trench**, like the Peru–Chile trench (Figure 19.21A). Trenches formed in this manner may be thousands of kilometres long, 8 to 12 kilometres deep, and between 50 and 100 kilometres wide (Figure 19.22). Destructive plate margins where oceanic crust is being consumed in the mantle are called **subduction** (*sub* = under, *duct* = lead) **zones**.

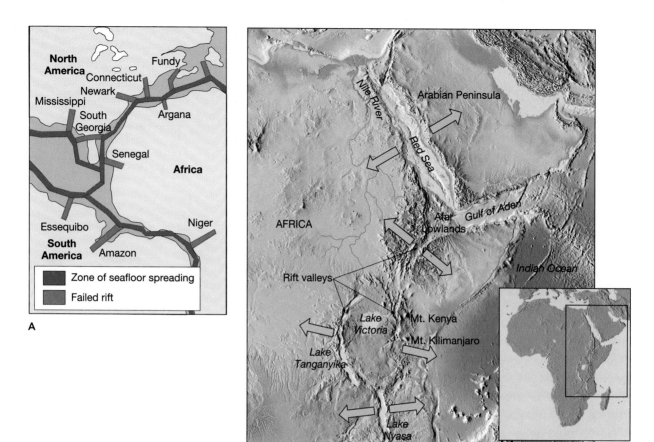

◆ **Figure 19.20   A.** Cracks produced by mantle plume–related doming of the lithosphere tend to produce rift zones that meet at triple junctions. Where two of the three rift arms in a particular triple junction zone meet rift arms of other triple junction systems, they form a composite rift zone, along which spreading can continue and ultimately form an ocean basin. The arms that do not connect with others eventually stop spreading, and therefore become failed rifts that simply become filled with sediment. Some of the major river systems follow the trends of ancient failed rifts. This diagram shows the rift arms that developed just prior to the opening of the Atlantic Ocean. **B.** East African rift valleys and associated features.

The average angle at which oceanic lithosphere descends into the mantle is about 45 degrees. However, depending on its buoyancy, a plate may descend at an angle as small as a few degrees or it may plunge vertically (90 degrees) into the mantle. When a spreading centre is located near a subduction zone, the lithosphere is young and, therefore, warm and buoyant. Therefore, the angle of descent is small. This is the situation along parts of the Peru–Chile trench. Low dip angles are usually associated with a strong coupling between the descending slab and the overriding plate. Consequently, these regions experience great earthquakes. By contrast, some subduction zones, such as the Mariana trench, have steep angles of descent and few strong earthquakes.

Although all convergent zones have the same basic characteristics, they are highly variable features. Each is controlled by the type of crustal material involved and the tectonic setting. Convergent boundaries can form between two oceanic plates, one oceanic and one continental plate, or two continental plates. All three situations are illustrated in Figure 19.21.

Oceanic–Continental Convergence

Whenever the leading edge of a continental plate converges with an oceanic plate, the buoyant continental plate remains "floating," while the denser oceanic slab sinks into the asthenosphere (Figure 19.21A). As the oceanic slab descends, some of the sediments carried on the subducting plate, as well as pieces of oceanic crust, are scraped off and plastered against the edge of the overriding continental block. This chaotic accumulation of deformed sediment and scraps of oceanic crust is called an **accretionary** (*ad* = toward, *crescere* = to grow) **wedge**. Studies conducted in the coastal regions of western Mexico, where the Cocos plate is being subducted, indicate that at least half of the sediment carried on the descending plate can be removed in this manner. Therefore, this process contributes to the already substantial accumulation of sediment deposited along continental margins.

When a descending plate reaches a depth of about 100 to 150 kilometres, heat drives water and other

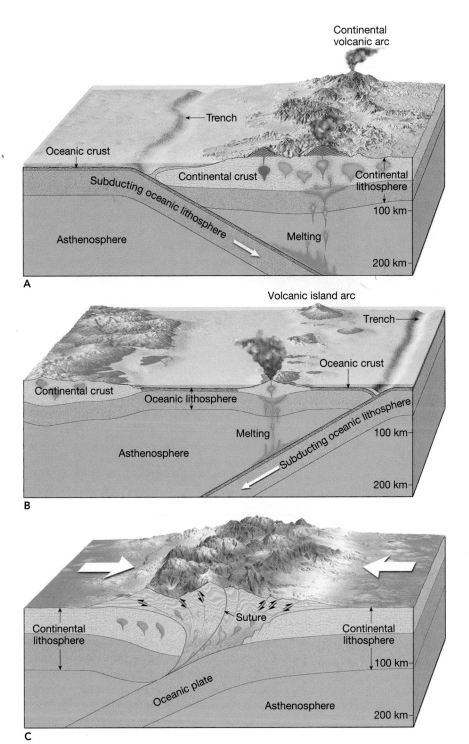

◆ **Figure 19.21** Zones of plate convergence. **A.** Oceanic-continental **B.** Oceanic-oceanic **C.** Continental-continental

volatile components from the subducted sediments into the overlying mantle. These substances act as a flux does at a foundry, inducing partial melting of mantle rocks at reduced temperatures. The partial melting of mantle rock generates magmas having a mafic or occasionally intermediate composition. (Recent research had shown that in some settings the subducted oceanic lithosphere might also partially melt, generating some magma). The newly formed magma, being less dense than the rocks of the mantle, will buoyantly rise. Often the magma, being more dense than continental rocks, will pool beneath the overlying continental crust, where it may melt some of the silica-enriched rocks. Eventually some of this silica-rich magma may migrate to the surface, where it can give rise to volcanic eruptions, some of which are explosive.

The volcanoes of the Andean arc located along the western flank of South America are the product

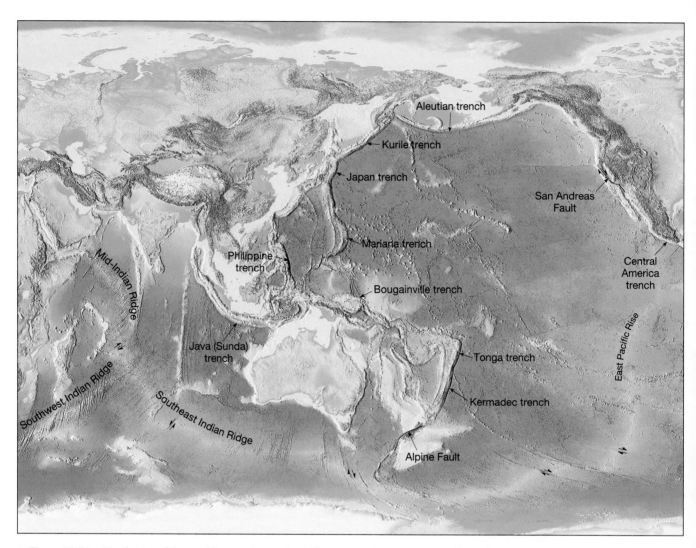

◆ **Figure 19.22** Distribution of the world's oceanic trenches, ridge system, fracture zones, and transform faults. Where transform faults offset ridge segments, they permit the ridge to change direction (curve) as can be seen in the Atlantic Ocean.

of magma generated as the Nazca plate descends beneath the continent (see Figure 19.17). In the central section of the southern Andes the subduction angle is very shallow, which probably accounts for the lack of volcanism in that area. As the South American plate moves westward, it overruns the Nazca plate. The result is a seaward migration of the Peru–Chile trench and a reduction in the size of the Nazca plate.

Mountains such as the Andes, which are produced in part by volcanic activity associated with the subduction of oceanic lithosphere, are called **continental volcanic arcs**. Another active continental volcanic arc is located in western North America. The Cascade Range of British Columbia, Washington, Oregon, and California consists of several well-known volcanic mountains, including Mount Rainier, Mount Shasta, and Mount St. Helens (see Figure 4.35, p. 120). As the continuing activity of Mount St. Helens testifies,

the Cascade Range is still active. The magma here arises from melting triggered by the subduction of a small remaining segment of the Farallon plate, of which the Juan de Fuca plate is the largest northern segment.

A remnant of a formerly extensive continental volcanic arc is California's Sierra Nevada, in which Yosemite National Park is located. The Sierra Nevada is much older than the Cascade Range, and it has been inactive for several million years, as evidenced by the absence of volcanic cones. Here erosion has stripped away most of the obvious traces of volcanic activity and left exposed the large, crystallized magma chambers that once fed lofty volcanoes.

Oceanic–Oceanic Convergence

When two oceanic slabs converge, one descends beneath the other, initiating volcanic activity in a manner similar

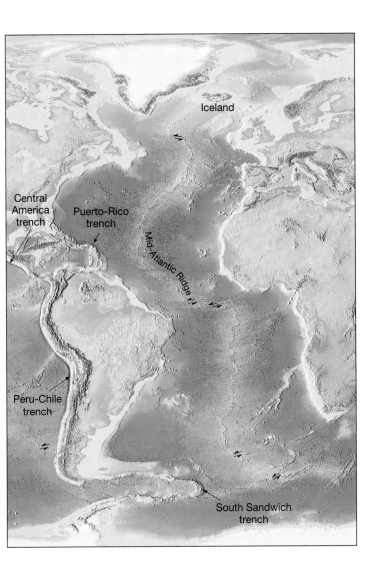

Some, such as the New Hebrides and Mariana arcs, have a small accretionary wedge or none at all. Either very little sediment is deposited in these trenches or most of the sediment is carried into the mantle by the subducting plate. At these sites the subducting Pacific crust is relatively old and dense and therefore will readily sink into the mantle. This is thought to account for the steep angle of descent (often approaching 90 degrees) common in the trenches of the western Pacific. Further, many of these subduction zones lack the large earthquakes that are associated with some other convergent zones, such as the Peru–Chile trench.

Only two volcanic island arcs are located in the Atlantic—the Lesser Antilles arc adjacent to the Caribbean Sea, and the Sandwich Islands in the South Atlantic. The Lesser Antilles are a product of the subduction of the Atlantic beneath the Caribbean plate. Located within this arc is the island of Martinique, where Mount Pelee erupted in 1902, destroying the town of St. Pierre and killing an estimated 28,000 people; and the island of Montserrat, where volcanic activity has occurred very recently.

Relatively young island arcs are fairly simple structures that are underlain by crust that is generally less than 20 kilometres thick. Examples include the arcs of Tonga, the Aleutians, and the Lesser Antilles. By contrast, older island arcs are more complex and are underlain by crust that ranges in thickness from 20 to 35 kilometres. Examples include the Japanese and Indonesian arcs, which are built upon material generated by earlier episodes of subduction. In a few places, volcanic island arcs are gradually transformed into an Andean-type volcanic chain. For example, the western section of the Aleutian arc consists of numerous volcanic islands built on oceanic crust, whereas the volcanoes at the eastern end of the chain are part of the Alaskan Peninsula—a piece of continental crust.

Continental–Continental Convergence

As you saw earlier, when an oceanic plate is subducted beneath continental lithosphere, an Andean-type volcanic arc develops along the margin of the continent. However, if the subducting plate also contains continental lithosphere, continued subduction eventually brings the two continents together (Figure 19.23). Whereas oceanic lithosphere is relatively dense and sinks into the asthenosphere, continental lithosphere is buoyant, which prevents it from being subducted to any great depth. The result is a collision between the two continental blocks (Figure 19.23).

Such a collision occurred when the subcontinent of India "rammed" into Asia and produced the Himalayas—the most spectacular mountain range on Earth (Figure 19.23). During this collision, the

to that which occurs at an oceanic–continental boundary. In this case, however, the volcanoes form on the ocean floor rather than on a continent (Figure 19.21B). If this activity is sustained, it will eventually build a chain of volcanic structures that emerge as islands. The volcanic islands are spaced about 80 kilometres apart and are built upon submerged ridges of volcanic material a few hundred kilometres wide. This newly formed land consisting of an arc-shaped chain of small volcanic islands is called a **volcanic island arc**. The Aleutian, Mariana, and Tonga islands are examples of volcanic island arcs. Island arcs such as these are generally located 200 to 300 kilometres from the trench axis. Located adjacent to the island arcs just mentioned are the Aleutian trench, the Mariana trench, and the Tonga trench (see Figure 19.22). Most volcanic island arcs are located in the western Pacific.

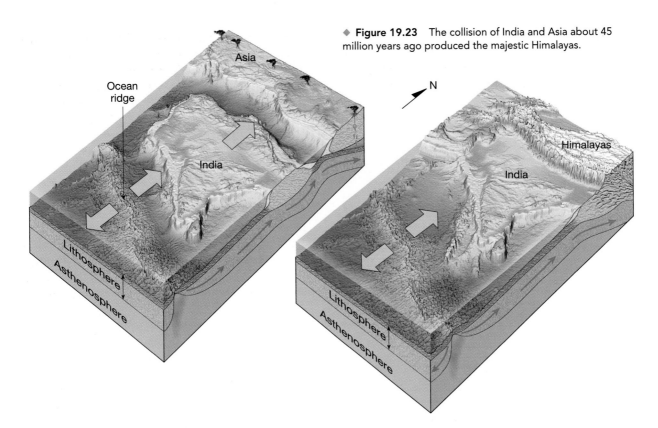

◆ **Figure 19.23** The collision of India and Asia about 45 million years ago produced the majestic Himalayas.

continental crust buckled, fractured, and was generally shortened and thickened. In addition to the Himalayas, several other major mountain systems, including the Alps, Appalachians, and Urals, formed during continental collisions.

Prior to a continental collision, the landmasses involved are separated by an ocean basin. As the continental blocks converge, the intervening seafloor is subducted beneath one of the plates. Subduction initiates partial melting in the overlying mantle rocks, which in turn results in the growth of a volcanic arc. Depending on the location of the subduction zone, the volcanic arc could develop on either of the converging landmasses, or if the subduction zone developed several hundred kilometres seaward from the coast, a volcanic island arc would form. Eventually, as the intervening seafloor is consumed, these continental masses collide. This folds and deforms the accumulation of sediments along the continental margin as if they had been placed in a gigantic vise. The result is the formation of a new mountain range composed of deformed and metamorphosed sedimentary rocks, fragments of the volcanic arc, and possibly slivers of oceanic crust.

After continents collide, the subducted oceanic plate may separate from the continental block and continue its downward movement. However, because of its buoyancy, continental lithosphere cannot be carried very far into the mantle. In the case of the Himalayas, the leading edge of the Indian plate was forced partially under Asia, generating an unusually great thickness of continental lithosphere. This accumulation accounts, in part, for the high elevation of the Himalayas and may help explain the elevated Tibetan Plateau to the north.

Transform Fault Boundaries

Internal Processes
↳ Plate Tectonics

The third type of plate boundary is the **transform** (*trans* = across, *forma* = form) **fault boundary**, which is characterized by strike-slip faulting where plates grind past one another without the production or destruction of lithosphere (*conservative plate margins*). Transform faults were first identified where they join offset segments of an ocean ridge (see Figure 19.22). At first it was erroneously assumed that the ridge system originally formed a long and continuous chain that was later offset by horizontal displacement along these large faults. However, the displacement along these faults was found to be in the exact opposite direction required to produce the offset ridge segments.

The true nature of transform faults was discovered in 1965 by J. Tuzo Wilson of the University of Toronto (Box 19.2). Wilson suggested that these large faults connect the global active belts (convergent boundaries, divergent boundaries, and other transform faults) into a continuous network that divides Earth's outer shell into several rigid plates. Thus, Wilson became the first to suggest that Earth was made of individual plates, while at the same time identifying the faults along which relative motion between the plates is made possible.

Most transform faults join two segments of a mid-ocean ridge (Figure 19.24). Here, they are part of prominent linear breaks in the oceanic crust known as **fracture zones**, which include both the transform faults and their inactive extensions into the plate interior. These fracture zones are present approximately every 100 kilometres along the trend of a ridge axis. As shown in Figure 19.24, active transform faults lie only *between* the two offset ridge segments. Here, seafloor produced at one ridge axis moves in the opposite direction from seafloor produced at an opposing ridge segment. Thus, between the ridge segments these adjacent slabs of oceanic crust are grinding past each other along a transform fault. Beyond the ridge crests are the inactive zones, where the fractures produced by strike-slip faulting are preserved as linear topographic

scars. These fracture zones tend to curve such that small segments roughly parallel the direction of plate motion at the time of their formation.

In another role, transform faults provide the means by which the oceanic crust created at ridge crests can be transported to a site of destruction: the deep-ocean trenches. Figure 19.25 illustrates this situation. Notice that the Juan de Fuca plate moves in a southeasterly direction, eventually being subducted under the west coast of the United States. The southern end of this relatively small plate is bounded by the Mendocino transform fault. This transform fault boundary connects the Juan de Fuca ridge to the Cascade subduction zone (Figure 19.25). Therefore, it facilitates the movement of the crustal material created at the ridge crest to its destination beneath the North American continent (Figure 19.25). Another example of a *ridge-trench* transform fault is found southeast of the tip of South America. Here transform faults on the north and south margins of the Scotia plate connect the trench to a short spreading axis (see Figure 19.17).

Although most transform faults are located within the ocean basins, a few cut through continental crust. Two examples are the earthquake-prone San Andreas Fault of California and the Alpine Fault of New Zealand. Notice in Figure 19.25 that the San Andreas Fault connects a spreading centre located in the Gulf of California to the Cascadia subduction zone and the Mendocino fault located along the northwest coast of the United States. Along the San Andreas Fault, the Pacific plate is moving toward the northwest, past the North American plate. If this movement continues, that part of California west of the fault zone, including the Baja Peninsula, will become an island off the Pacific Coast of the United States and Canada. It could eventually reach Alaska. However, a more immediate concern is the earthquake activity triggered by movements along this fault system.

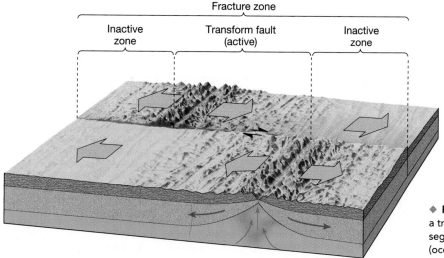

◆ **Figure 19.24** Diagram illustrating a transform fault boundary offsetting segments of a divergent boundary (oceanic ridge).

BOX 19.2

Canadian Profile
John Tuzo Wilson: Canada's Sponsor of Plate Tectonics

Vic Tyrer*

In the 1960s J. Tuzo Wilson (Figure 19.B) was one of a group of earth scientists who developed the theory of Plate Tectonics. His many contributions to the theory include an explanation for the opening and closing of ocean basins, now known as the "Wilson Cycle" (see Chapter 20). In 1965 he proposed the existence of "transform faults" to explain the numerous narrow fracture zones and earthquakes found in oceanic crust. His idea of plumes of magma rising through the Earth's crust creating local "hot spots" explained intraplate volcanoes such as the Hawaiian Island chain.

Born in Ottawa, J. Tuzo Wilson was the first graduate of geophysics from the University of Toronto in 1930. He went on to become an assistant geologist with the Geological Survey of Canada from 1936 to 1939. Following service in the Canadian Army, he became a Professor of Geophysics at the University of Toronto from 1946 to 1974.

Wilson was the Director General of the Ontario Science Centre from 1974 to 1985 where he combined his scientific vision with his commitment to public education. He was as comfortable in the Science Centre's classrooms as he was in the boardroom, frequently capturing the imagination of young students and adults alike with hands-on demonstrations of plate tectonics using his familiar "lighter and paper" explanation of the Hawaiian Islands.

◆ **Figure 19.B**   John Tuzo Wilson, Canadian Pioneer of Plate Tectonics.
(Photo courtesy of the Ontario Science Centre)

Wilson gave his last public lecture at the Ontario Science Centre in February 1992. His lecture, "With Zest to Go, in Quest to Know" recounted his lifetime travels of over three million kilometres and seven continents in his quest to understand and teach the inner workings of the Earth.

On May 3, 2001, the Ontario Science Centre unveiled a tribute to its former Director General in the form of a geodetic monument. Central to the monument is the sculpture's large iron spike. Visitors are asked to imagine that the spike has been driven into the Earth's core. The iron spike remains fixed, seemingly gouging a trough in the surrounding concrete plaza as the North American Tectonic Plate "drifts" westward a few centimetres per year. The trough marks a total of 2.3 metres of tectonic movement during the lifetime of Tuzo Wilson (1908-1993).

A notch on the spike marks a point calculated using Global Positioning System technology. Geomatics Canada's Geodetic Survey located this site by monitoring the broadcast from numerous orbiting GPS satellites over a period of 24 hours in July 1997.

———————

*Vic Tyrer was the Earth Science Programmer at the Ontario Science Centre from 1978 to 1999.

Testing the Plate Tectonics Model

With the development of plate tectonics, researchers from all Earth sciences began testing this model of how Earth works. Some of the evidence supporting continental drift and seafloor spreading has already been presented. In addition, some of the evidence that was instrumental in solidifying the support for this new idea follows. Note that much of this evidence was not new; rather, it was new interpretations of already existing data that swayed the tide of opinion.

Plate Tectonics and Earthquakes

By 1968 the basic outline of global tectonics was firmly established. In that year, three seismologists published papers demonstrating how successfully the new plate tectonics model accounted for the global distribution of earthquakes (Figure 19.26). In particular, these scientists were able to account for the close association between deep-focus earthquakes and subduction zones. Furthermore, the absence of deep-focus earthquakes along the oceanic ridge system was also shown to be consistent with the new theory.

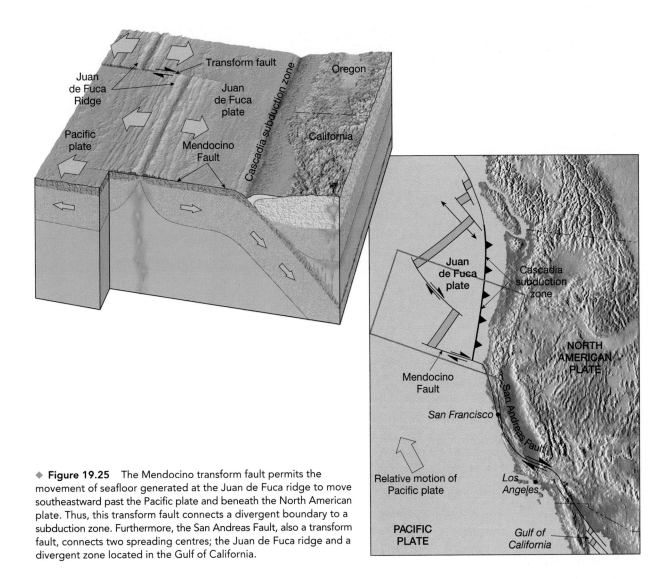

◆ **Figure 19.25** The Mendocino transform fault permits the movement of seafloor generated at the Juan de Fuca ridge to move southeastward past the Pacific plate and beneath the North American plate. Thus, this transform fault connects a divergent boundary to a subduction zone. Furthermore, the San Andreas Fault, also a transform fault, connects two spreading centres; the Juan de Fuca ridge and a divergent zone located in the Gulf of California.

Note the close association between plate boundaries and earthquakes by comparing the distribution of earthquakes shown in Figure 19.26 with the map of plate boundaries in Figure 19.17 (p. 528–529). In trench regions, where dense slabs of lithosphere plunge into the mantle, this association is especially striking. When the depths of earthquake foci and their locations within the trench systems are plotted, an interesting pattern emerges. Figure 19.27, which shows the distribution of earthquakes in the vicinity of the Japan trench, is an example. Here most shallow-focus earthquakes occur within or adjacent to the trench, whereas intermediate- and deep-focus earthquakes occur toward the mainland. A similar distribution pattern exists along the western margin of South America where the Nazca plate is being subducted beneath the South American continent.

In the plate tectonics model, deep-ocean trenches form where dense slabs of oceanic lithosphere plunge into the mantle (Figure 19.27). Shallow-focus earthquakes are produced as the descending plate interacts with the overriding lithosphere. As the slab descends farther into the asthenosphere, deeper-focus earthquakes are generated. Because the earthquakes occur within the rigid subducting plate rather than in the "plastic" mantle, they provide a method for tracking the plate's descent. Recall from Chapter 16 that the zones of inclined seismic activity that extend from the trench into the mantle are called *Wadati-Benioff zones* after two seismologists who conducted extensive studies on the distribution of earthquake foci. Very few earthquakes have been recorded below 700 kilometres, possibly because the slab has been heated sufficiently to lose its rigidity.

Evidence from Ocean Drilling

Some of the most convincing evidence confirming seafloor spreading has come from drilling directly into

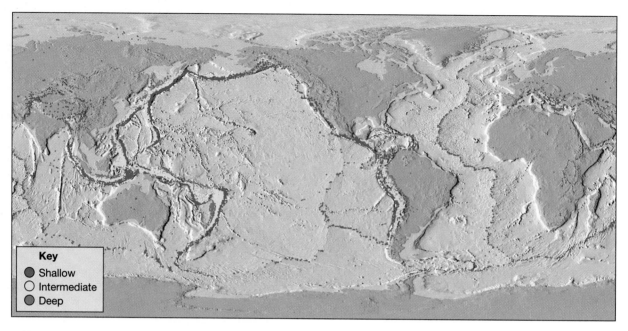

◆ **Figure 19.26**   Distribution of shallow-, intermediate-, and deep-focus earthquakes. Note that deep-focus earthquakes occur only in association with convergent plate boundaries and subduction zones. (Data from NOAA)

ocean-floor sediment. From 1968 until 1983 the source of these important data was the Deep Sea Drilling Project, an international program sponsored by several major oceanographic institutions and the U.S. National Science Foundation. The primary goal was to gather firsthand information about the age of and processes that formed the ocean basins. To accomplish this, a new drilling ship, the *Glomar Challenger*, was built.

Operations began in August 1968 in the South Atlantic. At several sites, holes were drilled through the entire thickness of sediments to the mafic rock below. An important objective was to gather samples of sediment from just above the igneous crust as a means of dating the seafloor at each site.* Because sedimentation begins immediately after the oceanic crust forms, remains of microorganisms found in the oldest sediments—those resting directly on the basalt—can be used to date the ocean floor at that site.

When the oldest sediment from each drill site was plotted against its distance from the ridge crest, the plot demonstrated that the age of the sediment increased with increasing distance from the ridge. This finding agreed with the seafloor-spreading hypothesis, which predicted that the youngest oceanic crust would be found at the ridge crest and that the oldest oceanic crust would be at the continental margins.

The data from the Deep Sea Drilling Project also reinforced the idea that the ocean basins are geologically youthful because no sediment with an age in excess of 180 million years was found. By comparison, continental crust that exceeds 4 billion years in age has been dated.

The thickness of ocean-floor sediments provided additional verification of seafloor spreading. Drill cores from the *Glomar Challenger* revealed that sediments are almost entirely absent on the ridge crest and that the sediment thickens with increasing distance from the ridge. Because the ridge crest is younger than the areas farther away from it, this pattern of sediment distribution should be expected if the seafloor-spreading hypothesis is correct.

The Ocean Drilling Project has succeeded the Deep Sea Drilling Project and, like its predecessor, is a major international program. The more technologically advanced drilling ship, the *JOIDES Resolution*, now continues the work of the *Glomar Challenger*.* The *JOIDES Resolution* can drill in water depths as great as 8200 metres and contains onboard laboratories equipped with the largest and most varied array of seagoing scientific research equipment in the world (Figure 19.28).

---

*Radiometric dates of the ocean crust itself are unreliable because of the alteration of basalt by seawater.

*JOIDES stands for Joint Oceanographic Institutions for Deep Earth Sampling.

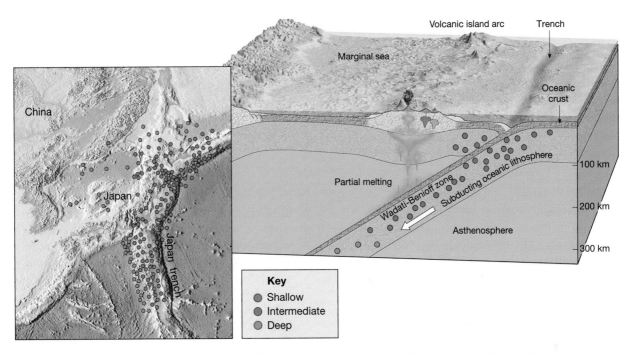

◆ **Figure 19.27** Distribution of earthquake foci in the vicinity of the Japan trench. Note that intermediate- and deep-focus earthquakes occur only within the sinking slab of oceanic lithosphere. (Data from NOAA)

Hot Spots

Mapping of seamounts in the Pacific Ocean revealed a chain of volcanic structures extending from the Hawaiian Islands to Midway Island and continuing northward toward the Aleutian trench (Figure 19.29). Radiometric dating in this chain showed that the volcanoes increase in age with increasing distance from Hawaii. Hawaii, the youngest volcano in the chain, rose from the seafloor less than a million years ago, whereas Midway Island is 27 million years old and Suiko Seamount, near the Aleutian trench, is 65 million years old (Figure 19.29).

Researchers are in agreement that a rising plume of mantle material is located beneath the island of Hawaii. Decompression melting of this hot rock as it enters the low-pressure environment near the surface generates a volcanic area known as a **hot spot**. As the Pacific plate moved over this hot spot, successive volcanic structures were built. The age of each volcano indicates the time when it was situated over the relatively stationary mantle plume.

This pattern is shown in Figure 19.29. Kauai is the oldest of the large islands in the Hawaiian chain. Five million years ago, when it was positioned over

◆ **Figure 19.28** The *JOIDES Resolution*, the drilling ship of the Ocean Drilling Program. This modern drilling ship has replaced the *Glomar Challenger* in the important work of sampling the floor of the world's oceans. (Photo courtesy of Ocean Drilling Program)

the hot spot, Kauai was the only Hawaiian Island (Figure 19.29). Visible evidence of the age of Kauai can be seen by examining its extinct volcanoes, which have been eroded into jagged peaks and vast canyons. By contrast, the south slopes of the comparatively young island of Hawaii consist of fresh lava flows, and two of Hawaii's volcanoes, Mauna Loa and Kilauea, remain active.

Two island groups parallel the Hawaiian Island–Emperor Seamount chain. One chain consists of the Tuamotu and Line islands, and the other includes the Austral, Gilbert, and Marshall islands. In each case, the most recent volcanic activity has occurred at the southeastern end of the chain, and the islands get progressively older to the northwest. Thus, like the Hawaiian Island–Emperor Seamount chain, these volcanic structures apparently formed by the same motion of the Pacific plate over fixed mantle plumes. Not only does this evidence support the fact that the plates do indeed move relative to Earth's interior but also the hot spot "tracks" trace the direction of the plate motion. Notice in Figure 19.29, for example, that the Hawaiian Island–Emperor Seamount chain bends. This particular bend in the track occurred about 40 million years ago when the motion of the Pacific plate changed from nearly due north to a northwesterly path. Similarly, hot spots found on the floor of the Atlantic have increased our understanding of the migration of landmasses following the breakup of Pangaea.

Students Sometimes Ask...

If the continents move, do other features like segments of the mid-ocean ridge also move?
That's a good observation, and yes, they do! It is interesting to note that very little is really fixed in place on Earth's surface. When we talk about movement of features on Earth, we must consider the question, "Moving relative to what?" Certainly, the mid-ocean ridge does move relative to the continents (which sometimes causes segments of the mid-ocean ridges to be subducted beneath the continents). In addition, the mid-ocean ridge is moving relative to a fixed location outside Earth. This means that an observer orbiting above Earth would notice, after only a few million years, that all continental and seafloor features—as well as plate boundaries—are indeed moving. The exception to this are hot spots, which seem to be relatively stationary and can be used to determine the motions of other features.

The existence of mantle plumes and their association with hot spots is well documented. Most mantle plumes are long-lived structures that appear to maintain relatively fixed positions within the mantle. Further, research suggests that at least some mantle plumes originate at great depth, perhaps at the mantle-core boundary. Others, however, may have a much shallower origin. Of the 40 or so hot spots that have been

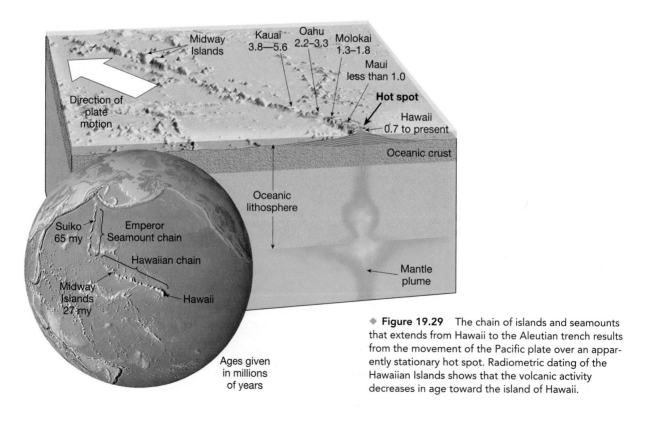

◆ **Figure 19.29** The chain of islands and seamounts that extends from Hawaii to the Aleutian trench results from the movement of the Pacific plate over an apparently stationary hot spot. Radiometric dating of the Hawaiian Islands shows that the volcanic activity decreases in age toward the island of Hawaii.

identified, over a dozen are located near spreading centres. For example, the mantle plume located beneath Iceland is responsible for the large accumulation of volcanic rocks found along the northern section of the Mid-Atlantic Ridge.

Measuring Plate Motion

A number of methods have been employed to establish the direction and rate of plate motion. As noted earlier, hot spot "tracks" like those of the Hawaiian Island–Emperor Seamount chain trace the movement of the Pacific Plate relative to the mantle below. Further, by measuring the length of this volcanic chain and the time interval between the formation of the oldest structure (Suiko Seamount) and youngest structure (Hawaii), an average rate of plate motion can be calculated. In this case the volcanic chain is roughly 3000 kilometres long and has formed over the past 65 million years—making the average rate of movement about 9 centimetres per year. The accuracy of this calculation hinges on the hot spot maintaining a fixed position in the mantle. Based on current evidence, this appears to be a reasonable assumption.

Recall that the magnetic stripes measured on the floor of the ocean also provide a method to measure rates of plate motion—at least as averaged over millions of years. Using paleomagnetism and other

indirect techniques, researchers have been able to work out relative plate velocities as shown on the map in Figure 19.30 (see Box 19.3).

It is currently possible, with the use of space-age technology, to directly measure the relative motion between plates. This is accomplished by periodically establishing the exact locations, and hence the distance between two observing stations situated on opposite sides of a plate boundary. Two of the methods used for this calculation are *Very Long Baseline Interferometry* (VLBI) and a satellite positioning technique that employs the *Global Positioning System* (GPS). The Very Long Baseline Interferometry system utilizes large

Students Sometimes Ask...

Will plate tectonics eventually "turn off" and cease to operate on Earth?
Because plate tectonic processes are powered by heat from within Earth (which is of a finite amount), the forces will slow sometime in the distant future to the point that the plates will cease to move. The work of external processes, however, will continue to erode Earth's features, most of which will be eventually eroded flat. What a different world it will be—an Earth with no earthquakes, no volcanoes, and no mountains. Flatness will prevail!

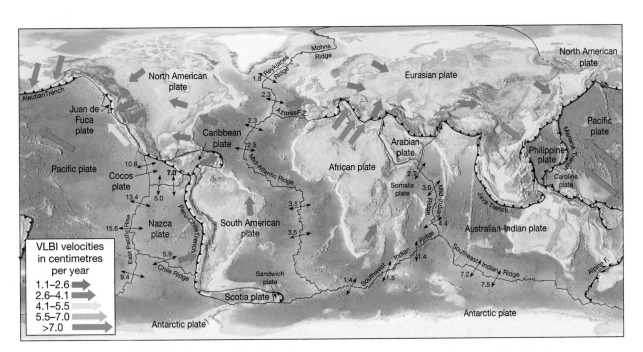

◆ **Figure 19.30** This map illustrates directions and rates of plate motion in centimetres per year. Seafloor-spreading velocities (as shown with black arrows and labels) are based on the spacing of dated magnetic stripes (anomalies). The coloured arrows show Very Long Baseline Interferometry (VLBI) data of plate motion at selected locations. The data obtained by these methods are typically consistent. (Seafloor data from DeMets and others, VLBI data from Ryan and others)

BOX 19.3

Understanding Earth
Plate Tectonics into the Future

Two geologists, Robert Dietz and John Holden, extrapolated present-day plate movements into the future. Figure 19.C illustrates where they envision Earth's landmasses will be 50 million years from now if present plate movements persist for this time span.

In North America we see that the Baja Peninsula and the portion of southern California that lies west of the San Andreas Fault will have slid past the North American plate. If this northward migration takes place, Los Angeles and San Francisco will pass each other in about 10 million years, and in about 60 million years Los Angeles will begin to descend into the Aleutian trench.

Significant changes are seen in Africa, where a new sea emerges as East Africa parts company with the mainland. In addition, Africa will have moved slowly into Europe, perhaps initiating the next major mountain-building stage on our dynamic planet. Meanwhile, the Arabian Peninsula continues to diverge from Africa, allowing the Red Sea to widen and causing the Persian Gulf to close.

In other parts of the world, Australia is now astride the equator and, along with New Guinea, is on a collision course with Asia. Meanwhile, North and South America are beginning to separate, while the Atlantic and Indian oceans continue to grow at the expense of the Pacific Ocean.

These projections into the future, although interesting, must be viewed with caution because many assumptions must be correct for these events to unfold as just described. Nevertheless, changes in the shapes and positions of continents that are equally profound will undoubtedly occur for millions of years to come.

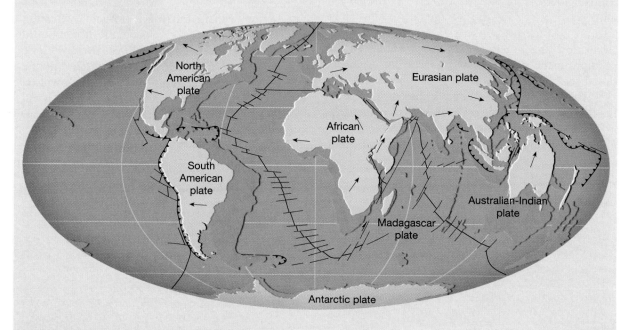

◆ **Figure 19.C**   The world as it may look 50 million years from now.
(From "The Breakup of Pangaea," Robert S. Dietz and John C. Holden. Copyright 1970 by Scientific American, Inc., All rights reserved)

radio telescopes to record signals from very distant quasars (quasi-stellar objects) (Figure 19.31). Quasars lie billions of light-years from Earth, so they act as stationary reference points. The millisecond differences in the arrival times of the same signal at different earthbound observatories provide a means of establishing the precise distance between receivers. A typical survey may take a day to perform and involves two widely spaced radio telescopes observing perhaps a dozen quasars, 5 to 10 times each. This scheme provides an estimate of the distance between these observatories, which is accurate to about 2 centimetres. By repeating this experiment at a later date, researchers can establish the relative motion of these sites. This method has been particularly useful in establishing large-scale plate motions, such as the separation that is occurring between North America and Europe.

You may be familiar with the Global Positioning System, which uses 21 satellites to accurately locate any individual who is equipped with a handheld receiver.

◆ **Figure 19.31**   Radio telescopes like this one located at Green Bank, West Virginia, are used to determine accurately the distance between two distant sites. Data collected by repeated measurements have detected relative plate motions of 1 to 15 centimetres per year between various sites worldwide. (Courtesy of National Radio Astronomy Observatory)

By using two spaced receivers, signals obtained by these instruments can be used to calculate their relative positions with considerable accuracy. Techniques using GPS receivers have been shown to be useful in establishing small-scale crustal movements such as those that occur along local faults in regions known to be tectonically active.

Confirming data obtained from these and other techniques leave little doubt that real plate motion has been detected. Calculations show that Hawaii is moving in a northwesterly direction and approaching Japan at 8.3 centimetres per year. A site located in Maryland is retreating from one in England at a rate of about 1.7 centimetres per year—a rate that is close to the 2.3-centimetres-per-year spreading rate that was established from paleomagnetic evidence.

The Driving Mechanism

The plate tectonics theory *describes* plate motion and the effects of this motion. Therefore, acceptance of this model does not rely on the mechanism that drives plate motion. This is fortunate, because none of the driving mechanisms yet proposed can account for all major facets of plate tectonics. Nevertheless, researchers generally agree on the following:

1. Convective flow in the rocky 2900-kilometre-thick mantle—in which warm, less dense rock rises and cooler, more dense material sinks—is the basic driving force for plate movement.

2. Mantle convection and plate tectonics are part of the same system. Oceanic plates represent the cold downward-moving portion of convective flow.
3. The slow movements of the plates and mantle are driven by the unequal distribution of heat within Earth's interior. This flow is the mechanism that transports heat away from Earth's interior.

What is not known with any large degree of certainty is the precise nature of this convective flow.

Some researchers have argued that the mantle is like a giant layer cake, divided at a depth of 660 kilometres. Convection operates in both layers, but mixing between layers is minimal. At the other end of the spectrum is a model that loosely resembles a pot of just barely boiling water, churning ever so slowly from top to bottom over eons of geologic time. Neither model fits all of the available data. We will first look at some of the processes that are thought to contribute to plate motion and then examine a few of the models that have been proposed to describe plate-mantle convection.

Slab-Pull, Ridge-Push, and Mantle Plumes

Several mechanisms generate forces that contribute to plate motion. One relies on the fact that old oceanic crust, which is relatively cool and dense, sinks into the asthenosphere and "pulls" the trailing lithosphere along. This mechanism, called **slab-pull**, is thought to be the primary downward arm of the convective flow operating in the mantle. By contrast, **ridge-push** results from the elevated position of the oceanic ridge system and causes oceanic lithosphere to gravitationally slide down the flanks of the ridge. Ridge-push, although apparently active in some spreading centres, is probably less important than slab-pull.

Most models suggest that hot, buoyant plumes of rock are the upward flowing arms in the convective mechanism at work in the mantle. These rising **mantle plumes** manifest themselves on Earth's surface as hot spots with their associated volcanic activity. Mapping of Earth's interior using seismic tomography indicates that at least some of these hot plumes extend upward from the vicinity of the mantle-core boundary (see Box 17.3, p. 486). Others, however, appear to originate higher in the mantle. Further, some upward flow in the mantle is not related to plumes and occurs just below the ridge crests as a result of seafloor spreading.

Models of Plate-Mantle Convection

Any model describing mantle convection must be consistent with the observed physical and chemical properties of the mantle. In particular, these models must explain why basalts that erupt along the oceanic ridge are fairly homogeneous in composition and depleted in certain trace elements. It is assumed that these ridge basalts are derived from rocks located in the upper mantle that experienced an earlier period of chemical differentiation, in which trace elements were removed. By contrast, higher concentrations of these elements are evident in mafic eruptions associated with hot-spot volcanism. Because basalts that erupted in different settings have different concentrations of trace elements, they are assumed to be derived from chemically distinct regions of the mantle. Basalts associated with mantle plumes are thought to come from a primitive (less differentiated) source, which more closely resembles the average chemical composition of the early mantle.

Layering at 660 kilometres   Earlier we referred to the "layer cake" version of mantle convection. As shown in Figure 19.32A, one of these layered models has two zones of convection—a thin, convective layer above 660 kilometres and a thick one located below. This model successfully explains why the basalts that erupt along the oceanic ridges have a somewhat different composition than those that erupt in Hawaii as a result of hot-spot activity. The mid-oceanic ridge basalts come from the upper convective layer, which is well mixed, whereas the mantle plume that feeds the Hawaiian volcanoes taps a deeper, more primitive source that resides in the lower convective layer.

Despite evidence that supports this model, recent seismic imaging has shown that subducting slabs of cold oceanic lithosphere are able to penetrate the 660-kilometre boundary. The subducting lithosphere serves to mix the upper and lower layers together. As a result, the layered mantle structure is lost.

Whole-Mantle Convection   Because of problems with the layered model, researchers began to favour whole-mantle convection. In a whole-mantle convection model, slabs of cold oceanic lithosphere descend into the lower mantle, providing the downward arm of convective flow (Figure 19.32B). Simultaneously, hot mantle plumes originating near the mantle-core boundary transport heat toward the surface. It was also suggested that at extremely slow rates of convection, primitive (undepleted) mantle rock would exist at depth in quantities sufficient to feed the rising mantle plumes.

Recent work, however, has shown that whole mantle mixing would cause the lower mantle to become appreciably mixed in a matter of a few hundred million years. This mixing would tend to eliminate the source of primitive magma observed in hot-spot volcanism.

Deep-Layer Model   A remaining possibility is layering deeper in the mantle. One deep-layer model has been described as analogous to a lava lamp on a low setting. As shown in Figure 19.32C, approximately the lower third of the mantle is like the coloured fluid in the bottom layer of a lava lamp. Like a lava lamp on low, heat from Earth's interior causes the two layers to slowly swell and shrink in complex patterns without substantial mixing. A small amount of material from the lower layer flows upward as mantle plumes to generate hot-spot volcanism at the surface.

This model provides the two chemically different mantle sources for basalt that are required by observational data. Further, it is compatible with seismic images that show cold lithospheric plates sinking deep into the mantle. Despite its attractiveness, there is little seismic evidence to suggest that a deep mantle layer of this nature exists, except for the very thin layer located at the mantle-core boundary.

Although there is still much to be learned about the mechanisms that cause plates to move, some facts are clear. The unequal distribution of heat in Earth generates some type of thermal convection that ultimately drives plate-mantle motion. Furthermore, the descending lithospheric plates are active components of downwelling, and they serve to transport cold material into the mantle. Exactly how this convective flow operates is yet to be determined.

Importance of Plate Tectonics

Plate tectonics is the first theory to provide a unified explanation of Earth's major surface features, including the continents and ocean basins. As such, it has linked many aspects of geology, which previously had been considered unrelated. Various branches of geology have joined to provide a better understanding of the workings of our dynamic planet. Within the framework of plate tectonics, geologists have found explanations for the geologic distribution of earthquakes, volcanoes, and mountain belts. Further, explanations for past distributions of plants and animals and the occurrence of economically significant mineral deposits have resulted from an understanding of plate tectonics.

Despite its usefulness in explaining many of the large-scale geologic processes operating on Earth, plate tectonics is not completely understood. The model that was set forth in 1968 was simply a basic framework that left much of the detail to future research. Through critical testing, this initial model has been modified and expanded to become the theory we know today. The current theory will certainly be refined as additional data and observations are obtained. The theory of plate tectonics, although a powerful tool, is nonetheless an evolving model of Earth's dynamic processes.

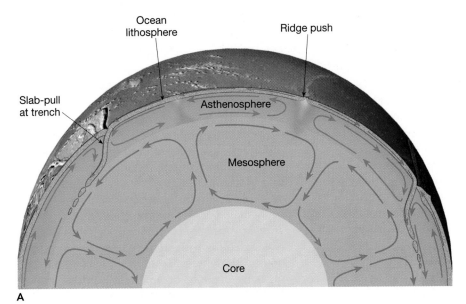

A

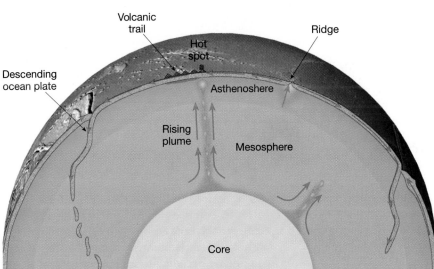

B

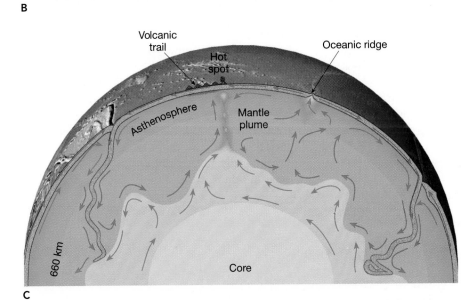

C

◆ **Figure 19.32** Proposed models for mantle convection. **A.** The model shown in this illustration consists of two convection layers—a thin, convective layer above 660 kilometres and a thick one below. **B.** In this whole-mantle convection model, cold oceanic lithosphere descends into the lowermost mantle while hot mantle plumes transport heat toward the surface. **C.** This deep-layer model suggests that the mantle operates similar to a lava lamp on a low setting. Earth's heat causes these layers of convection to slowly swell and shrink in complex patterns without substantial mixing. Some material from the lower layer flows upward as mantle plumes.

Chapter Summary

- In the early 1900s *Alfred Wegener* set forth the *continental drift* hypothesis. One of its major tenets was that a supercontinent called *Pangaea* began breaking apart into smaller continents about 200 million years ago. The smaller continental fragments then "drifted" to their present positions. To support the claim that the now-separate continents were once joined, Wegener and others used the *fit of South America and Africa, fossil evidence, rock types and structures*, and *ancient climates*. One of the main objections to the continental drift hypothesis was its inability to provide an acceptable mechanism for the movement of continents.

- From the study of *paleomagnetism*, researchers learned that the continents had wandered as Wegener proposed. In 1962, Harry Hess formulated the idea of *seafloor spreading*, which states that new seafloor is continually being generated at mid-oceanic ridges and old, dense seafloor is being consumed at the deep-ocean trenches. Support for seafloor spreading followed, with the discovery of alternating stripes of high- and low-intensity magnetism that parallel the ridge crests.

- By 1968, continental drift and seafloor spreading were united into a far more encompassing theory known as *plate tectonics*. According to plate tectonics, Earth's rigid outer layer (*lithosphere*) overlies a weaker region called the *asthenosphere*. Further, the lithosphere is broken into seven large and numerous smaller segments called *plates* that are in motion and continually changing in shape and size. Plates move as relatively coherent units and are deformed mainly along their boundaries.

- *Divergent plate boundaries* occur where plates move apart, resulting in upwelling of material from the mantle to create new seafloor. Most divergent boundaries occur along the axis of the oceanic ridge system and are associated with seafloor spreading, which occurs at rates of 2 to 15 centimetres per year. New divergent boundaries may form within a continent (for example, the East African rift valleys) where they may fragment a landmass and develop a new ocean basin.

- *Convergent plate boundaries* occur where plates move together, resulting in the subduction of oceanic lithosphere into the mantle along a deep oceanic trench. Convergence between an oceanic and continental block results in subduction of the oceanic slab and the formation of a *continental volcanic arc* such as the Andes of South America. Oceanic–oceanic convergence results in an arc-shaped chain of volcanic islands called a *volcanic island arc*. When two plates carrying continental crust converge, both plates are too buoyant to be subducted. The result is a "collision" resulting in the formation of a mountain belt such as the Himalayas.

- *Transform fault boundaries* occur where plates grind past each other without the production or destruction of lithosphere. Most transform faults join two segments of a mid-oceanic ridge. Others connect spreading centres to subduction zones and thus facilitate the transport of oceanic crust created at a ridge crest to its site of destruction, at a deep-ocean trench. Still others, like the San Andreas Fault, cut through continental crust.

- The theory of plate tectonics is supported by (1) the global distribution of *earthquakes* and their close association with plate boundaries; (2) the ages and thickness of *sediments* from the floors of the deep-ocean basins; and (3) the existence of island chains that formed over *hot spots* and provide a frame of reference for tracing the direction of plate motion.

- Three basic models for mantle convection are currently being evaluated. Mechanisms that contribute to this convective flow are slab-pull, ridge-push, and mantle plumes. *Slab-pull* occurs where cold, dense oceanic lithosphere is subducted and pulls the trailing lithosphere along. *Ridge-push* results when gravity sets the elevated slabs astride ocean ridges in motion. Hot, buoyant *mantle plumes* are considered the upward flowing arms of mantle convection. One model suggests that mantle convection occurs in two layers separated at a depth of 660 kilometres. Another model proposes whole-mantle convection that stirs the entire 2900-kilometre-thick rocky mantle. Yet another model suggests that the bottom third of the mantle gradually bulges upward in some areas and sinks in others without appreciable mixing.

Review Questions

1. What first led scientists such as Alfred Wegener to suspect that the continents were once joined?

2. What was Pangaea?

3. List the evidence that Wegener and his followers gathered to support the continental drift hypothesis.

4. Early in this century, what was the prevailing view of how land animals migrated across vast expanses of ocean?

5. Briefly explain why the recent acceptance of plate tectonics has been described as a scientific "revolution."

6. How does evidence for a late Paleozoic glaciation in the Southern Hemisphere support the continental drift hypothesis?

7. Explain how paleomagnetism can be used to establish the latitude of a specific place at some distant time.

8. What is meant by seafloor spreading? Who is credited with formulating the concept of seafloor spreading?

9. Describe how Fred Vine and D. H. Matthews related the seafloor-spreading hypothesis to magnetic reversals.

10. On what basis were plate boundaries first established?

11. Where is lithosphere being formed? Consumed? Why must the production and destruction of the lithosphere be going on at about the same rate?

12. Why is oceanic lithosphere subducted while continental lithosphere is not?

13. In what ways may the origin of the Japanese Islands be considered similar to the formation of the Andes Mountains? How do they differ?

14. Differentiate between transform faults and the two other types of plate boundaries.

15. Some people predict that California will sink into the ocean. Is this idea consistent with the concept of plate tectonics?

16. Applying the idea that hot spots remain fixed, in what direction was the Pacific plate moving while the Emperor Seamounts were being produced? (See Figure 19.29, p. 544) While the Hawaiian Islands were being produced?

17. With what type of plate boundary are the following places or features associated (be as specific as possible): Himalayas, Aleutian Islands, Red Sea, Andes Mountains, San Andreas Fault, Iceland, Japan, Mount St. Helens?

18. Briefly describe the three models proposed for mantle-plate convection. What is lacking in each of these models?

Key Terms

accretionary wedge (p. 534)
asthenosphere (p. 525)
continental drift (p. 515)
continental volcanic arc (p. 536)
convergent plate boundary (p. 533)

Curie point (p. 521)
divergent plate boundary (p. 527)
fracture zone (p. 539)
hot spot (p. 543)
lithosphere (p. 525)
magnetometer (p. 524)
mantle plume (p. 547)

normal polarity (p. 524)
oceanic ridge (p. 527)
paleomagnetism (p. 521)
Pangaea (p. 515)
plate (p. 525)
plate tectonics (p. 525)
reverse polarity (p. 524)
ridge-push (p. 547)

rift, or rift valley (p. 527)
seafloor spreading (p. 531)
slab-pull (p. 547)
subduction zone (p. 533)
transform fault boundary (p. 538)
trench (p. 533)
volcanic island arc (p. 537)

Web Resources

The *Earth* Web site uses the resources and flexibility of the Internet to aid in your study of the topics in this chapter. Written and developed by geology instructors, this site will help improve your understanding of geology. Visit **http://www.pearson.ca/tarbuck** and click on the cover of the text to find:

■ Online review quizzes.

■ Web-based critical thinking and writing exercises.

■ Links to chapter-specific Web resources.

■ Internet-wide key-term searches.

http://www.pearson.ca/tarbuck

Chapter 20
Mountain Building and Continental Frameworks

Mountains in Banff National Park, Alberta.
(Photo by Steve Hicock)

Few geological features of the Earth have captured the attention of humans as much as mountains. Throughout the ages, mountains have inspired legends, have appeared as prominent motifs in famous works of art, have influenced the placement of political boundaries, and have challenged the physical abilities of those bold enough to climb them.

To geologists, mountains are extremely important features for different reasons. First, mountains expose spectacular sedimentary rock sequences that are the basis for interpreting the sequence of events in Earth's history. Second, igneous and metamorphic rocks forced upward to the surface during mountain building provide information on processes that occur so deep in Earth's crust that they cannot be observed directly. Third, mountains supply geologists with information that is critical to the understanding of large-scale tectonic processes.

In the context of the Earth system, mountains influence drainage patterns of rivers, supply sediment that is the parent material for soil in lowland regions, and in some areas of the world, greatly influence regional climatic conditions.

The fact that mountains differ in shape, size, and arrangement reflects differences in the processes that produce them as well as in the character and duration of the processes that act to wear them down (Figure 20.1). In this chapter, the manner in which mountains are believed to form, as well as their significance to the formation of continents, is discussed.

Isostasy: Why the Earth Isn't Smooth

In Chapter 1, we noted that continents stand higher than ocean basins due to **isostasy**, the state of gravitational equilibrium achieved by the flotation of buoyant lithosphere on top of the more dense material of the asthenosphere. Recall that the lower density and greater thickness of continental crust, relative to oceanic crust, allows the upper surface of a continent to stand at a significantly greater elevation than the ocean floor. On a slightly smaller scale, isostasy also explains why regions within a continental landmass can vary considerably in elevation.

Perhaps the easiest way to grasp the concept of how isostasy affects continental topography is to envision a series of wooden blocks of different heights floating in water as shown in Figure 20.2. Note that the thin wooden blocks float only slightly above the water line, whereas the thick blocks stand very high. In a very general sense, the same principle applies to

◆ **Figure 20.1**   Comparison of two of Canada's mountain systems. **A.** The relatively youthful mountains in the Western Cordillera have high, sharp peaks. **B.** The much older mountains in the Appalachians, as seen here along the Cabot Trail of Cape Breton Island, Nova Scotia, have been worn down to rolling hills. (Photo A by Mike Schering; Photo B © J.A.Kraulis/Masterfile)

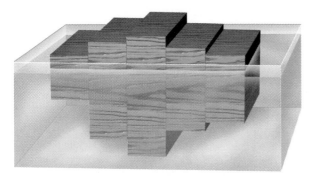

◆ **Figure 20.2**   This drawing illustrates how wooden blocks of different thicknesses float in water. In a similar manner, thick sections of crustal material float higher than do thinner crustal slabs.

elevation differences in continents; regions underlain by thick continental lithosphere stand higher than those underlain by thin continental lithosphere.

This simple model of isostasy is a good starting point for explaining why certain features of the Earth stand lower or higher than others on a regional scale. However, because it is a rather static depiction of the lithosphere, it does not give us insights as to *how, where,* or *when* such features are produced. Moreover, it does not account for the smaller scale deviations in elevation (e.g., high-standing volcanoes in rift zones) that complicate the overall appearance of certain physiographic regions. To address these aspects, it is necessary to examine mountains in the context of their tectonic history. The name for the processes that collectively produce mountains is **orogenesis** (*oros* = mountains, *genesis* = to come into being).

Mountains and Plate Tectonic Environments

A **mountain** can be loosely defined as a landform with an obvious peak that rises significantly above the land surrounding it, generally on the order of 300 metres or more. Such a definition, taken at face value, could imply that a mountain is no more than a big hill in the middle of nowhere. In the context of geology, however, this view could not be further from the truth. For when one considers mountains as part of the larger scale framework of continental features, it becomes obvious that the distribution of mountains is far from random (Figure 20.3).

Mountains tend to occur in linear patterns ranging in scale from small **mountain ranges**, consisting simply of a linked series of mountains, to regionally extensive **mountain belts** that can be extremely complex both in relief and internal structure. These linear features are generally oriented parallel to plate boundaries, where stresses generated by plate movement are the most intense. Due to isostasy, the modifications of crustal thickness that take place in extensional or compressional tectonic environments are accompanied by dramatic changes in land elevation.

In general, continental lithosphere subjected to extension is stretched and thinned, and effectively collapses as fault-bounded blocks. In such cases, mountainous relief is primarily due to the rotation of sinking lithospheric blocks, as observed in the formation of rift valleys. In contrast, continental lithosphere that has been compressed, deformed, and thickened, as observed along convergent margins, stands very high, producing

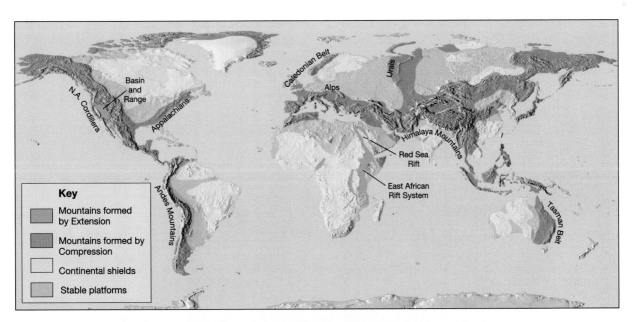

◆ **Figure 20.3**   The basic framework of continents consists of mountains associated with extensional settings (rifts), compressional mountain belts, continental shields, and stable platforms.

towering highlands. Local areas of compression generated in transform boundaries can also locally thicken the crust and produce mountains on a limited scale.

In addition to being important geologic entities in today's world, mountains have also left their mark in the formation and restructuring of continental masses through much of Earth's history. This is particularly true for mountains produced in convergent margins which, despite being worn down, have left evidence of their existence as deformed belts within the ancient shields of continental landmasses. These deformed belts, called *orogens*, indicate that large continents, such as North America, are the products of interactions with several crustal blocks. Before we explore this concept further, we will examine how various types of mountains fit into the plate tectonic scheme.

Mountains in Divergent Margin Settings

Significant topographic relief is produced in some continental regions by the downward sliding and rotation of fault-bounded blocks associated with tensional forces within the crust. As already discussed in Chapter 19, continental crust may be put into a state of tension when an ascending mantle plume pushes upward on base of the lithosphere, causing it to dome upward. In the process, the lithosphere thins and cracks to form rift valleys. Margins of rift valleys begin to collapse, causing blocks of lithosphere to slide downward relative to one another along normal faults. The resulting rift margins on either side of the valley have a step-like appearance, the blocks near the axis of the valley lying at lower elevations than those at the outer edges of the valley (Figure 20.4). The spectacular relief observed in the East African rift is largely the result of the downward sliding of such fault-bounded blocks (Figure 20.5).

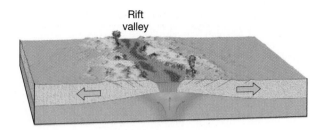

◆ **Figure 20.4**   Linear mountains mark the tops of tilted fault blocks produced by crustal extension. Volcanoes are built by the extrusion of magma, adding to the rugged topography of rift regions.

Students Sometimes Ask...

You mentioned that most mountains are the result of crustal deformation. Are there areas that exhibit mountainous topography but have been produced without crustal deformation?

Yes. Plateaus—areas of high-standing rocks that are essentially horizontal—are one example of a feature that can be deeply dissected by erosional forces into rugged, mountain-like landscapes. Although these highlands resemble mountains topographically, they lack the structures associated with orogenesis. The opposite situation also exists. For instance, the Piedmont section of the eastern Appalachians exhibits topography that is nearly as subdued as that seen on the Great Plains. Yet, because this region is composed of deformed metamorphic rocks, it is clearly part of the Appalachian Mountains.

Volcanism that accompanies the stretching and faulting of the crust contributes further to the high relief produced in continental rift margins. For example, Mount Kilimanjaro (5895 metres), located in Tanzania just adjacent to the Kenya border, was produced by mafic volcanism that accompanied the formation of the East African Rift.

◆ **Figure 20.5**   Mountainous terrain of Jebel Zabara along the Red Sea coast of Egypt. The highlands represent the exposed edges of tilted fault blocks of the Red Sea rift system. (Photo by W.R. Church)

block rose upward to achieve isostatic equilibrium (Figure 20.14). In such places as Gros Morne National Park, Newfoundland, the ophiolite sequence is so complete that the base of the oceanic crust can be observed (see Figure 18.14B).

Thick slices of ophiolites also occur in mountain ranges marking the collision of two continents, as seen in Alps and Himalayas. The scientific value of ophiolites preserved in the suture zones of such mountain ranges is obvious, for they provide tangible evidence of ocean basins that separated the continents prior to their collision.

Isostatic Adjustment after Active Convergence

At the beginning of this chapter, we used the example of wooden blocks of different heights floating in water to illustrate why some regions of Earth's surface stand higher than others. We noted that the thicker blocks have thicker roots and float higher than thinner blocks.

Applying the concept of isostasy, we should expect that when material is added to the top of a crustal block, such as the thrusted masses of rock piled up during mountain building, the root of the crustal block will subside to deeper level. Accordingly, we should expect the original block to rebound and the thickness of its root to be reduced when weight is removed. (Visualize what happens to a ship as cargo is being loaded and unloaded.)

Much of our understanding of how the crust responds to the application and removal of loads has emerged from the study of the after-effects of the last Ice Age. When continental ice sheets occupied portions of North America during the Pleistocene Epoch, the added weight of 3-kilometre-thick masses of ice caused downwarping of Earth's crust by hundreds of metres. In the 8000 years since the last ice sheet melted, uplifting of as much as 330 metres has occurred in the Hudson Bay region, where the thickest ice had accumulated (see Figure 12.30, p. 354).

Because of isostasy, deformed strata experience regional uplift both during mountain building and for an extended period afterward. Usually, little additional deformation is associated with the latter activity. As the crust rises, the processes of erosion accelerate, carving the deformed strata into a mountainous landscape.

One of the consequences of isostatic adjustment is that as erosion lowers the summits of mountains, the crust will rise in response to the reduced load (Figure 20.16A, B). The processes of uplifting and erosion will continue until the mountain block reaches "normal" crustal thickness. When this occurs, the mountains will be eroded to near sea level, and the once

deeply buried interior of the mountain will be exposed at the surface. In addition, as mountains are worn down, the eroded sediment is deposited on adjacent landscapes, causing these areas to subside (Figure 20.16C). Isostatic adjustment is responsible for both exposing the deep roots of the Appalachian mountains in eastern North America, but has also contributed to the large amounts of sediment deposited along the Atlantic and Gulf Coastal Plain that now cover the older bedrock of accreted terranes (Figure 20.15).

Mountains Produced at Transform Margins

As discussed in the previous chapters, transform plate boundaries involve the lateral sliding of one plate past another. Transform plate boundaries are not nearly as common in continental lithosphere as they are in oceanic plates, and as such, are not associated with mountain building in many continental areas. In addition, the stresses imposed on the crust at transform boundaries tend to be more localized than at divergent or convergent boundaries and therefore do not deform the crust on a regional scale. Nevertheless, mountains are associated with transform boundaries in some areas, including the San Andreas fault system of southwestern North America.

The formation of mountains along transform boundaries is linked to local areas of stress generated at bends in the fault system that deviate from the main trend of fault movement. Lateral movement along the fault produces local areas of compression that experience intense deformation. The Transverse Ranges just north of Los Angeles were formed as a result of such local compression along a westward-trending kink in the San Andreas fault system.

Broad Vertical Movements in Continents

Regions of Uplift

Although most young mountain belts, including the Alps, Andes, and Himalayas, have deep crustal roots and are supported by isostatic buoyancy, another important mechanism contributes to the elevated position of some landforms. With the development of the plate tectonics theory, it became clear that some form of up-and-down convective flow was occurring in the mantle. While the exact nature of this movement is still not fully understood, it is known that zones of hot mantle upwelling underlie some landmasses. The buoyancy of this hot rising material accounts for broad upwarping in the overlying lithosphere.

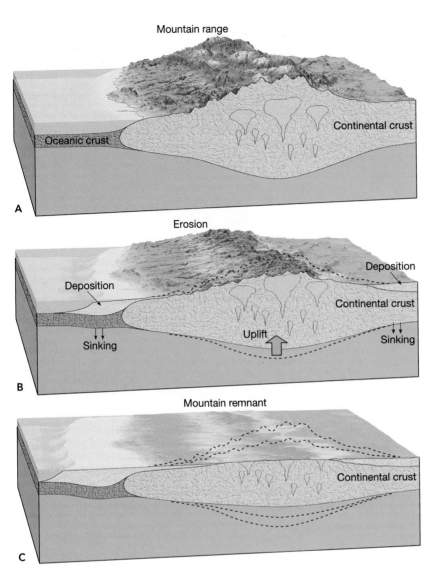

Figure 20.16 This sequence illustrates how the combined effect of erosion and isostatic adjustment results in a thinning of the crust in mountainous regions. A. When mountains are young, the continental crust is thickest. B. As erosion lowers the mountains, the crust rises in response to the reduced load. C. Erosion and uplift continue until the mountains reach "normal" crustal thickness.

Basin and Range Topography   In the American West, extending from the Front Range of the southern Rocky Mountains across the Colorado Plateau and through the Basin and Range province, the topography consists of lofty peaks and elevated plateaus (see Figure 20.10). According to the plate tectonics model, mountain belts are produced along continental margins in association with convergent plate boundaries. But this region of mountainous topography extends inland almost 1600 kilometres, far from the nearest plate boundary.

It was once assumed that, like other regions of mountainous topography, this area stood tall because the crust had been thickened by past geologic events. However, seismic studies conducted across the southern Rockies revealed a crustal thickness no greater than that found below Denver. These data ruled out crustal buoyancy as the cause for the abrupt 2-kilometre jump in elevation that occurs where the Great Plains meet the Rockies. Instead, this elevation difference must somehow be the result of mantle flow. In fact, seismic evidence indicates that the mantle beneath this region is warm and buoyant compared to the relatively cool, dense mantle beneath the eastern United States.

The event that may have triggered this period of uplift started with the nearly horizontal subduction of the Farallon plate eastward beneath North America as far inland as the Black Hills of South Dakota (Figure 20.17A). As the subducted slab scraped beneath the continent, compressional forces initiated a period of tectonic activity. About 65 million years ago the Farallon plate began to sink into the mantle. As this relatively cool plate gradually separated from the lithosphere above, it was replaced by hot rock that upwelled from the mantle (Figure 20.17B). According to this scenario, the hot mantle provided the buoyancy to raise the southern Rockies, as well as the Colorado Plateau and the Basin and Range province.

In the southern Rockies this event uplifted large blocks of ancient basement rocks along high-angle faults to produce mountainous topography separated by large sediment-filled basins. The upwelling that is associated with the Basin and Range province started about 50 million years ago and remains active today. Here the buoyancy of the warm material caused upwarping and rifting that elongated the overlying crust by 200 to 300 kilometres. The lower crust is ductile and easily stretched. The upper crust, on the other hand, is brittle and deforms by faulting (see Figure 15.22, p. 429). The extension and faulting broke the uplifted crust, causing individual blocks to slump. The high portions of these tilted blocks make up the mountain ranges, whereas their low areas form the basins, now partially filled with sediment.

It is important to point out that not all geologists studying this region agree with the model just presented. Another hypothesis suggests that the piecemeal addition of terranes to North America produced the uplift observed in the American West.

Continent-Scale Uplift   Southern Africa is one area where large-scale vertical motion is evident. Here much

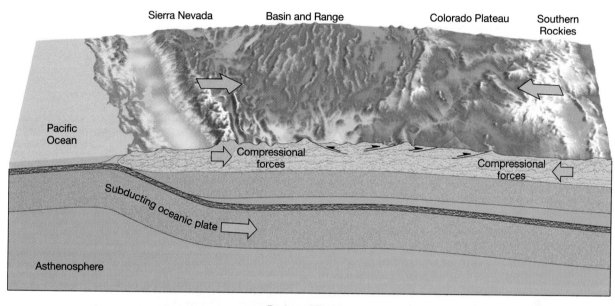

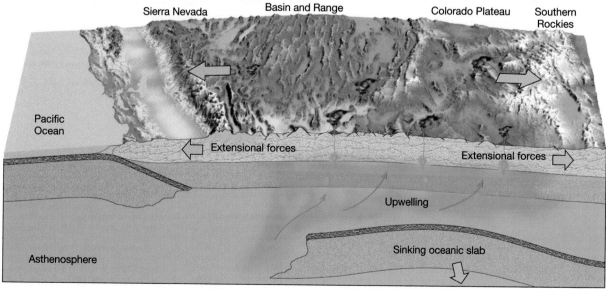

◆ **Figure 20.17**   One possible mechanism for the uplift that led to the formation of the southern Rockies, Colorado Plateau, and Basin and Range province. **A.** Nearly horizontal subduction of an oceanic plate initiated a period of tectonic activity. **B.** Sinking of this oceanic slab allowed for upwelling of hot mantle material that buoyantly raised the crust.

of the region exists as an expansive plateau having an average elevation of nearly 1500 metres. Geologic evidence shows that southern Africa and the surrounding seafloor have been slowly rising for the past 100 million years, even though it has not experienced a plate collision for nearly 400 million years. As mentioned in Chapter 17 (Box 17.3), such large-scale uplift might be related to the link between surface and deep internal Earth processes.

Regions of Subsidence

While some continental regions show evidence of broad uplift, a host of others show evidence of gentle downwarping. We know that some downwarping is simply the result of an extra load being added to the crust, such as occurred during the last Ice Age. Other examples of crustal loading occur along the continental margins where sediments are being deposited. This phenomenon is particularly dramatic in regions like the Gulf of Mexico, where the Mississippi River has dumped enormous quantities of sediment. In the case of broad regional uplift, however, some downwarped areas may be linked to flow dynamics of the mantle.

Foreland and Intracratonic Basins   Continental interiors, although having a more subdued topography than mountain belts along their margins, show a number of basins and intervening arches that manifest subtle deflections of the lithosphere. Of particular importance in North America are *foreland* and *intracratonic basins*, which have served as important catchment areas for sediments that now host significant volumes of oil and gas.

Foreland basins are elongate, asymmetrical, trough-like depressions (Figure 20.18) that are produced by the loading of a continental mass along one of its edges. Similar to how the free end of a diving board bends in response to the load of a diver, the lithosphere flexes downward under the increased mass of thickened crust. Among the best-known basins of this type in Canada is the Western Canada Foreland Basin.

This basin developed inland of the Rocky Mountains during the Mesozoic, in response to the crustal shortening and thickening that accompanied the accretion of terranes in British Columbia. Similarly, the Appalachian Foreland Basin in Eastern Canada was produced by crustal thickening that accompanied the accretion of terranes and uplift of the early Appalachian mountains during the Paleozoic.

Intracratonic basins are circular, bowl-shaped depressions that develop well within continental interiors. Two classic examples of intracratonic basins are the Williston Basin, which underlies the southern part of the Canadian prairies, and the Michigan Basin, the northern part of which underlies southwestern Ontario (Figure 20.19). The origin of intracratonic basins is somewhat obscure and remains a subject of significant debate. A few geologists believe that the formation of such basins is linked to far-field compressive forces related to plate convergence, but most assert that the circular basin shape must indicate a mantle-related phenomenon acting independently of plate interactions. In the latter camp, it has been suggested that an intracratonic basin forms by lithospheric thinning and subsequent sagging associated with the ascent of mantle plumes. Others have suggested that intracratonic basins represent circular dimples produced by blobs of dense mantle material that attached to the base of the lithosphere and later sank due to their increased density upon cooling.

Downwarping of Continental Margins and Interiors
There is also evidence to suggest that much of the North American interior experienced a major episode of subsidence during the Cretaceous Period. Although sea level was no more than 120 metres higher than it is today, vast seas invaded much of the continent. The Cretaceous seas eventually retreated, leaving behind thick accumulations of marine sedimentary rocks. Today these largely undeformed strata are found hundred of metres above sea level in the Canadian prairies (Figure 20.20).

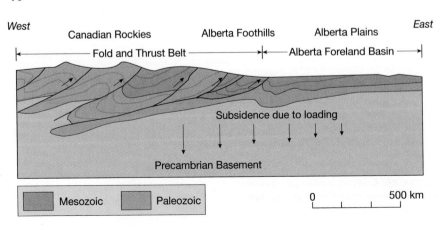

◆ **Figure 20.18**   Simplified cross section through Canadian Rocky Mountains, foothills, and Alberta Plains. A foreland basin underlies the plains region due to loading by thrust sheets that accompanied the shortening of crust that formed the Rockies. Note that crustal shortening continued long after the main phase of foreland basin development. Thus, Mesozoic sedimentary rocks that originally accumulated in the western part of the foreland basin have been disturbed by thrust faulting.

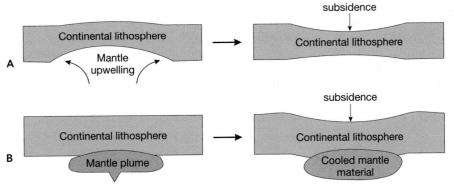

◆ **Figure 20.19** Two possible mechanisms responsible for forming a circular intracratonic basin. **A.** The lithosphere thins and sags due to localized mantle upwelling. **B.** The lithosphere is weighed down locally by a mass of mantle material that has attached to the base of the lithosphere and has sunk due to increased density upon cooling.

Similar episodes of large-scale downwarping are known on other continents, including Australia. The cause of these downward movements followed by rebound may be linked to subduction of oceanic lithosphere. One proposal suggests that when subduction ceases along a continental margin, the subducting slab detaches from the trailing lithosphere and continues its descent into the mantle. As this detached lithospheric slab sinks, it creates a downward flow in its wake that tugs on the base of the overriding continent. In some situations the crust is apparently pulled down enough to allow the ocean to extend inland. As the plate segment sinks deeper, the pull of the trailing wake weakens and the continent "floats" back into

isostatic balance. Although computer modelling supports this proposed mechanism, observational data to test this hypothesis has yet to be gathered.

The Origin and Evolution of Continental Crust

Earlier in this chapter, we learned that the theory of plate tectonics provides a model from which to examine the formation of Earth's major mountainous belts. But what roles have plate tectonics and mountain building played in the events that led to the origin and evolution of the continents? At this time no single answer to this question has met with overwhelming acceptance.

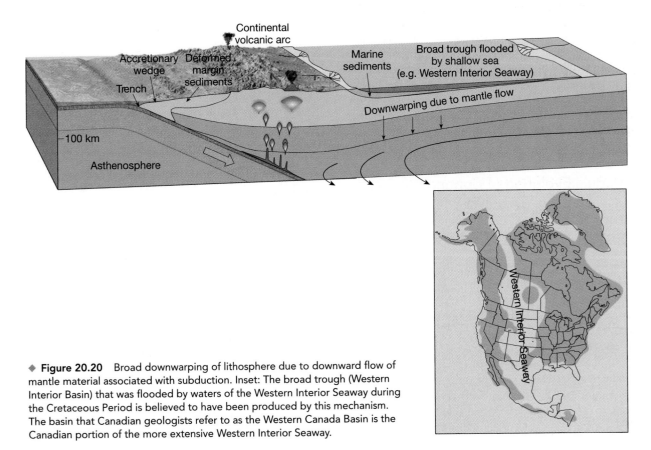

◆ **Figure 20.20** Broad downwarping of lithosphere due to downward flow of mantle material associated with subduction. Inset: The broad trough (Western Interior Basin) that was flooded by waters of the Western Interior Seaway during the Cretaceous Period is believed to have been produced by this mechanism. The basin that Canadian geologists refer to as the Western Canada Basin is the Canadian portion of the more extensive Western Interior Seaway.

The lack of agreement among geologists can be partly attributed to the complex nature and antiquity of most continental material, which makes deciphering its history very difficult. Whereas oceanic crust has a relatively simple layered structure and a rather uniform composition, the much older continental crust consists of a collage of highly deformed and metamorphosed igneous and sedimentary rocks. Moreover, in many places, such as the Canadian Prairies, the basement rocks are mantled by thick accumulations of much younger sedimentary rocks—a fact that inhibits detailed study. Nevertheless, during the last few decades, great strides have been made in unravelling the secrets held by the rocks that make up the stable continental interiors.

Early Evolution of the Continental Crust

At one extreme is a proposal that the continental crust formed early in Earth's history, perhaps in the first billion years. During this period, chemical differentiation resulted in the upward migration of the less-dense, silica-rich constituents of the mantle to produce a "scum" of continental-type rocks. The more dense silicate minerals, those enriched in iron and magnesium, remained in the mantle.

Shortly after this period of chemical differentiation, a mechanism that may have resembled plate tectonics continually reworked and recycled the continental crust. Through such activity, the continental crust was deformed, metamorphosed, and even remelted. However, because of their buoyancy, these silica-rich rocks either were never consumed into the mantle or, if they were subducted, they melted and became small or possibly nonexistent. Then, through gradual chemical differentiation of mantle material, the continents slowly grew over vast spans of geologic time.

Gradual Evolution of Continental Crust

An opposing view contends that the continents have grown larger through geologic time by the gradual accretion of material derived from the upper mantle. A main tenet of this hypothesis is that the primitive crust was of an oceanic type and the continents were returned to the crust as magma. Thus, the essence of this hypothesis is that the total volume of continental crust has not changed appreciably since its origin—only the distribution and shape of the landmasses have been modified by tectonic activity.

This view proposes that the formation of continental material takes place in multiple stages, as shown in Figure 20.21. The first step occurs in the upper mantle directly beneath the oceanic ridges. Here partial melting of the rock peridotite yields basaltic magma,

which rises to form oceanic crust. The rocks of the ocean floor are higher in silica, potassium, and sodium and lower in iron and magnesium than the rocks of the upper mantle from which they were derived.

As new ocean floor is generated at the ridge crests, older oceanic crust is being destroyed at the oceanic trenches. As an oceanic plate is thrust into the mantle, heat and pressure drives water from the subducting crustal rocks. These volatiles migrate into the wedge of hot mantle that lie above and trigger melting. Once enough molten rock forms, it will buoyantly rise toward the surface. During its ascent, the magma usually undergoes chemical differentiation (Figure 20.21). This gives rise to comparatively silica-rich magma, which is emplaced in a volcanic arc.

According to this view, the earliest continental rocks came into existence at a few isolated island arcs. Once formed, these island arcs coalesced to form larger continental masses, while deforming the volcanic and sedimentary rocks that were deposited in the intervening oceans. Eventually this process generated masses of continental crust having the size and thickness of modern continents.

Evidence supporting this view of continental growth comes from research in regions of plate subduction, such as Japan and the western flanks of the Americas. Equally important, however, has been the research conducted in the stable interiors of the continents, particularly in the shield areas (see Figure 20.3, p. 555). All continents have these vast, flat-lying expanses of highly deformed and metamorphosed igneous and sedimentary rocks.

Radiometric dating of rocks from shield areas, including those in Canada, Minnesota, and Greenland, has revealed that the oldest terranes formed about 3.8 billion years ago. This date represents one of the earliest periods of mountain building. At that time, possibly only 10 percent of the present continental crust existed. The next major period of continental evolution may have taken place between 3 billion and 2.5 billion years ago, as indicated by radiometric dates of similar terranes found in the shield areas of Canada, Africa, and western Australia. This period of mountain building generated the Superior and Churchill Cratons shown in Figure 20.22. The locations of these ancient continental blocks during their formation are not known. However, about 1.9 billion years ago, these two continental fragments collided to produce the Trans-Hudson orogen (Figure 20.22). This mountain-building episode was not restricted to North America, because ancient deformed strata of similar age are found on other continents.

It is not known with certainty how many periods of mountain building have occurred since the formation of Earth. The last major episode of global

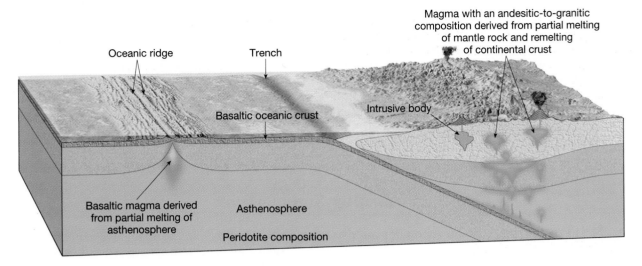

Oceanic ridge

Trench

Magma with an andesitic-to-granitic composition derived from partial melting of mantle rock and remelting of continental crust

Basaltic oceanic crust

Intrusive body

Basaltic magma derived from partial melting of asthenosphere

Asthenosphere

Peridotite composition

◆ **Figure 20.21**   The multistage process for transforming material from the asthenosphere into continental crust. Once continental crust is generated, its low density apparently keeps it afloat indefinitely.

scale evidently coincided with the closing of the proto-Atlantic and other ancient ocean basins during the formation of the supercontinent Pangea.

If the continents do in fact grow by accretion of material to their flanks, then the continents have grown larger at the expense of oceanic crust. Like the other proposal, this view assumes a buoyant and nonsubductable continental crust. Although continental crust remains afloat, continents have occasionally fragmented

and been carried along in a conveyor-belt fashion until they collided with other landmasses. Presently, Australia, which separated from Antarctica, is being rafted northward and will probably join Asia in much the same manner as India did in the recent geologic past. Thus, according to this view, fragmentation and the formation of new crustal rocks that accompanied the reshuffling of these fragments are responsible for the present volume, structure, and configuration of continents.

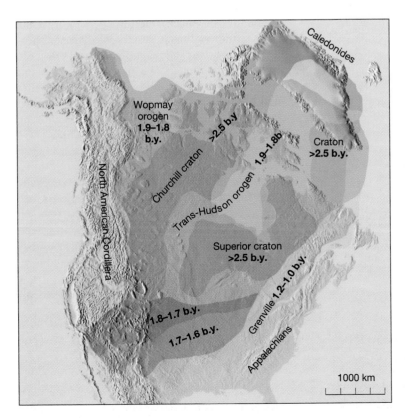

Caledonides

Wopmay orogen 1.9–1.8 b.y.

Churchill craton >2.5 b.y

1.9–1.8b

Craton >2.5 b.y.

North American Cordillera

Trans-Hudson orogen

Superior craton >2.5 b.y.

1.8–1.7 b.y.

1.7–1.6 b.y.

Grenville 1.2–1.0 b.y.

Appalachians

1000 km

◆ **Figure 20.22**   The map shows the major Precambrian mountain belts (orogens) and cores of ancient continental blocks (cratons) and their ages in billions of years (b.y.). It appears that North America was assembled by crustal blocks that were joined by processes very similar to modern plate tectonics. These ancient collisions produced mountainous belts that include remnant volcanic island arcs trapped by the colliding continental fragments.

A word of caution is in order. The views set forth in this section to explain the origin and evolution of the continents are speculative. Plate tectonics appears to be the major force in crustal evolution over the last 2 to 3 billion years. However, during the early history of Earth, the heat released by the decay of uranium, thorium, and potassium would have been more than twice as great as today. Was plate tectonics active early in Earth's history, only at a different rate, or were different processes in operation? Was the primitive crust composed primarily of continental rocks, or was it of the oceanic type? These are some of the questions that remain areas of active research.

Chapter Summary

- Earth's less dense crust floats on top of the denser and deformable rocks of the mantle, much like wooden blocks floating in water. The concept of a floating crust in gravitational balance is called *isostasy*. Most mountainous topography is located where the crust has been shortened and thickened. Therefore, mountains have deep crustal roots that isostatically support them. As erosion lowers the peaks, *isostatic adjustment* gradually raises the mountains in response. The processes of uplifting and erosion will continue until the mountain block reaches "normal" crustal thickness.

- The name for the processes that collectively produce a mountain system is *orogenesis*. Most mountains consist of roughly parallel ridges of folded and faulted sedimentary and volcanic rocks, portions of which have been strongly metamorphosed and intruded by younger igneous bodies.

- Major mountain systems form along *convergent plate boundaries*. *Andean-type mountain building* along continental margins involves the convergence of an oceanic plate and a plate whose leading edge contains continental crust. At some point in the formation of Andean-type mountains a *subduction zone* forms along with a *continental volcanic arc*. Sediment from the land, as well as material scraped from the subducting plate, becomes plastered against the landward side of the trench, forming an *accretionary wedge*. One of the best examples of an inactive Andean-type mountain belt is found in the western United States and includes the Sierra Nevada and the Coast Range in California.

- *Continental collisions*, in which both plates are carrying continental crust, have resulted in the formation of the Himalaya Mountains and the Tibetan Plateau. Continental collisions also formed many other mountain belts, including the Alps, Urals, and Appalachians.

- Recent investigations indicate that *accretion*, a third mechanism of orogenesis, takes place where *small crustal fragments collide and accrete to continental margins* along plate boundaries. The accreted crustal blocks are referred to as *terranes*. The mountainous topography of westernmost North America formed as the result of the accretion of terranes to North America.

- Convective flow in the mantle contributes to the up-and-down bobbing of the crust. The upward flow of a large superplume located beneath southern Africa is thought to have elevated this region during the last 100 million years. Crustal subsidence has produced large basins and may have allowed the ocean to invade the continents several times in the geologic past. In the western United States upwelling of hot mantle rock may be responsible for the elevated position of the southern Rockies, the Colorado Plateau, and the Basin and Range province.

- Geologists are trying to determine what role plate tectonics and mountain building play in the origin and evolution of continents. At one extreme is the view that most continental crust was formed early in Earth's history and has simply been reworked by the processes of plate tectonics. An opposing view contends that the continents have gradually grown larger by accretion of material derived from the mantle.

Review Questions

1. In the plate tectonics model, which type of plate boundary is most directly associated with mountain building?

2. Briefly describe the development of a volcanic island arc.

3. The formation of mountainous topography at a volcanic island arc, such as Japan, is considered just one phase in the development of a major mountain belt. Explain.

4. What is an accretionary wedge? Briefly describe its formation.

5. What is a passive margin? Give an example. Give an example of an active continental margin.

6. How does continental divergence produce mountains?

7. Suppose a sliver of oceanic crust was discovered in the interior of a continent. Would this support or refute the theory of plate tectonics? Explain.

8. Compare mountain building along an Andean-type convergent boundary and a continent-continent convergent boundary.

9. How can the Appalachian Mountains be considered a collision-type mountain range when the nearest continent is 5000 kilometres away?

10. How does the plate tectonics theory help explain the existence of fossil marine life in rocks atop the Ural Mountains?

11. In your own words, briefly describe the stages in the formation of a major mountain system according to the plate tectonics model.

12. Define the term *terrane*. How is it different from the term *terrain*?

13. List three structures that fall under the heading "oceanic plateaus." In addition to oceanic plateaus, what other features are thought to be carried by the oceanic lithosphere and eventually accreted to a continent?

14. On the basis of current knowledge, describe the major difference between the evolution of the Appalachian Mountains and the North American Cordillera.

15. Briefly describe the formation of the Canadian Rocky Mountains (see Box 20.1, p. 564).

16. Compare the forces of deformation associated with fault-block mountains to those associated with most other major mountain belts.

17. List some evidence that supports the concept of crustal uplift.

18. What evidence initially led geologists to conclude that mountains have deep crustal roots (see Box 20.2, p. 566)?

19. What happens to a floating object when weight is added? Subtracted? How does this principle apply to changes in the elevation of mountains? What term is applied to the adjustment that causes crustal uplift of this type?

20. How does the formation and melting of Pleistocene ice sheets support the idea that the lithosphere tries to remain in isostatic balance?

21. How do some researchers explain the elevated position of southern Africa?

22. What mechanism has been proposed to explain the subsidence in the interior of North America during the Cretaceous Period?

23. Briefly describe the difference between a foreland basin and an intracratonic basin.

24. Contrast the opposing views on the origin of the continental crust.

Key Terms

isostasy (p. 554)
isostatic adjustment (p. 557)

mountain (p. 555)
mountain belts (p. 555)

mountain ranges (p. 555)
orogenesis (p. 555)

passive margin (p. 557)
terrane (p. 562)

Web Resources

The *Earth* Web site uses the resources and flexibility of the Internet to aid in your study of the topics in this chapter. Written and developed by geology instructors, this site will help improve your understanding of geology. Visit **http://www.pearson.ca/tarbuck** and click on the cover of the text to find:

■ Online review quizzes.

■ Web-based critical thinking and writing exercises.

■ Links to chapter-specific Web resources.

■ Internet-wide key-term searches.

http://www.pearson.ca/tarbuck

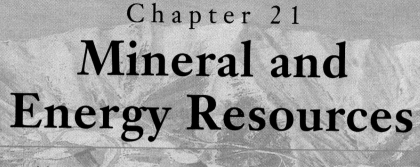

Chapter 21
Mineral and Energy Resources

Aerial view of the giant Bingham Canyon copper mine near Salt Lake City, Utah. (Photo by Michael Collier)

A popular saying among geologists is "if it can't be grown, it's gotta be mined." Materials extracted from the Earth are the basis for modern civilization (Figure 21.1). Everything we use in our lives that is not derived from living things comes from the solid Earth. We so depend on the Earth's resources that we generally take this fact for granted. Chances are that most of the metals, plastics, and energy sources that you are using today are derived from Earth resources that were ultimately discovered and developed with the help of geoscientists. For example, when you turn on your television or computer, you are actually using materials derived from the Earth. From the silicon chips to the copper in the wires, and from the glass screen to the energy that powers the computer, nearly all aspects of these high-tech devices are linked to mineral and fuel resources. The economic uses of several minerals, both metallic and nonmetallic, are indicated in Table 2.4 (page 54).

The mining industry is a vital contributor to the Canadian economy. The mining and mineral processing industries contributes over $35.1 billion to the Canadian economy, amounting to over $600 per person. Canada is also one of the world's largest exporters of minerals and mineral products. At least 80 percent of Canada's mineral and metal production is exported. Canada ranks first in the world for the production of potash and uranium, is the second largest producer of nickel and asbestos, and ranks in the top five for the production of zinc, cadmium, titanium concentrate, aluminum, platinum group metals, salt, gold, molybdenum, copper, gypsum, cobalt and lead. The five most important minerals in Canada with respect to production value are gold, nickel, potash, copper, and zinc.

The number of different **mineral resources** required by modern industries is immense (Figure 21.2), and although some countries, like Canada and the

◆ **Figure 21.1**   Toronto at night. As this scene reminds us, mineral and energy resources are the basis of modern civilization. (Photo © Lloyd Sutton/Masterfile)

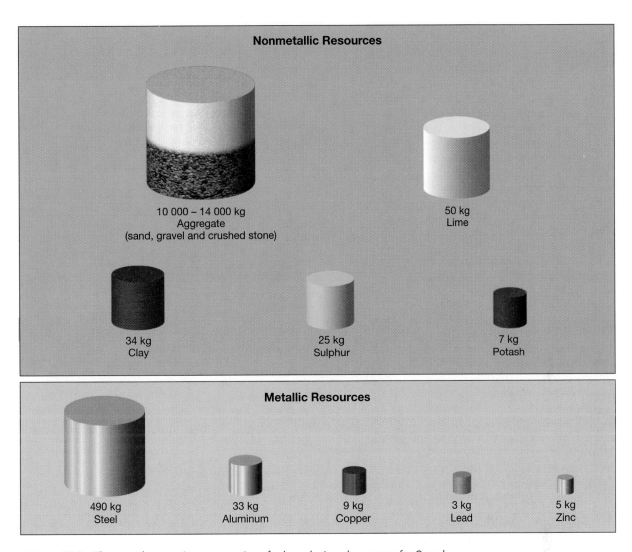

◆ **Figure 21.2**   The annual per capita consumption of selected mineral resources for Canada.
(Based on data from Statistics Canada, Aggregate Producers of British Columbia and Canadian Steel Producers Association)

United States, have rich supplies of many important minerals, no nation is completely self-sufficient. Consequently, resources of the Earth have long been the basis of world trade, have driven great waves of immigration, and have even generated wars.

Geoscientists are indispensable in many aspects of mineral and fuel resource discovery, measurement, and management. These include: predicting where an economically important resource might occur, determining the most effective way of extracting them, and determining the conditions of mineral deposit formation to aid in the discovery of similar deposits. At the same time, geoscientists are mindful of the great responsibilities nations face in their exploitation and utilization of natural resources (Box 21.1) and play crucial roles in assessing the possible environmental impacts of mineral extraction and developing techniques for environmental remediation.

Renewable and Nonrenewable Resources

Resources are the endowment of materials held in reserve that are useful to humans and are ultimately available commercially. Resources are commonly divided into two broad categories—renewable and nonrenewable.

Renewable resources can be replenished over relatively short time spans such as months, years, or decades. Common examples are plants and animals for food, natural fibres for clothing, and trees for lumber and paper. Energy from flowing water, wind, and the sun are also considered renewable. By contrast, **nonrenewable resources** continue to be formed in the Earth, but the processes that create them are so slow that significant deposits may take millions of years to accumulate. For human purposes, the Earth contains fixed quantities of these substances.

BOX 21.1

People and the Environment
Resource Consumption and Sustainability: A Global Perspective

Mike Powell*

In a global context, resources not only include minerals and fuel, but also anything else held in reserve that can be used in the future. For many of us, resources are those things that we use in excess to maintain our quality of life, or high standard of living (consumables, energy, etc.); this is a view embraced mainly by people living in "developed" countries. However, for most of the world's population, locally available resources are those things that simply maintain life on a daily basis (food, shelter, etc.). Many small countries possess a single resource (e.g., sandy white beaches or a pleasant climate), which translates into tourism dollars. From a global perspective, resources include the vast reservoirs of less tangible commodities such as air, water, and soil quality, as well as plant and animal species. Global resource consumption is not equitable or evenly distributed on a per capita basis. The vast majority of worldwide resources are channelled into economies that have the financial capacity to buy commodities.

Planet Earth can be thought of as a single mixed reservoir of all resources that are consumed mostly by a single compartment (humans). The rate of resource consumption is a product of population size and rate of consumption. The rate of population growth has accelerated exponentially throughout history. It took over 100,000 years for Earth to reach the 1 billion population mark, but it only took 13 years to add the 5th billion, and it will take only 12 years to reach the 6th billion. Increases in economic activity worldwide show that the richest 20 percent of people consume 85 percent of the resources while the poorest 25 percent use less than 1.5 percent. To put this into perspective, consider a scenario where China consumed resources at the same rate as North America. Conservatively, China would use the entire worldwide production of nearly all resources each year! Another interesting perspective is to consider how long the total static

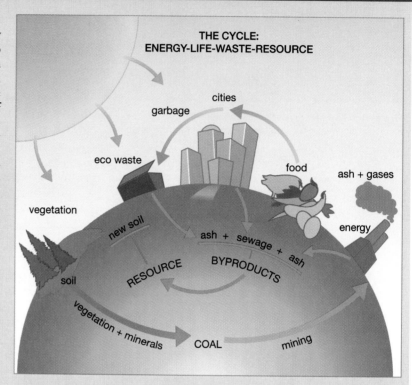

◆ **Figure 21.A**   Uses of "waste" resources from by-products. (Courtesy of Mike Powell)

resources would last if all the people in the world used resources at the same rate as North Americans—not very long, not even a generation.

Energy is the single most important resource. Per capita consumption is a measure of quality of life (and life expectancy) and industrial status worldwide. Approximately 25 percent of the global population use 75 percent of the total energy produced annually. North America (5 percent of the total population) uses twice as much energy as all developing countries combined and Western Europe (9 percent of the total population) uses about the same amount. On a per capita basis, people in developed countries use between 10 and 30 times as much energy as people in developing countries. Canada (6th largest consumer worldwide) uses 3.7 percent of the world's total generated electricity for a population of only 32 million. India (7th largest consumer worldwide), with a population of over 1 billion, uses less.

How long global resources will last depends on future population growth and consumption trends. Even at a conservative across-the-board increase of 1 percent, most of our proven mineral and energy reserves will last for less than 200 years. If the rate of consumption increases by 2 percent, static levels of resources will last for less than 100 years. So, whether or not we use resources in a "sustainable" way will be a complicated issue in the future, and one of the greatest problems facing future generations. The question remains: can we keep adding 100 million people to the surface of Planet Earth each year and still pretend that resource consumption can be tailored to "sustainability"? Figure 21.A shows the relationship between some types of waste and the need for new resources.

*Mike Powell is a professor in the Department of Earth Sciences at the University of Western Ontario.

When the present supplies have been mined or pumped from the ground, there will be no more. Examples are fuels (e.g., coal, oil, natural gas) and many important metals (e.g., iron, copper, uranium, gold). Some of these nonrenewable resources, such as aluminum, can be used over and over again; others, such as coal, cannot be recycled. Some resources can be placed in either category, depending on how they are used. Groundwater is one such example. Where it is pumped from the ground at a rate that can be replenished, groundwater can be classified as a renewable resource. However, in places where groundwater is withdrawn faster than it is replenished, the water table drops steadily. In this case, the groundwater is being "mined" just like other nonrenewable resources.

Resources include **reserves**, the already identified deposits from which minerals and fuel materials can be extracted profitably, and known deposits that are not yet economically or technologically recoverable. Resources may also include deposits that have not yet been discovered, but are inferred to exist by some evidence.

Metallic Mineral Deposits and Geologic Processes

An important point to be made about Earth's resources concerns the natural abundance of elements sought after; even the common metals such as aluminum and iron comprise less than 10 percent of average continental crust. Next, when one examines the distribution of mines and the primary mineral resources they extract, it becomes obvious that the various metals and other resources we rely on are not uniformly distributed across the globe—they are naturally concentrated only in particular areas. This uneven distribution is inherent to the Earth system. Not only does the composition of crust vary in different regions, but the effects of tectonic processes, weathering, sedimentation rates, the movement of fluids, and various other factors are expected to be subdued in some areas and enhanced in others. These circumstances, combined with the economics of extracting the resources, limit the localities on Earth where the mining of particular materials is practical.

Most of us have heard the term **ore**. This term is generally used to denote those useful metallic minerals (and a few nonmetallic minerals such as fluorite and sulphur) of sufficient concentrations to be mined at a profit. It is emphasized that a rock body may be considered an ore one day and not the next due to economic factors. These can include fluctuations in the price of the material sought after, the price of fuel required for the running of machinery and the efficiency of mining equipment. It should be noted that the term ore is not applied to all materials of economic interest. Most nonmetallic minerals used for such purposes as building stone, aggregate, abrasives, ceramics, and fertilizers are not usually called ores but are generally classified as **industrial minerals**.

In this section, a few of the most important types of metallic mineral deposits are discussed. The primary types of deposits discussed in this section and their contained metals are summarized in Table 21.1, and depicted in the context of plate tectonic environments in Figure 21.3. Canadian metallic mineral deposits of present and historical significance are shown in Figure 21.4.

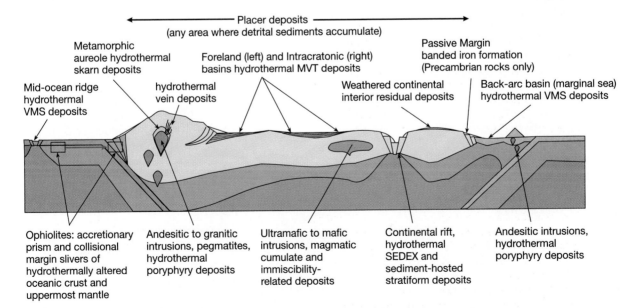

◆ **Figure 21.3** Distribution of common mineral deposit types and their contained metals with respect to plate tectonic environments. Abbreviations used for types of hydrothermal deposits: VMS = volcanogenic massive sulphide, SEDEX = Sedimentary − exhalative, MVT = Mississippi Valley type

◆ **Figure 21.19** Bauxite is the ore of aluminum and forms as a result of weathering processes under tropical conditions. Its colour varies from red or brown to nearly white. (Photo by E. J. Tarbuck)

vegetation for farming or forest growth. The long-term consequences of bauxite mining are obvious and continue to be points of concern for the many developing countries of the tropics.

Other Deposits Many copper and silver deposits result when weathering processes concentrate metals that are deposited through a low-grade primary ore. Usually such enrichment occurs in deposits containing pyrite (FeS_2), the most common and widespread sulphide mineral. Pyrite is important because when it chemically weathers, sulphuric acid forms, which enables percolating waters to dissolve the ore minerals. Once dissolved, the metals gradually migrate downward through the primary ore body until they are precipitated. Deposition takes place because of changes that occur in the chemistry of the solution when it reaches the groundwater zone (the zone beneath the surface where all pore spaces are filled with water). In this manner, the small percentage of dispersed metal can be removed from a large volume of rock and redeposited as a higher-grade ore in a smaller volume of rock. This enrichment process is responsible for the economic success of some metallic mineral deposits, particularly those containing copper.

Nonmetallic Resources

Earth materials that are not used as fuels or processed for metals are referred to as **nonmetallic mineral resources**. Realize the use of the world "mineral" is very broad in the economic context and is quite different from the geologist's strict definition of mineral found in Chapter 2. Nonmetallic mineral resources are extracted and processed either for the nonmetal-

lic elements they contain or for their desirable physical or chemical properties.

People often do not realize the importance of nonmetallic minerals because they see only the products that result from their use, not the minerals themselves. That is, many nonmetallics are used up in the process of creating other products. Examples include the fluorite and limestone that are part of the steelmaking process, the abrasives required to make a piece of machinery, and fertilizers needed to grow a food crop. Accordingly, the quantities of nonmetallics used in construction and other applications are huge.

Nonmetallic mineral resources are divided into two broad groups here—aggregate and stone, and industrial minerals. Because some substances have many uses, they can technically belong to either category. Limestone, for example, is used as crushed rock and building stone in its raw form, but its main mineral constituent, calcite, is important in its own right as the key ingredient for the manufacture of cement, and as a component of a score of other products including cosmetics and pharmaceuticals.

Aggregate and Stone

Because most building materials are widely distributed and present in almost unlimited quantities, they have little intrinsic value. Their economic worth comes only after the materials are removed from the ground and processed. Because their per-tonne value compared with metals and industrial minerals is low, mining and quarrying operations are usually undertaken to satisfy local needs. Except for special types of cut stone used for building and monuments, transportation costs greatly limit the distance most building materials can be moved. The aggregate and stone industry is very important in the lives of Canadians, as reflected in the thousands of sand and gravel pits and stone quarries that exist in our country.

Natural aggregate consists of crushed stone, sand, and gravel. From the standpoint of quantity and value, aggregate is a very important building material. It is produced in all provinces and territories in Canada and is used in nearly all road and building construction (Figure 21.20). Thanks to the large volume of sand and gravel in Canada deposited as glacial outwash during and at the end of the last ice age, every province and territory has an ample supply of aggregates for construction. Aggregate is used as-is in various construction applications, and is also important in the manufacture of concrete and road asphalt. To appreciate the importance of aggregate in our daily lives, one only needs to consider that just one kilometre of four-lane highway, such as Highway 401 in Ontario, requires more than 40 tonnes of aggregate.

◆ **Figure 21.20**   Aggregate pits are located in all provinces and territories of Canada. This aggregate pit is located near Aldergrove, British Columbia. (Photo by Steve Hicock)

Attesting to the diversity of Canada's geological makeup, many varieties of solid stone are quarried throughout the country, including rocks of igneous (e.g., granite), metamorphic (e.g., marble, gneiss, and slate), and sedimentary (e.g., sandstone, limestone, and dolostone) origin. Cut stone is largely used for aesthetic applications in construction, such as decorative facings, walkways, and countertops.

Industrial Minerals

Many nonmetallic resources are classified as industrial minerals. In some instances these materials are important because they contain specific chemical elements or compounds used in fertilizers or industrial chemicals. Some minerals possess useful physical properties. For example, garnet is both relatively hard and fairly common, making it a handy substance for the manufacture of abrasive products such as sandpaper. Other industrial minerals are used in the manufacture of products used in construction as fillers, coaters, and extenders in plastics, paints, and paper products and as heat-resistant materials, in addition to a host of other applications. Common industrial minerals and their uses are listed in Table 21.2. Some industrial minerals of particular importance in Canada, and their geologic occurrences, are indicated in Figure 21.21 and discussed next.

Diamonds   Diamonds are familiar as gemstones, but diamond is also extremely valuable as an abrasive by virtue of being the hardest natural substance known. Only about one quarter of all diamonds mined are used as gemstones.

The vast majority of diamonds are found in a unique ultramafic igneous rock called **kimberlite**. Magmas that produce kimberlite are believed to be generated by a small amount of partial melting in the asthenosphere at depths greater than 150 kilometres below Earth's surface. The diamonds themselves do not crystallize from kimberlitic magma. Rather, they are produced in the deep roots of the solid, lithospheric mantle beneath areas of old, stable, continental crust at depths of over 120 km (Figure 21.22). It is only here that high temperatures (between 900 and 1200°C) and pressures (over 40 kbar) remain stable for geologically long periods of time to allow the formation of diamonds.

Kimberlitic magma rapidly rises to the surface, acting as a sort of geological elevator that picks up chunks of lithospheric mantle material and crustal rock during its ascent. If kimberlitic magma passes through the deep areas of lithospheric mantle where diamonds reside, diamonds can be naturally sampled from this material as xenocrysts (foreign crystals) and as crystal components of xenoliths (foreign rock fragments) and can be carried upward by the magma.

TABLE 21.2   Occurrences and Uses of Nonmetallic Minerals

Mineral	Uses	Geological Occurrences
Apatite	Phosphorus fertilizers	Sedimentary deposits
Asbestos (chrysotile)	Incombustible fibres	Metamorphic alteration
Calcite	Aggregate; steelmaking; soil conditioning; chemicals; cement; building stone	Sedimentary deposits
Clay minerals (kaolinite)	Ceramics; china	Residual product of weathering
Corundum	Gemstones; abrasives	Metamorphic deposits
Diamond	Gemstones; abrasives	Kimberlite pipes; placers
Fluorite	Steelmaking; aluminum refining; glass; chemicals	Hydrothermal deposits
Garnet	Abrasives; gemstones	Metamorphic deposits
Graphite	Pencil lead; lubricant; refractories	Metamorphic deposits
Gypsum	Plaster of Paris	Evaporite deposits
Halite	Table salt; chemicals; ice control	Evaporite deposits; salt domes
Muscovite	Insulator in electrical applications	Pegmatites
Quartz	Primary ingredient in glass	Igneous intrusions; sedimentary deposits
Sulphur	Chemicals; fertilizer manufacture	Sedimentary deposits; hydrothermal deposits
Sylvite	Potassium fertilizers	Evaporite deposits
Talc	Powder used in paints, cosmetics, etc.	Metamorphic deposits

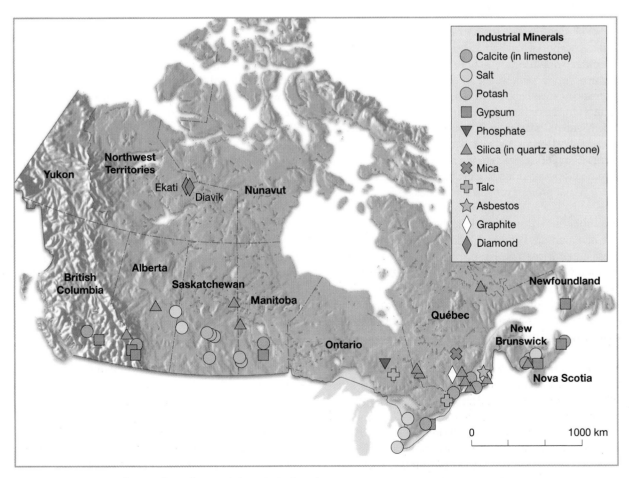

◆ **Figure 21.21**   Significant industrial mineral deposits in Canada.
(Data courtesy of Natural Resources Canada)

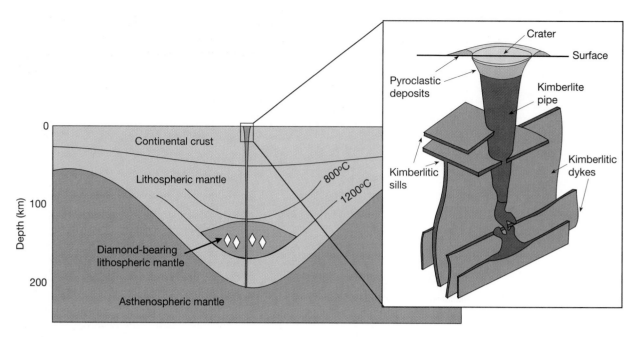

◆ **Figure 21.22**   Kimberlite pipes are carrot-shaped igneous bodies that are formed as mantle-sourced magma is violently extruded up to the Earth's surface. At the Ekati mine in the Northwest Territories, diamonds are mined from this rare type of igneous body.

During the last stage of magma ascent, and the rapid escape of carbon dioxide from the magma, together with the interaction of the hot kimberlitic magma with groundwater, an explosive volcanic eruption and a pipe-shaped vent is produced (Figure 21.22). The fractures and cavities within the vent are ultimately filled with kimberlite and shattered fragments of the surrounding rock. Associated with the main, carrot-shaped kimberlite pipe are ultramafic dykes and sills at depth and pyroclastic volcanic deposits at the surface.

Major producers of diamonds include South Africa (the hub of the early diamond industry), Australia, Namibia, and Russia. Canada has recently entered the diamond market through discoveries in the Northwest Territories. The Ekati mine, the first Canadian diamond mine, began production in 1998. It is projected that Canada may well supply 15 to 20 percent of the world's diamonds within the next decade.

Clays   Clays, the fine-grained, platy minerals formed by the weathering and hydration of aluminosilicate minerals, are important industrial minerals. Certain clay minerals can form economically important deposits at the place of origin in masses of weathered material. However, much larger quantities of clay minerals occur in sedimentary deposits of seas and lakes, having been transported long distances by wind or water from their site of formation. Major Canadian producers of clay include Saskatchewan, Ontario, and British Columbia.

The diverse physical and chemical properties of clays make them extremely useful in many applications. For example, sediment dominated by the clay mineral kaolinite (but also generally containing minor amounts of other minerals) is extensively mined for the production of bricks, sewer pipes, and pottery. As an industrial mineral, kaolinite is also used in the production of wall and floor tiles, porcelain, insulation products, ion-exchange products, as a filler (and natural whitener) for paint, adhesives, sealants, pharmaceuticals, cosmetics, and as a coater in paper. It might surprise you that the gloss finish on the pages of some books is actually a thin layer of kaolinite.

Clay minerals such as montmorillonite (a weathering product of volcanic ash) readily absorb moisture and are therefore useful as absorbents in products such as cat litter. Montmorillonite is also an important ingredient of lubricating drill mud in the petroleum industry. In addition, due to their high capacity for ion exchange, clays are used in oil refining and in waste treatment.

Carbonate Minerals   As discussed in Chapter 6, calcium carbonate in the form of calcite (and to a lesser extent, aragonite) is an important component of many sedimentary rocks. Vast quantities of limestone, most representing accumulations of calcareous skeletal material secreted by organisms, are quarried worldwide. While some limestone is quarried raw in the form of building stone, a great deal of it is extracted for its calcium carbonate content.

Large volumes of limestone are quarried and processed for the manufacture of cement. The principal raw materials used in the making of common cements are calcite (from limestone), with smaller amounts of silica (from sand), clay (from shale), and iron oxide. These provide the necessary ingredients—calcium, silica, alumina, and iron—that chemically interact to produce the desired qualities of cement. The raw materials are crushed into a powder and fed into the top of a tower containing a series of heating chambers. As the powdered substance drops through the series of chambers, it is heated to nearly 900°C, and then is fed through a kiln that heats the mixture further to about 1500–1600°C and converts the raw material into a substance called clinker. The key substance formed during this heating process is calcium silicate. The clinker is cooled, mixed with a small amount of gypsum (a regulator of cement setting time), and milled into a fine powder that is then ready for use as cement. When water is added to the cement powder, crystals of calcium silicate hydrate form an interlocking network that gives concrete its strength for construction purposes.

Calcium carbonate is also used in a great many other applications. It is used as a filler in asphalt, fertilizers, insecticides, paints, rubber and plastics, as a fluxing agent, as a mild abrasive in some toothpastes, as an acid neutralizer in a wide variety of cosmetic, medicinal, and paper products, and as a dietary supplement. In all likelihood, the very page you are reading in this text has a thin coat of calcium carbonate that has been added to the paper to provide a smooth, uniform surface for printing. Humans also use dolomite as a gentle abrasive and a dietary supplement.

Evaporite Salts   Evaporite deposits are important sources of many industrial minerals, particularly chloride and sulphate salts, and some carbonate minerals. During many times in Earth's past, widespread evaporite deposits were formed in shallow basins that were periodically flooded by seawater but dried up due to intense evaporation under arid conditions.

Normal seawater contains about 3.5 percent (by weight) dissolved matter that can be precipitated in mineral form when the water is evaporated. As observed in a glass of salty water left on the counter for several days, evaporation reduces the total volume of liquid, increasing the concentration of the dissolved matter to the point that salt can be precipitated. Partly due to differences in the relative concentrations of dissolved ions in seawater, minerals of different composition are precipitated sequentially as evaporation removes more and more water from the brine. In order of appearance, the principal minerals that form during progressive evaporation are: calcite,

gypsum, halite, and finally sylvite. Few evaporite deposits contain sylvite, probably because of the dilution of brines by episodic influxes of normal seawater from the adjacent ocean. It should also be noted that thick evaporite deposits represent several cycles of seawater influx, evaporation, and mineral precipitation rather than a single "drying-up" event in a basin.

Gypsum is primarily used for the manufacture of plaster products used in construction and other applications. Gypsum ($CaSO_4 \cdot 2H_2O$) is heated to about 180°C to drive off about three-quarters of its water (producing $CaSO_4 \cdot 1/2H_2O$), milled to a fine powder, and sold as plaster. When water is added to the powder, it rehydrates the material to produce tiny, interlocking crystals of gypsum that give plaster products their strength for interior construction applications. A plaster product familiar to most people is drywall (also called wallboard or gyprock), which consists of a sheet of plaster sandwiched between two sheets of paper. Physicians also commonly use plaster to make casts for patients who require their limbs to be stabilized as their fractured bones heal. Gypsum is also widely used as a filler or additive in many products, including cosmetics, pharmaceuticals, fertilizers, paints, and plastics. Significant deposits of gypsum occur in British Columbia, Manitoba, Ontario, Nova Scotia, and Newfoundland.

Halite, or salt, as it is more commonly known, is another important resource mined from evaporite deposits. The consumption of salt in Canada is very high, averaging about 360 kilograms of salt per person per year. This high figure reflects not only our liberal use of salt in the de-icing of roads, but also the extensive use of halite in the manufacture of important substances used in the chemical industry such as chlorine, caustic soda, baking soda and washing soda. It is also used to "soften" water, and of course, to season and preserve foods. Salt is a common evaporite and thick deposits are exploited using conventional underground mining and brining, which involves dissolving salt underground and re-precipitating it at the surface. In Canada, extensive Paleozoic salt deposits occur in the sedimentary basins of the Prairie Provinces, southwestern Ontario, Québec, and the Maritime Provinces. Globally, salt is extracted directly from seawater in arid, coastal areas, using the natural evaporating abilities of the Sun's heat.

The mining of potash, an evaporite rock containing the potassium-bearing minerals sylvite (KCl) and carnallite ($KClMgCl_2 \cdot 6H_2O$) is a huge industry in Canada. Most of Canada's potash is mined from immense deposits in Saskatchewan, where it occurs in horizontal layers up to three metres thick (Figure 2.26). Potash deposits also occur in the Maritime Provinces and are actively mined in New Brunswick. About 95 percent of all potash mined in Canada is exported to other countries, where it is primarily used in the production of fertilizers. It is also used in de-icing products, in the manufacture of soaps and detergents, and as an additive in glass and ceramic products.

Phosphate   Phosphate, contained in the mineral apatite, is mined primarily for the production of phosphoric acid for fertilizers, additives for livestock feed, industrial chemicals, and some home products. Although generally scarce in nature, phosphate can reach high concentrations in sedimentary rocks. The manner in which these deposits form is somewhat enigmatic, but appears to require a combination of high primary productivity, low oxygen, and low input rates of detrital sediment in a marine environment. Such conditions appear to be fairly common in oceanic environments influenced by the upwelling cold, nutrient-rich seawater. North America's major phosphate mines are restricted to the states of Florida, North Carolina, Idaho, and Utah, although sedimentary phosphate deposits are also known to exist in British Columbia. Canada's only active phosphate mine is located in Kapuskasing, Ontario, where the weathering of an igneous intrusion has left a rich residue of apatite crystals.

Sulphur   Because of its many uses, sulphur is an important nonmetallic resource. Sulphur is a major component of sulphuric acid and fertilizers: it is also featured in matches and gunpowder. Sources of sulphur include deposits of native sulphur associated with salt domes and volcanic areas and sulphides such as pyrite. In recent years, an increasingly important source has been the sulphur removed from coal, oil, and natural gas in order to reduce the sulphur dioxide emissions produced by their combustion.

Other Industrial Mineral Deposits   In addition to igneous and sedimentary deposits of industrial minerals, deposits of metamorphic origin also occur. Recall that, at convergent plate boundaries, the oceanic crust and sediments that have accumulated at the continental margins are carried to great depths. In these high-temperature, high-pressure environments, the mineralogy and texture of the subducted material are altered, producing deposits of nonmetallic minerals such as talc (used widely in cosmetic products) and graphite (used in the manufacture of steel products, lubricants, and pencils). Chrysotile asbestos is another product of metamorphism, resulting from the flushing of fluids through ultramafic rock. Concentrations of industrial minerals can also be produced in association with skarns during contact metamorphism. These minerals include garnet and corundum, used in abrasive products.

Nonrenewable Energy Resources
Petroleum

Petroleum is the group of hydrocarbons that includes oil and gas, and is presently a vital resource for the daily activities of humankind. It is mainly used as fuel, but it also comprises the main ingredients in plastics that find their way into synthetic fabrics, product packaging, casings, and a wide range of other products.

How Petroleum Is Formed   The origin of petroleum ultimately lies in organic matter contained in the tissues of living things. The vast majority of petroleum is derived from the remains of simple, aquatic organisms (mostly planktonic algae) that constitute most of the plankton in the seas and other large bodies of water, although some land-derived organic matter may also contribute to petroleum generation. Organic matter has a rather low preservation potential in most aquatic environments, since scavenging organisms and aerobic microbes naturally recycle it. However, it often accumulates in quiet-water environments of shallow seas and lakes, where low-oxygen conditions prevail. Accordingly, organic-rich sediments are typically fine-grained, and are represented in ancient deposits by shales and fine-grained carbonate rocks. Because they are the principal sources of organic matter from which petroleum is generated, these organic-rich rocks are called *source rocks*.

Although organic matter stands a better chance of survival under low-oxygen conditions, it is by no means immune to microbial attack. Organic-rich sediment can initially yield significant quantities of *biogenic gas* (principally methane) due to the breakdown of organic matter by anaerobic microbes (Figure 21.23). This microbial attack also removes much of the oxygen, nitrogen, phosphorus, and sulphur from the organic matter, leaving a residue that is enriched in hydrocarbons. With progressive burial, the production of biogenic gas ceases, the organic-rich sediment undergoes lithification, and heat becomes the main agent in the breakdown of the organic matter.

As the hydrocarbon-rich organic matter of a source rock is subjected to higher and higher temperatures, they are transformed into a solid waxy substance called *kerogen*. At temperatures above about 50°C, carbon-carbon bonds within the kerogen break, producing the lighter, less complex hydrocarbon molecules. This process is called cracking. As the temperature rises, more carbon-carbon bonds are broken, both in the kerogen and in the hydrocarbon molecules that formed previously, generating oil (Figure 21.23). Oil generation peaks at a temperature of about 100°C, beyond which the kerogen is depleted of its hydrocarbon supply, and significant amounts of

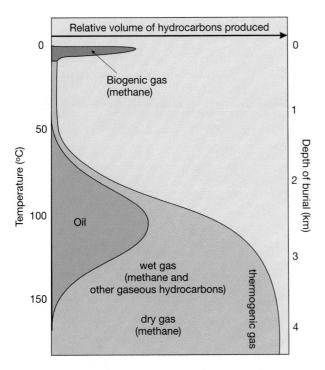

Relative volume of hydrocarbons produced

◆ **Figure 21.23** The temperatures and corresponding depths of burial at which oil and gas are generated. Note that biogenic gas is only produced at low temperatures, whereas themogenic gas produced by the thermal cracking of organic molecules dominates at higher temperatures. Oil generation is restricted to a temperature window between approximately 50 and 150°C. Note also the large amounts of natural gas produced at the expense of oil as hydrocarbons are cracked to progressively simpler molecular forms.

themogenic natural gas are produced (Figure 21.23). The natural gas produced at temperatures just above that of peak oil generation is commonly called *wet gas* because, in addition to its primary constituent, methane, it contains gaseous components such as ethane, propane, and butane, which can be separated by condensation. At temperatures above about 150°C, only *dry gas*, consisting almost entirely of methane, is generated.

It should be noted that the specific proportions of oil and gas generated in any given source rock depend not only on burial depth and duration of cooking, but can also reflect the composition of the organic matter from which the hydrocarbons are derived. Whereas organic matter from marine sources is *oil-prone*, land-derived organic matter tends to be more *gas-prone*. Because land-derived organic matter can be transported into aquatic environments, some source rocks can be expected to be more gas prone than others.

Petroleum Migration Unlike the organic matter from which they form, oil and gas are mobile. These fluids are gradually squeezed from the compacting, fine-grained source rocks where they originate into adjacent permeable beds such as sandstone, where openings between sediment grains are larger. Because this occurs underwater, the rock layers containing the oil and gas are saturated with water. Because oil and gas are less dense than water, they migrate upward through the water-filled pore spaces of the enclosing rocks. Unless something halts this upward migration, the fluids will eventually reach the surface, at which point the volatile components will evaporate.

Petroleum Traps Sometimes the upward migration of oil and gas is halted. A geologic environment that enables economically significant amounts of oil and gas to accumulate underground is called a **petroleum trap**. Several geologic structures may act as petroleum traps, but all have two basic conditions in common: a porous, permeable **reservoir rock** that will yield oil and natural gas in sufficient quantities to make drilling worthwhile; and a **cap rock**, such as shale, that is virtually impermeable to oil and gas. The cap rock halts the upwardly mobile oil and gas and keeps them from escaping to the surface. Petroleum traps can be produced in a number of ways, but they can be separated into two groups: structural traps and stratigraphic traps.

Structural Traps are petroleum traps that form due to the structural deformation of the rock layer that contains the hydrocarbons. One of the simplest structural traps is an anticline, an uparched succession of sedimentary strata (Figure 21.24A). As the strata are bent, the rising oil and gas collect at the apex (top) of the fold. Because of its lower density, the natural gas collects above the oil. In turn, the oil rests upon the denser water that saturates the reservoir rock. One of the world's largest oil fields, El Nala in Saudi Arabia, is the result of anticlinal traps, as is the famous Teapot Dome in Wyoming.

Fault traps form when strata are displaced in such a manner as to bring a dipping reservoir rock into position opposite an impermeable bed, as shown in Figure 21.24B. In this case the upward migration of the oil and gas is halted where it encounters the fault.

In the Gulf coastal plain region of the United States, important accumulations of oil occur in association with *salt domes*. Such areas have thick accumulations of sedimentary strata, including layers of rock salt. Salt occurring at great depths has been forced to rise in columns by the pressure of overlying beds. These rising salt columns gradually deform the overlying strata causing oil and gas to accumulate in the upturned sandstone beds adjacent to the salt column (Figure 21.24C).

Stratigraphic Traps are petroleum traps that are formed due to lateral and vertical changes in

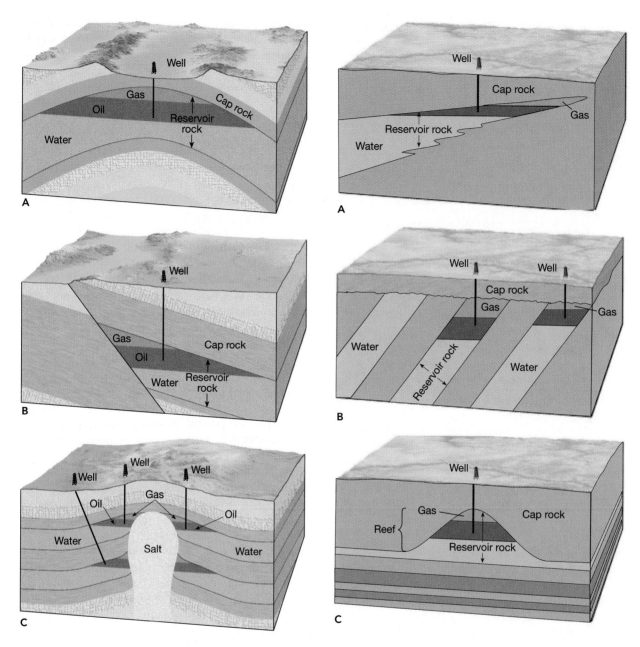

◆ **Figure 21.24** Common structural petroleum traps.
A. Anticline. **B.** Fault trap. **C.** Salt dome.

◆ **Figure 21.25** Common stratigraphic petroleum traps.
A. Pinch-out trap. **B.** Unconformity trap. **C.** Limestone reef.

rock type that reflect patterns of sediment deposition (Figure 21.25). A common type of stratigraphic trap is a pinch-out trap, characterized by the tapering off of a reservoir unit into a body of impermeable rock. An unconformity trap results from the truncation of a reservoir unit by erosion, followed by the deposition of impermeable cap rock sediments. Reefs also form important stratigraphic traps, by virtue of representing lens-shaped bodies of porous, permeable limestone that are commonly encased by impermeable shales.

Coal

Whereas petroleum is generally sourced from organic matter provided by aquatic plankton, coal is largely derived from the fibrous and woody remains of land plants. Similar to petroleum, the organic remains that form coal can only accumulate in appreciable quantities in oxygen-poor environments. Accordingly, accumulations of coal-forming organic matter tend to be restricted to areas such as swamps. In a swamp, the decomposition of organic matter consumes so much oxygen from the water that much of the organic matter escapes decomposition. The organic matter that survives decomposition is buried to progressively greater depths, where it is gradually cooked and compressed to form coal. This was discussed in greater detail in Chapter 6.

Coal has been an important fuel for centuries. In the nineteenth and early twentieth centuries, cheap and plentiful coal powered the Industrial Revolution. Until the 1950s, coal was an important domestic fuel as well as a power source for industry. However, its direct use in the home has largely been replaced by oil, natural gas, and electricity. These fuels are preferred because they are more readily available (delivered via pipes, tanks, or wiring) and cleaner to use. Nevertheless, coal remains the major fuel used by many nations.

Although coal is plentiful, its recovery and use present a number of problems. Surface mining can turn the countryside into a scarred wasteland if careful (and costly) reclamation is not carried out to restore the land. Although underground mining does not scar the landscape to the same degree, it has been costly in terms of human life and health.

Underground mining long ago ceased to be a pick-and-shovel operation and is today a highly mechanized and computerized process (Figure 21.26). However, the hazards of collapsing roofs, gas explosions, and working with heavy equipment remain.

Air pollution is a major problem associated with the burning of coal. Much coal contains significant quantities of sulphur. Despite efforts to remove sulphur before it is burned, the sulphur is converted into noxious sulphur oxide gases. Through a series of complex chemical reactions in the atmosphere, the sulphur oxides are converted to sulphuric acid, which then falls to the Earth's surface as rain or snow. This acid precipitation can have adverse ecological effects over widespread areas (see Box 5.1, p. 137).

As none of the problems just mentioned are likely to prevent the increased use of this important and abundant fuel, stronger efforts must be made to correct the problems associated with the mining and use of coal. Disposal of fly ash produced by coal burning also poses problems. However, as indicated in Figure 21.A, this material may well prove to be a future resource in itself.

Fossil Fuels in Canada

Canada is fortunate to host large quantities of coal and petroleum—so much that most of it is exported. The overwhelming majority of Canada's petroleum is produced in the southern areas of the Prairie Provinces and British Columbia that overlie strata of the Western Canada Sedimentary Basin (Figure 21.27). Through much of Paleozoic and Mesozoic time, warm, shallow seas, within which organic-rich sediments were deposited, occupied this area of Canada. Hydrocarbons were squeezed out of the organic-rich sediments and migrated to the permeable, low-pressure areas of the rock column such as Paleozoic reefs and units of Mesozoic sandstone. Similar circumstances allowed oil to be produced and trapped in basins and continental shelf areas of Newfoundland and Labrador (Hibernia project), Yukon, Northwest Territories,

◆ **Figure 21.26** Modern underground coal mining is highly mechanized. (Photo by Melvin Grubb/ Grubb Photo Services, Inc.) Strip mining (inset) of coal at Black Mesa, Arizona. Surface mining is common when coal seams are near the surface. (Photo by Richard W. Brooks/ Photo Researchers, Inc.)

◆ **Figure 21.27** Distribution of important oil and gas-bearing areas of North America.
(Source: Natural Resources Canada)

Nunavut, and to a lesser degree, British Columbia and Nova Scotia. Modest quantities of oil and gas are also extracted from Paleozoic sedimentary rocks of the Michigan Basin in Ontario (Figure 21.27). As discussed in Box 21.3, Canadians have played major roles in the development of the North American oil industry.

In the Prairie Provinces, the shallow seas that supplied marine organic matter to form oil were rimmed with coastal lowlands where land plants thrived and supplied their fibrous remains to form coal. About 98 percent of Canada's coal is produced in British Columbia, Alberta, and Saskatchewan and is of Mesozoic age (Figure 21.28). The famous coal of Nova Scotia is somewhat older, dating to the Late Paleozoic, and contributed greatly to the Canadian coal output for over 100 years. By 2001, Canada's coal production was becoming a multi-billion-dollar industry. Canada supplies over 20 countries (but mainly Japan and Korea) with coal and ranks as one of the world's leading coal exporters. A large proportion of coal remains in Canada to supply energy required for steel production.

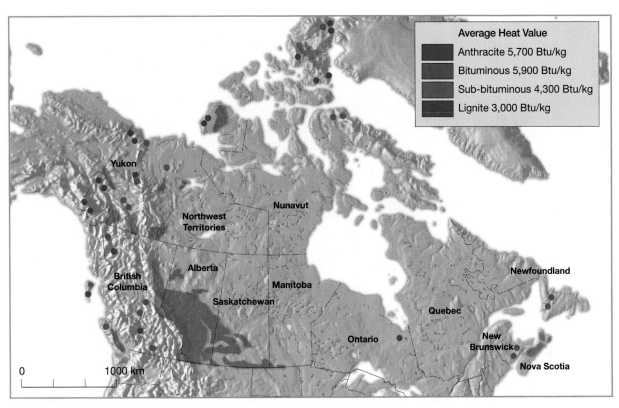

Average Heat Value
Anthracite 5,700 Btu/kg
Bituminous 5,900 Btu/kg
Sub-bituminous 4,300 Btu/kg
Lignite 3,000 Btu/kg

◆ **Figure 21.28**   Distribution of coal deposits in Canada. (Modified from an image courtesy of the Coal Association of Canada)

BOX 21.3

Canadian Profile
The Hard Oilers of Enniskillen Township and Canada's Petroleum Industry

Claudia Cochrane*

In the early 1850s, the industrial revolution was well underway in both Europe and the New World. As in ancient times, asphalt was needed for ship construction and for paving streets and sidewalks, and fuel was required for illumination. Although the Victorians had been using whale oil for their lamps, the great whales had been hunted almost to extinction. The world was desperate for a cheap source of petroleum. It was in the swampy hinterland of Lambton County's Enniskillen Township, Ontario that the North American oil industry was born.

In response, the enterprising Tripp Brothers, Henry and Charles, following their noses as well as reports from the Geological Survey of Canada, found their way to the site of the Lambton County oil seeps. In 1852 they started digging and manufacturing asphalt,

which they sold for ships and roads. "The International Mining & Manufacturing Company," chartered in 1854, was the first oil company in North America.

Initially the Tripps were immensely successful. Only a year after incorporation, their asphalt was displayed at the Paris Universal Exhibition and received an honourable mention from Emperor Napoleon III. More important, they received an order for asphalt to pave the streets of Paris. That winter the Enniskillen gum was transported overland by sled to Port Sarnia, and then by ship across the Atlantic Ocean to France. In 1854, another enterprising Canadian and government geologist, Abraham Gesner, was experimenting with oil seeps in the colony of Nova Scotia. He had just invented and patented a method of extracting kerosene lamp fuel from

the liquid oil component of natural seeps. When he set up a factory in New York the world suddenly lit up, and the search for crude oil was on.

Others, driven by dreams of profitable lighting, quickly followed the Tripps to Enniskillen Township and in 1858 James Miller Williams produced oil from sandy lenses at the base of the Tripp's gum bed. Initially pumped by hand and later by steam, his well yielded about 50 whiskey barrels a day. It was the first producing oil well in North America. By early winter, using a simple iron kettle distiller, he was refining kerosene. Although the resulting product was not very clean (it was called "skunk juice"), his was also the first oil refinery in North America. On the other side of Lake Erie, near Titusville in Pennsylvania, the Americans were exploiting gum beds as well, and a year later "Colonel" Edwin

Drake, drilling all the way through into bedrock, encountered oil.

By 1860 word got out in both countries and the twin oil rush was on. In Enniskillen Township workers flocked to the new village of Oil Springs, and within a year there were 400 wells in the field; some of them also drilled into bedrock, each producing 50–500 barrels a day. In 1862 one of these wells brought in a gusher, reputed to produce 2000 barrels a day, and the rush compounded. Oil Springs boomed and soon 1000 wells covered a square mile around Black Creek (Figure 21.C). A forest of shifting ash-pole service-derricks replaced the native woodland. The oil flowed freely and there was often no need to pump. Indeed sometimes it flowed too freely, running into Black Creek and into Lake St. Clair, where it produced an oil slick several centimetres thick. Oil Springs and Titusville have often produced conflicting claims for the position of "first in North America;" but neither has cared to take credit for the continent's first petroleum-related environmental debacle!

One of the more successful drillers, John Henry Fairbank, devised an ingenious method of pumping numerous wells efficiently and economically. The well pumps were linked each to another, and thereby all were connected up to a central power source by a series of parallel wooden rods. Known as the "jerker rod system," it could pump more than two hundred wells from one large central steam engine. Fairbank eventually expanded his operation over a large part of the Oil Springs field that is still operated by his descendants today. The Oil Springs boom was short-lived—less than eight years. The gushers "watered-out" and oil was discovered in Petrolia and further afield in Texas. Production went up, the oil prices went down as the market became saturated, and Oil Springs became a ghost town.

But for Enniskillen Township, the story was not finished. The producers and the party simply moved a few miles north to Petrolia, which produced its first gusher in 1866. This time, with a better infrastructure already in place, the boom was sustained for

◆ **Figure 21.C**   A photo taken in the late 1800s of drilling rigs in the Petrolia oilfield of southern Ontario, the birthplace of the Canadian oil industry. (Photo courtesy of Charles Fairbank)

four glorious decades and Petrolia thrived as the oil capital of Canada. The only intrusion on this refined lifestyle was the indelicate smell of sour crude that wafted over the town when the wind was blowing in the wrong direction—the townspeople referred to it as "the smell of money."

It was in Petrolia that the energetic drillers came to call themselves "hard oilers." Although the exact origin of the label is obscure, it has been said that a team of resolute Petrolia oil workers were threatened by their opponents at a ball game with a "hard oil finish" and that the term was born upon their victory. Many of the Petrolia Hard Oilers went on to open new oil fields abroad—in North Africa, Asia, and South America. Dubbed "the foreign drillers," they returned to Petrolia, full of exotic stories and souvenirs from strange lands. In all, they eventually went to 87 countries and were present in the 1920s and 1930s when giant oil fields were found in Iraq and other parts of the Middle East.

Price fluctuations were one of the greatest banes of the oil producers' lives. In an attempt to control supply,

Jacob Englehart bought up several refineries in Petrolia and the nearby city of London, where he and several other Ontario financiers established the Imperial Oil Company in 1880. Three years later, after a major fire in London, all of the operations were moved to Petrolia.

By 1893, Imperial Oil, still operating out of Petrolia, was a great refining success, with twenty-three branches all over Canada. But as demand outstripped supply, Imperial needed capital for expansion. As Canada was in a recession, the Directors initially approached British investors, but were unsuccessful in securing funds. The only option left was to sell to the Americans. Standard Oil had already bought many small refineries in the United States, including Abraham Gesner's New York kerosene factory, and had been casting a covetous eye on Canada's flagship firm for several years. The conditions of the takeover, effected in 1898, were unequivocal: the Imperial name was kept; Canadian headquarters and refineries were moved from Petrolia to Sarnia; and Imperial Oil went on to participate in developing the industry in other parts of Canada over the next century.

One of the new frontiers of petroleum exploration was western Canada. The first exploration well in Alberta was drilled in 1902 at the site of an oil seep near Cameron Brook, which is now part of the Waterton Lakes National Park. It produced a brief boomtown of Oil Springs that suffered the same fate as its namesake in Ontario. But the nearby Turner Valley field, discovered in 1914 and drilled with the assistance of Petrolia Hard Oilers, established that Alberta was indeed a fertile ground for petroleum production.

After 20 years and 133 dry holes, Imperial Oil encountered Canada's first giant oil field at Leduc, Alberta in 1947. Pipelines to Sarnia, Vancouver, and the U.S. were quickly built and Canada became a world-class oil-producing nation. But oil is not the only economically important hydrocarbon. In fact, natural gas is cleaner and requires no refining. Commercial natural gas production was pioneered in Alberta and Ontario. In 1883, CPR workers, drilling for water for their steam engines at Langevin in Southern Alberta, hit an unexpected flow of natural gas that ignited and destroyed their derrick. George Dawson of the Geological Survey of Canada, observing that the strata that contained the gas were continuous over much of Western Canada, predicted large volumes of gas from the Territory. Indeed, George Dawson's prediction was well-fulfilled in that the reserves of the Western Interior Basin are 80 percent gas prone.

Looking further afield for oil, Imperial pioneered exploration in the north with Canada's first Arctic discovery at Norman Wells in the Northwest Territories in 1920. The first well was drilled in the vicinity of yet another oil seep that had been originally noted by the intrepid explorer, Alexander Mackenzie. The field was produced only intermittently over the years until it was connected to the Canadian pipeline system in 1985.

The East Coast of Canada was also considered to be prime exploration territory and, at about the same time as the Leduc discovery, drillers were exploring underwater in the Atlantic Ocean near Prince Edward Island. This first effort at offshore drilling in Canada was not economically encouraging, but the technology was off to a good start. Half a century later, in 1997, the Hibernia Field was discovered 315 kilometres off the coast of Newfoundland.

With the dawn of the 21st century, Canada and its neighbours are encountering new petroleum challenges. Conventional sources of crude oil and natural gas are being depleted, and other more technologically sophisticated methods of extraction will be required for the future. The Canadian oil sands, North Atlantic offshore drilling, and Arctic reserves are all potential sources of hydrocarbon energy. However, their exploitation will require not only more advanced and expensive technology, but also recognition of environmental hazards and social concerns. In the Kyoto Accord, which was ratified by Ottawa in 2002, Canada pledged to reduce the emissions of greenhouse gases over the next 10 years. As well, Canada's First Nations are asserting their right to a share of the wealth, and long-ignored land claims must ultimately be negotiated. Therefore the future of the petroleum industry in Canada and elsewhere will depend not only upon rampant technological growth, as in the past, but upon the development of sober and responsible social and ecological policies as well.

At the start of the 21st century, the fourth generation of Fairbank Oil celebrated its 140th anniversary in tiny Oil Springs, Ontario (population 800). North America's first oil discovery field remains a commercially viable operation as the fourteenth-largest oil producer in Ontario. Victorian technology is still largely utilized, and remarkably, the original jerker rod system, designed by J.H. Fairbank early in the 1860s, remains in place. This is the oldest continuously operated oil company in the world.

*Claudia Cochrane ia a consulting geologist with Cairnlins Resources Limited and an adjunct professor in the Department of Earth Sciences at the University of Western Ontario.

Environmental Effects of Fossil Fuel Use

Humanity faces a broad array of human-caused environmental problems. Among the most serious are the impacts on the atmosphere that occur because of **fossil fuel** combustion. As human population continues its explosive growth (Figure 21.29), environmental problems associated with fossil fuel emissions will continue to grow unless drastic measures are taken to decrease them. Rural and urban air pollution, acid rain, and global greenhouse warming are all closely linked to the use of these basic energy resources.

Air pollution   Air *pollutants* are airborne particles and gases that occur in concentrations that endanger the health of organisms or disrupt the orderly function of the environment.

The human population constitutes a giant chemical factory producing a remarkable variety of undesirable products. Figure 21.30 shows the major primary pollutants and their sources. *Primary pollutants* are emitted directly from identifiable sources. They pollute the air immediately upon their emission. Much of the air pollution in Canada is particulate matter,most of which comes simply from dust blown from open sources such as roads and fields. For the remainder of pollution sources, the significance of the industrial and transportation categories is obvious. Industry (including fuel generating plants) and

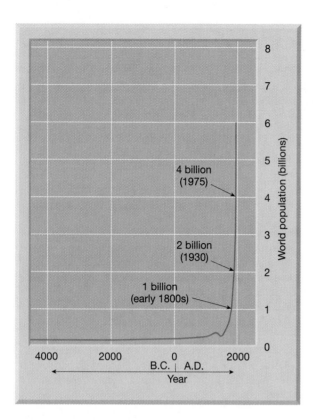

◆ **Figure 21.29**   Growth of world population. It took until 1800 for the number to reach 1 billion. By the year 2010, nearly 7 billion people may inhabit the planet. The demand for basic resources is growing faster than the rate of population increase. (Data from the Population Reference Bureau)

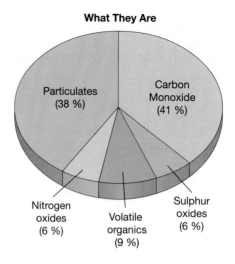

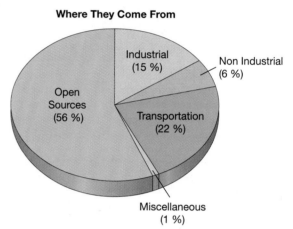

◆ **Figure 21.30**   Principal types and sources of air pollution in Canada per unit mass. While particulates from nonindustrial sources such as fields and roads comprise a large proportion of pollutants, the remaining pollutants can be linked largely to industrial and transportation-related emissions. (Pie charts based on data from *Criteria Air Contaminants Emission Summaries*, http://www.ec.gc.ca/pdb/ape/cape_home-e.cfm, Environment Canada)

the hundreds of millions of vehicles on the roads are the greatest contributors in this category.

When chemical reactions take place among primary pollutants, *secondary pollutants* are formed. The noxious mixture of gases and particles that make up urban smog is an important example that is created when volatile organic compounds and sulphur and nitrogen oxides react in the presence of sunlight. The burning of fossil fuels by industry and various modes of transportation are the greatest contributors of these compounds.

Carbon Dioxide and Global Warming   Warming of the lower atmosphere is global in scale. Unlike acid rain and urban air pollution, this issue is not associated with any of the primary pollutants in Figure 21.30. Rather, the connection between global warming and burning of fossil fuels relates to a basic product of combustion: carbon dioxide.

Carbon dioxide (CO_2) is a gas that occurs naturally in the atmosphere and is augmented by fuel combustion. Although CO_2 represents only about 0.036 percent of clean, dry air, it is nevertheless meteorologically significant. The importance of car-

bon dioxide lies in the fact that it is transparent to incoming, short-wave solar radiation, but is not transparent to some of the longer wavelength, outgoing radiation emitted by Earth. A portion of the energy leaving the ground is absorbed by carbon dioxide and subsequently re-emitted, part of it toward the surface, thereby keeping the air near the ground warmer than it would be without carbon dioxide. Thus, carbon dioxide is one of the gases responsible for warming the atmosphere. This process is called the *greenhouse effect* (Figure 21.31). Because carbon dioxide is an important heat absorber, any change in the air's carbon dioxide content could alter temperatures in the lower atmosphere.

Although the proportion of carbon dioxide in the air is relatively uniform at any given time, its percentage

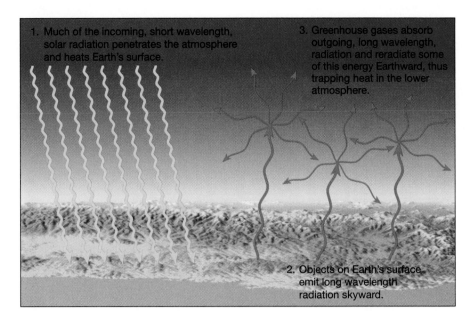

1. Much of the incoming, short wavelength, solar radiation penetrates the atmosphere and heats Earth's surface.

2. Objects on Earth's surface emit long wavelength radiation skyward.

3. Greenhouse gases absorb outgoing, long wavelength, radiation and reradiate some of this energy Earthward, thus trapping heat in the lower atmosphere.

◆ **Figure 21.31** The heating of the atmosphere. Most of the short-wavelength radiation from the Sun that is not reflected back to space passes through the atmosphere and is absorbed by Earth's land-sea surface. This energy is then emitted from the surface as longer-wavelength radiation, much of which is absorbed by certain gases in the atmosphere. Some of the energy absorbed by the atmosphere will be radiated Earthward. This so-called greenhouse effect is responsible for keeping Earth's surface much warmer than it would be otherwise.

Students Sometimes Ask ...

Is carbon dioxide the only gas that's responsible for global warming?

No. Although carbon dioxide is the most important, other gases also play a role. In recent years scientists have come to realize that the industrial and agricultural activities of people are causing a buildup of several trace gases that may also play a significant role. The substances are called *trace gases* because their concentrations are so much smaller than those of carbon dioxide. The trace gases that appear to be most important are methane (CH_4), nitrous oxide (N_2O), and chlorofluorocarbons (CFCs). These gases absorb wavelengths of outgoing radiation from Earth that would otherwise escape into space. Although individually their impact is modest, taken together the effects of these trace gases may be nearly as great as (CO_2) in warming the lower atmosphere.

has been rising steadily for more than a century (Figure 21.32). Much of this rise is the result of burning ever-increasing quantities of fossil fuels. From the mid-nineteenth century until 2000, there was an increase of more than 25 percent in the carbon dioxide content of the air.

Have global temperatures already increased as a result of the rising carbon dioxide level? This appears to be the case, according to the Intergovernmental Panel on Climate Change (IPCC). During the twentieth century the increase was about 0.6°C. Moreover,

recent years have been among the warmest on record. Globally, the 1990s were the warmest decade and 1998 the warmest year since 1861, when reliable instrument records began (Figure 21.33). Other studies indicate that the increase in temperature in the twentieth century is likely to have been the largest of any century during the past millennium. Are these temperatures caused by human activities or would they have occurred anyway? In January the IPCC declared that most of the observed warming over the last 50 years is likely due to the increase in carbon dioxide and other greenhouse gases that humans contribute to the atmosphere.

If fossil fuel use continues to increase at projected rates, the present carbon dioxide content of 360 parts per million (ppm) will climb to between 540 and 970 ppm by 2100. With such an increase, the greenhouse effect would be much more dramatic and measurable than in the past. The predicted result would be an average global surface temperature increase of 1.4 to 5.8°C. The warming of the lower atmosphere will not be the same everywhere. Rather, the temperature response in the polar regions could be much greater than the global average, while some low-latitude regions would experience much smaller fluctuations. According to the IPCC, the projected rate of warming during the next century will be much larger than the observed changes in the last century and is very likely without precedent during the last 10,000 years. However, the possible effects of aerosol emissions from fossil fuel combustion complicate this outlook considerably.

In December 1997, Canada and more than 160 other countries met in Kyoto, Japan, and agreed

◆ **Figure 21.32   A.** Carbon dioxide (CO_2) concentrations over the past 1000 years. Most of the record is based on data obtained from Antarctic ice cores. Bubbles of air trapped in the glacial ice provide samples of past atmospheres. The record since 1958 comes from direct measurements of atmospheric CO_2 taken at Mauna Loa Observatory, Hawaii. **B.** The rapid increase in CO_2 concentration since the onset of industrialization in the late 1700s is clear and has followed closely the rise in CO_2 emissions from fossil fuels.

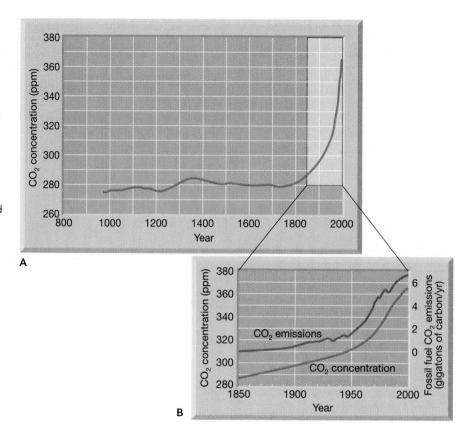

A

B

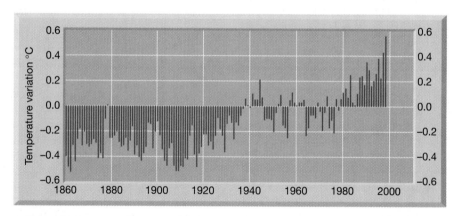

◆ **Figure 21.33**   Annual average global temperature variations for the period 1860–1998. The basis for comparison is the average for the 1961–90 period (the 0.0 line on the graph). Each narrow bar on the graph represents the departure of the global mean temperature from the 1961–1990 average for one year. For example, the global mean temperature for 1862 was more than 0.5°C *below* the 1961–90 average, whereas the global mean for 1998 was more than 0.5°C above. (Specifically, 1998 was 0.56°C warmer.) The bar graph clearly indicates that there can be *significant variations from year to year*. But the graph also shows a trend. Estimated global mean temperatures have been above the 1961–90 average every year since 1978. Also, the three warmest years in the 138-year record shown here were 1995, 1997, and 1998. *(After G. Bell, et al. "Climate Assessment for 1998," Bulletin of the American Meteorological Society, Vol. 80, No. 5, May 1999, p. 54.)*

on targets to reduce greenhouse gas emissions. The Kyoto Protocol includes the ratification of these targets and the proposed methods of achieving them. Canada's target is to reduce its greenhouse gas emis-

sions to 6 percent below 1990 levels by 2012. In December 2002, the federal government of Canada officially ratified Canada's commitment to the Kyoto Protocol.

People and the Environment
Aerosols from the "Human Volcano"

Increasing the levels of carbon dioxide and other greenhouse gases in the atmosphere is the most direct human influence on global climate. But it is not the only impact. Global climate is also affected by human activities that contribute to the atmosphere's aerosol content. *Aerosols* are the tiny, often microscopic, liquid and solid particles that are suspended in the air. Atmospheric aerosols are composed of many different materials, including soil, smoke, sea salt, and sulphuric acid. Natural sources are numerous and include such phenomena as dust storms and volcanoes. In Chapter 4 you learned that some explosive volcanoes (such as Mount Pinatubo) emit large quantities of sulphur dioxide gas high into the atmosphere. This gas combines with water vapour to produce clouds of tiny sulphuric acid aerosols that can lower air temperatures near the surface by reflecting solar energy back to space. So it is with sulphuric acid aerosols produced by human activities.

Presently the human contribution of aerosols to the atmosphere *equals* the quantity emitted by natural sources. Most human-generated aerosols come from the sulphur dioxide emitted during the combustion of fossil fuels and as a consequence of burning vegetation to clear agricultural land (Figure 21.A). Chemical reactions in the atmosphere convert the sulphur dioxide into aerosols, the same material that produces acid precipitation (see Box 5.1, p. 137).

The aerosols produced by human activity act directly by reflecting sunlight back to space and indirectly by making clouds "brighter" reflectors. The second effect relates to the fact that sulphuric acid aerosols attract water and thus are especially effective as cloud condensation nuclei (tiny particles upon which water vapour condenses). The

◆ **Figure 21.D**   One source of human-generated aerosols is power plants. (Photo by Bruce Forster/Tony Stone Images)

large quantity of aerosols produced by human activities (especially industrial emissions) trigger an increase in the number of cloud droplets that form within a cloud. A greater number of small droplets increases the cloud's brightness—that is, more sunlight is reflected back to space.

By reducing the amount of solar energy available to the climate system, aerosols have a net cooling effect. Studies indicate that the cooling effect of human-generated aerosols could offset a portion of the global warming caused by the growing quantities of greenhouse gases in the atmosphere. Unfortunately, the magnitude and extent of the cooling effect of aerosols is highly uncertain. This uncertainty is a significant hurdle in advancing our understanding of how humans alter Earth's climate.

It is important to point out some significant differences between global warming by greenhouse gases and

aerosol cooling. After being emitted, greenhouse gases such as carbon dioxide remain in the atmosphere for many decades. By contrast, aerosols released into the lower atmosphere remain there for only a few days or, at most, a few weeks before they are "washed out" by precipitation. Because of their short lifetime in the atmosphere, human-generated aerosols are distributed unevenly over the globe. As expected, they are concentrated near the areas that produce them, namely industrialized regions that burn fossil fuels and land areas where vegetation is burned.

Because the lifetime of human-generated aerosols in the atmosphere is short, the effect of the "human volcano" on today's climate is determined by the quantity of material emitted during the preceding couple of weeks. By contrast, the carbon dioxide released into the atmosphere remains for much longer spans and thus influences climate for many decades.

Unconventional Fossil Fuel Deposits

As supplies of petroleum and coal that can be extracted by conventional methods dwindle, we can expect that an increasing volume of fossil fuel will be produced from lower-grade deposits that require

innovative, unconventional methods (Figure 21.34). Two main types of deposits that are being seriously considered as candidates for future hydrocarbon extraction are oil shales and heavy oil sands.

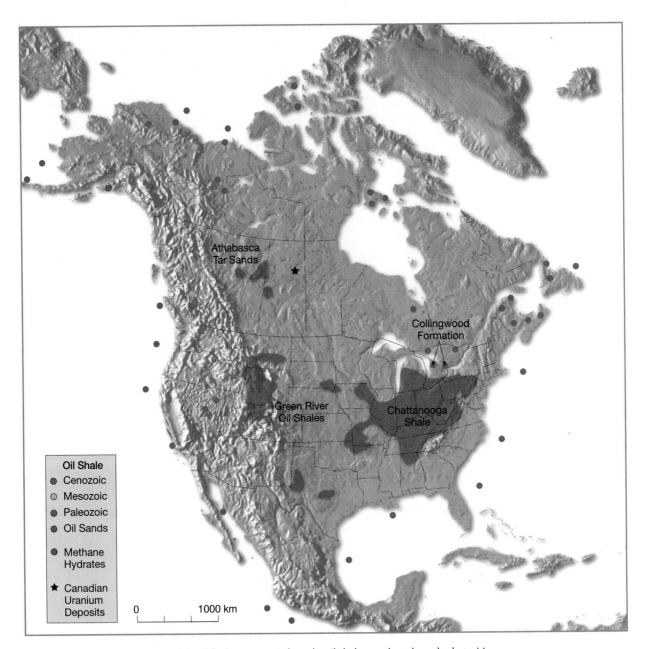

◆ **Figure 21.34**   Unconventional fossil fuel resources (oil sands, oil shales, and methane hydrates) in North America and uranium deposits of northern Saskatchewan.

Heavy Oil Sands   *Heavy oil sands*, also commonly called tar sands, are mixtures of sediment, water, and abundant, black, tar-like bitumen. It is thought that such deposits result from a very high level of inter-action between regular crude oil and groundwater linked to long-distance and near-surface oil migra-tion. A popular explanation for the origin of the tarry oil is that cool, oxygen-rich, groundwater provided optimal conditions for certain bacteria to preferen-tially consume light components of the oil, thus leav-ing behind a heavy residue. Exposure of petroleum to air at the surface of the Earth also allows its light

components to evaporate and could also contribute to enrichment of the heavier components of the petroleum. In any case, the hydrocarbons of the tar sands appear to represent degraded oil. Heavy oil sands occur in abundance in Alberta (Figure 21.34), the most famous of which are the Athabasca Tar Sands located northeast of Edmonton. Alberta's vast deposits are the source of about 15 percent of Canada's oil production.

Due to their high viscosity, hydrocarbons that occur in heavy oil sands cannot be pumped from a well like conventional oil. The tar sand is effectively

◆ **Figure 21.35** Outcrop of thinly bedded, kerogen-rich, Ordovician limestones ("oil shales") of the Collingwood Formation on the southern shore of Georgian Bay, Ontario. (Photo by C. Tsujita)

mined at the surface, heated to mobilize the hydrocarbons, and upgraded to a synthetic crude with the addition of hydrogen and removal of impurities. Projects have also been designed to bypass the mining process, whereby steam is injected directly into the tar sand deposit to mobilize the hydrocarbons, which are then recovered from pipes much like conventional crude oil. Much energy must be expended in the recovery process—about half as much as the end product yields. Relative to conventional petroleum resources, therefore, tar sands presently have a rather low profit margin per unit of liquid petroleum produced. Nonetheless, tar sands will undoubtedly play a major role in fossil fuel production as global supplies inevitably decrease.

Oil Shale   In some cases, fine-grained, organic-rich sedimentary rocks are not subjected to sufficiently high temperatures and pressures to liberate liquid and gas hydrocarbons, and they instead retain organic matter in the form of kerogen. A fine-grained sedimentary rock containing an abundance of kerogen is called an *oil shale*. Although widely used, this term is somewhat misleading since most oil shales are not shales at all, but fine-grained limestones (Figure 21.35). In any case, oil shales contain enormous amounts of hydrocarbons and can be artificially heated to crack their kerogen into liquid petroleum. Huge deposits of Cenozoic-age oil shale comprising the Green River Formation occur in Colorado, Utah, and Wyoming, and represent fine-grained lake sediments deposited during the Eocene Epoch. Paleozoic oil shales occur in a number of places in Canada, including New Brunswick, Nova Scotia, and southern Ontario, and a

few Mesozoic deposits have also been identified (Figure 21.34). Oil extraction from Canadian oil shales has been attempted, but is not yet profitable. Similar to the case of heavy oil sands, the extraction of a liquid product from oil shale involves a high input of energy, both because it must be mined, and also because it must be heated to liberate oil. Accordingly, a considerable increase in the price of oil will be necessary for a profitable industry to be built around this resource.

Methane Hydrate   *Methane hydrate* is a solid substance formed when water molecules form cages that surround single molecules of methane. Methane hydrates, once thought to be oddities in the natural world, are now known to be common in permafrost areas associated with methane seeps. Even more significant are vast methane hydrate deposits found in deep areas of continental shelves of nearly every continent on Earth, including North America (Figure 21.36). In those places, the methane-ice complexes are kept stable due to high pressures and cold temperatures at the seafloor. At least some of these deposits are associated with communities of microbes that produce methane as part of their normal metabolic activities. Methane hydrates are sufficiently high in methane content to produce a flame when ignited with a match.

While the thought of exploiting this large reservoir of methane is appealing, many concerns surround the use of this potential energy source. As with all unconventional hydrocarbon resources, the feasibility and cost of extraction is questionable at the present time. More serious, however, are the environmental impacts of extraction, both in terms of disturbing the seafloor, and the potential leakage of methane into the atmosphere during extraction. The warming effect of methane in the atmosphere is at least 20 times higher than that of carbon dioxide, and thus methane leakage from seafloor hydrates is a legitimate concern.

Coal Bed Methane   The slow heating of fibrous organic matter derived from land plants not only produces solid coal, but also liberates methane. This *coal bed methane* is trapped in the pore spaces of coal seams and associated detrital sediments. Because it is confined in the pores of sedimentary rocks, it is difficult to extract by conventional means and is therefore not currently exploited on a large scale. It has been found that coal bed methane can be released by pumping large quantities of water from a coal seam through several wells. The merits of this practice have, however, been recently questioned due to concerns about surface- and groundwater contamination, groundwater depletion, and the possible escape

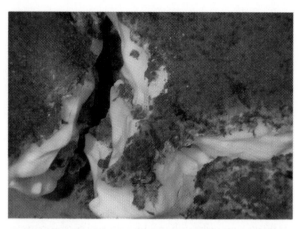

◆ **Figure 21.36**   A body of methane hydrate (yellowish-white) photographed on the seafloor some 850 metres below sea level, off the western coast of Vancouver Island, British Columbia. (Photo courtesy of Ross Chapman)

◆ **Figure 21.37**   A significant amount of electricity in Canada is produced by nuclear fission. Canadian nuclear reactors, such as those at the Pickering Nuclear Power Plant are of CANDU (Canadian Deuterium Uranium) design. (CP Photo/Kevin Frayer)

of the methane into the atmosphere. Still, coal bed methane is attractive as a clean-burning alternative to conventional fossil fuels such as oil and coal. As more efficient extraction technologies are developed, coal bed methane is likely to become more important as a fuel source in the future.

Nuclear Energy

Nuclear power plants generate roughly 14 percent of Canada's electricity supply. The fuel for these facilities comes from radioactive materials that release energy by the process of **nuclear fission**. Fission is accomplished by bombarding the nuclei of heavy atoms, commonly uranium-235, with neutrons. This causes the uranium nuclei to split into smaller nuclei and to emit neutrons and heat energy. The ejected neutrons, in turn, bombard the nuclei of adjacent uranium atoms, producing a chain reaction. If the supply of fissionable material is sufficient and if the reaction is allowed to proceed in an uncontrolled manner, an enormous amount of energy is released in the form of an atomic explosion. In most nuclear reactors, the chain reaction is regulated through the use of materials called *moderators*, which slow neutrons down to allow them to react with the uranium atoms in a controlled manner. Most reactors also use neutron-absorbing materials in control rods to prevent the participation of too many neutrons in the reaction at any one time.

CANDU Reactors   The production of nuclear power in Canada is centred on the CANDU (Canadian Deuterium-Uranium) design, which uses natural uranium-235 as fuel and deuterium oxide as a moderator and heat transport medium (Figure 21.37).

Deuterium is a stable isotope of hydrogen containing a proton and a neutron. Due to the higher mass of deuterium, relative to the more common "light" hydrogen that contains a single proton and no neutrons, deuterium oxide is heavier than regular hydrogen oxide (water). Deuterium oxide is therefore commonly called *heavy water*. The CANDU design allows on-power refuelling, does not require fuel reprocessing, and can potentially use spent fuel from other reactors. The versatility of CANDU reactors is a major factor that has contributed to their use in a number of countries outside North America, including Argentina, India, Korea, Pakistan, and Romania.

Uranium Occurrences   Although technically a metal, uranium is discussed in this section by virtue of being a fuel source. Uranium-235 is the only naturally occurring isotope that is readily fissionable and is therefore the primary fuel used in nuclear power plants. Although large quantities of uranium ore have been discovered, most contain les than 0.05 percent uranium. Of this small amount, 99.3 percent is the nonfissionable isotope uranium-238 and just 0.7 percent consists of the fissionable isotope uranium-235. Because most nuclear reactors operate with fuels that are at least 3 percent uranium-235, the two isotopes must be separated in order to concentrate the fissionable isotope. The process of separating the uranium isotopes is difficult, secretive, and substantially increases the cost of nuclear power.

In addition to being scarce, the size of the uranium atom generally prevents it from being incorporated into the crystal lattice of most rock-forming minerals. In addition, uranium is highly soluble in oxidizing environments and therefore requires unusual conditions to be precipitated in minerals at the Earth's surface. Uranium ores are extremely diverse in their geological settings, but it appears that a critical factor controlling the precipitation of uranium-bearing minerals is the absence of oxygen.

Uranium can be enriched in residual fluids of felsic magmas, reaching concentrations over a hundred parts per million and incorporated as a trace element in minerals such as zircon and apatite. In sufficient concentrations, however, uranium forms an oxide mineral (UO_2) called uraninite (pitchblende if very fine-grained), which can occur in veins associated with felsic igneous intrusions. Ores of uranium-bearing minerals also occur as placers in Precambrian rocks, reflecting the oxygen-free atmosphere of those times. The uranium ores of Elliot Lake, Ontario that formerly supplied a significant proportion of Canada's uranium represent such uranium placers.

The fact that uranium is soluble in oxidizing environments and insoluble in reducing environments can also contribute to the formation of uranium ores. For example, the rich uranium deposits of northern Saskatchewan's Athabasca Basin (Figure 21.34), which currently account for Canada's entire uranium production, appear to have been formed by the interaction of deep-sourced, oxygen-free fluid with shallow-sourced oxygenated fluid that carried dissolved uranium. In this case, this fluid interaction was focused in small areas where fault breccias intersect a nonconformity separating highly deformed metamorphic rocks from overlying sedimentary rocks. Lastly, uranium can be dissolved near the Earth's surface by oxygenated groundwater and carried through permeable, coarse-grained sedimentary rocks (e.g., sandstone). The uranium-charged fluid precipitates uranium in the mineral carnotite when it mixes with oxygen-poor water. Most of the uranium mined in Arizona, New Mexico, and Utah is contained in this type of ore. The close association of uranium and organic matter is also apparent in the slightly elevated concentration of uranium in organic-rich shales and phosphate deposits, relative to other sedimentary rock types.

Obstacles to Development   At one time, nuclear power was heralded as the clean, cheap source of energy to replace fossil fuels. However, several obstacles have emerged to hinder the development of nuclear power as a major energy source, not the least of which is the enormous cost of building the facilities that contain numerous safety features. More important, perhaps, is the concern over the possibility of a serious accident at one of the nearly 200 nuclear plants that exist worldwide. The Three Mile Island (Harrisburg, Pennsylvania) accident in 1979 brought this point home. Here, a malfunction led the plant operators to believe there was too much water in the

primary system rather than too little and allowed the reactor core to lie uncovered for several hours. Although the reactor sustained considerable damage, the public was fortunately spared great harm.

Unfortunately, the 1986 accident at Chernobyl in the former Soviet Union was far more serious. In this incident, the reactor ran out of control and two pressure-related explosions lifted the roof of the structure, allowing radioactive material to be strewn all over the immediate area. During the 10 days that it took to quench the fire that ensued, high levels of radioactive material were carried by the atmosphere and detected as far away as Norway. In addition to the 18 people who died within six weeks of the accident, many thousands more face an increased risk of death from radioactive fallout-related illnesses.

It should be emphasized that the concentrations of fissionable uranium-235 and the design of reactors are such that nuclear power plants cannot explode like an atomic bomb. The dangers arise from the possible escape of radioactive debris during a meltdown of the core or other malfunction. In addition, hazards such as the disposal of nuclear waste and the relationship that exists between nuclear energy programs and the proliferation of nuclear weapons must be considered as we evaluate the pros and cons of employing nuclear power.

Renewable Fuel Sources

Organic-Based Sources

Landfill Methane   Methane produced by the microbial decomposition of organic matter in landfills is a significant contributor of greenhouse gases to the atmosphere, although it usually escapes attention from the media. Thus, with the continued rise in global population, methane production from landfills can only increase in total volume and production rate. One way to deal with both the volume of methane released to the atmosphere and the demand for energy is to use this methane as fuel. About one quarter of all landfill methane is presently being collected in Canada, about 70 percent of which is used for energy production. While the use of landfill methane does significantly decrease total greenhouse emissions, it does not eliminate them, since carbon dioxide, although less potent than methane as a greenhouse gas, is produced from its combustion.

Biomass Energy   Biomass energy is derived from the tissues of organisms, specifically plants, and provided the first organic fuel used by humans. Plants, through photosynthesis, produce their structural tissues and sugars using solar energy, and this stored energy can later be released through combustion. You release this kind of biomass energy every time you burn wood in a campfire.

Biomass fuel is a primary source of energy in Third World countries in the form of wood and dung, but has been largely replaced in developed countries by fossil fuels. Recently, however, the use of biomass fuel has been seriously considered as a clean, renewable energy source.

One very important biomass fuel is ethanol, which is a product of anaerobic (oxygen-free) fermentation. In Canada, industries such as forestry generate lots of waste material that could support large ethanol-producing operations. Although not used as fuel in its pure form, ethanol can be used as an additive to gasoline and has been found to reduce carbon monoxide and carbon dioxide emissions. A form of biodiesel has been developed from vegetable oils and could also be used as an additive to conventional fuel. Although biomass fuels are far from replacing conventional fossil fuels, they are, at the very least, renewable and can reduce the total volume of greenhouse gas emissions.

Hydroelectric Power

Falling water has been an energy source for centuries. Through most of history, the mechanical energy produced by waterwheels was used to power mills and other machinery. Today, the power generated by falling water drives turbines that produce electricity, hence the term **hydroelectric power**. Canada is a world leader in the production of hydroelectricity, as it contributes to the production of over 60 percent of Canada's total electricity generation (Figure 21.38). Most of this energy is produced at large dams, which allow controlled flow of water (Figure 21.39). The water impounded in a reservoir is a form of stored (potential) energy that can be released at any time to produce electricity.

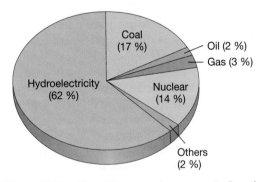

◆ **Figure 21.38** Electricity generation sources in Canada, 1997. Note the extremely high contribution of electricity from hydroelectric sources compared to fossil fuels and nuclear power generation.
(Source: Minister of Public Works and Government Services Canada and Natural Resources Canada, http://www2.nrcan. gc.ca/es/ener2000/online/html/chap3f-e.cfm#hydro)

Although waterpower is considered a renewable resource, the dams built to provide hydroelectricity have finite lifetimes and there are always negative environmental effects. All rivers carry suspended sediment that is deposited behind the dam as soon as it is built. Eventually the sediment will completely fill the reservoir. This takes 50 to 300 years, depending on the quantity of suspended material transported by the river. An example is Egypt's huge Aswan High Dam, which was completed in the 1960s. It is estimated that half of the reservoir will be filled with sediment from the Nile River by the year 2025.

The availability of appropriate sites is an important limiting factor in the development of large hydroelectric power plants. A good site provides a significant height for the water to fall and a high rate of flow. Most of the best U.S. sites have already been developed, limiting the future expansion of hydroelectric power. Canada still has many rivers that have not been dammed for hydroelectric power, but many of these are in remote areas that are particularly sensitive to environmental disturbance.

Solar Energy

The direct use of *solar energy* has gained much appeal as an alternative resource on a small scale. Many people use south-facing windows in their homes as *passive solar collectors* that allow objects in a room to absorb the energy of the Sun's rays and to re-emit the energy as heat. Together with more efficient insulation, south-facing windows are featured in many modern homes as a measure of heat conservation. More elaborate systems used in homes feature an active solar collector that utilizes large, glass-covered boxes to collect energy. This energy heats air or water that is, in turn, circulated through the house in pipes to supply heat where it is needed.

The use of solar energy on a large scale is becoming more practical as technology advances, but is still relatively insignificant compared to the energy used

◆ **Figure 21.39** Daniel Johnson Manic-v Hydroelectric Dam in Québec.
(Photo by Walter Muma)

◆ **Figure 21.40** Photograph of the EPCOR Solar Panel Array, located on EPCOR's corporate head office in downtown Edmonton, Alberta. The array has a generating capacity of 13.4 kilowatts. This building's integrated photovoltaic system, which converts sunlight directly into electricity, is connected to the Alberta grid. When installed in 1996, it was the largest of its kind in Canada, as well as the world's highest. (Photo used with permission of EPCOR Utilities Inc.)

from conventional fuel resources. So if solar energy is free, why isn't it used more? The reason lies in economics; while solar energy is free, the equipment required for its exploitation is expensive. As supplies of nonrenewable fuels decrease, and as solar collecting technology becomes more advanced, solar energy will surely increase in popularity and might even become cost effective in the near future. Another major impetus for the conversion to solar energy will be the continued quest for nonpolluting energy sources.

Research is currently underway to improve sunlight collecting technologies and experimental installations are becoming increasingly common throughout North America. One method currently under investigation in Barstow, California, involves the use of some 2000 sun-tracking mirrors that continually focus reflected light on a central receiving tower. The light energy collected by the receiver heats water in pressurized panels. This water is then transferred to turbines that generate electrical power. Another type of collector uses photovoltaic cells that collect sunlight and convert it directly into electricity. Examples of large experimental photovoltaic installations in Canada are the roof arrays of the Hugh MacMillan Rehabilitation Centre in Toronto, the CANMET Energy Technology Centre in Varennes, Québec, and the EPCOR office building in Edmonton (Figure 21.40).

Small rooftop photovoltaic systems have begun to be used in Third World countries such as the Dominican Republic, Sri Lanka, and Zimbabwe. The units are about as small as an open briefcase and use a battery to store electricity that is generated from sunlight during the day. These small photovoltaic systems can run a television or radio, plus a few light-bulbs for up to four hours. Although much cheaper than the construction of conventional electric generators, the units are still too expensive for poor families. Consequently, an estimated 2 billion people in developing countries still lack electricity.

Wind Energy

Approximately 0.25 percent of the solar energy that reaches the lower atmosphere is transformed into wind. Despite this deceptively minuscule figure, the total amount of energy that can be collected from the wind is enormous.

Wind has, of course, been used for centuries as an almost free and nonpolluting source of energy. Sailboats and windmills are two common forms of wind-powered devices used long ago. In addition, settlers of rural North America relied heavily on wind power for the pumping of water and other applications.

As technology has improved, efficiency has increased and the costs of wind-generated electricity have become more competitive. Between 1983 and 2000, technological advances cut the cost of wind power by about 85 percent. One area for the expansion of wind energy will likely be islands and other regions that are far from electrical grids and that presently must import fuel for power generation.

Wind turbines are becoming commonplace in Canada, with wind farms now well established in Alberta (Figure 21.41), Saskatchewan, Québec, and Prince Edward Island. At the time of writing, the largest modern wind turbine in North America is located next to Pickering nuclear plant in Ontario. Although wind generation is still modest in Canada, the amount of energy produced in 2002 was equivalent to that used by 50,000 Canadian homes in a year.

Although the future for wind power is promising, it is not without difficulties. In addition to technical advances that must continue to be made, noise pollution and costs of large tracts of land in populated areas present significant obstacles to development.

◆ **Figure 21.41**   Blue Ridge Wind Turbine, Waterton Wind Turbines Facility near Hillspring, Alberta. (Photo courtesy of Vision Quest Windelectric Inc.)

Geothermal Energy

Humans have taken advantage of water that has been heated by the Earth for centuries. In the modern age, such **geothermal energy** is now becoming popular for home and commercial heating, and is also being used in innovative applications such as the heating of greenhouses that can grow fruit and vegetable crops year-round. Another novel use of geothermal heat can be found in Springhill, Nova Scotia, where water that has been slightly warmed by the natural geothermal gradient in abandoned shafts of coalmines is pumped to provide space heating for an industrial development.

The use of geothermal energy for the generation of electricity is a relatively new development, and is focused on areas where groundwater is heated by near-surface volcanic activity. Geothermal power plants are presently well established in Iceland (Figure 21.42), Italy, New Zealand, and the United States. In these facilities, the steam and superheated water generated by hot springs is tapped and piped to turbines that generate electric power.

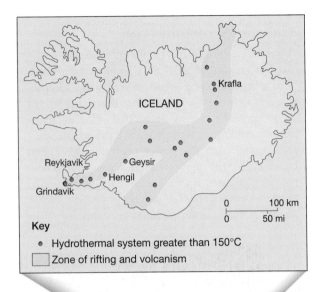

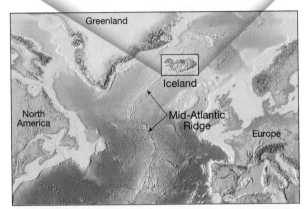

◆ **Figure 21.42**   Iceland straddles the Mid-Atlantic Ridge. This divergent plate boundary is the site of numerous active volcanoes and geothermal systems. Because the entire country consists of geologically young volcanic rocks, warm water can be encountered in holes drilled almost anywhere. More than 45 percent of Iceland's energy comes from geothermal sources. The photo shows a power station in southwestern Iceland. The steam is used to generate electricity. Hot (83°C) water from the plant is sent via an insulated pipeline to Reykjavik for space heating. (Photo by Simon Fraser/Science Photo Library)

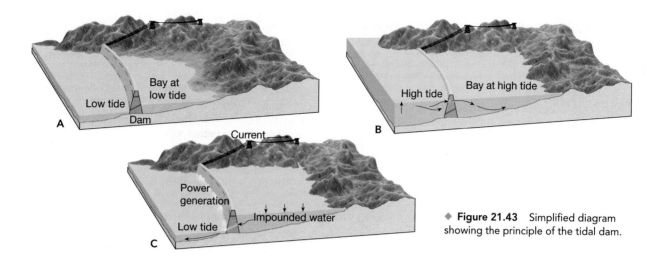

◆ **Figure 21.43** Simplified diagram showing the principle of the tidal dam.

Tidal Power

The ocean contains a huge amount of physical energy in the form of tides. Although harnessing this energy presently involves sophisticated technology, the principle of electricity generation from tidal energy is fairly simple. The most common method involves trapping ocean water in reservoirs at high tide, and later releasing it through hydroelectric turbines as the tide goes down (Figure 21.43). The largest tidal energy plant is located on the La Rance River in France, which generates 240 megawatts of power. The Annapolis Tidal Power Generation Station in the Bay of Fundy, Nova Scotia, is the only tidal plant in North America and generates about 20 megawatts. While tidal power is a good nonpolluting form of energy, its use is limited to areas where the tidal range is high and water turbidity is low. Thus, the focus of its further development will probably be on supplementing other forms of energy rather than replacing them.

Students Sometimes Ask ...

Is power from ocean waves a practical alternative energy source?

This possibility is being seriously explored. In November 2000 the world's first commercial wave power station began operating on the Scottish island of Islay, providing power to the United Kingdom power grid. The new 500-kilowatt power station uses a technology called the oscillating water column, in which the incoming waves push air up and down inside a concrete tube that is partially submerged in the ocean. The air rushing into and out of the top of the tube is used to drive a turbine to produce electricity. If this technology proves successful, it may open the door for wave power to be a significant contributor of renewable energy in appropriate coastal settings.

Chapter Summary

- *Mineral resources* are the endowment of useful minerals ultimately available commercially. Resources include already identified deposits from which minerals can be extracted profitably, called *reserves*, as well as known deposits that are not yet economically or technologically recoverable. Deposits inferred to exist but not yet discovered are also considered mineral resources. The term *ore* is used to denote those useful metallic minerals that can be mined for a profit, as well as some nonmetallic minerals, such as fluorite and sulphur, that contain useful substances.

- Some of the most important accumulations of metals, such as gold, silver, lead, and copper, are associated with igneous processes. The best-known and most important ore deposits are generated from *hydrothermal* (hot-water) *solutions*. Hydrothermal deposits can originate from hot, metal-rich fluids that are remnants of late-stage magmatic processes. These ion-rich solutions move along fractures or bedding planes, cool, and precipitate the metallic ions to produce *vein deposits*. In a *disseminated deposit* (e.g., much of the world's copper deposits) the ores from hydrothermal solutions are distributed as minute masses throughout the entire rock mass. Hydrothermal deposits can also be produced by mineral precipitation from fluids that picked up metals from volcanic or sedimentary rocks.

- Many of the most important metamorphic ore deposits are produced by contact metamorphism. Extensive aureoles of metal-rich deposits commonly surround igneous bodies where ions have invaded limestone strata. The most common metallic minerals associated with contact metamorphism are sphalerite (zinc), galena (lead), chalcopyrite (copper), magnetite (iron), and bornite (copper). Of equal economic importance are the metamorphic rocks themselves. In many regions, slate, marble, and quartzite are quarried for a variety of construction purposes.

- Weathering creates ore deposits by concentrating minor amounts of metals into economically valuable deposits. The process, often called *secondary enrichment*, is accomplished by either (1) removing undesirable materials and leaving the desired elements enriched in the upper zones of the soil, or (2) removing and carrying the desirable elements to lower zones where they are redeposited and become more concentrated. *Bauxite*, the principal ore of aluminum, is one important ore created as a result of enrichment by weathering processes. In addition, many copper deposits result when weathering processes concentrate metals that were formerly dispersed through low-grade primary ore.

- Earth materials that are not used as fuels or processed for the metals they contain are referred to as *nonmetallic resources*. Many are sediments or sedimentary rocks. The two broad groups of nonmetallic resources are *aggregate and stone* and *industrial minerals*. Limestone, perhaps the most versatile and widely used rock of all, is found in both groups.

- *Renewable resources* can be replenished over relatively short time spans. Examples include natural fibres for clothing, and trees for lumber. *Nonrenewable resources* form so slowly that, from a human standpoint, Earth contains fixed supplies. Examples include fuels such as oil and coal, and metals such as copper and gold. A rapidly growing world population and the desire for an improved living standard cause nonrenewable resources to become depleted at an increasing rate.

- Oil and natural gas, which commonly occur together in the pore spaces of some sedimentary rocks, consist of various *hydrocarbon compounds* (compounds made of hydrogen and carbon) mixed together. Petroleum formation is associated with the accumulation of sediment in ocean areas that are rich in plant and animal remains that become buried and isolated in an oxygen-deficient environment. As the mobile petroleum and natural gas form, they migrate and accumulate in adjacent permeable beds such as sandstone. If the upward migration is halted by an impermeable rock layer, referred to as a *cap rock*, a geologic environment that allows economically significant amounts of oil and gas to accumulate underground, called an *oil trap*, develops. The two basic conditions common to all oil traps are (1) a porous, permeable *reservoir rock* that will yield petroleum and/or natural gas in sufficient quantities, and (2) a cap rock.

- *Coal, petroleum*, and *natural gas*, the *fossil fuels* of our modern economy, are all associated with sedimentary rocks. Coal originates from large quantities of plant remains that accumulate in an oxygen-deficient environment, such as a swamp. Much of the present-day coal usage is for the generation of electricity. Air pollution produced by the sulphur oxide gases that form from burning most types of coal is a significant environmental problem.

- Environmental problems associated with burning fossil fuels include urban air pollution and global warming. The *primary pollutants* emitted by sources such as motor vehicles can react in the atmosphere to produce the *secondary pollutants* that make up urban smog. Combustion of fossil fuels is one of the ways that humans are increasing the atmosphere's carbon dioxide content. Greater quantities of this heat-absorbing gas contribute to global warming.

- When conventional petroleum resources are no longer adequate, fuels derived from *tar sands* and *oil shale* may become substitutes. Presently, tar sands from the province of Alberta are the source of about 15 percent of Canada's oil production. Oil from oil shale is presently uneconomical to produce. Oil production from both tar sands and oil shale has significant environmental drawbacks.

- Much of our energy is derived from non-fossil fuels. *Hydroelectric power* provides over 60 percent of Canada's electricity. Another important alternative energy source is *nuclear energy*. Other alternative energy sources are locally important but collectively provide little of the demand. These include *solar power, geothermal energy, wind energy*, and *tidal power*.

Review Questions

1. Contrast renewable and nonrenewable resources. Give one or more examples of each.

2. What is the estimated world population for the year 2010? How does this compare to the figures for 1930 and 1975? Is demand for resources growing as rapidly as world population?

3. Most coal in Canada is used for what purpose?

4. Describe two impacts on the atmospheric environment of burning fossil fuels.

5. What is an oil trap? List two conditions common to all oil traps.

6. List two drawbacks associated with the processing of tar sands recovered by surface mining.

7. North America has huge oil shale deposits but does not produce oil shale commercially. Explain.

8. What is the main fuel for nuclear fission reactors?

9. List two obstacles that have hindered the development of nuclear power as a major energy source.

10. Briefly describe two methods by which solar energy might be used to produce electricity.

11. Explain why dams built to provide hydroelectricity do not last indefinitely.

12. Is geothermal power considered an inexhaustible energy source? Explain.

13. What advantages does tidal power production offer? Is it likely that tides will ever provide a significant proportion of the world's electrical energy requirements?

14. Contrast *resource* and *reserve*.

15. What might cause a mineral deposit that had not been considered an ore to be reclassified as an ore?

16. List two general types of hydrothermal deposits.

17. Metamorphic ore deposits are often related to igneous processes. Provide an example.

18. Name the primary ore of aluminum and describe its formation.

19. Briefly describe the way in which minerals accumulate in placers. List four minerals that are mined from such deposits.

20. Describe the effects of metal mining in Canada on the natural environment.

21. Nonmetallic resources are commonly divided into two broad groups. List the two groups and some examples of materials that belong to each. Which group is most widely distributed?

Key Terms

banded iron formation (p. 593)
cap rock (p. 603)
disseminated deposit (p. 587)
fossil fuel (p. 609)
geothermal energy (p. 620)
hydroelectric power (p. 618)

hydrothermal solution (p. 587)
industrial minerals (p. 583)
kimberlite (p. 598)
mineral resource (p. 580)
nonmetallic mineral resource (p. 597)

nonrenewable resource (p. 581)
nuclear fission (p. 616)
ore (p. 583)
pegmatite (p. 586)
petroleum trap (p. 603)
placer (p. 594)

renewable resource (p. 581)
reserves (p. 583)
reservoir rock (p. 603)
resources (p. 581)
secondary enrichment (p. 596)
vein deposit (p. 587)

Web Resources

 The *Earth* Web site uses the resources and flexibility of the Internet to aid in your study of the topics in this chapter. Written and developed by geology instructors, this site will help improve your understanding of geology. Visit **http://www.pearson.ca/ tarbuck** and click on the cover of the text to find:

■ Online review quizzes.

■ Web-based critical thinking and writing exercises.

■ Links to chapter-specific Web resources.

■ Internet-wide key-term searches.

http://www.pearson.ca/tarbuck

Chapter 22
Planetary Geology

Barringer Crater, about 32 kilometres west of Winslow, Arizona. (Photo by Michael Collier)

The Sun is the hub of a huge rotating system consisting of nine planets, their satellites, and numerous small but interesting bodies, including asteroids, comets, and meteoroids. An estimated 99.85 percent of the mass of our solar system is contained within the Sun, while the planets collectively make up most of the remaining 0.15 percent. The planets, in order from the Sun, are Mercury, Venus, Earth, Mars, Jupiter, Saturn, Uranus, Neptune, and Pluto (Figure 22.1).

Under the control of the Sun's gravitational force, each planet maintains an elliptical orbit, and all of them travel in the same direction. The nearest planet to the Sun, Mercury, has the fastest orbital motion—48 kilometres per second—and the shortest period of revolution, 88 days. By contrast, the most distant planet, Pluto, has an orbital speed of 5 kilometres per second and requires 248 years to complete one revolution.

Imagine a planet's orbit drawn on a flat sheet of paper. The paper represents the planet's *orbital plane*. The orbital planes of all nine planets lie within 3 degrees of the plane of the Sun's equator, except for those of Mercury and Pluto, which are inclined 7 and 17 degrees, respectively.

When people first came to recognize that the planets are "worlds" much like Earth, a great deal of interest was generated. A primary concern has always been the possibility of intelligent life existing elsewhere in the universe. This expectation has not as yet come to pass. Nevertheless, as all of the planets most probably formed from the same primordial cloud of dust and gases, they should provide valuable information concerning Earth's history. Recent space explorations have been organized with this goal in mind. To date, Mercury, Venus, Mars, Jupiter, Saturn, Uranus, Neptune, and the Moon have been explored by space probes (Figure 22.2).

The Planets: An Overview

Careful examination of Table 22.1 shows that the planets fall quite nicely into two groups: the **terrestrial** (Earth-like) **planets** (Mercury, Venus, Earth, and Mars) and the **Jovian** (Jupiter-like) **planets** (Jupiter, Saturn, Uranus, and Neptune). Pluto is not included in either category, because its great distance from Earth and its small size make this planet's true nature a mystery (see Box 22.2).

The most obvious difference between the terrestrial and the Jovian planets is their size. The largest terrestrial planet (Earth) has a diameter only one-quarter as great as the diameter of the smallest Jovian planet (Neptune), and its mass is only one-seventeenth as great. Hence, the Jovian planets are often called *giants*. Also, because of their relative locations, the four Jovian planets are referred to as the *outer planets*, while the terrestrial planets are called the *inner planets*. As we shall see, there appears to be a correlation between the positions of these planets and their sizes.

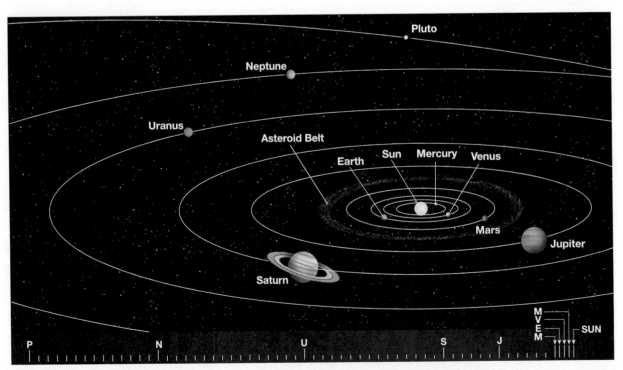

◆ **Figure 22.1**   Orbits of the planets (not to scale). Orbits are shown to scale along bottom of diagram.

TABLE 22.1   Planetary Data

Planet	Symbol	Mean Distance from Sun		Period of Revolution	Inclination of Orbit	Orbital Velocity	
		AU*	Millions of Kilometres			mi/s	km/s
Mercury	☿	0.39	58	88d	7°00′9	29.5	47.5
Venus	♀	0.72	108	225d	3°24′9	21.8	35.0
Earth	⊕	1.00	150	365.25d	0°00′0	18.5	29.8
Mars	♂	1.52	228	687d	1°51′9	14.9	24.1
Jupiter	♃	5.20	778	12yr	1°18′9	8.1	13.1
Saturn	♄	9.54	1427	29.5yr	2°29′9	6.0	9.6
Uranus	♅	19.18	2870	84yr	0°46′9	4.2	6.8
Neptune	♆	30.06	4497	165yr	1°46′9	3.3	5.3
Pluto	♇	39.44	5900	248yr	17°12′9	2.9	4.7

Planet	Period of Rotation	Diameter	Relative Mass (Earth = 1)	Average Density (g/cm³)	Polar Flattening (%)	Eccentricity	Number of Known Satellites†
		Kilometres					
Mercury	59d	4878	0.06	5.4	0.0	0.206	0
Venus	244d	12,104	0.82	5.2	0.0	0.007	0
Earth	23h56m04s	12,756	1.00	5.5	0.3	0.017	1
Mars	24h37m23s	6794	0.11	3.9	0.5	0.093	2
Jupiter	9h50m	143,884	317.87	1.3	6.7	0.048	28
Saturn	10h14m	120,536	95.14	0.7	10.4	0.056	30
Uranus	17h14m	51,118	14.56	1.2	2.3	0.047	21
Neptune	16h03m	50,530	17.21	1.7	1.8	0.009	8
Pluto	6.4d	~2300	0.002	1.8	0.0	0.250	1

*AU = astronomical unit, Earth's mean distance from the Sun.
†This number includes all satellites discovered as of April 2001. Additional small satellites are regularly being discovered around Jupiter, Saturn, and Uranus.

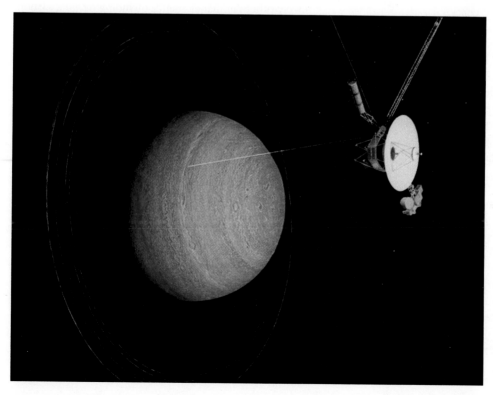

◆ **Figure 22.2** Painting of *Voyager 2* as it might have appeared when it encountered Uranus on January 24, 1986. (By Don Davis, courtesy of NASA)

Other dimensions along which the two groups markedly differ include density, composition, and rate of rotation. The densities of the terrestrial planets average about five times the density of water, whereas the Jovian planets have densities that average only 1.5 times that of water. One of the outer planets, Saturn, has a density only 0.7 that of water, which means that Saturn would float. Variations in the compositions of the planets are largely responsible for the density differences.

The substances that make up both groups of planets are divided into three groups—*gases*, *rocks*, and *ices*—based on their melting points.

1. The gases hydrogen and helium are those with melting points near absolute zero (0 Kelvin or −273°C), the lowest possible temperature.
2. The rocks are principally silicate minerals and metallic iron, which have melting points exceeding 700°C.
3. The ices have intermediate melting points (for example, H_2O has a melting point of 0°C) and include ammonia (NH_3), methane (CH_4), carbon dioxide (CO_2), and water (H_2O).

The terrestrial planets are mostly rock: dense rocky and metallic material, with minor amounts of gases. The Jovian planets, on the other hand, contain a large percentage of gases (hydrogen and helium), with varying amounts of ices (mostly water, ammonia, and methane). This accounts for their low densities. (The outer planets may contain as much rocky and metallic material as the terrestrial planets, but this material would be concentrated in their central cores.)

The Jovian planets have very thick atmospheres consisting of varying amounts of hydrogen, helium, methane, and ammonia. By comparison, the terrestrial planets have meagre atmospheres at best. A planet's ability to retain an atmosphere depends on its temperature and mass. Simply stated, a gas molecule can "evaporate" from a planet if it reaches a speed known as the **escape velocity**. For Earth, this velocity is 11 kilometres per second. Any material, including a rocket, must reach this speed before it can leave Earth and go into space.

The Jovian planets, because of their greater surface gravities, have escape velocities of 21–60 km/sec, much higher than the terrestrial planets. Consequently, it is more difficult for gases to "evaporate" from them. Also, because the molecular motion of a gas is temperature-dependent, at the low temperatures of the Jovian planets even the lightest gases are unlikely to acquire the speed needed to escape.

On the other hand, a comparatively warm body with a small surface gravity, like our Moon, is unable to hold even the heaviest gas and thus lacks an atmosphere. The slightly larger terrestrial planets of Earth,

Venus, and Mars retain some heavy gases like carbon dioxide, but even their atmospheres make up only an infinitesimally small portion of their total mass.

It is hypothesized that the primordial cloud of dust and gas from which all the planets are thought to have condensed had a composition somewhat similar to that of Jupiter. However, unlike Jupiter, the terrestrial planets today are nearly devoid of light gases and ices. Were the terrestrial planets once much larger? Did they contain these materials but lose them because of their relative closeness to the Sun? In the following section we will consider the evolutionary histories of these two diverse groups of planets in an attempt to answer these questions.

Origin and Evolution of the Planets

The orderly nature of our solar system leads most astronomers to conclude that the planets formed at essentially the same time and from the same primordial material as the Sun. This material formed a vast cloud of dust and gases called a *nebula* (*nebula* = a cloud). The **nebular hypothesis** suggests that all bodies of the solar system formed from an enormous nebular cloud consisting of approximately 80 percent hydrogen, 15 percent helium, and a few percent of all the other heavier elements known to exist (Figure 22.3). The heavier substances in this frigid cloud of dust and gases consisted mostly of elements such as silicon, aluminum, iron, and calcium—the substances of today's common rocky materials. Also prevalent were other familiar elements, including oxygen, carbon, and nitrogen.

About 5 billion years ago, and for reasons not yet fully understood, this huge cloud of minute rocky

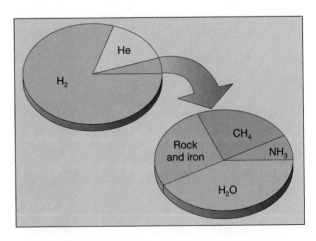

◆ **Figure 22.3** Composition of the primordial cloud of dust and gases from which the solar system is thought to have evolved. (Data from W. B. Hubbard)

Students Sometimes Ask...

Besides Earth, do any other bodies in the solar system have liquid water?

The planets closer to the Sun than Earth are considered too warm to contain liquid water and those farther from the Sun are generally too cold to have water in the liquid form (although some features on Mars suggest that it may have had abundant liquid water at some point in its history). However, the best prospects of finding liquid water within our solar system lie beneath the icy surfaces of some of Jupiter's moons. For instance, Europa is suspected to have an ocean of liquid water hidden under its outer covering of ice. Detailed images sent back to Earth from the *Galileo* spacecraft have revealed that Europa's icy surface is quite young and exhibits cracks apparently filled with dark fluid from below. This suggests that under its icy shell, Europa must have a warm, mobile interior—and perhaps an ocean. Because the presence of water in the liquid form is a necessity for life as we know it, there has been much interest in sending an orbiter to Europa—and eventually a lander capable of launching a robotic submarine—to determine if it too may harbour life.

The most prominent feature of Saturn is its system of rings (Figure 22.22), discovered by Galileo in 1610. Because he could not resolve them with his primitive telescope, they appeared to him as two smaller bodies adjacent to the planet. Their ring nature was revealed 50 years later by the Dutch astronomer Christian Huygens. Until the recent discovery that Jupiter, Uranus, and Neptune also have very faint ring systems, this phenomenon was thought to be unique to Saturn. When viewed from Earth, Saturn's rings appear as distinct bands.

Our view of Saturn slowly changes as both planets proceed along their orbits, continually shifting relative positions. This changes our angle of view of Saturn's rings. Once every 15 years, we see them edge-on, and they appear as an extremely fine line.

Saturn Close-up   In 1980 and 1981, fly-by missions of the nuclear-powered *Voyagers 1* and *2* space vehicles came within 100,000 kilometres of Saturn. More information was gained in a few days than had been acquired since Galileo first viewed this elegant planet telescopically (Figure 22.23):

1. Saturn's atmosphere is very dynamic, with winds roaring at up to 1500 kilometres per hour.

2. Large cyclonic "storms" similar to Jupiter's Great Red Spot, although much smaller, occur in Saturn's atmosphere.

3. Eleven additional moons were discovered.

4. The icy rings of Saturn were discovered to be more complex than expected. Each of the seven rings is made of numerous ringlets. The ringlets in one ring are intertwined in a braidlike configuration.

5. Satellite images reveal the thickness of the ring system to be no more than a few hundred metres. We easily see the thin rings from more than a billion kilometres' distance because they are highly reflective.

No image obtained so far can resolve the fine structures of the rings, but they undoubtedly are composed of small particles (moonlets) that orbit the planet much like any other satellite. Radar observations indicate that particles range in size from fine powder to objects as big as houses. The rings are close enough to Saturn that the tidal effect of the planet's gravity prevents similar-sized ring particles from coalescing under their own gravity into larger bodies.

Beyond the outermost bright ring (A ring), some moonlets have accreted to form very small satellites having diameters on the order of 100 kilometres. Five of these asteroid-size moons have been discovered orbiting within the faint outer rings, and others probably exist. Planetary geologists are very interested in the gravitational interaction in Saturn's ring system. They hope this will reveal how material from the primordial cloud of dust and gases condensed to produce the planets.

Saturn's Moons   The Saturnian satellite system consists of 30 named moons (Figure 22.23). (If you count the "moonlets" that comprise Saturn's rings, this planet has millions of satellites.) The largest, Titan, is bigger than Mercury and is the second-largest satellite in the solar system (after Jupiter's Ganymede). Titan and Neptune's Triton are the only satellites in the solar system known to have a substantial atmosphere. Because of its dense gaseous cover, the atmospheric pressure at Titan's surface is about 1.5 times that at Earth's surface. Another satellite, Phoebe, exhibits retrograde motion. It, like other moons with retrograde orbits, is most likely a captured asteroid or large planetesimal left over from the major episode of planetary formation.

Uranus and Neptune: The Twins

Earth and Venus have similar traits, but Uranus and Neptune are nearly twins. Only 1 percent different in diameter, both appear a pale greenish-blue, attributable to the methane in their atmospheres. Their structure and composition appear to be similar. Neptune, however, is colder, because it is half again as distant from the Sun's warmth as is Uranus.

◆ **Figure 22.22** A view of the dramatic ring system of Saturn.

◆ **Figure 22.23** Montage of the Saturnian satellite system. The moon Dione is in foreground; Tethys and Mimas are at lower right; Enceladus and Rhea are off ring's left; and Titan is upper right. (Photo courtesy of NASA)

Uranus: The Smaller Twin A feature of Uranus is that it rotates "on its side" (see Figure 22.2). Its axis of rotation, instead of being generally perpendicular to the plane of its orbit, like the other planets, lies nearly parallel to the plane of its orbit. Its rotational motion, therefore, has the appearance of rolling rather than the toplike spinning of the other planets. Because the axis of Uranus is inclined almost 90 degrees, the Sun is nearly overhead at one of its poles once each revolution, and then half a revolution later it is overhead at the other pole.

A surprise discovery in 1977 revealed that Uranus has rings, much like those encircling Jupiter. This find occurred as Uranus passed in front of a distant star and blocked its view, a process called *occultation* (the word *occult* means "hidden"). Observers saw the star "wink" briefly five times (meaning five rings) before the primary occultation and again five times afterward. Later studies indicate that Uranus has at least nine distinct belts of debris orbiting its equatorial region.

Spectacular views from *Voyager 2* of the five largest moons of Uranus show quite varied terrains. Some have long, deep canyons and linear scars, whereas others possess large, smooth areas on otherwise crater-riddled surfaces. The Jet Propulsion Laboratory described Miranda, the innermost of the five largest moons, as having a greater variety of landforms than any body yet examined in the solar system.

Neptune: The Windy Planet   Even when the most powerful telescope is focused on Neptune, it appears as a bluish fuzzy disk. Until *Voyager 2*'s 1989 encounter, astronomers knew very little about this planet. However, *Voyager 2*'s 12-year, nearly 5-billion-kilometre journey provided investigators with a great deal of new information about Neptune and its satellites.

Neptune has a dynamic atmosphere, much like those of Jupiter and Saturn (Figure 22.24). Winds exceeding 1000 kilometres per hour encircle the planet,

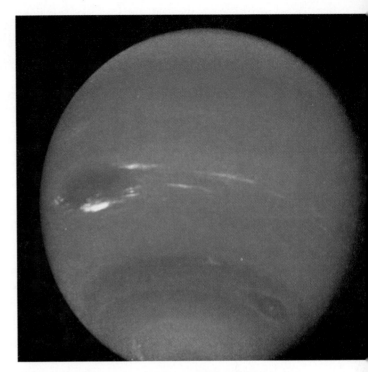

◆ **Figure 22.24** This image of Neptune shows the Great Dark Spot (left centre). Also visible are bright cirruslike clouds that travel at high speed around the planet. A second oval spot is at 54° south latitude on the east limb of the planet. (Courtesy of the Jet Propulsion Laboratory)

making it one of the windiest places in the solar system. It also has an Earth-size blemish called the *Great Dark Spot*, which is reminiscent of Jupiter's Great Red Spot, and is assumed to be a large rotating storm.

Perhaps most surprising are white, cirruslike clouds that occupy a layer about 50 kilometres above the main cloud deck, probably frozen methane. Six new satellites were discovered in the *Voyager* images, bringing Neptune's family to eight. All of the newly discovered moons orbit the planet in a direction opposite that of the two larger satellites. *Voyager* images also revealed a ring system around Neptune.

Triton, Neptune's largest moon, is a most interesting object. Its diameter is nearly that of Earth's moon. Triton is the only *large* moon in the solar system that exhibits retrograde motion. This suggests that Triton formed independently of Neptune and was gravitationally captured.

Triton also has the lowest surface temperature yet measured on any body in the solar system, −200°C. Its atmosphere is mostly nitrogen with a little methane. Despite low surface temperatures, Triton displays a volcanic-like activity. Two active plumes were discovered that extended to an altitude of 8 kilometres and were blown downwind for more than 100 kilometres. Presumably, the surface layers of darker methane ice

Students Sometimes Ask...

Why does Uranus spin on its side?
The most likely explanation for the unusual sideways spin of Uranus is that it started out spinning the same way as the other planets, but then its spin was altered by a giant impact—which was probably very common when the planets were first formed. However, a giant impact would be very difficult to verify because it would not have left any crater on Uranus, which has no solid surface. Like many events that occurred early on in the formation of our solar system, the reason for Uranus's sideways spin may never be known for sure.

BOX 22.2

Understanding Earth
Is Pluto Really a Planet?

Ever since Pluto's discovery in 1930, it has been a mystery on the edge of the solar system. At first, Pluto was thought to be about as large as Earth, but as better images were obtained, Pluto's diameter was estimated to be a little less than half that of Earth. Then, in 1978, astronomers discovered that Pluto has a moon (Charon), whose brightness when combined with its parent made Pluto appear much larger than it really was (Figure 22.B). Recent images obtained by the *Hubble Space Telescope* established the diameter of Pluto at only 2300 kilometres. This is about one-fifth that of Earth and less than half that of Mercury, long considered the runt of the solar system. In fact, seven moons in the solar system are larger than Pluto.

In addition to its diminutive size, during the last few years over 60 cometlike objects have been discovered in a belt that extends beyond the orbit of Neptune. One of these objects approaches 500 kilometres in diameter. It appears that Pluto and its moon Charon are just two of thousands of icy worlds orbiting in the far reaches of the solar system.

A growing number of astronomers assert that Pluto's small size and its location within a swarm of similar icy objects means that it should be reclassified as a "minor planet," like asteroids and comets. Others insist that regardless of how Pluto's identity changes, demoting Pluto to a minor planet would dishonour astronomical history and confuse the public.

For now it seems that the International Astronomical Union, a group that has the power to vote on whether or not Pluto is a planet, is content with the status quo. Nevertheless, Pluto's planetary status will never be the same. It is now clear that Pluto is unique among the planets, being very different from the four rocky innermost planets, and unlike the four gaseous giants. Perhaps Pluto is best described as the largest member of a belt of thousands of small, icy worlds that orbit in the outer reaches of our solar system.

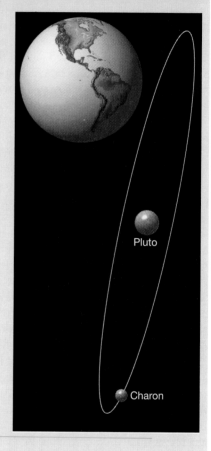

◆ **Figure 22.B** Pluto and its moon Charon. Earth is shown for scale.

readily absorb solar energy. Such surface warming vaporizes some of the underlying nitrogen ice. As subsurface pressures increase, explosive eruptions eventually result.

Pluto: Planet X

Pluto lies on the fringe of the solar system, almost 40 times as far from the Sun as Earth. It is 10,000 times too dim to be visible to the unaided eye. Because of its great distance and slow orbital speed, it takes Pluto 248 Earth-years to orbit the Sun. Since its discovery in 1930, it has completed about one-quarter of a revolution. Pluto's orbit is noticeably elongated (highly eccentric), causing it to occasionally travel inside the orbit of Neptune, where it resided from 1979 through February of 1999. There is no chance that Pluto and Neptune will ever collide, because their orbits are inclined to each other and do not actually cross (see Figure 22.1).

In 1978 the moon Charon was discovered orbiting Pluto. Because of its close proximity to the planet, the best ground-based images of Charon show it only as an elongated bulge. In 1990 the *Hubble Space Telescope* produced an image that clearly resolves the separation between these two icy worlds. Charon orbits Pluto once every 6.4 Earth-days at $1/20$ the distance from Pluto of our Moon from Earth (see Figure 22.B).

The discovery of Charon greatly altered earlier estimates of Pluto's size. Current data indicate that Pluto has a diameter of about 2300 kilometres, about one-fifth the size of Earth, making it the smallest planet in the solar system (see Box 22.2). Charon is about 1300 kilometres across, exceptionally large in proportion to its parent.

The average temperature of Pluto is estimated at −210°C, cold enough to solidify most gases that might be present. Thus, Pluto might best be described as a dirty iceball of frozen gases with lesser amounts of rocky substances.

BOX 22.3

Earth as a System
Comet Shoemaker-Levy Impacts Jupiter

In July 1994, Comet Shoemaker-Levy impacted Jupiter with the force equal to 6 million megatonnes of energy. (A megatonne is the equivalent of a million tonnes of the explosive TNT). Clearly, this was the most dramatic event in the solar system ever observed by people. Concern that a similar event might occur on Earth led NASA to establish the Near-Earth Object Search Committee to detect asteroids or comets with trajectories that might cross Earth's orbit.

It is possible that a few small fragments of a comet may have actually impacted Earth in 1908. That year, a strong explosion flattened more than 1000 square kilometres of a remote Siberian forest.

Comet Shoemaker-Levy was discovered at California's Mount Palomar Observatory barely a year before it impacted Jupiter. Careful observation showed that it consisted of two dozen fragments. Researchers concluded that the comet had broken up during an earlier pass by Jupiter. As the larger fragments penetrated Jupiter's outer

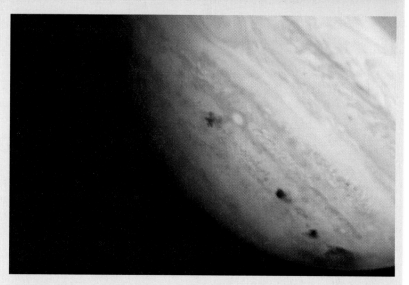

◆ **Figure 22.C**   Dark blemishes on Jupiter produced by the impact of fragments of Comet Shoemaker-Levy in July 1994. (Courtesy of NASA)

atmosphere, they produced brilliant impact flashes and debris plumes soaring thousands of kilometres above the planet. The result of these fiery impacts were dark zones in Jupiter's atmosphere that exceeded Earth's diameter in size.

The largest of these dark blemishes lasted for months (Figure 22.C). Investigators learned much about the dynamics of Jupiter's atmosphere by observing how these blemishes dispersed.

Minor Members of the Solar System

Asteroids: Microplanets

Asteroids are smaller bodies that have been likened to "flying mountains." The largest, Ceres, is about 1000 kilometres in diameter, but most of the 50,000 that have been observed are only about a kilometre across. The smallest asteroids are assumed to be no larger than grains of sand. Most lie between the orbits of Mars and Jupiter and have periods of three to six years (Figure 22.25). Some have very eccentric orbits and travel very near the Sun, and a few larger ones regularly pass close to Earth and its moon. Many of the most recent impact craters on the Moon and Earth were probably caused by collisions with asteroids. Inevitably, future Earth–asteroid collisions will occur (see Box 22.3).

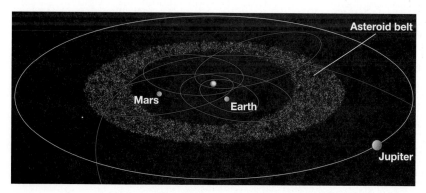

◆ **Figure 22.25**   The orbits of most asteroids lie between Mars and Jupiter. Also shown are the orbits of a few known near-Earth asteroids. Perhaps a thousand or more asteroids have near-Earth orbits. Luckily, only a few dozen are thought to be larger than 1 kilometre in diameter.

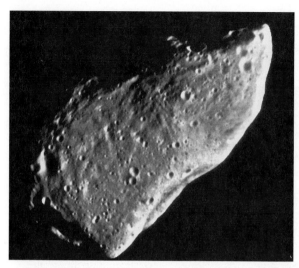

◆ **Figure 22.26** Image of asteroid 951 (Gaspra) obtained by the Jupiter-bound *Galileo* spacecraft. Like other asteroids, Gaspra is probably a collision-produced fragment of a larger body. (Courtesy of NASA)

Because many asteroids have irregular shapes, planetary geologists first speculated that they might be fragments of a broken planet that once orbited between Mars and Jupiter (Figure 22.26). However, the total mass of the asteroids is estimated to be only one-thousandth that of Earth, which itself is not a large planet. What happened to the remainder of the original planet? Others have hypothesized that several larger bodies once coexisted in close proximity and that their collisions produced numerous smaller ones. The existence of several "families" of asteroids has been used to support this explanation. However, no conclusive evidence has been found for either hypothesis.

In February 2001 an American spacecraft became the first visitor to an asteroid. Although it was not designed for landing, *NEAR Shoemaker* landed successfully and generated information that has planetary geologists intrigued and perplexed. Images obtained as the spacecraft drifted at the rate of 6 kilometres per hour toward the surface of Eros revealed a barren, rocky surface composed of particles ranging in size from fine dust to boulders up to 8 metres across. Researchers unexpectedly discovered that fine debris is concentrated in the low areas that form flat deposits that resemble ponds. Surrounding the low areas, the landscape is marked by an abundance of large boulders.

One of several hypotheses being considered as an explanation to the boulder-laden topography is seismic shaking, which would move the boulders upward. Analogous to what happens when a can of mixed nuts is shaken, the larger materials rise to the top while the smaller materials settle to the bottom.

Comets: Dirty Snowballs

Comets are among the most interesting and unpredictable bodies in the solar system. They have been compared to dirty snowballs, because they are made of frozen gases (water, ammonia, methane, carbon dioxide, and carbon monoxide) that hold together small pieces of rocky and metallic materials. Many comets travel very elongated orbits that carry them beyond Pluto.

When first observed, a comet appears very small; but as it approaches the Sun, solar energy begins to vaporize the frozen gases, producing a glowing head called the **coma** (Figure 22.27). The size of the coma varies greatly from one comet to another. Extremely rare ones exceed the size of the Sun, but most approximate the size of Jupiter. Within the coma a small glowing nucleus with a diameter of only a few kilometres can sometimes be detected. As comets approach the Sun, some, but not all, develop a tail that extends for millions of kilometres. Despite the enormous size of their tails and comas, they are comparatively small members of the solar system.

The fact that the tail of a comet points away from the Sun in a slightly curved manner (Figure 22.27) led early astronomers to propose that the Sun has a repulsive force that pushes the particles of the coma away, thus forming the tail. Today, two solar forces are known to contribute to this formation. One, *radiation pressure*, pushes dust particles away from the coma. The second, known as *solar wind*, is responsible for moving the ionized gases, particularly carbon monoxide. Sometimes a single tail composed of both dust and ionized gases is produced, but often two tails are observed (Figure 22.28).

As a comet moves away from the Sun, the gases forming the coma recondense, the tail disappears, and the comet returns to "cold storage." Material that was blown from the coma to form the tail is lost from the comet forever. Consequently, most comets cannot survive more than a few hundred close orbits of the Sun. Once all the gases are expelled, the remaining material—a swarm of unconnected metallic and stony particles—continues the orbit without a coma or a tail.

Little is known about the origin of comets. Millions are believed to orbit the Sun at distances greater than 10,000 times the Earth–Sun distance, in a structure called the *Oort cloud*. The gravitational effect of distant passing stars sends some comets into the highly eccentric orbits that carry them toward the centre of our solar system. Here, the gravitational pull of the larger planets, particularly Jupiter, alters a comet's orbit and accelerates its period of revolution.

The most famous short-period comet is Halley's comet. Its orbital period averages 76 years, and every one of its 29 appearances since 240 BC has been

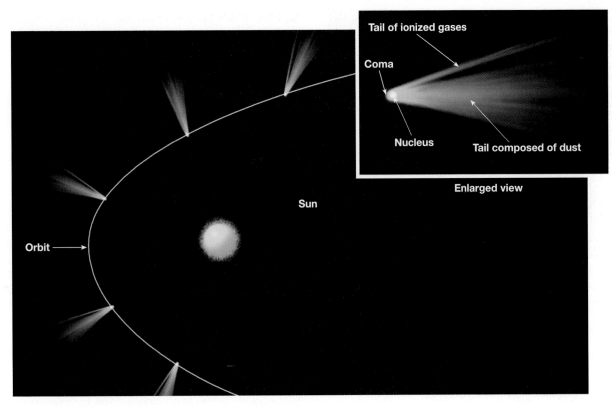

◆ **Figure 22.27** Orientation of a comet's tail as it orbits the Sun.

recorded by Chinese astronomers. This record is a testimonial to their dedication as astronomical observers and to the endurance of their culture. When seen in 1910, Halley's comet had developed a tail nearly 1.6 million kilometres long and was visible during the daylight hours.

In 1986 the unspectacular showing of Halley's comet was a disappointment to many people in the Northern Hemisphere. Yet it was during this most recent visit to the inner solar system that a great deal of new information was learned about this most famous of comets. The new data were gathered by space probes sent to rendezvous with the comet. Most notably, the European probe *Giotto* approached to within 600 kilometres of the comet's nucleus and obtained the first images of this elusive structure.

◆ **Figure 22.28** Comet Hale-Bopp. The two tails seen in the photograph are about 15 million to 25 million kilometres long. (A Peoria Astronomical Society Photograph by Eric Clifton and Greg Neaveill)

We now know that the nucleus is potato-shaped, 16 kilometres by 8 kilometres. The surface is irregular and full of craterlike pits. Gases and dust that vaporize from the nucleus to form the coma and tail appear to gush from its surface as bright jets or streams. Only about 10 percent of the comet's total surface was emitting these jets at the time of the rendezvous. The remaining surface area of the comet appeared to be covered with a dark layer that may consist of organic material.

In 1997 the comet Hale-Bopp made for spectacular viewing around the globe. As comets go, the nucleus of Hale-Bopp was unusually large, about 40 kilometres in diameter. As shown in Figure 22.28, two tails extended from this comet. The bluish gas-tail is composed of positively charged ions, and it points almost directly away from the Sun. The yellowish tail is composed of dust and other rocky debris. Because the rocky material is more massive than the ionized gases, it is less affected by the solar wind and follows a different trajectory away from the comet.

Meteoroids: Visitors to Earth

Nearly everyone has seen a **meteor**, popularly (but inaccurately) called a "shooting star." This streak of light lasts from an eye blink to a few seconds and occurs when a small solid particle, a **meteoroid**, enters Earth's atmosphere from interplanetary space. The friction between the meteoroid and the air heats both and produces the light we see. Most meteoroids originate from one of the following three sources: (1) interplanetary debris that was not gravitationally swept up by the planets during the formation of the solar system, (2) material that is continually being displaced from the asteroid belt, or (3) the solid remains of comets that once travelled near Earth's orbit. A few meteoroids are believed to be fragments of the Moon, or

possibly Mars, that were ejected when an asteroid impacted these bodies.

Although a rare meteoroid is as large as an asteroid, most are the size of sand grains and weigh less than 1/100 gram. Consequently, they vaporize before reaching Earth's surface. Some, called *micrometeorites*, are so tiny that their rate of fall becomes too slow to cause them to burn up, so they drift down as "space dust." Each day, the number of meteoroids that enter Earth's atmosphere must reach into the thousands. After sunset on a clear night, a half dozen or more are bright enough to be seen with the naked eye each hour from anywhere on Earth.

Occasionally, meteor sightings increase dramatically to 60 or more per hour. These displays, called **meteor showers**, result when Earth encounters a swarm of meteoroids traveling in the same direction and at nearly the same speed as Earth. The close association of these swarms to the orbits of some short-term comets strongly suggests that they represent material lost by these comets (Table 22.2). Some swarms not associated with orbits of known comets are probably the remains of the nucleus of a long-defunct comet. The notable Perseid meteor shower that occurs each year around August 12 is probably material from the Comet 1862 III, which has a period of 110 years.

Most meteoroids that are thought to be the remains of comets are small and only occasionally reach the ground (see Box 22.4). Most meteoroids large enough to survive the heated fall are thought to originate among the asteroids, where chance collisions modify their orbits and send them toward Earth. Earth's gravitational force does the rest.

The remains of meteoroids, when found on Earth, are referred to as **meteorites**. A few very large meteoroids have blasted out craters on Earth's surface that strongly resemble those on the lunar surface. The most famous is Meteor Crater in Arizona (see

TABLE 22.2	**Major Meteor Showers**	
Shower	Approximate Dates	Associated Comet
Quadrantids	January 4–6	—
Lyrids	April 20–23	Comet 1861 I
Eta Aquarids	May 3–5	Halley's comet
Delta Aquarids	July 30	—
Perseids	August 12	Comet 1862 III
Draconids	October 7–10	Comet Giacobini–Zinner
Orionids	October 20	Halley's comet
Taurids	November 3–13	Comet Encke
Andromedids	November 14	Comet Biela
Leonids	November 18	Comet 1866 I
Geminids	December 4–16	—

BOX 22.4

Earth as a System
Is Earth on a Collision Course?

The solar system is cluttered with meteoroids, asteroids, active comets, and extinct comets. These fragments travel at great speeds and can strike Earth with the explosive force of a powerful nuclear weapon.

In the last few decades, it has become increasingly clear that comets and asteroids have collided with Earth far more frequently than was previously known. The evidence is giant impact structures called *astroblemes*. More than 100 have been identified (Figure 22.D). Many were once mistakenly thought to result from some volcanic process. Although most astroblemes are so old they no longer resemble impact craters, evidence of their intense impact remains (Figure 22.E). One notable exception is a very fresh-looking crater near Winslow, Arizona, known as Meteor Crater—also known as Barringer Crater (see chapter-opening photo).

Evidence is mounting that about 65 million years ago a large asteroid about 10 kilometres in diameter collided with Earth. This impact may have caused the extinction of the dinosaurs, as well as nearly 50 percent of all plant and animal species. (For more on this, see Box 8.4, "Demise of the Dinosaurs," p. 238)

More recently, a spectacular explosion has been attributed to the collision of our planet with a comet or asteroid. In 1908, in a remote region of Siberia, a "fireball" that appeared more brilliant than the Sun exploded with a violent force. The shock waves rattled windows and caused sounds that were heard up to 1000 kilometres away. The "Tunguska event," as it is called, scorched, delimbed, and flattened trees to 30 kilometres from the epicentre. But expeditions to the

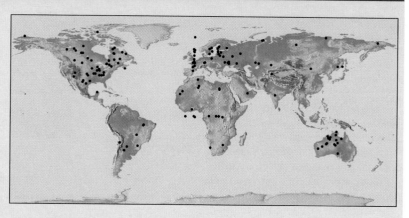

◆ **Figure 22.D** World map of major impact structures (astroblemes). Others are being identified every year. Many of these structures are quite old and have been identified from images obtained by satellites or by geologic mapping. (Data from Griffith Observatory)

area found no evidence of an impact crater or any fragments. Evidently, the explosion, which equalled at least a 10-megaton nuclear bomb, occurred a few kilometres above Earth's surface. Most likely it was the demise of a comet or perhaps a stony asteroid. Why it exploded prior to impact is uncertain.

The dangers of living with these small but deadly objects from space again came to public attention in 1989 when an asteroid nearly a kilometre across shot past Earth. It was a near miss, about twice the distance to the Moon. Travelling at 70,000 kilometres per hour, it could have produced a crater 10 kilometres in diameter and perhaps 2 kilometres deep. As an observer noted, "Sooner or later it will be back." As it was, it crossed our orbit just six hours ahead of Earth. Statistics show that collisions of this tremendous magnitude should take place every few hundred million years and could have drastic consequences for life on Earth.

◆ **Figure 22.E** Manicouagan, Quebec, is a 200-million-year-old eroded impact crater. The lake outlines the crater remnant, which is 70 kilometres across. Fractures related to this event extend outward for an additional 30 kilometres. (Courtesy of U.S. Geological Survey)

chapter-opening photo). This huge cavity is about 1.2 kilometres across, 170 metres deep, and has an upturned rim that rises 50 metres above the surrounding countryside. Over 30 tonnes of iron fragments have been found in the immediate area, but attempts to locate a main body have been unsuccessful.

Based on erosion, the impact likely occurred within the last 20,000 years.

Prior to Moon rocks brought back by lunar explorers, meteorites were the only extraterrestrial materials that could be directly examined (Figure 22.29). Meteorites are classified by their composition:

BOX 22.5

Canadian Profile
Fall and Recovery of the Tagish Lake Meteorite:
A Messenger from the Early Solar System

Philip McCausland*

In the predawn of January 18, 2000, an exceptionally bright fireball was widely seen travelling south-southeast throughout Yukon and northern B.C. Many reported seeing a terminal flash as bright as the sun, hearing loud booms, and noted a smoke trail that persisted for nearly half an hour. Some people took photos and video of the smoke trail which, along with U.S. defence satellite infrared and optical data, proved invaluable for estimating the size and pre-collision orbit of the object. A week after the fall, some dark fusion-crusted pieces of the meteorite were found and carefully collected from the frozen surface of Tagish Lake, B.C. by local resident Jim Brook. His find was remarkable: carbonaceous chondrite meteorites, dark, charcoal-like material that comes from the earliest history of the solar system. Much more of the meteorite was collected from the frozen lake surface by a team of Canadian researchers led by Alan Hildebrand (University of Calgary) in April and early May 2000, before the spring thaw (Figure 22.F). In all, some 500 fragments totalling about 6 kilograms of carbonaceous chondrite were collected from a large strewnfield approximately 16 kilometres long by 3 kilometres wide.

The Tagish Lake meteorite fall is an exceptional event: the meteorites are a particularly fragile kind of carbonaceous chondrite, and may be the most primitive solar system material ever sampled. Chondrite meteorites comprise about 85 percent of all meteorite falls. They are considered to represent primitive solar system material because they contain glassy beads called chondrules as well as calcium-aluminum inclusions (CAIs) and hardy pre–solar system grains (silicon carbide, microdiamonds, and C60 buckyballs),

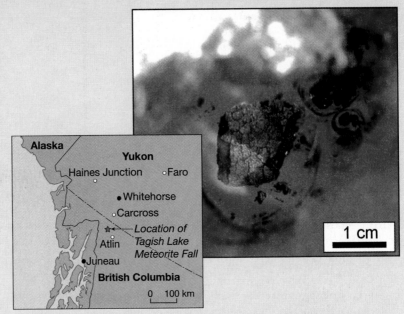

◆ **Figure 22.F**   Top-down view of a Tagish Lake meteorite fragment found in a 4 centimetre deep melt-hole in the ice surface of the lake. Note the fragmented brown fusion crust from the meteorite's passage through the Earth's atmosphere and the otherwise dark appearance of the meteorite, bespeckled with white CAIs. Inset: Location of Tagish Lake, British Columbia. (Photo by Phil McCausland)

all of which indicate that the meteorite has never been heated or altered by planetary processes. In bulk composition and mineralogy Tagish Lake is most similar to the CI-type carbonaceous chondrites, which are deemed to be closest to the original solar nebula composition and are used in chondrite-normalized rare and trace element plots. Like other carbonaceous chondrites, Tagish Lake bears amino acids and other simple organic compounds that may have provided the building blocks for life. Surprisingly, the Tagish Lake material has greater bulk porosity—30 percent—and higher carbon content (about 5 weight percent) than any other meteorite. In reflected sunlight, the Tagish Lake

material most closely resembles dark asteroids from the outer asteroid belt, the first meteorite to do so. It seems likely that the Tagish Lake meteorite represents a portion of the solar system that has not previously provided samples. Its unprecedented features, the pristine nature of the still frozen material collected by Jim Brook, and the large amount of material collected makes the Tagish Lake meteorite a premiere scientific event, the subject of essential early solar system research for years to come.

*Philip McCausland is a Research Scientist at the University of Michigan.

◆ **Figure 22.29**   Iron meteorite found near Barringer Crater, Arizona.
(Courtesy of Meteor Crater, Northern Arizona, USA)

(1) **iron**—mostly iron with 5 to 20 percent nickel, (2) **stony**—silicate minerals with inclusions of other minerals, and (3) **stony-iron**—mixtures. Although stony meteorites are probably more common, people find mostly irons. This is understandable, for irons withstand the impact better, weather more slowly, and are much easier for a layperson to distinguish from terrestrial rocks. Iron meteorites are probably fragments of once-molten cores of large asteroids or small planets.

One rare kind of meteorite, called a *carbonaceous chondrite* (see Box 22.5), was found to contain simple amino acids and other organic compounds, which are the basic building blocks of life. This discovery confirms similar findings in observational astronomy, which indicate that numerous organic compounds exist in the frigid realm of outer space.

If meteorites represent the makeup of Earth-like planets, as some planetary geologists suggest, then Earth must contain a much larger percentage of iron than is indicated by surface rocks. This is one reason why geologists suggest that Earth's core may be mostly iron and nickel. In addition, meteorite dating indicates that our solar system's age certainly exceeds 4.5 billion years. This "old age" has been confirmed by data from lunar samples.

Chapter Summary

- The planets can be arranged into two groups: the *terrestrial* (Earth-like) *planets* (Mercury, Venus, Earth, and Mars) and the *Jovian* (Jupiter-like) *planets* (Jupiter, Saturn, Uranus, and Neptune). Pluto is not included in either group. When compared to the Jovian planets, the terrestrial planets are smaller, more dense, contain proportionally more rocky material, and lesser amounts of gases (hydrogen and helium) and ices (water, ammonia, methane, and carbon dioxide).

- The *nebular hypothesis* describes the formation of the solar system. The planets and Sun began forming about 5 billion years ago from a large cloud of dust and gases called a *nebula*. As the nebular cloud contracted, it began to rotate and flatten into a disk. Material that was gravitationally pulled toward the centre became the *protosun*. Within the rotating disk, small centres formed, called *protoplanets*, sweeping up more and more of the nebular debris. Because of their high temperatures and weak gravitational fields, the inner planets (Mercury, Venus, Earth, and Mars) were unable to accumulate much of the lighter components (hydrogen, ammonia, methane, and water) of the nebula. However, because of the very cold temperatures existing far from the Sun, the fragments from which the Jovian planets formed contained a high percentage of ices—water, carbon dioxide, ammonia, and methane.

- The lunar surface exhibits several types of features. Most *craters* were produced by the impact of rapidly moving debris (meteoroids). Bright, densely cratered highlands make up much of the lunar surface. Dark, fairly smooth lowlands are called *maria*. Maria basins are enormous impact craters that have been flooded with layer upon layer of very fluid mafic lava. All lunar terrains are mantled with a soil-like layer of grey, unconsolidated debris, called *lunar regolith*, which has been derived from a few billion years of meteoric bombardment.

- Closest to the Sun, *Mercury* is small, dense, has no atmosphere, and exhibits the greatest temperature extremes of any planet. *Venus*, the brightest planet in the sky, has a thick, dense, carbon dioxide atmosphere, a surface of relatively subdued plains and inactive volcanoes, a surface atmospheric pressure 90 times that of Earth, and surface temperatures of 475°C. *Mars*, the red planet, has a carbon dioxide atmosphere only 1 percent as dense as Earth's, extensive winds and dust storms, numerous inactive volcanoes, many large canyons, and several valleys of debatable origin resembling drainage patterns similar to stream valleys on Earth. *Jupiter*, the largest planet, rotates rapidly, has a banded appearance, a Great Red Spot that varies in size, a ring system, and at least 16 moons (one of which, Io, is volcanically active). *Saturn*, best known for its rings, also has a dynamic atmosphere with winds up to 1500 kilometres per hour and "storms" similar to Jupiter's Great Red Spot. *Uranus* and *Neptune* are often called the twins because of similar structure and composition. Uranus is unique in rotating "on its side." Neptune has white cirruslike clouds above its main cloud deck and an Earth-size Great Dark

Spot, assumed to be a large rotating storm similar to Jupiter's Great Red Spot. *Pluto*, a small frozen world with one moon, has an elongated orbit causing it to occasionally travel inside the orbit of Neptune but with no chance of collision.

- The minor members of the solar system include *asteroids*, *comets*, and *meteoroids*. No conclusive evidence has been found to explain the origin of the asteroids. Comets are made of ices with small pieces of rocky and metallic material. Many travel in very elongated orbits that carry them beyond Pluto. Meteoroids, small solid particles that travel through interplanetary space, become *meteors* when they enter Earth's atmosphere and vaporize with a flash of light. *Meteor showers* occur when Earth encounters a swarm of meteoroids, probably fragments of a comet. *Meteorites* are the remains of meteoroids. The *three types of meteorites* are iron, stony, and stony-iron.

Review Questions

1. By what criteria are the planets placed into either the Jovian or the terrestrial group?
2. What are the three types of materials that make up the planets? How are they different? How does their distribution account for the density differences between the terrestrial and Jovian planetary groups?
3. Explain why different planets have different atmospheres.
4. How is crater density used in the relative dating of lunar features?
5. Briefly outline the history of the Moon.
6. How are the maria of the Moon similar to the Columbia Plateau?
7. Why has Mars been the planet most studied telescopically?
8. What surface features does Mars have that are also common on Earth?
9. What evidence supports a water cycle on Mars? What evidence refutes the possibility of a wet Martian climate?
10. Why are astrobiologists intrigued about evidence that groundwater has seeped onto the surface of Mars?
11. The two "moons" of Mars were once suggested to be artificial. What characteristics do they have that would cause such speculation?
12. What is the nature of Jupiter's Great Red Spot?
13. Why are the Galilean satellites of Jupiter so named?
14. What is distinctive about Jupiter's satellite Io?

15. Why are the four *outer* satellites of Jupiter thought to have been captured rather than having been formed with the rest of the satellite system?

16. How are Jupiter and Saturn similar?

17. How are Saturn's satellite Titan and Neptune's Triton similar?

18. What three bodies in the solar system exhibit volcanic-like activity?

19. Where are most asteroids found?

20. What do you think would happen if Earth passed through the tail of a comet?

21. Where are most comets thought to reside? What eventually becomes of comets that orbit close to the Sun?

22. Compare meteoroid, meteor, and meteorite.

23. What are the three main sources of meteoroids?

24. Why are meteorite craters more common on the Moon than on Earth, even though the Moon is a much smaller target?

25. It has been estimated that Halley's comet has a mass of 100 billion tonnes. Further, this comet is estimated to lose 100 million tonnes of material during the few months that its orbit brings it close to the Sun. With an orbital period of 76 years, what is the maximum remaining life span of Halley's comet?

Key Terms

asteroid (p. 647)	Jovian planet (p. 626)	meteoroid (p. 650)	stony meteorite (p. 653)
coma (p. 648)	lunar regolith (p. 633)	meteor shower (p. 650)	terrestrial planet (p. 626)
comet (p. 648)	maria (p. 630)	nebular hypothesis (p. 628)	
escape velocity (p. 628)	meteor (p. 650)	stony-iron meteorite	
iron meteorite (p. 653)	meteorite (p. 650)	(p. 653)	

Web Resources

 The *Earth* Web site uses the resources and flexibility of the Internet to aid in your study of the topics in this chapter. Written and developed by geology instructors, this site will help improve your understanding of geology. Visit **http://www.pearson.ca/tarbuck** and click on the cover of the text to find:

■ Online review quizzes.

■ Web-based critical thinking and writing exercises.

■ Links to chapter-specific Web resources.

■ Internet-wide key-term searches.

http://www.pearson.ca/tarbuck

Glossary

Aa flow A type of lava flow that has a jagged, blocky surface.

Ablation A general term for the loss of ice and snow from a glacier.

Abrasion The grinding and scraping of a rock surface by the friction and impact of rock particles carried by water, wind, and ice.

Abyssal plain Very level area of the deep-ocean floor, usually lying at the foot of the continental rise.

Accretionary wedge A large wedge-shaped mass of sediment that accumulates in subduction zones. Here sediment is scraped from the subducting oceanic plate and accreted to the overriding crustal block.

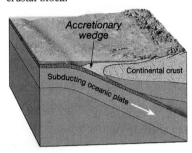

Active continental margin Usually narrow and consisting of highly deformed sediments. Such margins occur where oceanic lithosphere is being subducted beneath the margin of a continent.

Active layer The zone above the permafrost that thaws in summer and refreezes in winter.

Aftershock A smaller earthquake that follows the main earthquake.

Alluvial fan A fan-shaped deposit of sediment formed when a stream's slope is abruptly reduced.

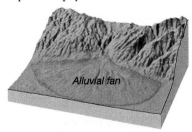

Alluvium Unconsolidated sediment deposited by a stream.

Alpine glacier A glacier confined to a mountain valley, which in most instances had previously been a stream valley.

Andesitic composition See *Intermediate composition*.

Angle of repose The steepest angle at which loose material remains stationary without sliding downslope.

Angular unconformity An unconformity in which the older strata dip at an angle different from that of the younger beds.

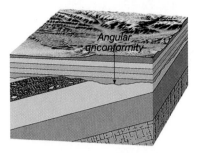

Antecedent stream A stream that continued to downcut and maintain its original course as an area along its course was uplifted by faulting or folding.

Anthracite A hard, metamorphic form of coal that burns cleanly and hot.

Anticline A fold in sedimentary strata that resembles an arch.

Aphanitic texture A texture of igneous rocks in which the crystals are too small for individual minerals to be distinguished without the aid of a microscope.

Aquifer Rock or sediment through which groundwater moves easily.

Aquitard An impermeable bed that hinders or prevents groundwater movement.

Archean Eon The second eon of Precambrian time. The eon following the Hadean and preceding the Proterozoic. It extends between 3.8 and 2.5 billion years ago.

Arête A narrow, knifelike ridge separating two adjacent glaciated valleys.

Arkose A feldspar-rich sandstone.

Artesian well A well in which the water rises above the level where it was initially encountered.

Assimilation In igneous activity, the process of incorporating country rock into a magma body.

Asteroid One of thousands of small planetlike bodies, ranging in size from a few hundred kilometres to less than one kilometre across. Most asteroids' orbits lie between those of Mars and Jupiter.

Asthenosphere A subdivision of the mantle situated below the lithosphere.

This zone of weak material exists below a depth of about 100 kilometres and in some regions extends as deep as 700 kilometres. The rock within this zone is easily deformed.

Atmosphere The gaseous portion of a planet, the planet's envelope of air. One of the traditional subdivisions of Earth's physical environment.

Atoll A coral island consisting of a nearly continuous ring of coral reef surrounding a central lagoon.

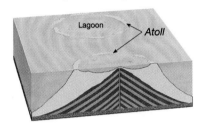

Atom The smallest particle that exists as an element.

Atomic mass unit A mass unit equal to exactly one-twelfth the mass of a carbon-12 atom.

Atomic number The number of protons in the nucleus of an atom.

Atomic weight The average of the atomic masses of isotopes for a given element.

Aureole A zone or halo of contact metamorphism found in the country rock surrounding an igneous intrusion.

Backswamp A poorly drained area on a floodplain resulting when natural levees are present.

Bajada An apron of sediment along a mountain front created by the coalescence of alluvial fans.

Banded iron formation A chemical sedimentary rock consisting of alternating bands of iron oxide minerals and chert. Restricted to the Precambrian and most widespread in Proterozoic rock sequences.

Bar Common term for sand and gravel deposits in a stream channel.

Barchan dune A solitary sand dune shaped like a crescent with its tips pointing downwind.

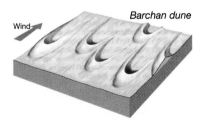

Barchanoid dune Dunes forming scalloped rows of sand oriented at right angles to the wind. This form is intermediate between isolated barchans and extensive waves of transverse dunes.

Barrier island A low, elongate ridge of sand that parallels the coast.

Basal slip A mechanism of glacial movement in which the ice mass slides over the surface below.

Basalt A fine-grained igneous rock of mafic composition.

Basaltic composition See *Mafic composition*.

Base level The level below which a stream cannot erode.

Basin A circular downfolded structure.

Batholith A large mass of igneous rock that formed when magma was emplaced at depth, crystallized, and was subsequently exposed by erosion.

Baymouth bar A sandbar that completely crosses a bay, sealing it off from the main body of water.

Beach drift The transport of sediment in a zigzag pattern along a beach, caused by the uprush of water from obliquely breaking waves.

Beach nourishment Process in which large quantities of sand are added to the beach system to offset losses caused by wave erosion. Building beaches seaward improves beach quality and storm protection.

Bed See *Strata*.

Bedding plane A nearly flat surface separating two beds of sedimentary rock. Each bedding plane marks the end of one deposit and the beginning of another having different characteristics.

Bed load Sediment moved along the bottom of a stream by moving water, or particles moved along the ground surface by wind.

Belt of soil moisture A zone in which water is held as a film on the surface of soil particles and may be used by plants or withdrawn by evaporation. The uppermost subdivision of the zone of aeration.

Benioff zone See *Wadati-Benioff zone*.

Biochemical A type of chemical sediment that forms when material dissolved in water is precipitated by water-dwelling organisms. Shells are common examples.

Biogenous sediment Seafloor sediments consisting of material of marine-organic origin.

Biosphere The totality of life forms on Earth.

Bituminous coal The most common form of coal, often called soft, black coal.

Blowout A depression excavated by wind in easily eroded materials.

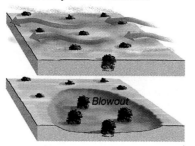

Body wave A seismic wave that travels through Earth's interior.

Bottomset bed A layer of fine sediment deposited beyond the advancing edge of a delta and then buried by continued delta growth.

Bowen's reaction series A concept proposed by N. L. Bowen that illustrates the relationships between magma and the minerals crystallizing from it during the formation of igneous rocks.

Braided stream A stream consisting of numerous intertwining channels.

Breakwater A structure protecting a nearshore area from breaking waves.

Breccia A sedimentary rock composed of angular fragments that were lithified.

Brittle deformation Deformation that involves the fracturing of rock. Associated with rocks near the surface.

Burial metamorphism Low-grade metamorphism that occurs in the lowest layers of very thick accumulations of sedimentary strata.

Caldera A large depression typically caused by collapse or ejection of the summit area of a volcano.

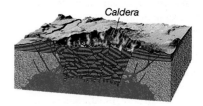

Caliche A hard layer, rich in calcium carbonate, that forms beneath the *B* horizon in soils of arid regions.

Calving Wastage of a glacier that occurs when large pieces of ice break into the water.

Capacity The total amount of sediment a stream is able to transport.

Capillary fringe A relatively narrow zone at the base of the zone of aeration. Here water rises from the water table in tiny threadlike openings between grains of soil or sediment.

Cap rock A necessary part of an oil trap. The cap rock is impermeable and hence keeps upwardly mobile oil and gas from escaping at the surface.

Cassini gap A wide gap in the ring system of Saturn between the *A* ring and the *B* ring.

Cataclastic metamorphism A type of metamorphism caused by extreme shearing of rock, generally associated with fault systems.

Catastrophism The concept that Earth was shaped by catastrophic events of a short-term nature.

Cavern A naturally formed underground chamber or series of chambers most commonly produced by solution activity in limestone.

Cementation One way in which sedimentary rocks are lithified. As material precipitates from water from that percolates through the sediment, open spaces are filled and particles joined into a solid mass.

Cenozoic Era A time span on the geologic time scale beginning about 65 million years ago, following the Mesozoic Era.

Chemical sedimentary rock Sedimentary rock consisting of material that was precipitated from water by either inorganic or organic means.

Chemical weathering The processes by which the internal structure of a mineral is altered by the removal and/or addition of elements.

Chemically active fluid A fluid composed mainly of water but also containing chemical components that can react with- a rock body through which the fluid penetrates.

Cinder cone A rather small volcano built primarily of ejected lava fragments that consist mostly of pea- to walnut-size lapilli.

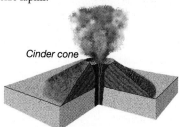

Cinder cone

Cirque An amphitheatre-shaped basin at the head of a glaciated valley produced by frost wedging and plucking.

Clastic A sedimentary rock texture consisting of broken fragments of preexisting rock.

Cleavage The tendency of a mineral to break along planes of weak bonding.

Col A pass between mountain valleys where the headwalls of two cirques intersect.

Colour A phenomenon of light by which otherwise identical objects may be differentiated.

Column A feature found in caves that is formed when a stalactite and stalagmite join.

Columnar joints A pattern of cracks that forms during cooling of molten rock to generate columns.

Coma The fuzzy, gaseous component of a comet's head.

Comet A small body that generally revolves about the Sun in an elongated orbit.

Compaction A type of lithification in which the weight of overlying material compresses more deeply buried sediment. It is most important in the fine-grained sedimentary rocks such as shale.

Competence A measure of the largest particle a stream can transport; a factor dependent on velocity.

Composite cone A volcano composed of both lava flows and pyroclastic material.

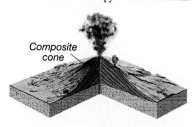

Composite cone

Compound A substance formed by the chemical combination of two or more elements in definite proportions and usually having properties different from those of its constituent elements.

Compressional stress Differential stress that shortens a rock body.

Compressional stress

Concordant A term used to describe intrusive igneous masses that form parallel to the bedding of the surrounding rock.

Conduction The transfer of heat through matter by molecular activity.

Conduit A pipelike opening through which magma moves toward Earth's surface. It terminates at a surface opening called a vent.

Cone of depression A cone-shaped depression in the water table immediately surrounding a well.

Cone of depression

Confining pressure See *Uniform stress*.

Conformable layers Rock layers that were deposited without interruption.

Conglomerate A sedimentary rock composed of rounded gravel-size particles.

Contact metamorphism Changes in rock caused by the heat from a nearby magma body.

Continental drift A hypothesis, credited largely to Alfred Wegener, that suggested all present continents once existed as a single supercontinent. Further, beginning about 200 million years ago, the supercontinent began breaking into smaller continents, which then "drifted" to their present positions.

Continental margin That portion of the seafloor adjacent to the continents. It may include the continental shelf, continental slope, and continental rise.

Continental rise The gently sloping surface at the base of the continental slope.

Continental shelf The gently sloping submerged portion of the continental margin, extending from the shoreline to the continental slope.

Continental slope The steep gradient that leads to the deep-ocean floor and marks the seaward edge of the continental shelf.

Continental volcanic arc Mountains formed in part by igneous activity associated with the subduction of oceanic lithosphere beneath a continent. Examples include the Andes and the Cascades.

Convection The transfer of heat by the mass movement or circulation of a substance.

Convergent plate boundary A boundary in which two plates move together, resulting in oceanic lithosphere being thrust beneath an overriding plate, eventually to be reabsorbed into the mantle. It can also involve the collision of two continental plates to create a mountain system.

Convergent boundary

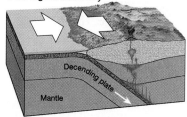

Coral reef Structure formed in a warm, shallow, sunlit ocean environment that consists primarily of the calcite-rich remains of corals as well as the limy secretions of algae and the hard parts of many other small organisms.

Core The innermost layer of Earth based on composition. It is thought to be largely an iron-nickel alloy with minor amounts of oxygen, silicon, and sulphur.

Correlation Establishing the equivalence of rocks of similar age in different areas.

Covalent bond A chemical bond produced by the sharing of electrons.

Crater The depression at the summit of a volcano, or that which is produced by a meteorite impact.

Creep The slow downhill movement of soil and regolith.

Crevasse A deep crack in the brittle surface of a glacier.

Cross-bedding Structure in which relatively thin layers are inclined at an angle to the main bedding. Formed by currents of wind or water.

Cross-cutting A principle of relative dating. A rock or fault is younger than any rock (or fault) through which it cuts.

Crust The very thin outermost layer of Earth.

Crystal An orderly arrangement of atoms.

Crystal form The external appearance of a mineral as determined by its internal arrangement of atoms.

Crystal settling During the crystallization of magma, the earlier-formed minerals are denser than the liquid portion and settle to the bottom of the magma chamber.

Crystallization The formation and growth of a crystalline solid from a liquid or gas.

Curie point The temperature above which a material loses its magnetization.

Cut bank The area of active erosion on the outside of a meander.

Cutoff A short channel segment created when a river erodes through the narrow neck of land between meanders.

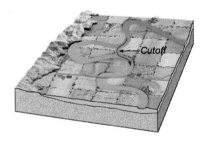
Cutoff

Darcy's law When permeability is uniform, the velocity of groundwater increases as the slope of the water table increases. It is expressed by the formula: $V = \dfrac{K}{n}\dfrac{h}{l}$, where V is velocity, n the porosity, h the head, l the length of flow, and K the coefficient of permeability.

Dark silicate Silicate minerals containing ions of iron and/or magnesium in their structure. They are dark in colour and have a higher specific gravity than nonferromagnesian silicates.

Daughter product An isotope resulting from radioactive decay.

Debris flow A flow of soil and regolith containing a large amount of water. Most common in semiarid mountainous regions and on the slopes of some volcanoes.

Decompression melting Melting that occurs as rock ascends due to a drop in confining pressure.

Deep-ocean basin The portion of seafloor that lies between the continental margin and the oceanic ridge system. This region comprises almost 30 percent of Earth's surface.

Deep-ocean trench A narrow, elongated depression of the seafloor.

Deep-sea fan A cone-shaped deposit at the base of the continental slope. The sediment is transported to the fan by turbidity currents that follow submarine canyons.

Deflation The lifting and removal of loose material by wind.

Deformation General term for the processes of folding, faulting, shearing, compression, or extension of rocks as the result of various natural forces.

Delta An accumulation of sediment formed where a stream enters a lake or an ocean.

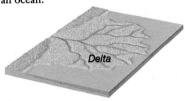
Delta

Dendritic pattern A stream system that resembles the pattern of a branching tree.

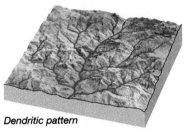
Dendritic pattern

Density The weight per unit volume of a particular material.

Desalination The removal of salts and other chemicals from seawater.

Desert One of the two types of dry climate; the driest of the dry climates.

Desert pavement A layer of coarse pebbles and gravel created when wind removed the finer material.

Detrital sedimentary rocks Rocks that form from the accumulation of materials that originate and are transported as solid particles derived from both mechanical and chemical weathering.

Diagenesis A collective term for all the chemical, physical, and biological changes that take place after sediments are deposited and during and after lithification.

Differential stress Forces that are unequal in different directions.

Differential weathering The variation in the rate and degree of weathering caused by such factors as mineral make-up, degree of jointing, and climate.

Dip The angle at which a rock layer or fault is inclined from the horizontal. The direction of dip is at a right angle to the strike.

Dip-slip fault A fault in which the movement is parallel to the dip of the fault.

Discharge The quantity of water in a stream that passes a given point in a period of time.

Disconformity A type of unconformity in which the beds above and below are parallel.

Discontinuity A sudden change with depth in one or more of the physical properties of the material making up Earth's interior. The boundary between two dissimilar materials in Earth's interior as determined by the behaviour of seismic waves.

Discordant A term used to describe plutons that cut across existing rock structures, such as bedding planes.

Disseminated deposit Any economic mineral deposit in which the desired mineral occurs as scattered particles in the rock but in sufficient quantity to make the deposit an ore.

Dissolution A common form of chemical weathering, it is the process of dissolving into a homogeneous solution, as when an acidic solution dissolves limestone.

Dissolved load That portion of a stream's load carried in solution.

Distributary A section of a stream that leaves the main flow.

Divergent plate boundary A boundary in which two plates move apart, resulting in upwelling of material from the mantle to create new seafloor.

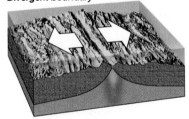
Divergent boundary

Divide An imaginary line that separates the drainage of two streams, often found along a ridge.

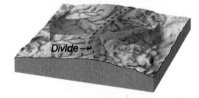
Divide

D″ layer A region in roughly the lowermost 200 kilometres of the mantle where P-waves experience a sharp decrease in velocity.

Dome A roughly circular upfolded structure.

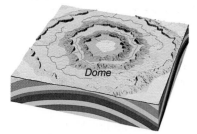

Drainage basin The land area that contributes water to a stream.

Drawdown The difference in height between the bottom of a cone of depression and the original height of the water table.

Drift See Glacial drift.

Drumlin A streamlined symmetrical hill composed of glacial till. The steep side of the hill faces the direction from which the ice advanced.

Dry climate A climate in which yearly precipitation is less than the potential loss of water by evaporation.

Ductile deformation A type of solid-state flow that produces a change in the size and shape of a rock body without fracturing. Occurs at depths where temperatures and confining pressures are high.

Dune A hill or ridge of wind-deposited sand.

Dyke A tabular-shaped intrusive igneous feature that cuts through the surrounding rock.

Earthflow The downslope movement of water-saturated, clay-rich sediment. Most characteristic of humid regions.

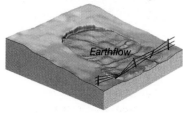

Earthquake Vibration of Earth produced by the rapid release of energy.

Ebb current The movement of tidal current away from the shore.

Echo sounder An instrument used to determine the depth of water by measuring the time interval between emission of a sound signal and the return of its echo from the bottom.

Elastic rebound The sudden release of stored strain in rocks that results in movement along a fault.

Electron A negatively charged subatomic particle that has a negligible mass and is found outside an atom's nucleus.

Element A substance that cannot be decomposed into simpler substances by ordinary chemical or physical means.

Eluviation The washing out of fine soil components from the *A* horizon by downward-percolating water.

Emergent coast A coast where land formerly below sea level has been exposed by crustal uplift or a drop in sea level or both.

End moraine A ridge of till marking a former position of the front of a glacier.

Energy-levels or shells Spherically shaped, negatively charged zones that surround the nucleus of an atom.

Environment of deposition A geographic setting where sediment accumulates. Each site is characterized by a particular combination of geologic processes and environmental conditions.

Eon The largest time unit on the geologic time scale, next in order of magnitude above era.

Ephemeral stream A stream that is usually dry because it carries water only in response to specific episodes of rainfall. Most desert streams are of this type.

Epicentre The location on Earth's surface that lies directly above the focus of an earthquake.

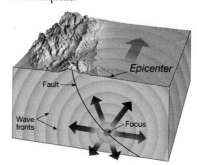

Epoch A unit of the geologic time scale that is a subdivision of a period.

Equilibrium line The lower limit of perennial snow accumulation on a glacier.

Era A major division on the geologic time scale; eras are divided into shorter units called periods.

Erosion The incorporation and transportation of material by a mobile agent, such as water, wind, or ice.

Eruption column Buoyant plumes of hot ash-laden gases that can extend thousands of metres into the atmosphere.

Escape velocity The initial velocity an object needs to escape from the surface of a celestial body.

Esker Sinuous ridge composed largely of sand and gravel deposited by a stream flowing in a tunnel beneath a glacier near its terminus.

Estuary A funnel-shaped inlet of the sea that formed when a rise in sea level or subsidence of land caused the mouth of a river to be flooded.

Evaporite A sedimentary rock formed of material deposited from solution by evaporation of the water.

Evapotranspiration The combined effect of evaporation and transpiration.

Exfoliation dome Large, dome-shaped structure, usually composed of granite, formed by sheeting.

Exotic stream A permanent stream that traverses a desert and has its source in well-watered areas outside the desert.

External process Process such as weathering, mass wasting, or erosion that is powered by the Sun and contributes to the transformation of solid rock into sediment.

Extrusive Igneous activity that occurs at Earth's surface.

Facies A portion of a rock unit that possesses a distinctive set of characteristics that distinguishes it from other parts of the same unit.

Fall A type of movement common to mass-wasting processes that refers to the freefalling of detached individual pieces of any size.

Fault A break in a rock mass along which movement has occurred.

Fault-block mountain A mountain formed by the displacement of rock along a fault.

Fault creep Slow gradual displacement along a fault. Such activity occurs relatively smoothly and with little noticeable seismic activity.

Fault scarp A cliff created by movement along a fault. It represents the exposed surface of the fault prior to modification by weathering and erosion.

Felsic composition A compositional group of igneous rocks indicating the rock is composed almost entirely of light-coloured silicates.

Ferromagnesian silicate See *Dark silicate*.

Fetch The distance that the wind has traveled across the open water.

Fiord A steep-sided inlet of the sea formed when a glacial trough was partially submerged.

Firn Granular recrystallized snow. A transitional stage between snow and glacial ice.

Fissility The property of splitting easily into thin layers along closely spaced, parallel surfaces, such as bedding planes in shale.

Fission (nuclear) The splitting of a heavy nucleus into two or more lighter nuclei, caused by the collision with a neutron. During this process a large amount of energy is released.

Fissure A crack in rock along which there is a distinct separation.

Fissure eruption An eruption in which lava is extruded from narrow fractures or cracks in the crust.

Flood The overflow of a stream channel that occurs when discharge exceeds the channel's capacity. The most common and destructive geologic hazard.

Flood basalts Flows of basaltic lava that issue from numerous cracks or fissures and commonly cover extensive areas to thicknesses of hundreds of metres.

Floodplain The flat, low-lying portion of a stream valley subject to periodic inundation.

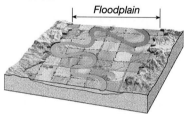

Flood current The tidal current associated with the increase in the height of the tide.

Flow A type of movement common to mass-wasting processes in which water-saturated material moves downslope as a viscous fluid.

Flowing artesian well An artesian well in which water flows freely at Earth's surface because the pressure surface is above ground level.

Fluorescence The absorption of ultraviolet light, which is reemitted as visible light.

Focus (earthquake) The zone within Earth where rock displacement produces an earthquake.

Fold A bent layer or series of layers that were originally horizontal and subsequently deformed.

Foliated texture A texture of metamorphic rocks that gives the rock a layered appearance.

Foliation A term for a linear arrangement of textural features often exhibited by metamorphic rocks.

Foreset bed An inclined bed deposited along the front of a delta.

Foreshocks Small earthquakes that often precede a major earthquake.

Force That which tends to put stationary objects in motion or change motions of moving bodies.

Fossil The remains or traces of organisms preserved from the geologic past.

Fossil fuel General term for any hydrocarbon that may be used as a fuel, including coal, oil, natural gas, bitumen from tar sands, and shale oil.

Fossil succession Fossil organisms succeed one another in a definite and determinable order, and any time period can be recognized by its fossil content.

Fracture Any break or rupture in rock along which no appreciable movement has taken place.

Fracture zone Linear zone of irregular topography on the deep-ocean floor that follows transform faults and their inactive extensions.

Fragmental texture See *Pyroclastic texture*.

Frost wedging The mechanical breakup of rock caused by the expansion of freezing water in cracks and crevices.

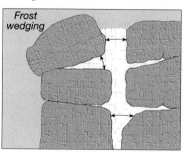

Fumarole A vent in a volcanic area from which fumes or gases escape.

Gaining stream Streams that gain water from the inflow of groundwater through the streambed.

Geologic time scale The division of Earth history into blocks of time—eons, eras, periods, and epochs. The time scale was created using relative dating principles.

Geology The science that examines Earth, its form and composition, and the changes that it has undergone and is undergoing.

Geothermal energy Natural steam used for power generation.

Geothermal gradient The gradual increase in temperature with depth in the crust. The average is 30°C per kilometre in the upper crust.

Geyser A fountain of hot water ejected periodically from the ground.

Glacial budget The balance, or lack of balance, between ice formation at the upper end of a glacier, and ice loss in the zone of wastage.

Glacial drift An all-embracing term for sediments of glacial origin, no matter how, where, or in what shape they were deposited.

Glacial erratic An ice-transported boulder that was not derived from the bedrock near its present site.

Glacial striations Scratches and grooves on bedrock caused by glacial abrasion.

Glacial trough A mountain valley that has been widened, deepened, and straightened by a glacier.

Glacier A thick mass of ice originating on land from the compaction and recrystallization of snow that shows evidence of past or present flow.

Glass (volcanic) Natural glass produced when molten lava cools too rapidly to permit recrystallization. Volcanic glass is a solid composed of unordered atoms.

Glassy A term used to describe the texture of certain igneous rocks, such as obsidian, that contain no crystals.

Gneissic texture A texture of metamorphic rocks in which dark and light silicate minerals are separated, giving the rock a banded appearance.

Gondwanaland The southern portion of Pangaea consisting of South America, Africa, Australia, India, and Antarctica.

Graben A valley formed by the downward displacement of a fault-bounded block.

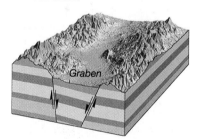

Graded bed A sediment layer characterized by a decrease in sediment size from bottom to top.

Graded stream A stream that has the correct channel characteristics to maintain exactly the velocity required to transport the material supplied to it.

Gradient The slope of a stream, generally expressed as the vertical drop over a fixed distance.

Granitic composition See *Felsic composition*.

Greenhouse effect Carbon dioxide and water vapour in a planet's atmosphere absorb and reradiate infrared wavelengths, effectively trapping solar energy and raising the temperature.

Groin A short wall built at a right angle to the seashore to trap moving sand.

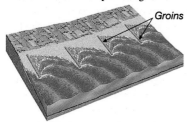

Groins

Groundmass The matrix of smaller crystals within an igneous rock that has porphyritic texture.

Ground moraine An undulating layer of till deposited as the ice front retreats.

Groundwater Water in the zone of saturation.

Guyot A submerged flat-topped seamount.

Hadean Eon The first eon on the geologic time scale. The eon ending 3.8 billion years ago that preceded the Archean Eon.

Half-life The time required for one-half of the atoms of a radioactive substance to decay.

Hanging valley A tributary valley that enters a glacial trough at a considerable height above the floor of the trough.

Hanging valley

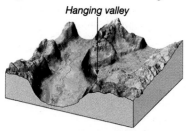

Hardness A mineral's resistance to scratching and abrasion.

Head groundwater The vertical distance between the recharge and discharge points of a water table. Also the source area or beginning of a valley.

Head (stream) The beginning or source area for a stream. Also called the headwaters.

Headward erosion The extension upslope of the head of a valley due to erosion.

Historical geology A major division of geology that deals with the origin of Earth and its development through time. Usually involves the study of fossils and their sequence in rock beds.

Hogback A narrow, sharp-crested ridge formed by the upturned edge of a steeply dipping bed of resistant rock.

Horizon A layer in a soil profile.

Horn A pyramid-like peak formed by glacial action in three or more cirques surrounding a mountain summit.

Horst An elongate, uplifted block of crust bounded by faults.

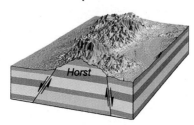

Horst

Hot spot A concentration of heat in the mantle, capable of producing magma that, in turn extrudes onto Earth's surface. The intraplate volcanism that produced the Hawaiian Islands is one example.

Hot spring A spring in which the water is 6–9°C warmer than the mean annual air temperature of its locality.

Hummocky cross-stratification Sedimentary structure characterized by draped laminations that dip away from a central high (hummock). Hummocks are typically about one metre in diameter. Believed to be produced by storm waves.

Humus Organic matter in soil produced by the decomposition of plants and animals.

Hydraulic gradient The slope of the water table. Expressed as h/l, where h is the head and l the length of flow.

Hydroelectric power Electricity generated by falling water that is used to drive turbines.

Hydrogenous sediment Seafloor sediment consisting of minerals that crystallize from seawater. An important example is manganese nodules.

Hydrologic cycle The unending circulation of Earth's water supply. The cycle is powered by energy from the Sun and is characterized by continuous exchanges of water among the oceans, the atmosphere, and the continents.

Hydrolysis A chemical weathering process in which minerals are altered by chemically reacting with water and acids.

Hydrosphere The water portion of our planet; one of the traditional subdivisions of Earth's physical environment.

Hydrothermal metamorphism Chemical alterations that occur as hot, ion-rich water circulates through fractures in rock.

Hydrothermal solution The hot, watery solution that escapes from a mass of magma during the latter stages of crystallization. Such solutions may alter the surrounding country rock and are frequently the source of significant ore deposits.

Hypothesis A tentative explanation that is then tested to determine if it is valid.

Ice cap A mass of glacial ice covering a high upland or plateau and spreading out radially.

Ice-contact deposit An accumulation of stratified drift deposited in contact with a supporting mass of ice.

Ice sheet A very large, thick mass of glacial ice flowing outward in all directions from one or more accumulation centres.

Ice shelf Forming where glacial ice flows into bays, it is a large, relatively flat mass of floating ice that extends seaward from the coast but remains attached to the land along one or more sides.

Igneous rock Rock formed from the crystallization of magma.

Immature soil A soil lacking horizons.

Impact metamorphism Metamorphism that occurs when meteorites strike Earth's surface.

Incised meander Meandering channel that flows in a steep, narrow valley. These features form either when an area is uplifted or when base level drops.

Inclusion A piece of one rock unit contained within another. Inclusions are used in relative dating. The rock mass adjacent to the one containing the inclusion must have been there first in order to provide the fragment.

Index fossil A fossil that is associated with a particular span of geologic time.

Index mineral A mineral that is a good indicator of the metamorphic environment in which it formed. Used to distinguish different zones of regional metamorphism.

Inertia Objects at rest tend to remain at rest, and objects in motion tend to stay in motion unless either is acted upon by an outside force.

Infiltration The movement of surface water into rock or soil through cracks and pore spaces.

Infiltration capacity The maximum rate at which soil can absorb water.

Inner core The solid innermost layer of Earth, about 1216 kilometres in radius.

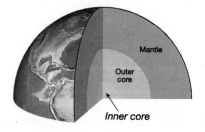

Inner core

Inselberg An isolated mountain remnant characteristic of the late stage of erosion in a mountainous arid region.

Intensity (earthquake) A measure of the degree of earthquake shaking at a given locale based on the amount of damage.

Interior drainage A discontinuous pattern of intermittent streams that do not flow to the ocean.

Intermediate composition A compositional group of igneous rocks, indicating that the rock contains at least 25 percent dark silicate minerals. The other dominant mineral is plagioclase feldspar.

Internal process A process such as mountain building or volcanism that derives its energy from Earth's interior and elevates Earth's surface.

Intraplate volcanism Igneous activity that occurs with in a tectonic plate away from plate boundaries.

Intrusive rock Igneous rock that formed below Earth's surface.

Ion An atom or molecule that possesses an electrical charge.

Ionic bond A chemical bond between two oppositely charged ions formed by the transfer of valence electrons from one atom to the other.

Iron meteorite One of the three main categories of meteorites. This group is composed largely of iron with varying amounts of nickel (5–20 percent). Most meteorite finds are irons.

Island arc See *Volcanic island arc.*

Volcanic island arc

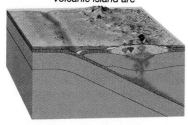

Isostasy The concept that Earth's crust is "floating" in gravitational balance upon the material of the mantle.

Isotatic adjustment Compensation of the lithosphere when weight is added or removed. When weight is added, the lithosphere will respond by subsiding, and when weight is removed, there will be uplift.

Isotopes Varieties of the same element that have different mass numbers; their nuclei contain the same number of protons but different numbers of neutrons.

Jetties A pair of structures extending into the ocean at the entrance to a harbour or river that are built for the purpose of protecting against storm waves and sediment deposition.

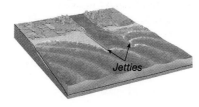

Joint A fracture in rock along which there has been no movement.

Jovian planet One of the Jupiter-like planets, Jupiter, Saturn, Uranus, and Neptune. These planets have relatively low densities.

Kame A steep-sided hill composed of sand and gravel, originating when sediment collected in openings in stagnant glacial ice.

Kame terrace A narrow, terrace-like mass of stratified drift deposited between a glacier and an adjacent valley wall.

Karst A type of topography formed on soluble rock (especially limestone) primarily by dissolution. It is characterized by sinkholes, caves, and underground drainage.

Kettle holes Depressions created when blocks of ice become lodged in glacial deposits and subsequently melt.

Klippe A remnant or outlier of a thrust sheet that was isolated by erosion.

Laccolith A massive igneous body intruded between preexisting strata.

Laccolith

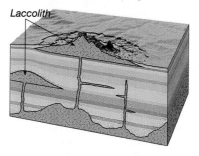

Lag time The amount of time between a rainstorm and the occurrence of flooding.

Lahar Debris flows on the slopes of volcanoes that result when unstable layers of ash and debris become saturated and flow downslope, usually following stream channels.

Laminar flow The movement of water particles in straight-line paths that are parallel to the channel. The water particles move downstream without mixing.

Lateral accretion surfaces Gently dipping sedimentary structures preserved within point bar deposits of meandering streams. These surfaces mark former positions of an accreting point bar.

Lateral moraine A ridge of till along the sides of a valley glacier composed primarily of debris that fell to the glacier from the valley walls.

Laterite A red, highly leached soil type found in the tropics that is rich in oxides of iron and aluminum.

Laurasia The northern portion of Pangaea consisting of North America and Eurasia.

Lava Magma that reaches Earth's surface.

Lava dome A bulbous mass associated with an old-age volcano, produced when thick lava is slowly squeezed from the vent. Lava domes may act as plugs to deflect subsequent gaseous eruptions.

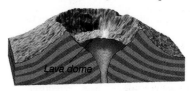

Lava tube Tunnel in hardened lava that acts as a horizontal conduit for lava flowing from a volcanic vent. Lava tubes allow fluid lavas to advance great distances.

Law A formal statement of the regular manner in which a natural phenomenon occurs under given conditions; e.g., the "law of superposition."

Law of superposition In any undeformed sequence of sedimentary rocks, each bed is older than the one above and younger than the one below.

Leaching The depletion of soluble materials from the upper soil by downward-percolating water.

Light silicate Silicate minerals that lack iron and/or magnesium. They are generally lighter in colour and have lower specific gravities than dark silicates.

Liquefaction The transformation of a stable soil into a fluid that is often unable to support buildings or other structures.

Lithification The process, generally cementation and/or compaction, of converting sediments to solid rock.

Lithosphere The rigid outer layer of Earth, including the crust and upper mantle.

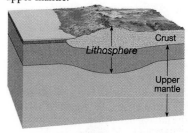

Local base level See *Temporary base level.*
Loess Deposits of windblown silt, lacking visible layers, generally buff-colored, and capable of maintaining a nearly vertical cliff.
Longitudinal dunes Long ridges of sand oriented parallel to the prevailing wind; these dunes form where sand supplies are limited.

Longitudinal dunes

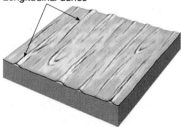

Longitudinal profile A cross-section of a stream channel along its descending course from the head to the mouth.
Longshore current A nearshore current that flows parallel to the shore.
Long (L) waves These earthquake-generated waves travel along the outer layer of Earth and are responsible for most of the surface damage. L waves have longer periods than other seismic waves.
Losing stream Streams that lose water to the groundwater system by outflow through the streambed.
Lower mantle See *Mesosphere.*
Low-velocity zone A subdivision of the mantle located between 100 and 250 kilometres and discernible by a marked decrease in the velocity of seismic waves. This zone does not encircle Earth.
Lunar breccia A lunar rock formed when angular fragments and dust are welded together by the heat generated by the impact of a meteoroid.
Lunar regolith A thin, gray layer on the surface of the moon, consisting of loosely compacted, fragmented material believed to have been formed by repeated meteoritic impacts.
Lustre The appearance or quality of light reflected from the surface of a mineral.

Mafic composition A compositional group of igneous rocks indicating that the rock contains substantial dark silicate minerals and calcium-rich plagioclase feldspar.
Magma A body of molten rock found at depth, including any dissolved gases and crystals.
Magma mixing The process of altering the composition of a magma through the mixing of material from another magma body.

Magmatic differentiation The process of generating more than one rock type from a single magma.
Magnetometer A sensitive instrument used to measure the intensity of Earth's magnetic field at various points.
Magnitude (earthquake) An estimate of the total amount of energy released during an earthquake, based on seismic records.
Manganese nodules A type of hydrogenous sediment scattered on the ocean floor, consisting mainly of manganese and iron and usually containing small amounts of copper, nickel, and cobalt.
Mantle One of Earth's compositional layers. The solid rocky shell that extends from the base of the crust to a depth of 2900 kilometres.
Mantle plume A mass of hotter-than-normal mantle material that ascends toward the surface, where it may lead to igneous activity. These plumes of solid yet mobile material may originate as deep as the core-mantle boundary.
Maria The smooth areas on our moon's surface that were incorrectly thought to be seas.
Massive An igneous pluton that is not tabular in shape.
Mass number The sum of the number of neutrons and protons in the nucleus of an atom.
Mass wasting The downslope movement of rock, regolith, and soil under the direct influence of gravity.
Meander A loop like bend in the course of a stream.

Meander

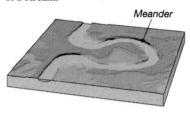

Meander scar A floodplain feature created when an oxbow lake becomes filled with sediment.
Mechanical weathering The physical disintegration of rock, resulting in smaller fragments.
Medial moraine A ridge of till formed when lateral moraines from two coalescing alpine glaciers join.
Melt The liquid portion of magma excluding the solid crystals.
Mercalli intensity scale See *Modified Mercalli intensity scale.*
Mesosphere The part of the mantle that extends from the core-mantle boundary to a depth of 660 kilometres. Also known as the lower mantle.

Mesozoic Era A time span on the geologic time scale between the Paleozoic and Cenozoic Eras—from about 248 to 65 million years ago.
Metallic bond A chemical bond present in all metals that may be characterized as an extreme type of electron sharing in which the electrons move freely from atom to atom.
Metamorphic facies A distinctive assemblage of minerals that can be used to designate a particular set of metamorphic environmental conditions recorded in a metamorphic rock.
Metamorphic rock Rock formed by the alteration of preexisting rock deep within Earth (but still in the solid state) by heat, pressure, and/or chemically active fluids.
Metamorphism The changes in mineral composition and texture of a rock subjected to high temperatures and pressures within Earth.
Meteor The luminous phenomenon observed when a meteoroid enters Earth's atmosphere and burns up; popularly called a "shooting star."
Meteorite Any portion of a meteoroid that survives its traverse through Earth's atmosphere and strikes the surface.
Meteoroid Any small, solid particle that has an orbit in the solar system.
Meteor shower Numerous meteoroids traveling in the same direction and at nearly the same speed. They are thought to be material lost by comets.
Micrometeorite A very small meteorite that does not create sufficient friction to burn up in the atmosphere, but slowly drifts down to Earth.
Mid-ocean ridge A continuous mountainous ridge on the floor of all the major ocean basins and varying in width from 500 to 5000 kilometres. The rifts at the crests of these ridges represent divergent plate boundaries.

Mid-ocean ridge

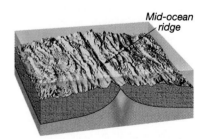

Migmatite A rock exhibiting both igneous and metamorphic rock characteristics. Such rocks may form when light-coloured silicate minerals melt and then crystallize, while the dark silicate minerals remain solid.

Mineral A naturally occurring, inorganic crystalline material with a unique chemical structure.

Mineralogy The study of minerals.

Mineral resource All discovered and undiscovered deposits of a useful mineral that can be extracted now or at some time in the future.

Modified Mercalli intensity scale A 12-point scale developed to evaluate earthquake intensity based on the amount of damage to various structures.

Mohorovičič discontinuity (Moho) The boundary separating the crust and the mantle, discernible by an increase in seismic velocity.

Mohs scale A series of 10 minerals used as a standard in determining hardness.

Moment magnitude A more precise measure of earthquake magnitude than the Richter scale that is derived from the amount of displacement that occurs along a fault zone.

Monocline A one-limbed flexure in strata. The strata are usually flat-lying or very gently dipping on both sides of the monocline.

Mountain A landform with an obvious peak that rises significantly above the land surrounding it, generally on the order of 300 metres or more.

Mountain belt A regionally extensive group of mountains composed of multiple mountain ranges.

Mountain range A linked series of mountains.

Mouth The point downstream where a river empties into another stream or water body.

Mud crack A feature in some sedimentary rocks that forms when wet mud dries out, shrinks, and cracks.

Mudflow See *Debris flow*.

Mudrock A fine-grained sedimentary rock consisting primarily of clay- to silt-sized particles.

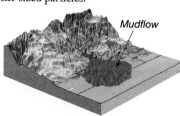
Mudflow

Natural levee An elevated landform composed of alluvium that parallels some streams and acts to confine their waters, except during floodstage.

Neap tide The lowest tidal range, occurring near the times of the first and third quarters of the moon.

Nebular hypothesis A model for the origin of the solar system that supposes a rotating nebula of dust and gases that contracted to form the Sun and planets.

Neutron A subatomic particle found in the nucleus of an atom. The neutron is electrically neutral, with a mass approximately equal to that of a proton.

Nonclastic A term for the texture of sedimentary rocks in which the minerals form a pattern of interlocking crystals.

Nonconformity An unconformity in which older metamorphic or intrusive igneous rocks are overlain by younger sedimentary strata.

Nonferromagnesian silicate See *Light silicate*.

Nonflowing artesian well An artesian well in which water does not rise to the surface, because the pressure surface is below ground level.

Nonfoliated Metamorphic rocks that do not exhibit foliation.

Nonmetallic mineral resource A mineral resource that is not a fuel or processed for the metals it contains.

Nonrenewable resource A resource that forms or accumulates over such long time spans that it must be considered as fixed in total quantity.

Normal fault A fault in which the rock above the fault plane has moved down relative to the rock below.

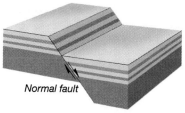
Normal fault

Normal polarity A magnetic field the same as that which presently exists.

Nuclear fission The splitting of atomic nuclei into smaller nuclei, causing neutrons to be emitted and heat energy to be released.

Nucleus The small, heavy core of an atom that contains all of its positive charge and most of its mass.

Nuée ardente Incandescent volcanic debris buoyed up by hot gases that moves downslope in an avalanche fashion.

Numerical date The number of years that have passed since an event occurred.

Occultation The disappearance of light resulting when one object passes behind an apparently larger one. For example, the passage of Uranus in front of a distant star.

Oceanic ridge system See *Mid-ocean ridge*.

Octet rule Atoms combine in order that each may have the electron arrangement of a noble gas; that is, the outer energy level contains eight neutrons.

Oil trap A geologic structure that allows for significant amounts of oil and gas to accumulate.

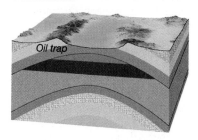
Oil trap

Ophiolite complex The sequence of rocks that make up the oceanic crust. The three-layer sequence includes an upper layer of pillow basalts, a middle zone of sheeted dykes, and a lower layer of gabbro.

Ore Usually a useful metallic mineral that can be mined at a profit. The term is also applied to certain nonmetallic minerals such as fluorite and sulphur.

Original horizontality Layers of sediment that are generally deposited in a horizontal or nearly horizontal position.

Orogenesis The processes that collectively result in the formation of mountains.

Outer core A layer beneath the mantle about 2270 kilometres thick, which has the properties of a liquid.

Outlet glacier A tongue of ice normally flowing rapidly outward from an ice cap or ice sheet, usually through mountainous terrain to the sea.

Outwash plain A relatively flat, gently sloping plain consisting of materials deposited by meltwater streams in front of the margin of an ice sheet.

Oxbow lake A curved lake produced when a stream cuts off a meander.

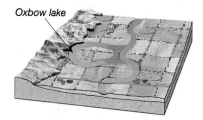
Oxbow lake

Oxidation The removal of one or more electrons from an atom or ion. So named because elements commonly combine with oxygen.

Pahoehoe Flow A lava flow with a smooth-to-ropy surface.

Paleomagnetism The natural remnant magnetism in rock bodies. The permanent magnetization acquired by rock that can be used to determine the location of the magnetic poles and the latitude of the rock at the time it became magnetized.

Paleontology The systematic study of fossils and the history of life on Earth.

Paleozoic Era A time span on the geologic time scale between the Precambrian and Mesozoic Eras—from about 540 million to 248 million years ago.

Pangaea The proposed supercontinent that 200 million years ago began to break apart and form the present landmasses.

Parabolic dune A sand dune similar in shape to a barchan dune except that its tips point into the wind. These dunes often form along coasts that have strong onshore winds, abundant sand, and vegetation that partly covers the sand.

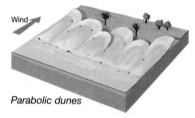

Parabolic dunes

Paradigm Theory that is held with a very high degree of confidence and is comprehensive in scope.

Parasitic cone A volcanic cone that forms on the flank of a larger volcano.

Parent material The material upon which a soil develops.

Parent rock The rock from which a metamorphic rock formed.

Partial melting The process by which most igneous rocks melt. Since individual minerals have different melting points, most igneous rocks melt over a temperature range of a few hundred degrees. If the liquid is squeezed out after some melting has occurred, a melt with a higher silica content results.

Passive continental margin A margin that consists of a continental shelf, continental slope, and continental rise. They are not associated with plate boundaries and therefore experience little volcanism and few earthquakes.

Pater noster lakes A chain of small lakes in a glacial trough that occupies basins created by glacial erosion.

Pedalfer Soil of humid regions characterized by the accumulation of iron oxides and aluminum-rich clays in the *B* horizon.

Pedocal Soil associated with drier regions and characterized by an accumulation of calcium carbonate in the upper horizons.

Pegmatite A very coarse-grained igneous rock (typically granite) commonly found as a dyke associated with a large mass of plutonic rock that has smaller crystals. Crystallization in a water-rich environment is believed to be responsible for the very large crystals.

Pegmatitic texture A texture of igneous rocks in which the interlocking crystals are all larger than one centimetre in diameter.

Perched water table A localized zone of saturation above the main water table, created by an impermeable layer (aquiclude).

Peridotite An igneous rock of ultramafic composition thought to be abundant in the upper mantle.

Period A basic unit of the geologic time scale that is a subdivision of an era. Periods may be divided into smaller units called epochs.

Permafrost Any permanently frozen subsoil. Usually found in the subarctic and arctic regions.

Permeability A measure of a material's ability to transmit water.

Phaneritic texture An igneous rock texture in which the crystals are roughly equal in size and large enough so the individual minerals can be identified without the aid of a microscope.

Phanerozoic Eon That part of geologic time represented by rocks containing abundant fossil evidence. The eon extending from the end of the Proterozoic Eon (540 million years ago) to the present.

Phenocryst Conspicuously large crystal embedded in a matrix of finer-grained crystals.

Physical geology A major division of geology that examines the materials of Earth and seeks to understand the processes and forces acting beneath and upon Earth's surface.

Piedmont glacier A glacier that forms when one or more alpine glaciers emerge from the confining walls of mountain valleys and spread out to create a broad sheet in the lowlands at the base of the mountains.

Pillow lava Basaltic lava that solidifies in an underwater environment and develops a structure that resembles a pile of pillows.

Pipe A vertical conduit through which magmatic materials have passed.

Placer Deposit formed when heavy minerals are mechanically concentrated by currents, most commonly streams and waves. Placers are sources of gold, tin, platinum, diamonds, and other valuable minerals.

Plastic flow A type of glacial movement that occurs within the glacier, below a depth of approximately 50 metres, in which the ice is not fractured.

Plate One of numerous rigid sections of the lithosphere that moves as a unit over the material of the asthenosphere.

Plate tectonics The theory that proposes that Earth's outer shell consists of individual plates that interact in various ways and thereby produce earthquakes, volcanoes, mountains, and the crust itself.

Playa The flat central area of an undrained desert basin.

Playa lake A temporary lake in a playa.

Playfair's law A well-known and oft-quoted statement by John Playfair that states that a valley is the result of the work of the stream that flows in it.

Pleistocene Epoch An epoch of the Quaternary Period beginning about 1.8 million years ago and ending about 10,000 years ago. Best known as a time of extensive continental glaciation.

Plucking The process by which pieces of bedrock are lifted out of place by a glacier.

Pluton A structure that results from the emplacement and crystallization of magma beneath the surface of Earth.

Plutonic rock Igneous rocks that form at depth. After Pluto, the god of the lower world in classical mythology.

Pluvial lake A lake formed during a period of increased rainfall. For example, this occurred in many nonglaciated areas during periods of ice advance elsewhere.

Point bar A crescent-shaped accumulation of sand and gravel deposited on the inside of a meander.

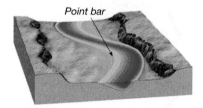

Point bar

Polymorphs Two or more minerals having the same chemical composition but different crystalline structures. Exemplified by the diamond and graphite forms of carbon.

Porosity The volume of open spaces in rock or soil.

Porphyritic texture An igneous rock texture characterized by two distinctively different crystal sizes. The larger crystals are called phenocrysts, whereas the matrix of smaller crystals is termed the groundmass.

Porphyroblastic texture A texture of metamorphic rocks in which particularly large grains (porphyroblasts) are surrounded by a fine-grained matrix of other minerals.

Porphyry An igneous rock with a porphyritic texture.

Pothole A depression formed in a stream channel by the abrasive action of the water's sediment load.

Precambrian All geologic time prior to the Paleozoic Era.

Primary (P) wave A type of seismic wave that involves alternating compression and expansion of the material through which it passes.

Principle of faunal succession Fossil organisms succeed one another in a definite and determinable order, and any time period can be recognized by its fossil content.

Principle of original horizontality Layers of sediment are generally deposited in a horizontal or nearly horizontal position.

Proterozoic Eon The eon following the Archean and preceding the Phanerozoic. It extends between 2500 and 540 million years ago.

Proton A positively charged subatomic particle found in the nucleus of an atom.

Pumice A light-coloured glassy vesicular rock commonly having a felsic composition.

P wave The fastest earthquake wave, which travels by compression and expansion of the medium.

Pyroclastic texture An igneous rock texture resulting from the consolidation of individual rock fragments that are ejected during a violent volcanic eruption.

Pyroclastic flow A highly heated mixture, largely of ash and pumice fragments, traveling down the flanks of a volcano or along the surface of the ground.

Pyroclastic material The volcanic rock ejected during an eruption. Pyroclastics include ash, bombs, and blocks.

Radial pattern A system of streams running in all directions away from a central elevated structure, such as a volcano.

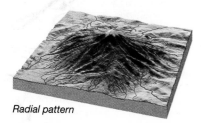

Radial pattern

Radioactive decay The spontaneous decay of certain unstable atomic nuclei.

Radiocarbon (carbon-14) The radioactive isotope of carbon, which is produced continuously in the atmosphere and used in dating events from the very recent geologic past (the last few tens of thousands of years).

Radiometric dating The procedure of calculating the absolute ages of rocks and minerals that contain certain radioactive isotopes.

Rainshadow desert A dry area on the lee side of a mountain range. Many middle-latitude deserts are of this type.

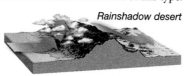

Rainshadow desert

Rapids A part of a stream channel in which the water suddenly begins flowing more swiftly and turbulently because of an abrupt steepening of the gradient.

Rays Bright streaks that appear to radiate from certain craters on the lunar surface. The rays consist of fine debris ejected from the primary crater.

Recessional moraine An end moraine formed as the ice front stagnated during glacial retreat.

Rectangular pattern A drainage pattern characterized by numerous right angle bends that develops on jointed or fractured bedrock.

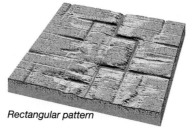

Rectangular pattern

Refraction See *Wave refraction*.

Regional metamorphism Metamorphism associated with large-scale mountain building.

Regolith The layer of rock and mineral fragments that nearly everywhere covers Earth's land surface.

Rejuvenation A change in relation to base level, often caused by regional uplift, which causes the forces of erosion to intensify.

Relative dating Rocks and structures are placed in their proper sequence or order. Only the chronological order of events is determined.

Renewable resource A resource that is virtually inexhaustible or that can be replenished over relatively short time spans.

Reserve Already identified deposits from which minerals can be extracted profitably.

Reservoir rock The porous, permeable portion of an oil trap that yields oil and gas.

Residual soil Soil developed directly from the weathering of the bedrock below.

Reverse fault A fault in which the material above the fault plane moves up in relation to the material below.

Reverse fault

Reverse polarity A magnetic field opposite to that which presently exists.

Richter scale A scale of earthquake magnitude based on the amplitude of the largest seismic wave.

Ridge push A mechanism that may contribute to plate motion. It involves the oceanic lithosphere sliding down the oceanic ridge under the pull of gravity.

Rift A long, narrow trough bounded by normal faults. It represents a region where divergence is taking place.

Rift valley See *Rift*.

Rills Tiny channels that develop as unconfined flow begins producing threads of current.

Ripple marks Small waves of sand that develop on the surface of a sediment layer by the action of moving water or air.

Roche moutonnée An asymmetrical knob of bedrock formed when glacial abrasion smoothes the gentle slope facing the advancing ice sheet and plucking steepens the opposite side as the ice overrides the knob.

Rock A consolidated mixture of minerals.

Rock avalanche The very rapid downslope movement of rock and debris. These rapid movements may be aided by a layer of air trapped beneath the debris, and they have been known to reach speeds in excess of 200 kilometres per hour.

Rock cleavage The tendency of rocks to split along parallel, closely spaced surfaces. These surfaces are often highly inclined to the bedding planes in the rock.

Rock cycle A model that illustrates the origin of the three basic rock types and the interrelatedness of Earth materials and processes.

Rock flour Ground-up rock produced by the grinding effect of a glacier.

Rockslide The rapid slide of a mass of rock downslope along planes of weakness.

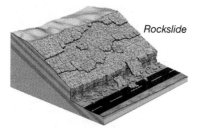

Rockslide

Rock structure All features created by the processes of deformation from minor fractures in bedrock to a major mountain chain.

Runoff Water that flows over the land rather than infiltrating into the ground.

Salinity The proportion of dissolved salts to pure water, usually expressed in parts per thousand (0/000).

Saltation Transportation of sediment through a series of leaps or bounces.

Salt flat A white crust on the ground produced when water evaporates and leaves its dissolved materials behind.

Schistosity A type of foliation characteristic of coarser-grained metamorphic rocks. Such rocks have a parallel arrangement of platy minerals such as the micas.

Scoria Vesicular ejecta that is the product of basaltic magma.

Scoria cone See *Cinder cone.*

Sea arch An arch formed by wave erosion when caves on opposite sides of a headland unite.

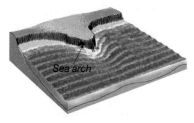

Sea arch

Seafloor metamorphism Metamorphism that occurs when seafloor basalt is invaded by circulating seawater associated with igneous activity at spreading ridges. Widespread hydration of minerals in basalt results from this process.

Seafloor spreading The hypothesis first proposed in the 1960s by Harry Hess that suggested that new oceanic crust is produced at the crests of mid-ocean ridges, which are the sites of divergence.

Seamount An isolated volcanic peak that rises at least 1000 metres above the deep-ocean floor.

Sea stack An isolated mass of rock standing just offshore, produced by wave erosion of a headland.

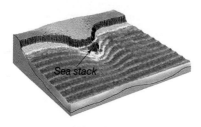

Sea stack

Seawall A barrier constructed to prevent waves from reaching the area behind the wall. Its purpose is to defend property from the force of breaking waves.

Secondary enrichment The concentration of minor amounts of metals that are scattered through unweathered rock into economically valuable concentrations by weathering processes.

Secondary (S) wave A seismic wave that involves oscillation perpendicular to the direction of propagation.

Sediment Unconsolidated particles created by the weathering and erosion of rock, by chemical precipitation from solution in water, or from the secretions of organisms, and transported by water, wind, or glaciers.

Sedimentary environment See *Environment of deposition.*

Sedimentary facies The characteristics of a rock unit that reflect its environment of deposition and differentiate the rock unit from time-equivalent units adjacent to it.

Sedimentary rock Rock formed from the weathered products of preexisting rocks that have been transported, deposited, and lithified.

Seismic gap A segment of an active fault zone that has not experienced a major earthquake over a span when most other segments have. Such segments are probable sites for future major earthquakes.

Seismic sea wave A rapidly moving ocean wave generated by earthquake activity, which is capable of inflicting heavy damage in coastal regions.

Seismogram The record made by a seismograph.

Seismograph An instrument that records earthquake waves.

Seismology The study of earthquakes and seismic waves.

Settling velocity The speed at which a particle falls through a still fluid. The size, shape, and specific gravity of particles influence settling velocity.

Shadow zone The zone between 105 and 140 degrees distance from an earthquake epicentre, which direct waves do not penetrate because of refraction by Earth's core.

Shatter cones Sets of conical fractures present in rocks affected by shock waves of an explosion. Closely associated with meteorite impact structures.

Shear Stress that causes two adjacent parts of a body to slide past one another.

Sheeted dykes A large group of nearly parallel dykes.

Sheet flow Runoff moving in unconfined thin sheets.

Sheeting A mechanical weathering process characterized by the splitting off of slablike sheets of rock.

Shelf break The point at which a rapid steepening of the gradient occurs, marking the outer edge of the continental shelf and the beginning of the continental slope.

Shield A large, relatively flat expanse of ancient metamorphic rock within the stable continental interior.

Shield volcano A broad, gently sloping volcano built from fluid basaltic lavas.

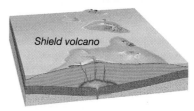

Shield volcano

Silicate mineral Any one of numerous minerals that have the silicon-oxygen tetrahedron as their basic structure.

Silicon-oxygen tetrahedron A structure composed of four oxygen atoms surrounding a silicon atom that constitutes the basic building block of silicate minerals.

Sill A tabular igneous body that was intruded parallel to the layering of preexisting rock.

Sinkhole A depression produced in a region where soluble rock has been removed by groundwater.

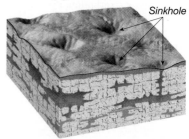

Sinkhole

Slab-pull A mechanism that contributes to plate motion in which cool, dense oceanic crust sinks into the mantle and "pulls" the trailing lithosphere along.

Slaty cleavage The type of foliation characteristic of slates in which there is a parallel arrangement of fine-grained metamorphic minerals.

Slide A movement common to mass-wasting processes in which the material moving downslope remains fairly coherent and moves along a well-defined surface.

Slip face The steep, leeward surface of a sand dune that maintains a slope of about 34 degrees.

Slump The downward slipping of a mass of rock or unconsolidated material moving as a unit along a curved surface.

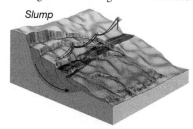
Slump

Snowfield An area where snow persists throughout the year.

Soil A combination of mineral and organic matter, water, and air; that portion of the regolith that supports plant growth.

Soil horizon A layer of soil that has identifiable characteristics produced by chemical weathering and other soil-forming processes.

Soil profile A vertical section through a soil, showing its succession of horizons and the underlying parent material.

Solifluction Slow, downslope flow of water-saturated materials common to permafrost areas.

Solum The *O, A,* and *B* horizons in a soil profile. Living roots and other plant and animal life are largely confined to this zone.

Sorting The degree of similarity in particle size in sediment or sedimentary rock.

Specific gravity The ratio of a substance's weight to the weight of an equal volume of water.

Speleothem A collective term for the dripstone features found in caverns.

Spheroidal weathering Any weathering process that tends to produce a spherical shape from an initially blocky shape.

Spit An elongate ridge of sand that projects from the land into the mouth of an adjacent bay.

Spring A flow of groundwater that emerges naturally at the ground surface.

Spring tide The highest tidal range. Occurs near the times of the new and full moons.

Stalactite The icicle-like structure that hangs from the ceiling of a cavern.

Stalagmite The columnlike form that grows upward from the floor of a cavern.

Star dune An isolated hill of sand that exhibits a complex form and develops where wind directions are variable.

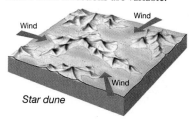
Star dune

Steppe One of the two types of dry climate. A marginal and more humid variant of the desert that separates it from bordering humid climates.

Stock A pluton similar to but smaller than a batholith.

Stony-iron meteorite One of the three main categories of meteorites. This group, as the name implies, is a mixture of iron and silicate minerals.

Stony meteorite One of the three main categories of meteorites. Such meteorites are composed largely of silicate minerals with inclusions of other minerals.

Strain An irreversible change in the shape and size of a rock body caused by stress.

Strata Parallel layers of sedimentary rock.

Stratified drift Sediments deposited by glacial meltwater.

Stratovolcano See *Composite cone.*

Streak The colour of a mineral in powdered form.

Stream A general term to denote the flow of water within any natural channel. Thus, a small creek and a large river are both streams.

Stream piracy The diversion of the drainage of one stream, resulting from the headward erosion of another stream.

Stress The force per unit area acting on any surface within a solid.

Striations (glacial) Scratches or grooves in a bedrock surface caused by the grinding action of a glacier and its load of sediment.

Strike The compass direction of the line of intersection created by a dipping bed or fault and a horizontal surface. Strike is always perpendicular to the direction of dip.

Strike-slip fault A fault along which the movement is horizontal.

Subduction The process by which oceanic lithosphere plunges into the mantle along a convergent zone.

Subduction

Subduction zone A long, narrow zone where one lithospheric plate descends beneath another.

Submarine canyon A seaward extension of a valley that was cut on the continental shelf during a time when sea level was lower, or a canyon carved into the outer continental shelf, slope, and rise by turbidity currents.

Submergent coast A coast whose form is largely the result of the partial drowning of a former land surface due to a rise of sea level or subsidence of the crust, or both.

Subsoil A term applied to the *B* horizon of a soil profile.

Superposed stream A stream that cuts through a ridge lying across its path. The stream established its course on uniform layers at a higher level without regard to underlying structures and subsequently downcut.

Superposition, law of In any undeformed sequence of sedimentary rocks, each bed is older than the one above and younger than the one below.

Surf A collective term for breakers; also the wave activity in the area between the shoreline and the outer limit of breakers.

Surface waves Seismic waves that travel along the outer layer of Earth.

Surge A period of rapid glacial advance. Surges are typically sporadic and short-lived.

Suspended load The fine sediment carried within the body of flowing water or air.

S wave An earthquake wave, slower than a P wave, that travels only in solids.

Swells Wind-generated waves that have moved into an area of weaker winds or calm.

Syncline A linear downfold in sedimentary strata; the opposite of anticline.

Syncline

System A group of interacting or interdependent parts that form a complex whole.

Tablemount See *Guyot.*
Tabular Describing a feature such as an igneous pluton having two dimensions that are much longer than the third.

Talus An accumulation of rock debris at the base of a cliff.

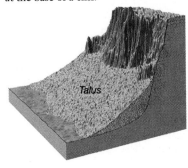

Talus

Tarn A small lake in a cirque.

Tectonics The study of the large-scale processes that collectively deform Earth's crust.

Temporary (local) base level The level of a lake, resistant rock layer, or any other base level that stands above sea level.

Tensional stress The type of stress that tends to pull a body apart.

Terminal moraine The end moraine marking the farthest advance of a glacier.

Terrace A flat, benchlike structure produced by a stream, which was left elevated as the stream cut downward.

Terrane A crustal block bounded by faults, whose geologic history is distinct from the histories of adjoining crustal blocks.

Terrestrial planet One of the Earthlike planets: Mercury, Venus, Earth, and Mars. These planets have similar densities.

Terrigenous sediment Seafloor sediments derived from terrestrial weathering and erosion.

Texture The size, shape, and distribution of the particles that collectively constitute a rock.

Theory A well-tested and widely accepted view that explains certain observable facts.

Thermal metamorphism See *Contact metamorphism*.

Thrust fault A low-angle reverse fault.

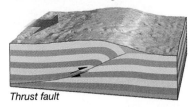

Thrust fault

Tidal current The alternating horizontal movement of water associated with the rise and fall of the tide.

Tidal delta A deltalike feature created when a rapidly moving tidal current emerges from a narrow inlet and slows, depositing its load of sediment.

Tidal flat A marshy or muddy area that is alternately covered and uncovered by the rise and fall of the tide.

Tide Periodic change in the elevation of the ocean surface.

Till Unsorted sediment deposited directly by a glacier.

Tillite A rock formed when glacial till is lithified.

Tombolo A ridge of sand that connects an island to the mainland or to another island.

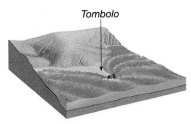

Tombolo

Topset bed An essentially horizontal sedimentary layer deposited on top of a delta during floodstage.

Transform fault A major strike-slip fault that cuts through the lithosphere and accommodates motion between two plates.

Transform fault boundary A boundary in which two plates slide past one another without creating or destroying lithosphere.

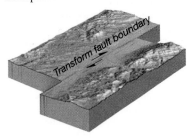

Transform fault boundary

Transpiration The release of water vapour to the atmosphere by plants.

Transported soil Soils that form on unconsolidated deposits.

Transverse dunes A series of long ridges oriented at right angles to the prevailing wind; these dunes form where vegetation is sparse and sand is very plentiful.

Travertine A form of limestone ($CaCO_3$) that is deposited by hot springs or as a cave deposit.

Trellis drainage pattern A system of streams in which nearly parallel tributaries occupy valleys cut in folded strata.

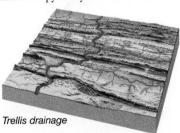

Trellis drainage

Trench An elongate depression in the seafloor produced by bending of oceanic crust during subduction.

Trigger A factor that initiates an event such as an episode of mass-wasting.

Truncated spurs Triangular-shaped cliffs produced when spurs of land that extend into a valley are removed by the great erosional force of a valley glacier.

Tsunami The Japanese word for a seismic sea wave.

Turbidite Turbidity current deposit characterized by graded bedding.

Turbidity current A downslope movement of dense, sediment-laden water created when sand and mud on the continental shelf and slope are dislodged and thrown into suspension.

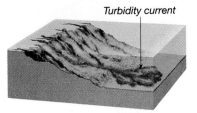

Turbidity current

Turbulent flow The movement of water in an erratic fashion often characterized by swirling, whirlpool-like eddies. Most streamflow is of this type.

Ultimate base level Sea level; the lowest level to which stream erosion could lower the land.

Ultramafic composition A compositional group of igneous rocks containing mostly olivine and pyroxene.

Unconformity A surface that represents a break in the rock record, caused by erosion and nondeposition.

Uniform stress Forces that are equal in all directions.

Uniformitarianism The concept that the processes that have shaped Earth in the geologic past are essentially the same as those operating today.

Valence electron The electrons involved in the bonding process; the electrons occupying the highest principal energy level of an atom.

Valley glacier See *Alpine glacier*.

Valley train A relatively narrow body of stratified drift deposited on a valley floor by meltwater streams that issue from the terminus of an alpine glacier.

Vein deposit A mineral filling a fracture or fault in a host rock. Such deposits have a sheetlike, or tabular, form.

Vent The surface opening of a conduit or pipe.

Ventifact A cobble or pebble polished and shaped by the sandblasting effect of wind.

Vesicles Spherical or elongated openings on the outer portion of a lava flow that were created by escaping gases.

Vesicular texture A term applied to aphanitic igneous rocks that contain many small cavities called vesicles.

Viscosity A measure of a fluid's resistance to flow.

Volatiles Gaseous components of magma dissolved in the melt. Volatiles will readily vaporize (form a gas) at surface pressures.

Volcanic Pertaining to the activities, structures, or rock types of a volcano.

Volcanic bomb A streamlined pyroclastic fragment ejected from a volcano while still semimolten.

Volcanic island arc A chain of volcanic islands generally located a few hundred kilometres from a trench where there is active subduction of one oceanic plate beneath another.

Volcanic neck An isolated, steep-sided, erosional remnant consisting of lava that once occupied the vent of a volcano.

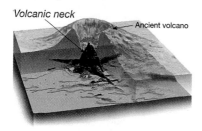

Volcano A mountain formed from lava and/or pyroclastics.

Wadati-Benioff zone The narrow zone of inclined seismic activity that extends from a trench downward into the asthenosphere.

Walther's Law The concept that sedimentary units of a vertical succession represent the deposits of laterally adjacent sedimentary environments (sedimentary facies) that have migrated over one another through time (provided that unconformities are not present in the vertical succession).

Waterfall A precipitous drop in a stream channel that causes water to fall to a lower level.

Water gap A pass through a ridge or mountain in which a stream flows.

Water table The upper level of the saturated zone of groundwater.

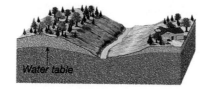

Wave-cut cliff A seaward-facing cliff along a steep shoreline formed by wave erosion at its base and mass wasting.

Wave-cut platform A bench or shelf along a shore at sea level, cut by wave erosion.

Wave height The vertical distance between the trough and crest of a wave.

Wavelength The horizontal distance separating successive crests or troughs.

Wave of oscillation A water wave in which the wave form advances as the water particles move in circular orbits.

Wave of translation The turbulent advance of water created by breaking waves.

Wave period The time interval between the passage of successive crests at a stationary point.

Wave refraction A change in direction of waves as they enter shallow water. The portion of the wave in shallow water is slowed, which causes the waves to bend and align with the underwater contours.

Weathering The disintegration and decomposition of rock at or near the surface of Earth.

Welded tuff A pyroclastic deposit composed of particles fused together by the combination of heat still contained in the deposit after it has come to rest and the weight of overlying material.

Well An opening bored into the zone of saturation.

Wind gap An abandoned water gap. These gorges typically result from stream piracy.

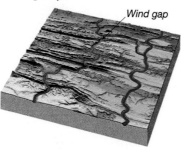

Xenolith An inclusion of unmelted country rock in an igneous pluton.

Xerophyte A plant highly tolerant of drought.

Yardang A streamlined, wind-sculpted ridge having the appearance of an inverted ship's hull that is oriented parallel to the prevailing wind.

Yazoo tributary A tributary that flows parallel to the main stream because a natural levee is present.

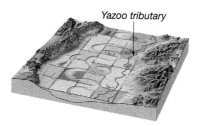

Zone of accumulation The part of a glacier characterized by snow accumulation and ice formation. The outer limit of this zone is the snowline.

Zone of aeration The area above the water table where openings in soil, sediment, and rock are not saturated but filled mainly with air.

Zone of fracture The upper portion of a glacier consisting of brittle ice.

Zone of saturation The zone where all open spaces in sediment and rock are completely filled with water.

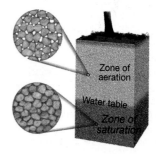

Zone of wastage The part of a glacier beyond the snowline where annually there is a net loss of ice.

Index

License Agreement
GEODe III

READ THIS LICENSE CAREFULLY BEFORE OPENING THE DISKETTE PACKAGE. BY OPENING THIS PACKAGE, YOU ARE AGREEING TO THE TERMS AND CONDITIONS OF THIS LICENSE. IF YOU DO NOT AGREE, DO NOT OPEN THE PACKAGE. PROMPTLY RETURN THE UNOPENED PACKAGE AND ALL ACCOMPANYING ITEMS TO THE PLACE YOU OBTAINED THEM. THESE TERMS APPLY TO ALL LICENSED SOFTWARE ON THE DISK EXCEPT THAT THE TERMS FOR USE OF ANY SHAREWARE OR FREEWARE ON THE DISKETTES ARE AS SET FORTH IN THE ELECTRONIC LICENSE LOCATED ON THE DISK:

1. GRANT OF LICENSE and OWNERSHIP: The enclosed computer programs <<and any data>> ("Software") are licensed, not sold, to you by Pearson Education Canada Inc. ("We" or the "Company") in consideration of your adoption of the accompanying Company textbooks and/or other materials, and your agreement to these terms. You own only the disk(s) but we and/or our licensors own the Software itself. This license allows instructors and students enrolled in the course using the Company textbook that accompanies this Software (the "Course") to use and display the enclosed copy of the Software for academic use only, so long as you comply with the terms of this Agreement. You may make one copy for back up only. We reserve any rights not granted to you.

2. USE RESTRICTIONS: You may not sell or license copies of the Software or the Documentation to others. You may not transfer, distribute or make available the Software or the Documentation, except to instructors and students in your school who are users of the adopted Company textbook that accompanies this Software in connection with the course for which the textbook was adopted. You may not reverse engineer, disassemble, decompile, modify, adapt, translate or create derivative works based on the Software or the Documentation. You may be held legally responsible for any copying or copyright infringement which is caused by your failure to abide by the terms of these restrictions.

3. TERMINATION: This license is effective until terminated. This license will terminate automatically without notice from the Company if you fail to comply with any provisions or limitations of this license. Upon termination, you shall destroy the Documentation and all copies of the Software. All provisions of this Agreement as to limitation and disclaimer of warranties, limitation of liability, remedies or damages, and our ownership rights shall survive termination.

4. DISCLAIMER OF WARRANTY: THE COMPANY AND ITS LICENSORS MAKE NO WARRANTIES ABOUT THE SOFTWARE, WHICH IS PROVIDED "AS-IS." IF THE DISK IS DEFECTIVE IN MATERIALS OR WORKMANSHIP, YOUR ONLY REMEDY IS TO RETURN IT TO THE COMPANY WITHIN 30 DAYS FOR REPLACEMENT UNLESS THE COMPANY DETERMINES IN GOOD FAITH THAT THE DISK HAS BEEN MISUSED OR IMPROPERLY INSTALLED, REPAIRED, ALTERED OR DAMAGED. THE COMPANY DISCLAIMS ALL WARRANTIES, EXPRESS OR IMPLIED, INCLUDING WITHOUT LIMITATION, THE IMPLIED WARRANTIES OF MERCHANTABILITY AND FITNESS FOR A PARTICULAR PURPOSE. THE COMPANY DOES NOT WARRANT, GUARANTEE OR MAKE ANY REPRESENTATION REGARDING THE ACCURACY, RELIABILITY, CURRENTNESS, USE, OR RESULTS OF USE, OF THE SOFTWARE.

5. LIMITATION OF REMEDIES AND DAMAGES: IN NO EVENT, SHALL THE COMPANY OR ITS EMPLOYEES, AGENTS, LICENSORS OR CONTRACTORS BE LIABLE FOR ANY INCIDENTAL, INDIRECT, SPECIAL OR CONSEQUENTIAL DAMAGES ARISING OUT OF OR IN CONNECTION WITH THIS LICENSE OR THE SOFTWARE, INCLUDING, WITHOUT LIMITATION, LOSS OF USE, LOSS OF DATA, LOSS OF INCOME OR PROFIT, OR OTHER LOSSES SUSTAINED AS A RESULT OF INJURY TO ANY PERSON, OR LOSS OF OR DAMAGE TO PROPERTY, OR CLAIMS OF THIRD PARTIES, EVEN IF THE COMPANY OR AN AUTHORIZED REPRESENTATIVE OF THE COMPANY HAS BEEN ADVISED OF THE POSSIBILITY OF SUCH DAMAGES. SOME JURISDICTIONS DO NOT ALLOW THE LIMITATION OF DAMAGES IN CERTAIN CIRCUMSTANCES, SO THE ABOVE LIMITATIONS MAY NOT ALWAYS APPLY.

6. GENERAL: THIS AGREEMENT SHALL BE CONSTRUED AND INTERPRETED ACCORDING TO THE LAWS OF THE PROVINCE OF ONTARIO. This Agreement is the complete and exclusive statement of the agreement between you and the Company and supersedes all proposals, prior agreements, oral or written, and any other communications between you and the company or any of its representatives relating to the subject matter.

Should you have any questions concerning this agreement or if you wish to contact the Company for any reason, please contact in writing: Customer Service, Pearson Education Canada, 26 Prince Andrew Place, Don Mills, Ontario M3C 2T8.

Should you have any questions concerning technical support, you may write to:
Tasa Graphic Arts, Inc.
9301 Indian School Road NE
Suite 208
Albuquerque, NM 87112-2861